Lecture Notes in Electrical Engineering

Volume 1229

The book series *Lecture Notes in Electrical Engineering* (LNEE) publishes the latest developments in Electrical Engineering—quickly, informally and in high quality. While original research reported in proceedings and monographs has traditionally formed the core of LNEE, we also encourage authors to submit books devoted to supporting student education and professional training in the various fields and applications areas of electrical engineering. The series cover classical and emerging topics concerning:

- Communication Engineering, Information Theory and Networks
- Electronics Engineering and Microelectronics
- Signal, Image and Speech Processing
- Wireless and Mobile Communication
- Circuits and Systems
- Energy Systems, Power Electronics and Electrical Machines
- Electro-optical Engineering
- Instrumentation Engineering
- Avionics Engineering
- Control Systems
- Internet-of-Things and Cybersecurity
- Biomedical Devices, MEMS and NEMS

For general information about this book series, comments or suggestions, please contact leontina.dicecco@springer.com.

To submit a proposal or request further information, please contact the Publishing Editor in your country:

China

Jasmine Dou, Editor (jasmine.dou@springer.com)

India, Japan, Rest of Asia

Swati Meherishi, Editorial Director (Swati.Meherishi@springer.com)

Southeast Asia, Australia, New Zealand

Ramesh Nath Premnath, Editor (ramesh.premnath@springernature.com)

USA, Canada

Michael Luby, Senior Editor (michael.luby@springer.com)

All other Countries

Leontina Di Cecco, Senior Editor (leontina.dicecco@springer.com)

**** This series is indexed by EI Compendex and Scopus databases. ****

Phillip G. Bradford · S. Andrew Gadsden ·
Shiban K. Koul · Kamakhya Prasad Ghatak
Editors

Proceedings of IEMTRONICS 2024

International IoT, Electronics and Mechatronics Conference, Volume 1

Springer

Editors
Phillip G. Bradford
University of Connecticut
Stamford, CT, USA

S. Andrew Gadsden
McMaster University
Hamilton, ON, Canada

Shiban K. Koul
Centre for Applied Research in Electronics
Indian Institute of Technology Delhi
New Delhi, Delhi, India

Kamakhya Prasad Ghatak
Institute of Engineering and Management
Kolkata, West Bengal, India

ISSN 1876-1100 ISSN 1876-1119 (electronic)
Lecture Notes in Electrical Engineering
ISBN 978-981-97-4779-5 ISBN 978-981-97-4780-1 (eBook)
https://doi.org/10.1007/978-981-97-4780-1

This Springer imprint is published by the registered company Springer Nature Singapore Pte Ltd.
The registered company address is: 152 Beach Road, #21-01/04 Gateway East, Singapore 189721, Singapore

Organizing Committee

General Chair

Prof. Kenneth T. V. Grattan OBE (Officer of the Order of the British Empire) FREng, Royal Academy of Engineering—George Daniels Professor of Scientific Instrumentation, University of London, London, UK

Technical Co-chair

Prof. Bob Gill, British Columbia Institute of Technology, Burnaby, Canada

Convener

Prof. Malay Gangopadhyay, Institute of Engineering and Management, Kolkata, India

Finance Chair

Prof. Sanghamitra Poddar, Institute of Engineering and Management, Kolkata, India

Sponsors

SMART Society, USA
IEM-UEM Group

Advisory Committee

Dr. Chuck Easttom, University of Dallas, USA and Georgetown University, USA
Dr. Ronald F. DeMara, University of Central Florida, USA
Dr. Ashutosh Datta, Johns Hopkins University, USA
Dr. Yang Hao, Queen Mary University, London
Dr. Vien Van, University of Alberta, Canada
Dr. Omar Ramahi, University of Waterloo, Canada
Dr. Yahia Antar, Royal Military College, Canada
Dr. Zhizhang (David) Chen, Dalhousie University, Canada
Dr. Detlef Streitferdt, Technische Universitat Ilmenau, Germany
Prof. Shahab Tayeb, California State University, Fresno
Dr. Bhaskar Gupta, Jadavpur University, India
Dr. Chinmoy Saha, Indian Institute of Space Science and Technology, India
Dr. Jawad Y. Siddiqui, Institute of Radio Physics and Electronics, University of Calcutta, India

Technical Committee

Dr. Nabeeh Kandalaft, Grand Valley State University, USA
Dr. Alex "Sandy" Antunes, Capitol Technology University, USA
Dr. Izzat Alsmadi, Texas A&M, San Antonio, USA
Dr. Lo'ai Tawalbeh, Texas A&M University-San Antonio, USA
Dr. Pratik Chattopadhyay, Indian Institute of Technology (BHU), Varanasi
Dr. Doina Bein, California State University, Fullerton, USA
Dr. Hasan Yasar, Carnegie Mellon University, USA
Dr. Moises Levy, West Texas A&M University, USA
Dr. Christian Trefftz, Grand Valley State University, USA
Dr. Mrinal Sen, Indian Institute of Technology (ISM), Dhanbad
Dr. Petros Spachos, University of Guelph, Canada
Dr. Kanika Sood, California State University, Fullerton
Dr. Ke-Lin Du, Concordia University, Canada
Dr. Wenlin Han, California State University, Fullerton
Dr. Ashiq Adnan Sakib, Florida Polytechnic University, USA
Dr. Morteza Modarresi Asem, Islamic Azad University, Iran

Dr. Md. Liakat Ali, Rider University, USA
Dr. Tarek El Salti, Sheridan College, Canada
Dr. Sukomal Dey, Indian Institute of Technology, Palakkad
Dr. Maysam Chamanzar, Carnegie Mellon University, USA
Dr. Kean Boon Lee, Sheffield University, UK
Dr. Ali Abdullah S. AlQahtani, North Carolina A&T State University

Foreword

We are delighted to present the proceedings of the 4th International IoT, Electronics and Mechatronics Conference (IEMTRONICS 2024), held in a hybrid format from April 3rd to 5th, 2024. We are grateful to the appropriate authorities of the Imperial College London, UK for hosting the physical component of the conference.

IEMTRONICS 2024 witnessed a vibrant response from the global research community, receiving a total of 353 research paper submissions from a diverse range of 25 countries. Following a meticulous triple-blind review process, 77 high-quality papers were accepted for publication, resulting in a competitive acceptance rate of approximately 26%.

We would like to express our sincere gratitude to the authors for their valuable contributions and the reviewers for their dedication and expertise in ensuring the high quality of the accepted papers. Special thanks are extended to the IEM-UEM Group and SMART Society USA for their generous sponsorship, which significantly contributed to the success of the conference. We are extremely grateful to the respected President and Hon'ble Director of IEM-UEM Group, India for inspiration and involvement behind this project.

These proceedings capture the cutting-edge research advancements presented at IEMTRONICS 2024, spanning the vast and dynamic fields of the Internet of Things (IoT), Electronics, and Mechatronics respectively. We are confident that this collection will serve as a valuable resource for researchers, practitioners, engineers, scientists, and graduate students alike, fostering further innovation and collaboration within these critical and highly relevant disciplines.

April 2024

Dr. Malay Gangopadhyaya
Convener
IEMTRONICS—2024

Preface

The present volume compiles a selection of research papers presented at the International IoT, Electronics and Mechatronics Conference (IEMTRONICS) 2024, held in hybrid mode at Imperial College London, UK, from April 3rd to 5th, 2024. The conference aimed to foster a platform for scholars to share their latest research advancements in the rapidly evolving fields of IoT, electronics and mechatronics, respectively.

The conference program encompassed a diverse array of technical sessions, special sessions, tutorials, and demonstration sessions. Papers were rigorously peer-reviewed to ensure their quality and relevance to the conference theme. The proceedings are organized into five primary tracks, namely IoT and Data Science, Electronics, Mechatronics, Information Technology, and Mobile Communication, reflecting the broad spectrum of topics covered.

This collection of papers provides valuable insights into the current state-of-the-art and emerging trends in these critical fields. It serves as a resource for researchers, industry professionals, and academia, stimulating further exploration and innovation in this context.

We wish all attendees of IEMTRONICS 2024 an enjoyable scientific gathering in London, UK. We look forward to seeing all of you next year at the conference.

Stamford, USA — Phillip G. Bradford
Hamilton, Canada — S. Andrew Gadsden
New Delhi, India — Shiban K. Koul
Kolkata, India — Kamakhya Prasad Ghatak

Contents

About the Editors

Dr. Phillip G. Bradford is Associate Professor-in-Residence, Computer Science, University of Connecticut, Stamford, USA. His research interests include graph theory, Bayes methods, automobile industry, automotive electronics, calculus, complex networks, computational complexity, directed graphs, fault diagnosis, formal languages, genetics, matrix multiplication and network.

Dr. S. Andrew Gadsden is the Associate Chair, Graduate Studies, Department of Mechanical Engineering and Associate Professor, Department of Mechanical Engineering, McMaster University, Canada. His expertise includes Artificial Intelligence; Cognitive Systems; Control Systems; Data Science; Estimation Theory; Machine Learning; Mechatronics; Smart Systems. Dr. Gadsden's research background and expertise includes a broad consideration of control and estimation theory, artificial intelligence and machine learning, and smart or cognitive systems.

Dr. Shiban K. Koul is a Professor at the Indian Institute of Technology Delhi. He is also the Chairman of Astra Microwave Products Limited, Hyderabad, a major company involved in the Development of RF and Microwave systems in India. His current research interests include RF MEMS, Device modeling, Microwave and Millimeter wave IC design and Reconfigurable microwave circuits including antennas. He is currently a serving ADCOM member and a Member of IEEE MTT society's technical committees on Microwave and Millimeter Wave Integrated Circuits (MTT-6) and RF MEMs (MTT-21), Member of India Initiative team of IEEE MTT-S, Membership Services Regional Co-coordinator India, Vice Chair MGA and MTT-S Speaker bureau lecturer. Dr. Koul is the author/co-author of 199 Research Papers and seven state-of-the art books and holds six patents and six copyrights. He is a Fellow of the Institution of Electrical and Electronics Engineers, USA (IEEE), Fellow of the Indian National Academy of Engineering (INAE) and Fellow of the Institution of Electronics and Telecommunication Engineers (IETE).

Dr. Kamakhya Prasad Ghatak obtained his Ph.D. Degree of the University of Calcutta, India, in 1988. In accordance with the analysis of World Ranking of top 2% Scientists as prepared by Stanford University in 2020, USA, (shorturl.at/qHIJ4), Prof. K. P. Ghatak of IEM UEM group stays within top 1% in the field of Applied Physics. Professor Ghatak is the principal co-author of more than 200 research papers in SCI Journals and 385 research publications in SCOPUS, respectively. He is the Principal author of 20 research monographs from Springer-Verlag, Germany, De Gruyter, Nova and World Scientific Publishing Company respectively. He has guided successfully more than 100 Ph.D. Students in different topics of Nano Technology. He was Assistant Professor in the Department of Radio Physics and Electronics of the University of Calcutta from 1983 to 1987, Associate Professor in the Department of ETCE of Jadavpur University from 1987 to 1994, Professor and HOD of Electronic Science of the University of Calcutta from 1994 to 2011, HOD and Dean of the National Institute of Technology, Agartala from 2012 to

2014 and from January 2015 he has joined in the IEM/UEM group, Kolkata as Senior Professor and research Director. He was at the top of the merit lists in all the said cases.

Fast Software Implementation of Edwards Curve Cryptography for IoT Secure Communication Purposes

Ihor Koriakov, Yevgen Kotukh, and Maksym Korobchynskyi

Abstract The article examines Ukraine's national standard DSTU 9041:2020 for encrypting short messages using Twisted Edwards Elliptic Curves. It focuses on the standard's performance enhancements and compatibility with existing cryptographic tools. The implementation of this standard in C programming language is explored, addressing challenges such as computational efficiency, bandwidth limitations, and energy consumption. The article serves as a resource for the scientific and engineering community, providing mathematical explanations, code snippets, and examples for the implementation and cryptographic transformations of the DSTU 9041:2020 standard.

Keywords Edwards curves · Encryption · IoT · Fast cryptography

I. Koriakov
Krypton Research and Development Company Co., Ltd., Kyiv, Ukraine
e-mail: ikor@i.ua

Y. Kotukh (✉) · M. Korobchynskyi
School of Cybersecurity, Yevhenii Bereznyak Military Academy, Kyiv, Ukraine
e-mail: yevgenkotukh@gmail.com

M. Korobchynskyi
e-mail: maks_kor@ukr.net

P. G. Bradford et al. (eds.), *Proceedings of IEMTRONICS 2024*, Lecture Notes in Electrical Engineering 1229, https://doi.org/10.1007/978-981-97-4780-1_1

1 Introduction

In 2020, a new National Standard, DSTU 9041:2020, was introduced, entitled "Information Technology. Cryptographic Information Protection. Algorithm for Encrypting Short Messages Based on Twisted Edwards Elliptic Curves." This algorithm is designed for the encryption of short messages (up to 616 bits) and is compatible with encryption algorithms, including those defined by the national standards of Ukraine [1].

Like the digital signature standard DSTU 4145:2002, the new algorithm employs cryptographic transformations within the group of elliptic curve points, utilizing the latest developments in elliptic curve cryptography—Edwards curves. This innovation provides significant performance advantages, improving speed by more than threefold. The new standard is developed with the latest demands for cryptographic algorithm resilience in mind. For instance, the lower bound of cryptosystem resilience in this standard equals 2 127 (≈ 10.42), which is more than one and a half times higher than in DSTU 4145, and other levels such as 2 255 (≈ 10.85), 2 383 (≈ 10.127), and 2767 (≈ 10.255) can be selected. Moreover, its resilience to both attacks for plaintext recovery and distinguishing attacks is rigorously substantiated [2].

The encryption algorithm project, which forms the basis of this standard, has been tested both within Ukraine and internationally (Central European Conference on Cryptography in June 2020)—a forum for leading cryptologists worldwide. DSTU-9041 is harmonized with all current national standards in Ukraine. The novelty of the standard lies in its application scope—key encapsulation, state-of-the-art mathematical apparatus, and a new algorithm for generating pseudo-random sequences, which, unlike the analogous algorithm from DSTU 4145, uses exclusively national cryptographic algorithms and does not reference the corresponding post-Soviet standards whose validity has nearly expired.

While the new standard is not categorized as post-quantum, its resilience will only be threatened once quantum computers with 700 or more qubits emerge (currently, the number of "working" qubits achieved is about 50). Its advantage over post-quantum algorithms lies in the relatively small key size (dozens or even hundreds of times smaller than in post-quantum algorithms).

Incorporating the importance of the DSTU 9041:2020 standard for Internet of Things (IoT) devices in this article highlights how critical secure encryption and decryption algorithms are for the rapidly expanding IoT domain. IoT devices often transmit sensitive data, making them vulnerable to cyber threats. The DSTU 9041:2020 standard, with its focus on Edwards curves over prime fields and advanced block cipher encryption algorithms like Kalina-l/k-KW and Kalina-l/k-KW-p, offers robust security solutions. Additionally, the use of the Kupyna hash function further enhances data integrity and security. These features are crucial for IoT environments, where the reliability and security of communication between numerous interconnected devices are paramount.

Despite the growing interest in asymmetric encryption [3, 4] resilient to quantum attacks, the demand for rapid encryption solutions, particularly for applications in the Internet of Things (IoT), continues to escalate. This trend underscores a dual focus in cryptographic research and development: on the one hand, addressing the imminent threat posed by quantum computing to current cryptographic standards, and, on the other, catering to the immediate need for efficient, high-speed encryption technologies that can be seamlessly integrated into IoT devices. These devices, characterized by their limited computational resources and power constraints, necessitate lightweight yet secure encryption methods to ensure data integrity and confidentiality in real-time communication and data processing within the IoT ecosystem.

The aim of this work is to present the implementation of DSTU 9041:2020 in a form understandable for general comprehension, providing practical examples of source code implementation in C++. This article is targeted at the scientific and engineering community interested in exploring the possibilities of modern cryptographic primitives in applied fields. Special attention is given to the implementation of secure information exchange protocols, considering limitations on computational speed and efficiency, bandwidth of physical telecommunication channels, energy consumption requirements, and other key parameters.

2 Background

Let's consider the objects and parameters that the DSTU 9041:2020 algorithm operates on. This is of practical importance for understanding further operations related to elliptic curves, which we perform at the source code level. The Edwards curve is defined by the set of (x, y) pairs that satisfy the equation

$$x^2 + ay^2 = dx^2y^2 + 1,$$

where $a = 2$ and d is determined from the application with the parameters of the recommended elliptic curves. This equation is for illustrative purposes only; we will not use it directly; the calculations will be determined by specific procedures. Operations with curve points are defined by a simple field Fp, which is a set of numbers $\{0, 1, \ldots, p - 1\}$ where p is a prime number.

When choosing a specific recommended curve, the base point P is specified by the values of its coordinates (xp, yp). Two main operations are defined on curve points: addition of points and multiplication of a point by a scalar. The addition of points A and B is denoted as $A + B$, and the multiplication of point A by the scalar s is denoted as sA. All numbers that we will have to operate with will have a large capacity (in the examples below, 512 bits) and will be presented as arrays of integers. Moreover, to effectively perform operations with carries and expansion of bit depth during multiplication, we will use 32-bit signed integers, in the lower 16 bits of which 512-bit numbers will be placed, divided into 16-bit portions. For calculations, a convenient order is when the low-order digits of a number are placed at the beginning of the array. For example, the xp coordinate as a 512-bit number (in hexadecimal format):

5230A1EE747050A072BD7319741586EA520388B6B5309
4571C821A2FC9A9E83D
56665346B5DB04C43E75261DBDA512728FAAFAC48A
E9260A5A184E2933E3A400

will be represented as an array (see Sect. 4).

Note that to the array, which was supposed to contain 32 integers ($512/16 = 32$), another 33rd element was added, which is necessary to take into account carries and extensions of bit depth during arithmetic operations. Next, we will consider the basis of the algorithm for generating a public key on elliptic curves.

Let's consider the general principle of the exchange between Bob and Alice.

1. Alice wants to send an encrypted message to Bob.
2. Bob generates a random number e (random numbers in these protocols are secret and destroyed at the end of the session).
3. Bob calculates the point $Q = eP$, which is the public decryption key.
4. Bob passes the Q to Alice.
5. Alice generates a random number ε.
6. Alice calculates the point $R = \varepsilon P$, which is the public encryption key.
7. Alice calculates the point $T = \varepsilon Q = (xT, yT)$ and uses xT to generate the secret encryption key.
8. Alice encrypts the message and gives Bob the public encryption key R and the encrypted message.
9. Bob calculates point $T' = (xT', yT')$ as $T' = eR$ and uses xT' to generate the secret decryption key.

Note that $T = \varepsilon\, Q = \varepsilon\, eP$ and $T' = eR = e\varepsilon P$ are commutative operations, which is equivalent to the commutative exponentiation operation in the Diffie-Hellman protocol.

Let's consider the general principle of the exchange between Bob and Alice, which is implemented by the DSTU 9041:2020 protocol. A feature of the implementation is that Alice, as part of the cryptogram, does not transmit to Bob the full coordinates of the point R, but only the x coordinate R, since Bob can uniquely restore the second coordinate yR along the xR coordinate.

1. Bob generates a random number e.
2. Bob calculates the point $Q = eP$, which is the public decryption key.
3. Bob passes Q to Alice.
4. Alice generates a random number ε.
5. Alice calculates the point $R = \varepsilon P$, which is the public encryption key, and stores $r = xR$.
6. Alice calculates the point $T = \varepsilon Q = (xT, yT)$ and uses xT to generate the secret

encryption key.
7. Alice encrypts message S and sends Bob the cryptogram $C = (r \parallel t)$, where t is the ciphertext.
8. Bob calculates the yR coordinate of point R' as

$$\sqrt{(1 - r^2)(a - dr^2)^{-1}}$$

9. Bob calculates point $T' = (xT', yT')$ as $T' = eR'$ and uses xT' to generate the secret decryption key.
10. Bob decrypts the ciphertext t and recovers the message S.

3 Our Solution

First, let's define a couple of constants and types (see Sect. 4),

Further, in the process of presentation, we will declare the variables used and explain their purpose.

3.1 *Multiplying a Point by a Scalar*

As we have seen, the main operation is the multiplication of a point by a scalar (number), which is implemented in the form of a binary algorithm: the bits of the number are sequentially analyzed and when the value is "1", the addition of certain points is performed and, regardless of the fulfillment of the condition, the point is doubled, that is also addition, but with itself.

Despite the basic formula for adding points in affine coordinates, DSTU 9041:2020 provides a formula for adding points in three-dimensional projective coordinates, which is calculated much faster, which requires translation from affine coordinates to projective at the beginning of the calculation and reverse translation from projective to affine at the end. Consider the algorithm for multiplying a point by a scalar:

1. Receive as input the value of the scalar number fk and the coordinates of the point R: RX and RY.
2. Construct a singular point O in projective coordinates as Q: ($QX = 1$, $QY = 0$, $QZ = 1$).
3. Complete the third coordinate $RZ = 1$ for the input point R.
4. Iterate through all the words and bits in the words of the scalar fk.

5. If bit = 1, then add the points $Q = Q + R$.
6. Double the point $R = 2R$.
7. Return point Q back to affine coordinates, for which the inverse value QZ^{-1} for QZ is obtained and calculate $kX = QX * QZ^{-1}$ and $kY = QY * \mathrm{QZ}^{-1}$
8. Return kX and kY.

A practical implementation of this algorithm looks like this (see Sect. 4).

There are calls to the following functions: LoOne—loads the number 0…1 into a long variable, LoZero—loads the number 0, Mov—rewrites the first variable into the second, AddPointPro—adds the point specified by the first three coordinates to the point specified by the second three coordinates and returns the third triple of coordinates of the total point (doubling the point—the point is added to itself), InvMod—converts the first variable to the inverse value (multiplicative inversion), MulMod—multiplies the coordinate by the inverse value.

Next, we will analyze only the main functions, since some functions are quite clear, but the full text of the example contains definitions of all the functions used.

3.2 Addition of Points

Addition is implemented in projective coordinates in accordance with the algorithm given in DSTU 9041:2020 in Appendix B, formula B.4:

$$XR = AG(C - aD);\ YR = AF((XT + YT)(XS + YS) - C - D);\ ZR = FG,$$

where $A = ZTZS$; $B = A\ 2$; $C = XTXS$; $D = YTYS$; $E = dCD$; $F = B - E$; $G = B + E$.

These formulas determine the coordinates of point $R = (XR\!: YR\!: ZR)$, which represents the sum of two points $T = (XT\!: YT\!: ZT)$ and $S = (XS\!: YS\!: ZS)$.

The practical implementation of this function looks as shown in Sect. 4.

There are calls to the following functions: MulMod.

- the already mentioned multiplication modulo p, Addm and
- Subm, respectively, addition and subtraction modulo the same.

3.3 Modulo Multiplication

The so-called binary modulo multiplication algorithm is used, which performs sequential multiplications of the elements of the first factor by the entire second factor and forms a modulo sum of partial products, analyzing the values of the bits of the highest element of the sum of partial products. If at this position the bit is equal to "1", then addition modulo prime p is performed with p's 2's complement shifted by the appropriate number of bits.

Let's describe the algorithm:

1. Obtain the values of factors A and B at the input.
2. Reset Mu result.
3. For all elements $A[i]$ do:
4. Shift the result Mu to the left by $k - 1$ digit (in our case by one word int).
5. Fold product of the current word $A[i]$ by B with result Mu.
6. For all bits of the high-order word of the result Mu, do the following:
7. If the bit is equal to "1", then add the modulo result of Mu with the shifted complement of p to 2.
8. Output Mu.

The practical implementation of the modulo multiplication function looks as shown in Sect. 4.

Here additional functions MulaiB have appeared—calculating the partial product of an element of the first factor by the second factor, and AddL—long addition (addition of all 33 elements) without taking the modulus.

Shifted p's complements to 2 are pre-tabulated when initialized in a table Ptb containing k number of length L.

3.4 Partial Product Calculation

This is a simple function that accumulates a partial product as a result of Mul, taking into account the expansion of bit depth during multiplication. The practical implementation of the function looks as shown in Sect. 4.

3.5 Restoring the Ordinate of Point R

Bob calculates the yR coordinate of point R' using the formula: $\sqrt{(1 - r^2)(a - dr^2)^{-1}}$ according to the following function (the calculated elements of the formula are marked in the comments). The practical implementation of this function looks as shown in Sect. 4. Here we use already familiar functions, of which the multiplicative inversion InvMod and the extraction of the square root Scroot have not yet been described.

3.6 Multiplicative Inversion

Multiplicative inversion or reciprocal of q^{-1} modulo p is the value of h such that $hq^{-1} = 1$. In accordance with the property (q^{-1} mod p) = (q^{p-2} mod p), to calculate the multiplicative inversion, it is enough to raise q to the power $p - 2$.

The practical implementation is shown in Sect. 4.

There is a new function for exponentiation modulo PowMod.

3.7 Exponentiation _

Exponentiation modulo is implemented by the well-known binary method, in which the degree value is initially set to one, then the bits of the exponent are analyzed and, if the bit is equal to "1", then the current value of the degree is multiplied by the base of the degree; in addition, the current value is squared degrees.

The practical implementation of the modulo exponentiation function is shown in Sect. 4.

3.8 Square Root

The modulo square root is calculated as follows. Based on the input radical value v, $f = v^{(p-1)/4}$ is calculated, then $z = v^{(p+3)/8}$ and the value of the root $u = z$, but if $f = p - 1$, then $u = (w * z)$ mod p. The w parameter is given in Appendix B of DSTU 9041:2020.

A practical implementation of calculating the square root modulo function is shown in Sect. 4.

The constants pd 4, pd 8, p 1 are formed during initialization.

Simple arithmetic operations and their implementation are designed in Sect. 4.

4 Source Code

A practical implementation of calculating the square root modulo function is shown in Section VI:
The constants pd 4, pd 8, p 1 are formed during initialization.
Simple arithmetic operations and their implementation are designed in Section VI:

IV. SOURCE CODE

```
// Elliptic arithmetica DSTU 9041:2020 Appl.B, tabl. B.3,
E512/1
#include <iostream>
#include <array>

#define k 17 // Z=2^(k-1)=65536 (16 bit word)
#define L 32 // num words in 512 bit

typedef std::array<int, L+1> Lng;// Lng[L+1] = 512+16 bit
;

// Parameters from DSTU 9041:2020 tabl. B.3 E512/1
Lng p = {

0xFC95,0xFFFF,0xFFFF,0xFFFF,0xFFFF,0xFFFF,0xFFFF,
0xFFFF,

0xFFFF,0xFFFF,0xFFFF,0xFFFF,0xFFFF,0xFFFF,0xFFFF,
0xFFFF,

0xFFFF,0xFFFF,0xFFFF,0xFFFF,0xFFFF,0xFFFF,0xFFFF,
0xFFFF,

0xFFFF,0xFFFF,0xFFFF,0xFFFF,0xFFFF,0xFFFF,0xFFFF,
0xFFFF,0};
Lng Xp = {

0xA400,0x33E3,0x4E29,0x5A18,0x260A,0x8AE9,0xFAC4
,0x8FAA,

0x1272,0xBDA5,0x261D,0x3E75,0x04C4,0xB5DB,0x5346
,0x5666,

0xE83D,0xC9A9,0x1A2F,0x1C82,0x9457,0xB530,0x88B6,
0x5203,

0x86EA,0x7415,0x7319,0x72BD,0x50A0,0x7470,0xA1EE,
0x5230,0};
Lng Yp = {



0x0A85,0xAE7D,0x8F0B,0x9C50,0x4F52,0x2C44,0xE442,
0xB675,

0x40D6,0x27A5,0xDA2E,0xFDD1,0xB930,0xA0E6,0x272
4,0x1A83,

0x2B3A,0x86B4,0x991A,0x7124,0x5EF2,0xA5CB,0xFCC
3,0x4DBC,

0xEF21,0x978A,0xF451,0x9762,0xC921,0xCC63,0x0D50,
0x053A,0};
Lng w = {
```

```
0x96E3,0x8CBA,0xE8FE,0x278B,0xB43A,0x39A5,0x4FA
5,0x0401,

0x41AB,0xBE20,0x8350,0xBB8D,0xFDF4,0x98EB,0x4B1
E,0xCA1E,

0xB910,0xA8BA,0xA043,0xC423,0xD8F2,0x1EA5,0x56F
C,0xEAE5,

0x2A62,0x95AC,0x8679,0x12E6,0x7479,0xD202,0xDAB5,
0x658D,0};

 Lng const_a,const_d; //elliptic coeff a=2, d=0x10D
 Lng np; // 2-p = 0x036B,0x0...
 Lng p1,p2; // p-1=0xFC94, p-2=0xFC93,0xFFFF...
 Lng pd4,pd8; // (p-1)/4=0xFF25...0x3FFF;
(p+3)/8=0xFF93...0x1FFF
 Lng F,A,B; // work
 Lng Ptb[k]; // p << j, j=0..k-1
 Lng r,XQ,YQ,XR,YR,XT,YT; // notation from DSTU
 Lng r2,v,y; // notation from DSTU

void PrntLng(Lng Lo){ // Console write Lng L+1
int i,j;
 for (i=0; i<(L/8); i++) {
  for (j=0; j<(L/4); j++) {
   std::cout << " " << std::hex << Lo[L-1-(i*8+j)];
  }
  std::cout << std::endl;
 }
}

void LoZero(Lng Lo){ // Lo = 0
int i;
 for (i=0; i<=L; i++) {
  Lo[i] = 0;
 }
}

void LoOne(Lng Lo){ // Lo = 1
int i;
 Lo[0] = 1;
 for (i=1; i<=L; i++) {
  Lo[i] = 0;
 }
}

int Comp(Lng A, Lng B){ //(A>B)=>Comp>0;
(A<B)=>Comp<0; (A=B)=>Comp=0
int i,s;
 for (i=L; i>=0; i--) {
  s = A[i] - B[i];
  if (s!=0) {break;}
 }
 return s;
}

void Add(Lng A, Lng B, Lng AB){ // A + B
int i,c,s;
 c = 0;
 for (i=0; i<L; i++) {
```

```
    s = A[i] + B[i] + c;
    AB[i] = s & 0xFFFF;
    c = s >> 16;
  }
  AB[L] = c;
}

void Sub(Lng A, Lng B, Lng AB){ // A - B
int i,c,s,b;
  c = 1;
  for (i=0; i<L; i++) {
    b = B[i];
    s = A[i] + ((~b) & 0xFFFF) + c;
    AB[i] = s & 0xFFFF;
    c = s >> 16;
  }
  AB[L] = c;
}

void Mov(Lng A, Lng B){ // A -> B
int i;
  for (i=0; i<=L; i++) {
    B[i] = A[i];
  }
}

void Addm(Lng A, Lng B, Lng AB){ // AB = (A + B)mod
p
  Add(A,B,AB);
  if (Comp(AB,p)>=0) {
    Add(AB,np,AB);
  }
  AB[L] = 0;
}

void Subm(Lng A, Lng B, Lng AB){ // AB = (A - B)mod p
  Sub(p,B,AB);
  Addm(A,AB,AB);
  AB[L] = 0;
}

void Left1(Lng X, Lng AB){ // shift Lng << 1
int i,c,s;
  c = 0;
  for (i=0; i<=L; i++) {
    s = (X[i] << 1) | c;
    AB[i] = s & 0xFFFF;
    c = s >> 16;
  }
}

void Left16(Lng X, Lng AB){ // shift Lng << 16
int i;
  for (i=L-1; i>=0; i--) {
    AB[i+1] = X[i];
  }
  AB[0] = 0;
}

void AddL(Lng A, Lng B, Lng AB){ // A + B ext
int i,c,s;
  c = 0;
```

```
  for (i=0; i<=L; i++) {
    s = A[i] + B[i] + c;
    AB[i] = s & 0xFFFF;
    c = s >> 16;
  }
  AB[L] = AB[L] + (c << 16);
}

void SubL(Lng A, Lng B, Lng AB){ // A - B ext
int i,c,s,b;
  c = 1;
  for (i=0; i<=L; i++) {
    b = B[i];
    s = A[i] + ((~b) & 0xFFFF) + c;
    AB[i] = s & 0xFFFF;
    c = s >> 16;
  }
}

void MulaiB(int ai, Lng B, Lng Mul) { // ai * B
int j,carr;
  LoZero(Mul);
  carr = 0;
  for (j=0; j<L; j++) {
    carr = Mul[j]+ai*B[j]+((carr >> 16)&0xFFFF);
    Mul[j] = carr & 0xFFFF;
  }
  Mul[L] = ((carr >> 16)&0xFFFF);
}

void MulMod(Lng A, Lng B, Lng Mu){// (A * B) mod p =>
Mu
int i,j,tt; Lng T1,T2;
  LoZero(Mu);
  for (i=L-1; i>=0; i--) {
    Left16(Mu,T1);
    MulaiB(A[i],B,T2);
    AddL(T1,T2,Mu);
    tt = Mu[L];
    Mu[L] = 0;
    for (j=k-1; j>=0; j--) {
      if (((tt >> j) & 1) != 0) {
        Addm(Mu,Ptb[j],Mu);
      }
    }
  }
}

void PowMod(Lng UU, Lng S, Lng V) { // V = U^S mod p
int i,j; Lng VV,Pow,U;
  LoOne(VV);
  Mov(UU,U);
  for (i=0; i<L; i++) {
    for (j=0; j<=15; j++) {
      if (((S[i] >> j) & 1) != 0) {
        MulMod(VV,U,Pow); // V*U if S[i,j]=1
        Mov(Pow,VV);
      }
      MulMod(U,U,Pow); // U^2
      Mov(Pow,U);
    }
  }
```

```
 Mov(VV,V);
}

void InvMod(Lng num, Lng MuInv) { // Multiply inversion
 PowMod(num,p2,MuInv);
}

void Scroot(Lng v, Lng u) { // Square Root
Lng f,z;
 PowMod(v,pd4,f); //f=v^(p-1)/4
 PowMod(v,pd8,z); //z=v^(p+3)/8
 Mov(z,u);
 if (Comp(f,p1)==0) MulMod(w,z,u);
}

void AddPointPro(Lng Xt, Lng Yt, Lng Zt, Lng Xs, Lng Ys,
Lng Zs, Lng Xr, Lng Yr, Lng Zr) { // Add Points
Lng A,B,C,D,E,F,G,V,U,H,q;
 MulMod(Zt,Zs,A);
 MulMod(A,A,B);
 MulMod(Xt,Xs,C);
 MulMod(Yt,Ys,D);
 MulMod(C,D,V);
 MulMod(const_d,V,E);
 Subm(B,E,F);
 Addm(B,E,G);
 MulMod(const_a,D,V);
 Subm(C,V,U);
 MulMod(G,U,V);
 MulMod(A,V,q);
 Addm(Xt,Yt,V);
 Addm(Xs,Ys,U);
 MulMod(U,V,H);
 Subm(H,C,V);
 Subm(V,D,U);
 MulMod(A,U,V);
 Mov(q,Xr);
 MulMod(F,V,Yr);
 MulMod(F,G,Zr);
}

void MulPointPro(Lng fk, Lng RXin, Lng RYin, Lng kX,
Lng kY) { // scalar * Point
int i,j; Lng RX,RY,RZ,QX,QY,QZ,AZ; // in Projective
coordinates
 LoOne(QX); // Projective coordinate Point "O" = (1,0,1)
!!!
 LoZero(QY);
 LoOne(QZ);
 Mov(RXin,RX);
 Mov(RYin,RY);
 LoOne(RZ); // Projective coordinate our Point =
(RX,RY,1)
 for (i=0; i<L; i++) {
  for (j=0; j<=15; j++) {
   if (((fk[i] >> j) & 1) != 0) {
    AddPointPro(QX,QY,QZ,RX,RY,RZ,QX,QY,QZ);
   }
   AddPointPro(RX,RY,RZ,RX,RY,RZ,RX,RY,RZ);
  }
 }
 InvMod(QZ,AZ);
```

```
 MulMod(QX,AZ,kX);
 MulMod(QY,AZ,kY);
}

void RecYR() {
 MulMod(r,r,r2);
 LoOne(F);
 Subm(F,r2,A); // A=(1-r^2)
 MulMod(const_d,r2,B);
 Subm(const_a,B,F); // F=(a-dr^2)
 InvMod(F,B); // B=(a-dr^2)^-1
 MulMod(A,B,v); // v=(a-dr^2)(a-dr^2)^-1
 Scroot(v,y);
 LoZero(F);
 Subm(F,y,y); // RecYR
}

void Init(){
int i;
 LoOne(A);
 Sub(p,A,p1); // p-1 = FC94
 p1[L] = 0;
 A[0] = 2;
 Sub(p,A,p2); // p-2 = FC93
 p2[L] = 0;
 Mov(A,const_a); // a=2
 A[0] = 0x010D;
 Mov(A,const_d); // d=10D
 LoZero(A);
 Sub(A,p,np); // 2 - p = 036B
 Mov(p,pd4); // p
 pd4[0] = 0xFF25; //p-1 div 4
 pd4[L-1] = 0x3FFF; // div 4
 Mov(p,pd8); // p
 pd8[0] = 0xFF93; //p+3 div 8
 pd8[L-1] = 0x1FFF; // div 8
 Mov(np,Ptb[0]);
 for (i=1; i<k; i++) { // Mtb init
  Left1(Ptb[i-1],Ptb[i]);
 }
}

int main(){
 Init();
 std::cout << "BOB: Open Key for Decrypt Q = eP ---------
------------------" << std::endl;
 LoZero(F);
 F[0] = 0x25;//25;// e - decrypt key
 MulPointPro(F,Xp,Yp,XQ,YQ); // Q = eP
 std::cout << "XQ = " << std::endl;
 PrntLng(XQ);
 std::cout << "YQ = " << std::endl;
 PrntLng(YQ);
 std::cout << "ALICE: Encrypt Key = XT -------------------
-----------------" << std::endl;
 F[0] = 0x7; // eps - encrypt key
 MulPointPro(F,Xp,Yp,XR,YR); // R = epsP
 std::cout << "R = epsP (eps=7):" << std::endl;
 std::cout << "r = XR = " << std::endl;
 PrntLng(XR);
 F[0] = 0x7; // eps - encrypt key
 MulPointPro(F,XQ,YQ,XT,YT); // T = epsQ
```

```
 std::cout << "T = epsQ (eps=7):" << std::endl;
 std::cout << "XT = Encrypt Key =" << std::endl;
 PrntLng(XT);
 std::cout << "YT = " << std::endl;
 PrntLng(YT);
 Mov(XR,r);
 std::cout << "BOB: Decrypt Key = XT' ---------------------
-----------------" << std::endl;
 RecYR();
 std::cout << "r^2 = " << std::endl;
 PrntLng(r2);
 std::cout << "Recovery y = " << std::endl;
 PrntLng(y);
 LoOne(F);
 F[0] = 0x25; // e - decrypt key
 MulPointPro(F,r,y,XT,YT); // Q = eP
 std::cout << "T' = eR' (e=$25):" << std::endl;
 std::cout << "XT' = Decrypt Key =" << std::endl;
 PrntLng(XT);
 std::cout << "OK" << std::endl;
 std::cin.get();
 return 0;
}
```

5 Conclusion and Future Work

This article provided an in-depth examination of the DSTU 9041:2020, Ukraine's national standard for encrypting short messages using Twisted Edwards Elliptic Curves. It emphasized the standard's innovative aspects, such as enhanced performance and compatibility with existing cryptographic tools. The practical implementation of the standard in the C programming language was also explored, addressing challenges like computational efficiency, bandwidth constraints, and energy consumption. This study serves as a valuable resource for the scientific and engineering community, particularly those invested in cutting-edge cryptographic techniques and secure information exchange protocols. Through mathematical explanations, code snippets, and practical examples, the article elucidated the implementation and cryptographic transformations of the DSTU 9041:2020 standard.

Future research could focus on optimizing the standard's implementation for various platforms, particularly for IoT devices, considering their unique constraints. Another avenue for exploration is the potential integration of the standard with emerging technologies like blockchain and quantum computing, which could revolutionize data security. Additionally, conducting comparative studies with other elliptic curve cryptography standards could provide deeper insights into performance and security efficacy. Continuous updates and improvements in line with the evolving landscape of cybersecurity threats will ensure the standard's relevance and robustness in secure communication protocols [5].

References

1. Shchur N, Pokotylo O, Bailiuk Ye (2023) Cryptography on elliptic curves and its practical application. Cybersecur Educ Sci Tech 1(21):48–64

2. Olefir PY (2021) Estimates of the effectiveness of methods for adding points on an elliptic curve in Edwards form. Master's thesis, Igor Sikorsky Kyiv Polytechnic Institute
3. Kotukh Y, Khalimov G, Korobchinskiy M (2023) Method of security improvement for MST3 cryptosystem based on automorphism group of Ree function field. Theor Appl Cybersecur 5(2)
4. Kotukh Y, Khalimov H (22023) Method of directed encryption based on the functional field of the Hermitian group enhanced by homomorphic transformation. Ukrainian Sci J Inf Secur 29(3):124–129
5. Krishna HV, Sekhar KR (2024) Enhancing security in IIoT applications through efficient quantum key exchange and advanced encryption standard. Soft Comput, 1–11

Lunar Robotics Evolution and Innovative Design

Serdar Kalaycioglu, Anton de Ruiter, Zachary Xie, Jeric Jiang, Angelina Tseng, and Haipeng Xie

Abstract The frontier of space exploration continues to expand with the Moon as a critical focus, offering insights into our universe's origins and potential resources for future missions. This paper presents a comprehensive overview of the advancements in autonomous lunar exploration systems, emphasizing the integration of innovative design and control mechanisms in rover technology. We explore the evolution from early lunar missions to the development of sophisticated Multi-Rover Systems equipped with advanced autonomy and mobility to navigate the challenging lunar terrain. This manuscript presents an innovative approach to lunar exploration with the development of the Lunar Exploration Rover System (LERS), an advanced prototype crafted for navigating and executing tasks across the Moon's varied landscape. The LERS integrates a mobile exploration unit and attached manipulator arms for detailed operations, aiming to facilitate a wide array of lunar research activities. The paper details the unique design principles, notable characteristics, and the experimental platform utilized for the system's iterative testing and enhancement. This initiative marks a critical leap forward in developing dynamic and efficient robotic solutions for the challenges of lunar exploration.

Keywords Design of lunar rovers · Multi-rover robotics · Modular design

1 Introduction

Space exploration has long transcended beyond mere curiosity, undertaking the ambitious task of unraveling the mysteries of the cosmos, with the Moon acting as a pivotal step in this journey. The development of Lunar Exploration Rover Systems

S. Kalaycioglu (✉) · A. de Ruiter
Toronto Metropolitan University, Toronto, ON M5B 2K3, Canada
e-mail: skalay@torontomu.ca

S. Kalaycioglu · Z. Xie · J. Jiang · A. Tseng · H. Xie
Dr. Robot Inc., Richmond Hill, ON L4B 1G8, Canada

P. G. Bradford et al. (eds.), *Proceedings of IEMTRONICS 2024*, Lecture Notes in Electrical Engineering 1229, https://doi.org/10.1007/978-981-97-4780-1_2

(LERS) has become instrumental in these endeavors, driven by the dual objectives of enhancing scientific research and proving technologies for future space exploration.

The lunar surface presents a complex array of challenges, from its diverse geological landscapes to the extreme environmental conditions. This necessitates the deployment of rover systems capable of extensive mobility and robust operational capabilities. Historical missions have laid the groundwork, while contemporary efforts focus on leveraging technological advancements for improved exploration outcomes [1–3].

The heterogeneity and complexity of the Moon's terrain require the deployment of mobile research platforms. Stationary instruments, although effective in geologically stable zones, are insufficient for investigating the Moon's diverse mineralogical formations [3]. It is estimated that exhaustive lunar exploration endeavors may span distances up to 500 km. This expansive scope, coupled with sophisticated geological operations such as core drilling and material sampling, necessitates the development of durable and adaptable roving vehicles. These vehicles are tasked with traversing extensive areas, negotiating challenging landscapes, preserving the condition of collected specimens, and executing field-based scientific analyses.

The progression of lunar rover technology has experienced notable enhancements over time. The initial models, exemplified by the USSR's Lunokhod and NASA's Apollo Lunar Rovers, laid the groundwork for the sophisticated designs we see today. In recent years, the emphasis in rover development has transitioned toward enhancing their mobility, flexibility, and self-governing capabilities [4, 5].

Contemporary models, such as NASA's Curiosity rover and the European Space Agency's ExoMars rover, boast state-of-the-art movement systems capable of navigating the moon's diverse landscapes, from its fine-grained soils to its rugged terrains [6, 7].

Despite these advancements, current rover models encounter significant obstacles related to energy conservation, the negotiation of physical barriers, and their ability to adapt to the extreme environmental conditions found on the lunar surface.

The design of lunar rovers has evolved significantly, influenced by the need for high mobility, energy efficiency, and the ability to perform complex scientific tasks. In the exploration of the complex issues related to rover movement across lunar terrains, various research efforts have investigated the principles of dynamics and design governing vehicle motion in these environments. The work of Irani and Bauer, for instance, has significantly contributed to the field by examining the terra-mechanics of lightweight vehicles equipped with solid wheels and grousers on sandy substrates [8]. Their findings lay crucial groundwork in comprehending how vehicles engage with loose, granular surfaces, which are prevalent on celestial bodies beyond Earth.

Leveraging foundational work in rover dynamics, a detailed examination of the performance of flexible wheels, focusing on the effects of wheel deflection, was conducted in [9]. This study provided key insights into the adaptability and operational efficiency of flexible wheels in lunar rover applications, offering a significant contribution to the understanding of wheel technology in extraterrestrial exploration. Further research [10] analyzed the mobility of wheeled lunar robots on soft terrain, emphasizing the importance of innovative wheel design in overcoming the challenges presented by the lunar environment [10]. These investigations highlighted the critical role of wheel adaptation in navigating the moon's surface, underscoring the need for ongoing advancements in rover locomotion technology.

Advancing the practical implications of these research outcomes, the study by Patel, Slade, and Clemmet [11] explored the complexities of the ExoMars rover's mobility subsystem. This investigation offered a comprehensive analysis of its architectural and functional specifications, aligning them with the principles of terramechanics [11]. In a subsequent development, Torre [12] introduced the conceptual framework for the Team ITALIA Rover, a contender in the Google Lunar X Prize competition. This work illustrated the effective translation of theoretical concepts into practical designs for lunar exploration vehicles, demonstrating the feasibility of applying advanced scientific theories to the engineering of rover systems for lunar missions [12].

The foundational basis for these research endeavors is established by the pioneering contributions of Wong [13, 14] offering an exhaustive overview of ground vehicle theory and the multifaceted discipline of terra-mechanics. The in-depth investigations conducted by Wong into the dynamics of terrain interaction, the efficacy of off-road vehicles, and their design principles stand as critical resources for grasping the intricate dynamics at play between vehicles and diverse surface conditions. This body of work significantly enhances the knowledge pool, guiding both the conceptualization and refinement processes involved in the engineering of lunar rover systems, ensuring their effective operation across the moon's challenging landscapes.

A wide array of characteristics distinguishes lunar rovers, encompassing their methods of movement, which range from legged locomotion and wheeled travel to limbless sliding. The design of their suspension systems also varies widely, incorporating configurations from the Rocker-bogie setup to independent and flexible arrangements. Steering functionalities among these vehicles differ, including Skid-steer and Ackerman steering systems, among more specialized techniques. The level of human oversight required is dictated by the rover's control algorithms, enabling operations from fully autonomous to semi-autonomous modes. In terms of structure, lunar rovers can be designed with either a unified body or a segmented architecture, each tailored to navigate specific types of terrain, from rugged landscapes to flat plains. Additionally, their ability to find their way is supported by a spectrum of technologies, ranging from celestial navigation using stars and the sun to Earth-based systems like GPS and advanced sensor arrays [15].

Recent advancements in design approaches have focused on improving the functionality and efficiency of lunar rovers, including the refinement of wheel configurations and the adoption of sophisticated suspension technologies such as the rocker-bogie system [16–18].

These innovations prioritize the rovers' stability, energy conservation, and flexibility across different landscapes. Moreover, the incorporation of cutting-edge artificial intelligence and machine learning techniques has significantly enhanced the autonomy of these vehicles, facilitated complex navigational tactics, and enabled instantaneous analytical processing.

With the escalation of goals in lunar exploration, there is a growing focus on the deployment of multi-rover systems. This collective strategy, involving the synchronized operation of several rovers, is aimed at enhancing the efficiency of exploration activities, spreading out scientific inquiries, and enabling the coverage of vast lunar expanses within reduced durations [19, 20]. Nonetheless, the development of these integrated systems presents additional challenges, including the need for sophisticated communication protocols, precise coordination mechanisms, and efficient energy utilization strategies.

Beyond the intricate design features, the autonomy of lunar rovers has been remarkably advanced through the adoption of advanced robust control algorithms, as highlighted in recent literature by Kalaycioglu and De Ruiter [21].

The study of wheeled locomotion mechanics has garnered significant interest within the scientific community [22–27]. Extensive research efforts have been dedicated to exploring the dynamics and kinematics control associated with the Mecanum wheel, a specialized type of omnidirectional wheel, highlighting its unique properties and applications [28–33]. Additionally, considerable attention has been paid to control of systems featuring a four-wheel independent drive and independent steering mechanism, commonly referred to as "Swerve" drive. This configuration offers enhanced maneuverability and precision in navigation, further broadening the scope of investigation into efficient locomotion strategies for mobile robots.

Substantial research has been conducted on the multifaceted aspects of wheeled locomotion, yet a comprehensive understanding of the control mechanisms governing the coordinated movement of multiple rovers and their integrated arms remains elusive. Specifically, the design of systems equipped with multiple rovers, each featuring dual manipulators capable of executing real-time trajectories while jointly handling a shared load, is in its nascent phase. This highlights a significant gap in the current state of robotics, pointing toward the complexity of developing a robust Lunar Exploration Rover System (LERS) carrying out these types of synchronized operations.

This manuscript explores the sophisticated design and utility of LERS, shedding light on their multifaceted applications. In Sect. 2, we detail the missions envisioned for LERS and outline the specific Strawman Tasks, offering an in-depth analysis of their intended purposes and capabilities. Section 3 is dedicated to an extensive examination of the Testbed System, encompassing the Remote Subsystem that includes the Rover—detailing its mechanical framework as well as its electrical and electronic systems, such as Command, Control, and Communication (C^3)—alongside the Robotic Arms, which are similarly categorized. Concurrently, we discuss the Local Subsystem, emphasizing the Control Workstation and the integration of a High-fidelity Unity Simulator equipped with Virtual Reality features. Central to this discussion is the Communication Interfaces, crucial for facilitating cohesive operation between the remote and local subsystems. The concluding section, Sect. 4, synthesizes our investigation, highlighting the principal insights and the broader implications of our research on lunar exploration technologies.

2 LERS Missions and Possible Strawman Tasks

The Lunar Exploration Rover System (LERS) plays a crucial role in this arena, offering a wide array of operational capabilities and missions that are specifically crafted to further the aims of lunar exploration.

A key aspect of the LERS mission framework is its dual functionality: the rover, optimized for navigation and discovery, alongside the robotic arm, tailored for meticulous and skillful execution of tasks. This integration facilitates a broad spectrum of activities, encompassing

- **Exploration and Cartography**: At the core of the rover's objectives lies its exploratory mission, through which it generates detailed cartographic representations of lunar terrains, supporting subsequent exploration endeavors.
- **Deployment of Scientific Instruments**: Equipped with its robotic arm, the LERS transforms into a versatile science laboratory on wheels, adept at positioning diverse instruments for extensive experimental activities.
- **Sample Gathering**: The lunar surface, rich in historical narratives encapsulated within its rocks and soil, offers a window into its past. The LERS, equipped with a comprehensive set of instruments, is adept at harvesting and examining these samples, shedding light on the Moon's developmental history and prospective resources.
- **Support for Astronauts**: Serving as a terrestrial ally for astronauts, the LERS is designed to transport equipment and provisions, thereby enhancing their exploratory capabilities and extending their operational reach across the lunar landscape.
- **Autonomous Lunar Exploration**: The rover extends its utility beyond astronaut assistance through its self-governing features, allowing it to conduct solo missions aimed at charting the Moon's topography and examining its geological composition.
- **Construction and Infrastructure Development**: The robotic arm's capabilities transcend exploration, proving essential for construction efforts. It plays a significant role in erecting structures, establishing lunar habitats, and performing upkeep on existing facilities.
- **Exploitation of Lunar Resources**: Leveraging the rover's navigational abilities and the robotic arm's accuracy, this system is well-equipped for extracting lunar resources, including water ice and precious minerals, thereby tapping into the Moon's wealth of assets.
- **Remote Operations and Sensing**: The functionality of the LERS extends beyond direct lunar interaction, allowing for remote control from Earth. This capability ensures safe and accurate maneuvers across the Moon's challenging landscapes, even from vast distances.
- **Collaboration and International Missions**: In the spirit of global cooperation, the LERS can integrate into collaborative missions, working synergistically with assets from international space agencies to further our collective knowledge of the Moon.

Building upon the multifaceted capabilities of the LERS, the system is designed for extensive locomotive missions, penetrating subsurface drilling to acquire lunar core samples, routine collection of surface regolith, and the conduct of a diverse array of scientific investigations encompassing gravimetry, magnetometry, and spectrometry. Furthermore, the LERS is equipped to establish enduring observational outposts and oversee the conveyance and manipulation of gathered samples to lunar transport vehicles at predetermined assembly locations.

3 LERS System Architecture Overview

The LERS testbed, crafted to simulate potential lunar exploration missions, forms a critical base for the evolution and confirmation of both rover and robotic arm capabilities. This segment offers a detailed examination of the testbed architecture, highlighting its key elements, the connectivity between these components, and their collective contribution to the overarching goals of lunar exploration.

The architectural design of the LERS system is composed of remote and local subsystems, interconnected through a communication interface, as delineated in Fig. 1a, b. The process of designing the LERS physical system is comprehensively detailed and visualized in Figs. 2 and 3, illustrating the step-by-step approach and methodologies involved. This detailed depiction provides a clear understanding of the workflow and key components that contribute to the development of the LERS system, showcasing the thoughtful consideration and engineering principles applied throughout the design phase.

3.1 Remote Subsystem

The remote subsystem is fundamentally composed of the rover, the robotic arm, and the Command, Control, and Communication (C^3) components. These elements are crucial to the operational capabilities of the LERS, playing a pivotal role in its functionality and effectiveness.

- **Rover**: The rover, depicted in Fig. 4, is engineered for traversing diverse lunar landscapes, establishing itself as a critical component for exploration endeavors. It amalgamates mechanical, electrical, and electronic parts, seamlessly integrated with Command, Control, and Communication (C^3) functionalities. This integration is vital for optimizing energy usage, supporting autonomous movement, and ensuring continuous communication with the control station, thereby enhancing its operational efficiency and effectiveness in lunar exploration missions.

The chassis serves as the rover's structural backbone, enveloping its sides and integrating the swerve modules, coupled with a central joint that links the corresponding modules on each side. Opting for a swerve design enhances the rover's

(a)

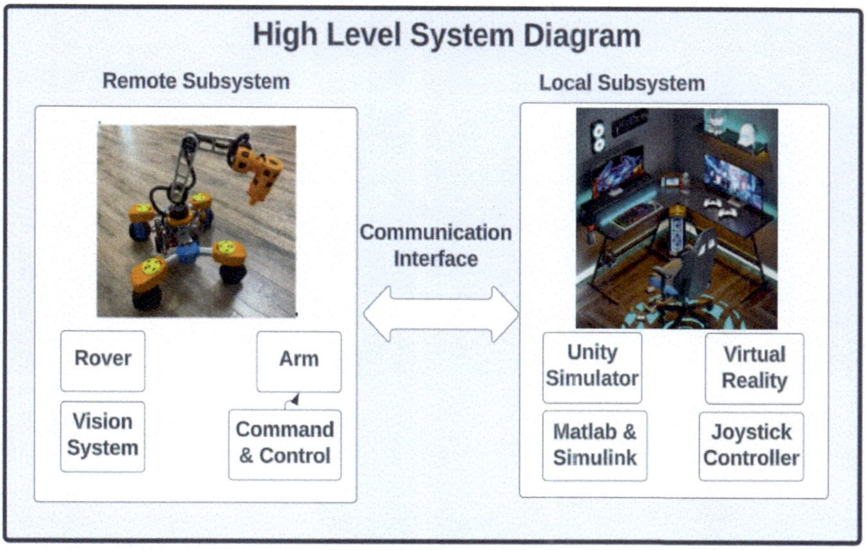

Fig. 1 **a** LERS high-level system architecture. **b**. Matlab/Simulink—control and unity interfaces

(b)

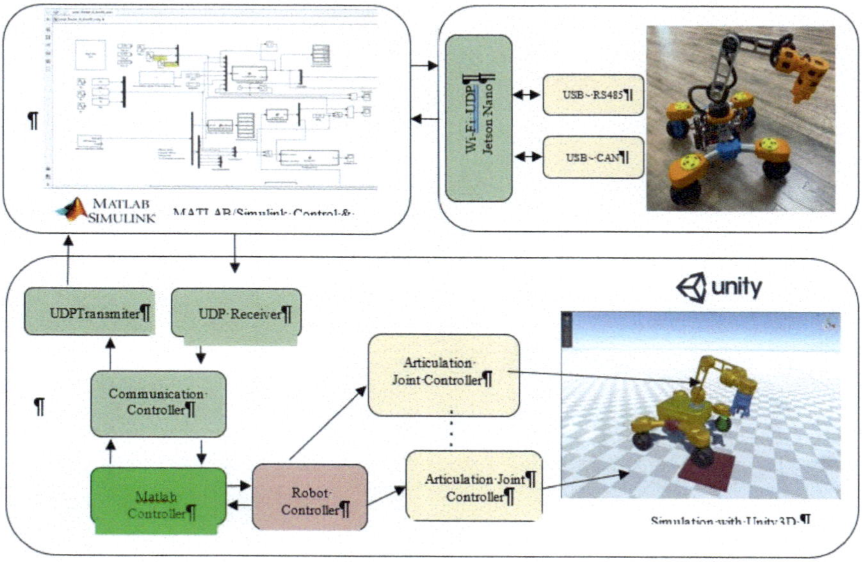

Fig. 1 (continued)

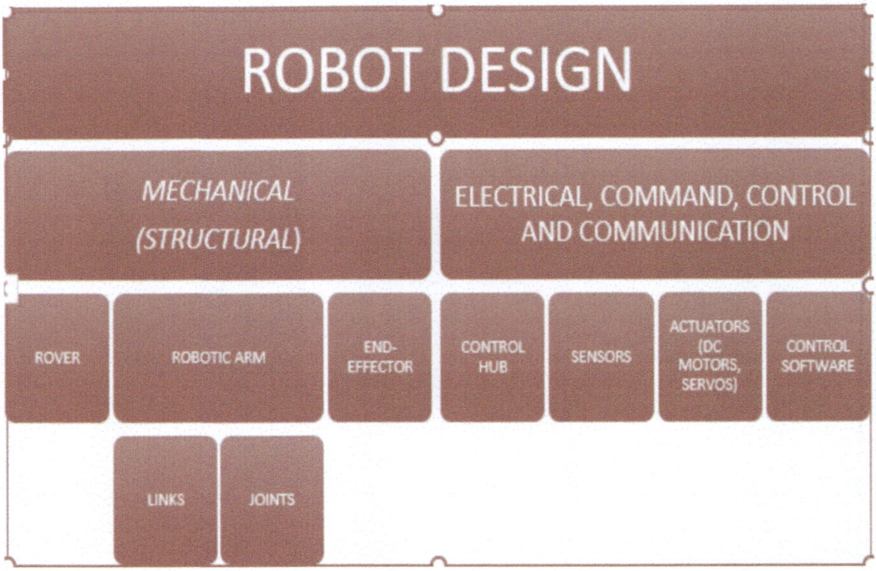

Fig. 2 LERS physical system design diagram

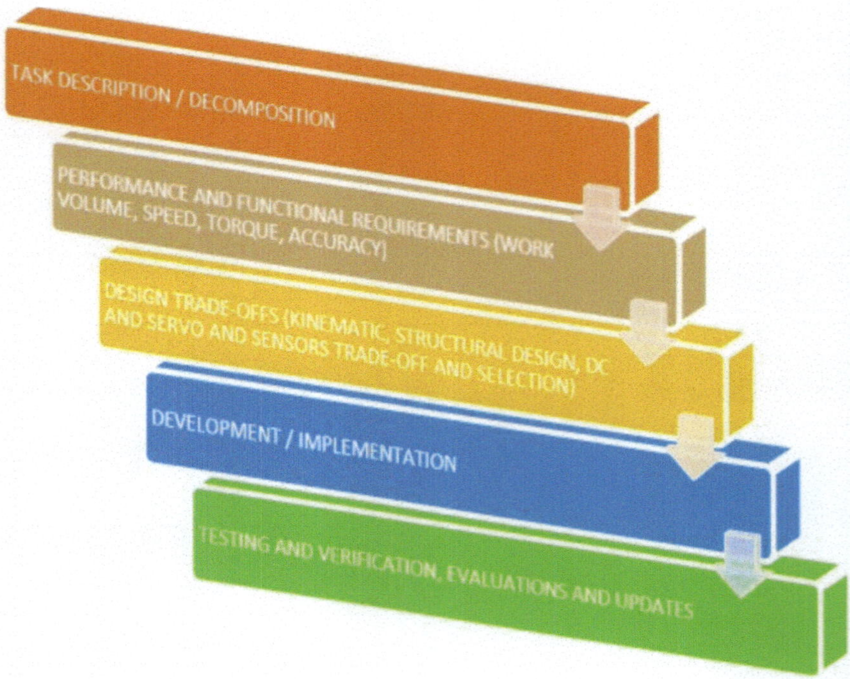

Fig. 3 Process of LERS systems engineering design cycle

Fig. 4 Rover CAD chassis design

agility, permitting each wheel to pivot independently around an axis that is perpendicular to its main rotation. This innovative design empowers the chassis to perform dynamic turns, execute strafing maneuvers, and preserve stability over uneven terrain, thereby optimizing the rover's maneuverability across various lunar surfaces.

The rover employs four RMD-L-5005 brushless DC motors for its steering mechanism, chosen for their high precision and programmable torque capabilities. These servos are instrumental in enhancing the rover's maneuverability over a wide range of lunar terrains. Their compact size, robust strength, and compatibility with a CAN bus system significantly bolster the rover's adaptability and performance across various exploration scenarios. This integration of advanced servos ensures that the rover can navigate the challenges of lunar exploration with greater efficiency and reliability.

Furthermore, the rover is outfitted with four DDSM115TM hub electric motors, tasked with managing wheel rotation. These motors stand out for their direct-drive system, ability to operate at low speeds with high torque, and their quiet operation. Such characteristics are pivotal for ensuring the rover maintains stability and control when navigating rough terrains, while also keeping noise disturbances to a minimum.

This combination of features makes these motors an ideal choice for the demanding conditions of lunar exploration, facilitating smooth and efficient movement across the moon's surface.

Figure 4 offers a detailed visualization of the chassis integrated with the robotic arm, with further elaboration provided in the following subsection. This illustration highlights the seamless integration of the chassis design with the robotic arm, showcasing the intricate engineering and design considerations that enable their effective collaboration in operational scenarios.

In conclusion, the rover's design, as illustrated in Fig. 4, integrates a range of components and subsystems for precise and adaptable navigation across lunar terrains. It is also equipped with an advanced vision system, tasked with determining the position and orientation of obstacles and targets. This enhancement significantly improves the rover's navigational capabilities, enabling a more informed and effective strategy for navigating the varied landscapes of the moon. This holistic approach to design and functionality ensures the rover is well-prepared to meet the challenges of lunar exploration.

- **Robotics arm**: Mounted on the rover, robotic arms are essential for precision tasks, combining mechanical components with advanced electrical and electronic systems, including a dedicated Command, Control, and Communication (C^3) module. These arms are designed for a variety of operations, from sample collection to equipment deployment. Figure 5a–d shows the arm, including the CAD model, illustrating their design and capabilities. The dimensions are provided in mm.

The robotic arm exhibits a structural configuration characterized by six Degrees of Freedom (DOF), with its dimensions specified in millimeters (mm). Notably, this design incorporates brushless gear DC motors, which are controlled by CAN Bus and recognized for their precision in torque programming, as well as their consistent performance and reliability in scientific applications.

(a)

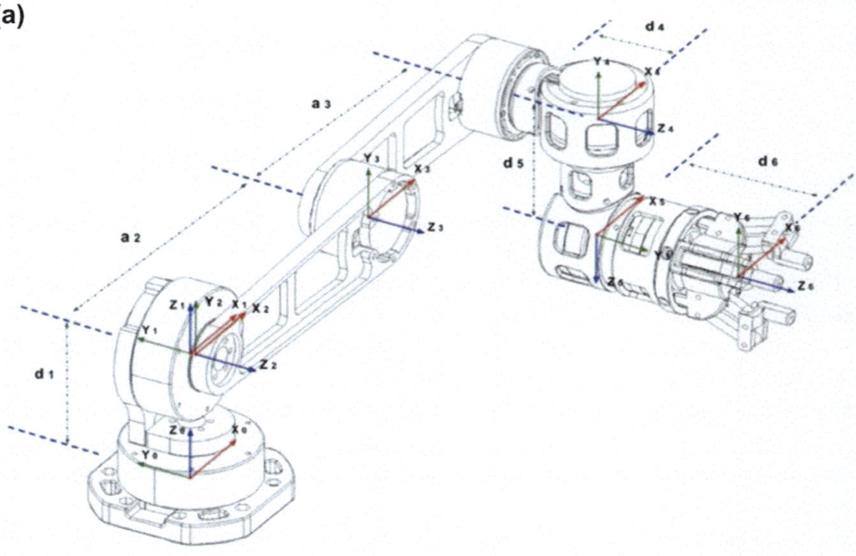

Fig. 5 **a** 6-DOF robotics arm—zero-home configuration. **b** 6-DOF robotics arm engineering drawings. **c** 6-DOF robotics arm engineering drawings—front and side views. **d** 6-DOF robotics arm CAD model. **e** LERS command, control, and communication C^3 architecture

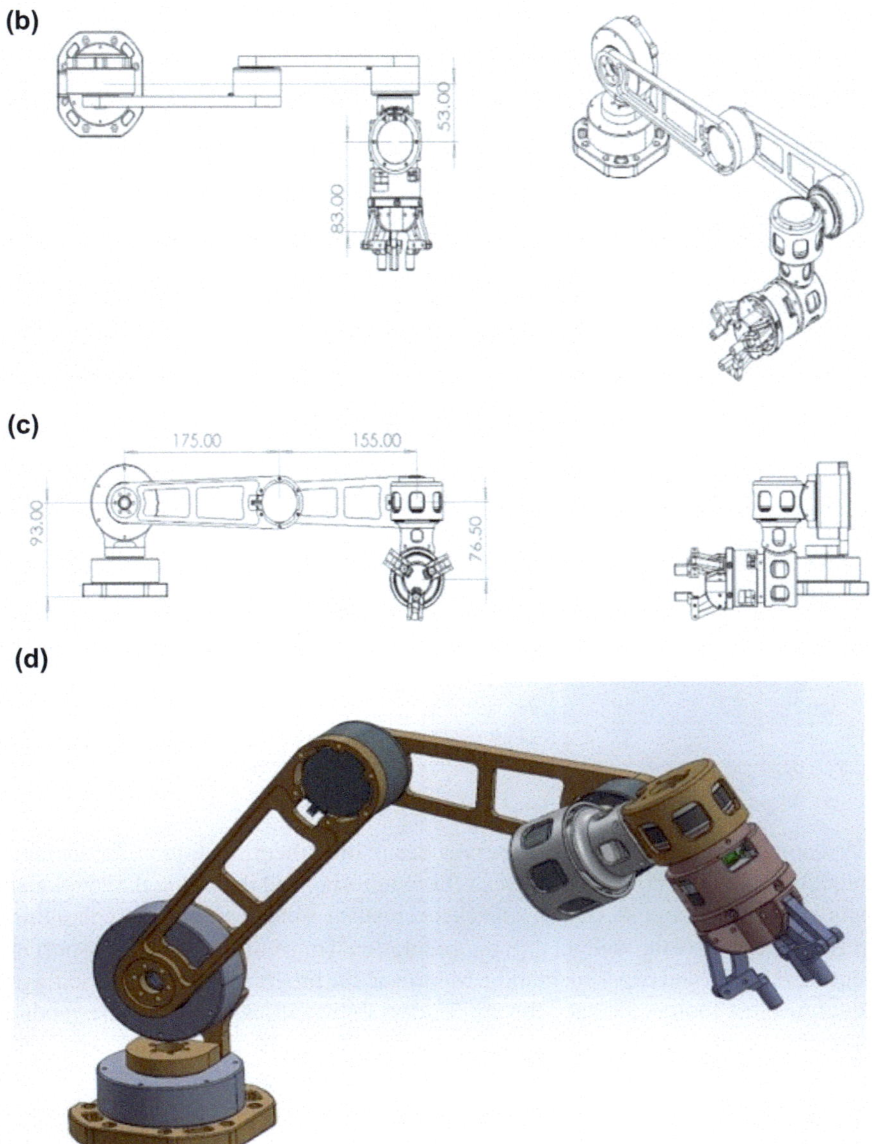

Fig. 5 (continued)

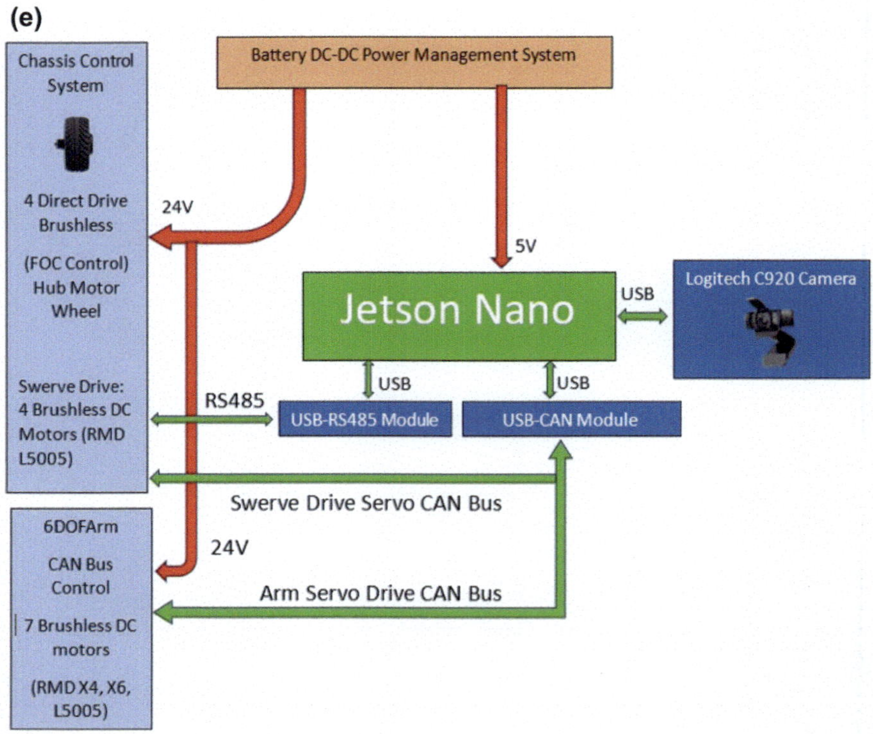

Fig. 5 (continued)

A distinguishing feature of these servos lies in their incorporation of 360-degree magnetic encoders, a critical component for monitoring and recording the rotational position of the servo output. This feedback mechanism is indispensable for upholding precision and control throughout the arm's operational movements, a crucial aspect in scientific research and experimentation. Moreover, the programmable torque settings in the brushless motors enhance the arm's adaptability, allowing for precise modulation of exerted force to align with the specific demands of scientific tasks and experiments.

- **Command, Control, and Communication C³**: The Command, Control, and Communication (C^3) system plays a crucial role in the rover's operations and interactions. At its core, we find the Jetson Nano processor, serving as the central processing unit (CPU). This CPU is constructed on the quad-core ARM Cortex-A57 with MPCore processor architecture, providing robust computational power and efficient multitasking capabilities, which are indispensable for the rover's diverse functionalities. Figure 5e illustrates the LERS Command, Control, and Communications (C^3) Architecture for reference.

The system is equipped with a Graphics Processing Unit (GPU) that leverages the NVIDIA Maxwell™ architecture and boasts 128 NVIDIA CUDA cores. This advanced GPU architecture guarantees high-performance graphics processing, empowering the rover to execute intricate graphical tasks and simulations. Consequently, this enhances its visual and navigational capabilities significantly. The incorporation of CUDA technology further amplifies the rover's computational abilities by enabling parallel computing and accelerating scientific simulations.

In addition to the CPU and GPU, the system is equipped with 4 GB of 64-bit LPDDR4 memory, operating at a frequency of 1600 MHz. This generous memory configuration provides a significant memory bandwidth of 25.6 GB/second, playing a crucial role in managing extensive datasets and ensuring the seamless execution of memory-intensive applications. A noteworthy feature is the support for CUDA and OpenCV, enabling the development and deployment of computer vision applications that are essential for the rover's navigation and observation tasks.

Furthermore, the C^3 system includes an ESP32 Servo Driver™ expansion board, which extends the rover's connectivity capabilities. This board comes equipped with integrated Wi-Fi and Bluetooth capabilities, allowing for wireless communication and data transfer. These connectivity features play a crucial role in maintaining real-time communication with the control station, ensuring remote operability of the rover, and facilitating the seamless transmission of data and telemetry.

In summary, the Command, Control, and Communication (C^3) system, with its core based on the Jetson Nano™ processor and enhanced by advanced memory and connectivity components, serves as the technological foundation of the rover. This integrated system guarantees robust computational power, streamlined communication, and versatile control, enabling the rover to navigate and operate effectively across various lunar environments.

3.2 *Local Subsystem and Communication Interface*

As the Remote Subsystem operates directly on the lunar surface, the Local Subsystem remains stationed at the base, serving as an interface for human operators and facilitating the oversight and management of LERS operations.

- **Control workstation**: The Control Workstation serves as the primary interface for human operators. Outfitted with cutting-edge software and hardware, it empowers operators to monitor, guide, and intervene in LERS operations as required. The workstation offers a real-time feed of the rover's activities, the status of its robotic arms, and essential system diagnostics.
- **Unity simulation and VR environment**: This state-of-the-art simulation platform plays a vital role within the testbed. Before actual lunar tasks commence, the LERS' operations can be replicated in a virtual setting that faithfully imitates lunar conditions. This not only assists in operator training but also permits the examination and enhancement of operational procedures. The Unity models for our

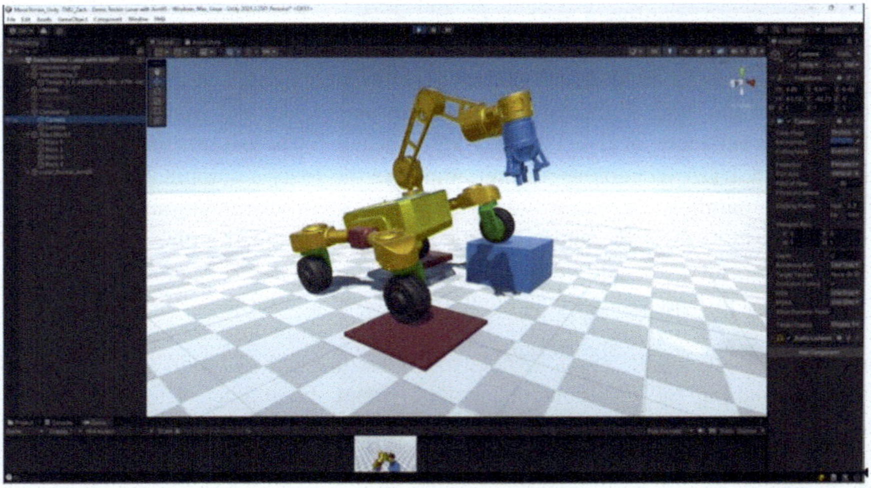

Fig. 6 Unity model and simulation

rocker-bogie lunar rover and 6-DOF manipulator arm are constructed using Unity Articulation Body and Rigid Body components. These completed models simulate full physics dynamics, with adjustable joint stiffness and damping, allowing for further fine-tuning of the model (Fig. 6) The Virtual Reality (VR) component enhances realism by providing a three-dimensional immersive experience, aiding in the understanding of challenges and the optimization of the robot's functions.

- **Communication Interface**: Bridging the Remote and Local Subsystems is a resilient Communication and Interface framework. This guarantees that directives from the Control Workstation are delivered to the LERS promptly and that data from the LERS, encompassing visuals, diagnostics, and other essential details, is transmitted in real time. In light of the vast interstellar distances and distinctive communication challenges, this system incorporates multiple safeguards and is engineered to manage potential disruptions while upholding a stable connection with the LERS.

In conclusion, the testbed system represents a thorough configuration tailored to simulate, assess, and oversee the Moon Rover Robotics System within a controlled setting. Its purpose is to guarantee that when the MRRS is deployed on the lunar surface, it functions at peak efficiency, thereby ensuring the success of its missions and the safety of any accompanying astronauts. An illustration of the system prototype is available in Fig. 7.

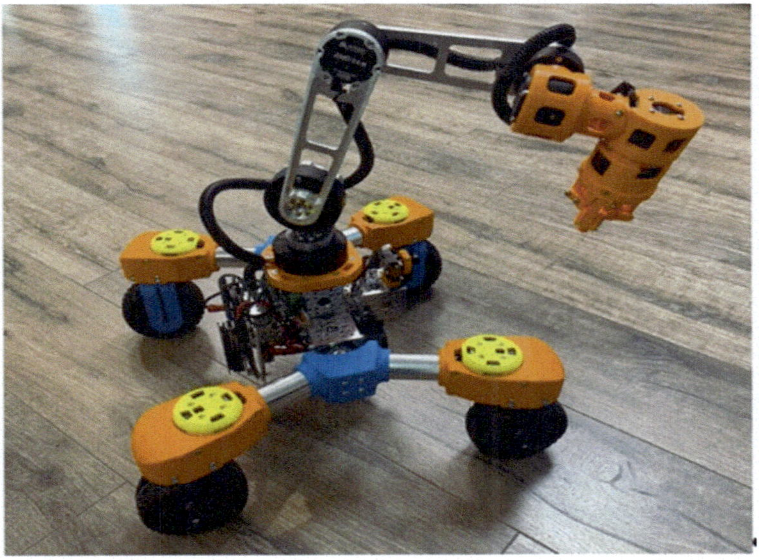

Fig. 7 A Rover system prototype

4 Conclusion

In summary, the Lunar Exploration and Robotic System (LERS) represents a significant leap forward in space exploration technology, especially for lunar missions. This system seamlessly combines the adaptability of a rover with the precision of robotic arms, equipping the LERS to address the multifaceted challenges presented by the rugged lunar terrain. Every component, from the rover's mobility mechanisms to the capabilities of its robotic arms, has been meticulously tailored to meet specific operational demands on the lunar surface.

Complementing the LERS development is our testbed system, which provides a structured framework for evaluating, refining, and validating the LERS's capabilities. Through high-fidelity simulations and an interactive control interface, we ensure that the system can be thoroughly assessed under controlled conditions before its eventual deployment in real-world scenarios.

However, it's essential to underscore the preliminary nature of this prototype. While the LERS displays promising features and functionalities, it is still undergoing rigorous testing and validation. The current system iteration requires the development of robust control algorithms to ensure optimal performance across diverse lunar conditions. Calibration, testing, and validation of hardware components will play a pivotal role in the subsequent research phases. Furthermore, as part of our future research endeavors, we plan to implement robust control techniques in the hardware implementation phase, further solidifying their practical application for the lunar rover's operations.

References

1. Siddiqi A (2018) Beyond Earth: a chronicle of deep space exploration, 1958–2016 (PDF), The NASA history series, 2nd edns. NASA History Program Office, Washington, DC, p 1. ISBN 9781626830424. LCCN 2017059404. SP2018-4041
2. Raykunav GG (2014) Fundamental space research. Fismatlit, Moscow, p 457. ISBN 978-5-9221-1559-9
3. Slyuta EN, Vasilev AV, Dalyaev IY (2017) "Robot-Geologist": scientific tasks and technical configuration. In: Proceedings of 48th Lunar and Planetary Science Conference, 20–24 March 2017
4. Wilhelms DE (1987) The Geologic history of the moon. United States Government Printing Office, Washington, p 329. ISBN 1495919854
5. Wilson J (2005) How we'll get back to the moon. GoddardView 1(223)
6. Putz P (1998) Space robotics in Europe: a survey. Robot Auton Syst 23:3–16
7. Siegwart R, Lamon P, Estier T, Lauria M, Piguet R (2002) Innovative design for wheeled locomotion in rough terrain. Robot Auton Syst 40:151–162
8. Irani R, Bauer R, Warkentin A (2011) A dynamic terramechanic model for small lightweight vehicles with rigid wheels and grousers operating in sandy soil. J Terramech 48:307–318
9. KojiroIizuka I, Kubota L, Takashi M (2009) Study of flexible wheels for lunar exploration rovers: running performance of flexible wheels and various amount of deflection. J Asian Electric Veh 7(2):2009
10. Iizuka K, Kunii Y, Takashi M (2008) Study on Wheeled forms of lunar robots for traversing soft terrain. In: 2008 IEEE/RSJ international conference on intelligent robots and systems. Acropolis Convention Center, Nice, France, 22–26 Sept 2008
11. Patel N, Slade R, Richard C, Clemmet J (2010) The Exomars rover locomotion subsystem. J Terramech 47:227–242
12. Della Torre A, Finzi AE, Genta G (2010) AMALIA mission lunar rover—the conceptual design of the team ITALIA Rover. Acta Astronaut 67(7–8):961–978
13. Wong JY (2001) Theory of ground vehicles, 3rd edn. Wiley
14. Wong JY (2010) Terramechanics & off-road vehicle engineering, terrain behaviour, off-road vehicle performance and design, 2nd edn. Butterworth-Heinemann, UK
15. Barlas F (2004) Design of a Mars Rover suspension mechanism. M.S. thesis, Izmir Institute of Technology, Izmir, pp 11–18
16. Carrier D (1992) Soviet rover systems. In: Proceedings of the AIAA space programs and technologies conference, pp 24–27, Huntsville, AL
17. Kemurdjian A, Gromov V, Mishkinyuk V, Kucherenko V, Sologub P (1992) Small Marsokhod configuration. In: International conference on robotics and automation, vol 1. IEEE, Nice, France, pp 165–168
18. Kim D, Hong H, Kim H, Kim J (2012) Optimal design and kinetic analysis of a stair-climbing mobile robot with rocker-bogie mechanism. Mech Mach Theory 50:90–108. ISSN 0094-114X. https://doi.org/10.1016/j.mechmachtheory.2011.11.013
19. Kalaycioglu S, de Ruiter A (2022) Coordinated motion and force control of multi-rover robotics system with Mecanum wheels. In: 2022 IEEE international IOT, electronics and mechatronics conference (IEMTRONICS). IEEE, pp 1–9. https://doi.org/10.1109/IEMTRONICS55184.2022.9795804
20. Kalaycioglu S, de Ruiter A (2023) Nonlinear model predictive control of rover robotics system. Adv Sci Technol Eng Syst J 8(1):44–56
21. Kalaycioglu S, De Ruiter A (2023) Passivity-based nonlinear model predictive control (PNMPC) of multi-robot systems for space applications. Front Robot AI 10:1181128. https://doi.org/10.3389/frobt.2023.1181128
22. Wada M, Mori S (1996) Holonomic and omnidirectional vehicle with conventional tires. In: Proceedings of IEEE international conference on robotics and automation. IEEE, pp 3671–3676. https://doi.org/10.1109/ROBOT.1996.509272

23. Ostrowski J, Burdick J (1998) The geometric mechanics of undulatory robotic locomotion. Int J Robot Res 17(7):683–701. https://doi.org/10.1177/027836499801700701
24. Stöger C, Müller A, Gattringer H (2018) Parameter identification and model-based control of redundantly actuated, non-holonomic, omnidirectional vehicles, pp 207–229. https://doi.org/10.1007/978-3-319-55011_4
25. Muir PF, Neuman CP (1987) Kinematic modeling of wheeled mobile robots. J Robot Syst 4(2):281–340. https://doi.org/10.1002/rob.4620040209
26. Pin FG, Killough SM (1994) A new family of omnidirectional and holonomic wheeled platforms for mobile robots. IEEE Trans Robot Autom 10(4):480–489. https://doi.org/10.1109/70.313098
27. Campion G, Bastin G, D'Andrea-Novel B (1993) Structural properties and classification of kinematic and dynamic models of wheeled mobile robots. In: Proceedings IEEE international conference on robotics and automation. IEEE Computer Society Press, pp 462–469. https://doi.org/10.1109/ROBOT.1993.292023
28. Wampfler G, Salecker M, Wittenburg J (1989) Kinematics, dynamics, and control of omnidirectional vehicles with Mecanum wheels. Mech Struct Mach 17(2):165–177. https://doi.org/10.1080/15397738909412814
29. Gfrerrer A (2008) Geometry and kinematics of the Mecanum wheel. Comput Aided Geometric Des 25(9):784–791. https://doi.org/10.1016/j.cagd.2008.07.008
30. Lin L-C, Shih H-Y (2013) Modeling and adaptive control of an omni-mecanum-wheeled robot. Intell Control Autom 04(02):166–179. https://doi.org/10.4236/ica.2013.42021
31. Shimada A, Yajima S, Viboonchaicheep P, Samura K (2005) Mecanum-wheel vehicle systems based on position corrective control. In: 31st Annual conference of IEEE Industrial Electronics Society, 2005. IECON 2005. IEEE, 6 pp. https://doi.org/10.1109/IECON.2005.1569224
32. Wang Y, Chang D (2009) Motion performance analysis and layout selection for motion system with four Mecanum wheels. J Mech Eng 45(5):307–316
33. Tatar MO, Popovici C, Mandru D, Ardelean I, Plesa A (2014) Design and development of an autonomous omni-directional mobile robot with Mecanum wheels. In: 2014 IEEE international conference on automation, quality and testing, robotics. IEEE, pp 1–6. https://doi.org/10.1109/AQTR.2014.6857869

Synthetic Datasets for Hand Gesture Recognition

Pranav Vaidik Dhulipala, Samuel Oncken, Steven Claypool, and Stavros Kalafatis

Abstract The ability to train a robot to recognize human gestures is critical in enabling close proximity to Human–Robot Interaction (HRI). To that end, generating the appropriate dataset for the corresponding Machine Learning (ML) algorithm is essential. In this paper, we introduced new datasets for hand gesture recognition. Given the complexity of generating thousands of physical hand gestures, we started with the basic hand gestures and developed additional synthetic gestures thus creating a comprehensive set.

Keywords Synthetic datasets · Deep learning · Computer vision · Domain randomization

1 Introduction

Over the years, deep learning and machine learning techniques have become essential tools for computer vision applications such as object recognition, localization, and segmentation [25]. Deep learning practitioners are often found to spend a significant amount of time finding a suitable dataset for their use, followed by the necessary cleaning and preprocessing steps before applying any of the deep learning techniques [11, 18].

Moreover, for human-centric computer vision applications such as hand gesture recognition, pose detection, and localization, finding the right datasets that satisfy all the annotation requirements is often challenging, due to the human body having many degrees of freedom with a lot of joints, unlike most solid objects. Furthermore,

P. V. Dhulipala (✉) · S. Oncken · S. Claypool · S. Kalafatis
Texas A&M University, College Station, TX 77845, USA
e-mail: pranav.d1993@tamu.edu

S. Claypool
e-mail: svclaypool@tamu.edu

S. Kalafatis
e-mail: skalafatis-tamu@tamu.edu

P. G. Bradford et al. (eds.), *Proceedings of IEMTRONICS 2024*, Lecture Notes in Electrical Engineering 1229, https://doi.org/10.1007/978-981-97-4780-1_3

with activity and gesture recognition applications, annotations also can drastically vary with project requirements, with the same pose or gesture conveying different interpretations, especially in the case of sign language. Consequently, human-centric computer vision applications have to often build custom datasets and annotate them according to their needs.

Common practices for building custom datasets involve acquiring pictures or videos of poses from either the practitioners themselves or a set of limited subjects under controlled conditions, usually indoors. This can affect the variability and scale of the resulting datasets, depending on the complexity of the application, making it impractical for reuse in other applications.

Other practices include a collection of video snippets from a video platform like YouTube, or web scraping from results of an image search engine, followed by manual annotation and cleaning of the acquired data. The manual annotation also poses huge challenges, as it involves labeling datasets on a large scale, with at least 10,000 images or video frames. While computer vision tasks frequently require such large-scale datasets, annotations can consequently be quite time-consuming [8].

Domain randomization is a recently emerged technique that addresses the mentioned challenges by suggesting the training of the deep learning models with simulated data that transfer to real data [26]. The technique focuses on randomizing rendering during training, aiming to add enough variability in the simulation so that the real world may appear as just another variation to the model. However, there is also a reality gap that emerges with the use of synthetic datasets for training, due to the simulators not being able to produce the photorealistic textures and lighting that matches their real counterpart. There have been several publications in the literature addressing the reality gap problem [12, 14, 23, 27, 29]. Recently, it has also been established that the synthetic datasets are easy to render and can be generated on a scale cheaper and faster [30].

Several synthetic datasets have been developed and published over the years for various computer vision applications such as object recognition [4, 13, 15, 19], gait detection [9], and semantic segmentation [5, 6]. There also have been methods and pipelines developed to build synthetic data for some applications. Unity Technologies provides Unity Perception toolkit [28] for building synthetic datasets.

In this work, we have introduced a new pipeline for building synthetic image dataset for hand gesture recognition. In order to evaluate our work, we built three synthetic gesture datasets based on existing datasets, namely the Sign Language for Numbers (Digits Dataset) [17], American Sign Language dataset [20], and HANDS dataset [24], and discussed the simulation environment, generation parameters, and synthetic dataset comparison with their real counterparts. In this document, we discuss some related works in Sect. 2, and the methodology in Sect. 3. The further discussion is mentioned in Sect. 4, followed by conclusion (Sect. 5).

2 Related Work

Over recent years, several human-centric synthetic datasets and pipelines have been published, taking advantage of the underlying principle of domain randomization [16].

Varol et al. [31] presented the dataset SURREAL, containing more than 6 million frames of synthetic human activity. Their work used motion capture data from CMU motion capture dataset [1] and applied them on to various human models generated from an SMPL body model [21]. The dataset is annotated for semantic segmentation of body parts and human pose estimation.

Unity Technologies has introduced the Sans People generator [10] for generating human-centric datasets for detection, localization, segmentation, and pose estimation. The generator consists of a unity scene with a virtual background screen. Annotated data is generated while the background wallpapers rapidly change while human models and obstructive objects spawn and disappear with random poses and orientations.

While using readily available motion capture datasets helps speed up the generation of the dataset, the datasets may not always contain the custom hand gestures one may want to utilize in their project. Furthermore, Sans People generator is also limited where it is not easy to restrict the human models to a specific set the poses.

Miura et al. [22] introduced a synthetic sign language dataset and a pipeline for gesture recognition. Their work consisted of applying a set of sign language poses onto a set of synthetic human models from the SURREAL dataset. The dataset, however, is limited to full-body images in contrast to just hand signs, which is more common with hand gesture datasets. Furthermore, the dataset does not use changing background textures to increase the variability of the dataset.

3 Methods

The sizes and description of the datasets generated are described below

- Digits dataset is 31.2 GB with 132,000 images consisting of 11 classes, 12,000 images per class. Figure 2 shows some sample images from the dataset, compared to its real dataset counterpart.
- ASL dataset is 88.6 GB with 336,000 images consisting of 28 classes, 12,000 images per class. Figure 3 shows some sample images from the dataset, compared to its real dataset counterpart.
- HANDS dataset is 37.1 GB with 144,000 images consisting of 12 classes, 12,000 images per class. Figure 1 shows some sample images from the dataset, compared to its real dataset counterpart. In order to only focus on one-handed gestures, we have omitted the simulation of the two-handed gestures from the real counterpart.

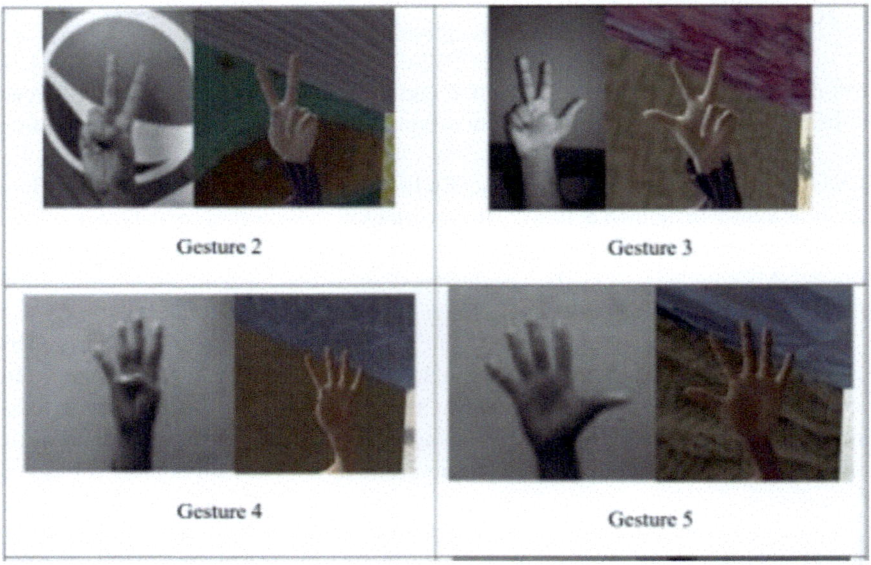

Fig. 1 Sample images from the HANDS dataset, shown in contrast to their real counterparts

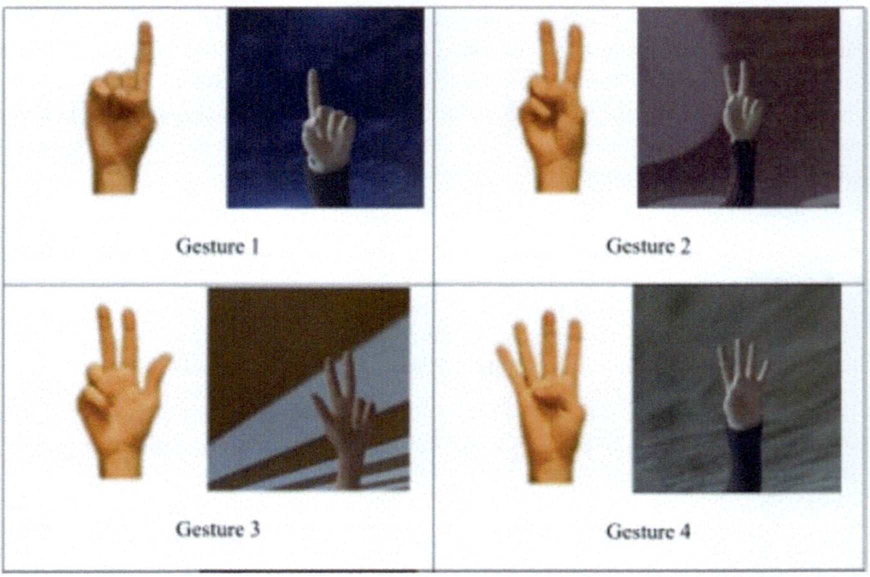

Fig. 2 Sample images from the Digits dataset, shown in contrast to their real counterparts

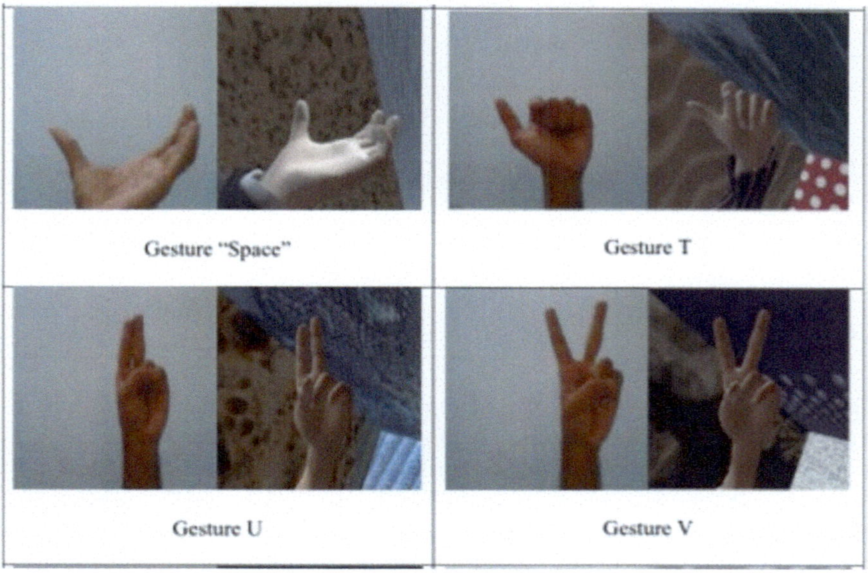

Fig. 3 Sample images from the ASL dataset, shown in contrast to their real counterparts

In order to build the synthetic datasets, a pipeline was developed as described in Fig. 4. We describe each step as follows.

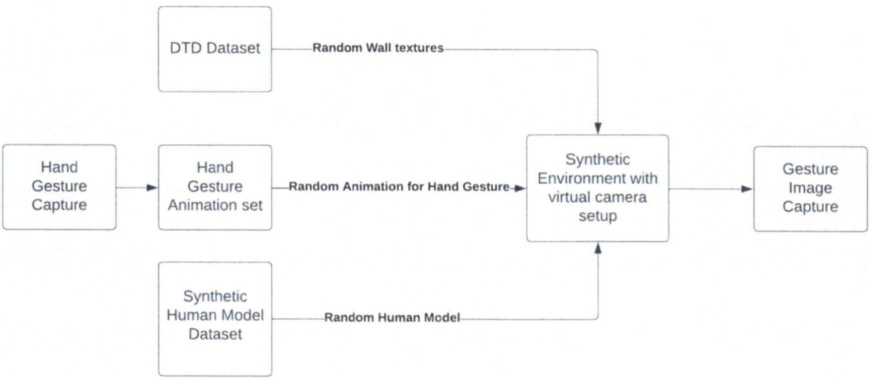

Fig. 4 Pipeline used for building the synthetic datasets

Fig. 5 Sample scenes with virtual human models

3.1 Hand Gesture Capture

Hand gestures were recorded using a Leap Motion Controller (LMC), allowing us to use its precise finger position tracking to map the hand gestures and record them as animation files. We have obtained recordings for the 3 datasets, 12 for the Digits dataset [17], 28 for ASL dataset [20], and 12 for the HANDS dataset [24] respectively. These animations, however, are mapped to only one of the hands, due to the fact that deep learning practitioners often tend to augment their datasets by flipping images, leading to duplicates for chiral (non-symmetric) images. Finger positions for some hand gestures, however, were animated using Blender [3], due to accuracy issues faced while capturing using the LMC.

3.2 Virtual Humans

Virtual human models are built using the MakeHuman software [2], which uses human models according to various characteristics such as size, skin and hair colors, facial structures such as Asian, African, and Caucasian, gender, hairstyles, rigging, and clothing. We modified its mass-produce plugin to generate 100 rigged virtual models, allowing us to add the recorded animations in the synthetic environment. The height of the virtual human models has been fixed during their generation, since their height, while being irrelevant for the hand gesture dataset, also helps with fixing the camera positions in the virtual environment to focus on the hand gestures (see sample scenes from Fig. 5).

3.3 Synthetic Environment

The virtual environment was built using Unity version 2020.3.22f. The background environment is modeled in an enclosed room, consisting of 4 walls, a ceiling, and a floor.

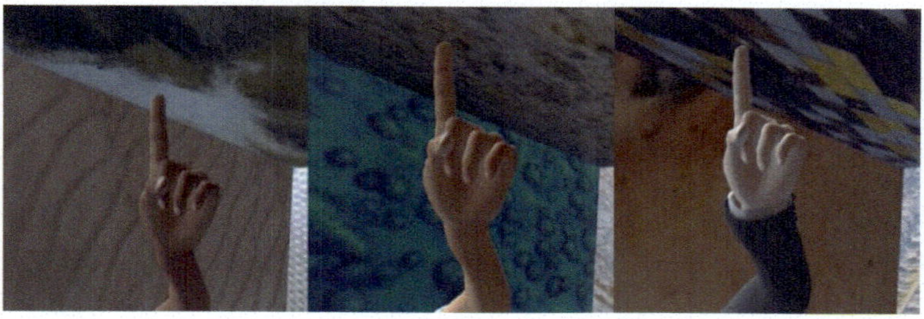

Fig. 6 Example images from Digits dataset showing different images of same gesture with different backgrounds

Background Textures: The environment is programmed to change all the surface textures at random, sourced from the Describable Texture Dataset (DTD) [7], which is composed of 5640 texture images across 40 different classes. Figure 6 shows different backgrounds for the same gesture.

Virtual Human Spawning: Virtual human models are set to spawn one at a time with a random hand gesture pose from the recorded animations. The model is replaced with the next model once the image capture is complete.

Camera Placement and image capture Camera placement was set according to the datasets, with multiple cameras set up according to each gesture. Each camera only captures images according to the specific gestures that it is programmed to capture. Images are captured using the image synthesis plugin for Unity. A total of 12,000 images are captured for each hand sign, for all the datasets, with a resolution of 400 $\times$ 400 pixels.

4 Discussion

4.1 Pose Capture Differences

While Leap motion controllers are excellent devices for capturing hand poses, the accuracy of the poses was largely dependent on their mounting and relative finger orientations. Initial tests were done using a desktop mount, which was not satisfactory. We have hence switched its mounting to the user's head for better accuracy. Table 1 shows an aggregated summary of gestures captured for each dataset for both desktop and head mounts and their respective number of gestures that were captured correctly. Figure 7 shows the differences in captured poses for the same pose with different mounts. It was also observed that the gestures with overlapping fingers were difficult to capture with the LMC. Some gestures were affected by this lack of accuracy, with an example shown in Fig. 8. Consequently, we have simulated the said poses using the blender software. Another possible solution for a higher accuracy pose capture

Table 1 Comparison of gesture accuracy capture for desktop versus head mounts

Dataset	Total number of gestures	Number of correctly captured gestures for Desktop mount	Number of correctly captured gestures for head mount
Digits	10	8	10
Alphabet	27	20	25
HANDS	12	10	12

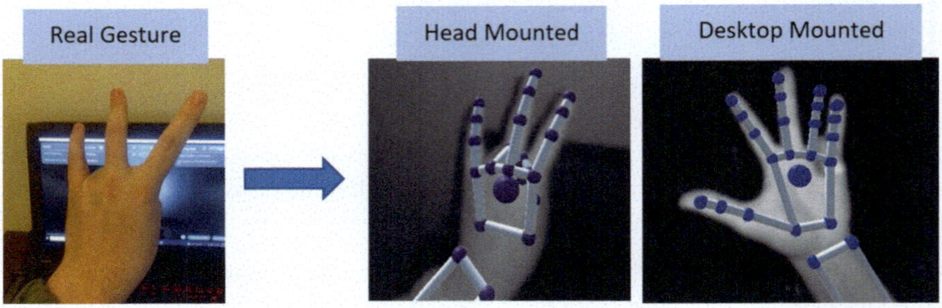

Fig. 7 Tracking accuracy of head-mounted LMC versus desktop-mounted LMC

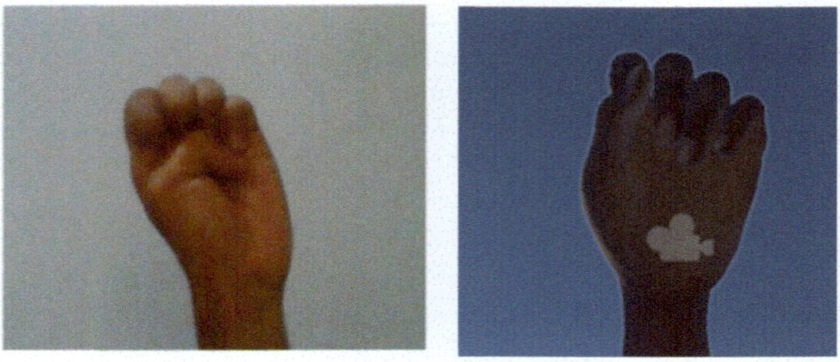

Fig. 8 Recording differences example

is to use high-fidelity motion-capture gloves, which are significantly more expensive but more reliable solutions.

4.2 *Dataset Differences*

By building the datasets, some changes and improvements were introduced to the synthetic datasets as compared to their real counterparts. Some of the differences are illustrated in Table 2. This subsection highlights some of the common and specific differences that are significant.

Table 2 Comparison of real versus synthetic datasets

Attributes	Real datasets	Synthetic datasets
Background variation	No	Yes
Variable contrast	No	Yes
Pose errors	No	Yes
Scalability	No	Yes
Skin color variation	No	Yes
Lighting variation	No	Yes
Gloves	No	Yes

All images captured from the virtual environment were from the left hand of the model. This decision was taken because the real datasets have all collected their samples from different hands—Digits and HANDS datasets used right-handed gestures, while the ASL dataset used left-handed gestures. Since the flipping of the images is a common step with data augmentation during training, we have decided to capture all the gestures for the datasets with the gestures from the left hand.

All the images from the real Digits and ASL datasets have a white or homogeneous background; the real HANDS dataset had somewhat variable backgrounds. All of the synthetic datasets used a variable background, as described in Sect. 3, as evident from Figs. 1, 2, and 3. This was introduced to help with image augmentation and add more variations in the images, thus helping build more robust models. The synthetic datasets have also added more variation for skin colors, which, along with the changing backgrounds, create a variance in contrast between the hands and the background. Another variance we have addressed is the use of gloves of various colors by human models in the environment. With most hand gesture datasets using the gesture images of bare hands, models trained with these datasets may not be able to detect the hand gestures with gloves present, as their texture and colors were not trained for. Figure 9 shows some gesture images with human models wearing gloves. Long sleeves from some of the virtual human models also add to the variations in the datasets.

The original HANDS dataset consisted of images of 15 gestures. We have omitted three two-handed gestures while building our synthetic version, in order to focus on only one-handed gestures for this work. Another significant difference is the use of

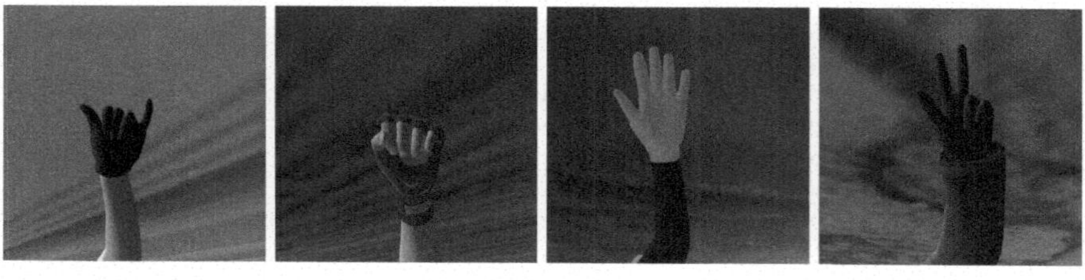

Fig. 9 Gesture images with gloves

colored images in the synthetic version in contrast to the use of grayscale images in the original version. This change was introduced to add more variance, while also accounting for the traditional use of grayscaling of images during the preprocessing stage of training the models using the datasets.

5 Conclusion

In this paper, we have developed and shared 3 synthetic hand gesture datasets. We have also described the process of building the datasets, including the environment and other tools used. Some key issues with building the datasets were also discussed in detail. The reality gap and its possible effects were also mentioned. A pipeline for building synthetic datasets for gesture recognition has also been established in this work. While the method of capturing gestures using motion capture is more accurate, our approach is cheaper and easier for the acquisition of less complex gesture poses due to not needing expensive motion capture equipment.

The differences between the synthetic and real datasets were compared and discussed in detail. The synthetic datasets accommodated more variation in the image data while preserving the features present in the original datasets. Some of the key features such as additional skin and glove colors, background variations, and increased variation in background contrast were mentioned.

Since the human models used were not very realistic, there exists a reality gap that needs to be addressed regarding the usage of the synthetic datasets generated. The difference in the training effectiveness of the datasets compared to their real counterparts needs to be addressed by training various models and evaluating their performances with the real and our synthetic datasets. In our future work, we aim to present and discuss them in detail.

References

1. Carnegie-mellon mocap database. http://mocap.cs.cmu.edu/
2. Makehuman community: open source tool for making 3D characters. http://www.makehuman community.org
3. Blender—a 3D modelling and animation software. https://www.blender.org/ (Year of the version you used), accessed: 12 MArch 2024
4. Akar CA, Tekli J, Jess D, Khoury M, Kamradt M, Guthe M (2022) Synthetic object recognition dataset for industries. In: 2022 35th SIBGRAPI conference on graphics, patterns and images (SIBGRAPI), vol 1, pp 150–155. https://doi.org/10.1109/SIBGRAPI55357.2022.9991784
5. Armstrong WS, Drakontaidis S, Lui N (2022) Synthetic data for semantic image segmentation of imagery of unmanned spacecraft
6. Chen Y, Li W, Chen X, Gool LV (2019) Learning semantic segmentation from synthetic data: a geometrically guided input-output adaptation approach. In: Proceedings of the IEEE/CVF conference on computer vision and pattern recognition, pp 1841–1850
7. Cimpoi M, Maji S, Kokkinos I, Mohamed S, Vedaldi A (2014) Describing textures in the wild. In: Proceedings of the IEEE conference on computer vision and pattern recognition (CVPR)

8. Deng J, Dong W, Socher R, Li LJ, Li K, Fei-Fei L (2015) Imagenet large scale visual recognition challenge. Int J Comput Vis (IJCV). https://doi.org/10.1007/s11263-015-0816-y
9. Dou H, Zhang W, Zhang P, Zhao Y, Li S, Qin Z, Wu F, Dong L, Li X (2021) Versatilegait: a large-scale synthetic gait dataset with fine-grained attributes and complicated scenarios. arXiv preprint arXiv:2101.01394
10. Erfanian Ebadi S, Jhang YC, Zook A, Dhakad S, Crespi A, Parisi P, Borkman S, Hogins J, Ganguly S (2021) Peoplesanspeople: a synthetic data generator for human-centric computer vision
11. Gulli A (2019) Deep learning illustrated
12. Hagelskjær F, Buch AG (2021) Bridging the reality gap for pose estimation networks using sensor-based domain randomization. In: Proceedings of the IEEE/CVF international conference on computer vision, pp 935–944
13. He B, Li X, Huang B, Gu E, Guo W, Wu L (2021) Unityship: a large-scale synthetic dataset for ship recognition in aerial images. Remote Sens 13(24):4999
14. Horváth D, Erdős G, Istenes Z, Horváth T, Földi S (2022) Object detection using sim2real domain randomization for robotic applications. IEEE Trans Robot 39(2):1225–1243
15. Jalal M, Spjut J, Boudaoud B, Betke M (2019) Sidod: a synthetic image dataset for 3D object pose recognition with distractors. In: Proceedings of the IEEE/CVF conference on computer vision and pattern recognition (CVPR) workshops
16. Joshi I, Grimmer M, Rathgeb C, Busch C, Bremond F, Dantcheva A (2022) Synthetic data in human analysis: a survey. arXiv preprint arXiv:2208.09191
17. Khalid M (2019) Sign language for numbers. https://www.kaggle.com/datasets/muhammadkhalid/sign-language-for-numbers
18. Kuzhahiev M, Cakir Y (2020) A survey of public datasets for computer vision tasks. Comput Vis Image Underst. https://doi.org/10.1016/j.cviu.2020.103015
19. Kvalsvik S, Rasmussen I, Hagen D, Nilsen Aune T, Andersen PA (2022) Synthetic data generated in unreal engine 4 (2022). https://doi.org/10.21227/sh9d-9t96
20. Londhe K (2021) American sign language. https://doi.org/10.34740/KAGGLE/DSV/2184214; https://www.kaggle.com/dsv/2184214
21. Loper M, Mahmood N, Romero J, Pons-Moll G, Black MJ (2015) Smpl: a skinned multi-person linear model. ACM Trans Graph (TOG) 34(6):1–16
22. Miura T, Sako S (2021) Synslag: synthetic sign language generator. In: Proceedings of the 23rd international ACM SIGACCESS conference on computers and accessibility, pp 1–4
23. Muratore F, Eilers C, Gienger M, Peters J (2021) Data-efficient domain randomization with Bayesian optimization. IEEE Robot Autom Lett 6(2):911–918
24. Nuzzi C, Pasinetti S, Pagani R, Coffetti G, Sansoni G (2021) Hands: an RGB-D dataset of static hand-gestures for human-robot interaction. Data Brief 35:106791
25. Schmidhuber J (2015) Deep learning in neural networks: an overview. Neural Netw. https://doi.org/10.1016/j.neunet.2014.09.003
26. Tobin J, Fong R, Ray A, Schneider J, Zaremba W, Abbeel P (2019) Domain randomization for transferring deep neural networks from simulation to the real world. In: 2017 IEEE/RSJ international conference on intelligent robots and systems (IROS). IEEE, pp 23–30
27. Tremblay J, Prakash A, Acuna D, Brophy M, Jampani V, Anil C, To T, Cameracci E, Boochoon S, Birchfield S (2018) Training deep networks with synthetic data: bridging the reality gap by domain randomization. In: Proceedings of the IEEE conference on computer vision and pattern recognition workshops, pp 969– 977
28. Unity technologies: unity perception package (2020). https://github.com/Unity-Technologies/com.unity.perception
29. Valtchev SZ, Wu J (2021) Domain randomization for neural network classification. J Big Data 8(1):94

30. Varol G, Laptev I, Schmid C et al (2021) Synthetic humans for action recognition from unseen viewpoints. Int J Comput Vis 129:2264–2287 (2021). https://doi.org/10.1007/s11263-021-01467-7
31. Varol G, Romero J, Martin X, Mahmood N, Black MJ, Laptev I, Schmid C (2017) Learning from synthetic humans. In: Proceedings of the IEEE conference on computer vision and pattern recognition, pp 109–117

Optimizing k-Nearest Neighbor for Color Detection

Omer Al Sumeri, Luis Villanueva, Drake Robert Sheela, Hitender Singh Sidhu, Jacob Luscomb, and Shahab Tayeb

Abstract This paper presents an improvement to the efficiency of the k-Nearest Neighbor (KNN) algorithm by reducing its time and space complexity while keeping accuracy high. KNN is a popular classification algorithm in machine learning that classifies a new data point by considering the k-number of nearest neighbors surrounding it. However, the naive implementation of KNN requires computing the distance of the entire dataset to the unknown point, resulting in significant time and space consumption. To address this issue, a new implementation of KNN is proposed for datasets that have distinct clusters of condensed data points. The new implementation reduces the clusters of condensed data points into a single data point. The k-value is improved from the square root of n to 1; where n is the total number of points in the dataset and 1 comes from the single closest data point to the unknown point. This new implementation significantly reduces the time and space complexity of the KNN algorithm. The clusters of data now represented by a single point reduce the number of comparisons and the amount of data the machine needs to store to use the algorithm. A color detection application was used to analyze the results from the naive to the improved implementation. Time complexity, space complexity, and an accuracy experiment were done to test the improved implementation from the naive. It is concluded that the improved implementation outperforms the naive in both space and time complexity while keeping the same accuracy.

O. Al Sumeri · L. Villanueva (✉) · D. R. Sheela · H. S. Sidhu · J. Luscomb · S. Tayeb
Electrical and Computer Engineering, California State University, Fresno, Fresno, CA, USA
e-mail: villanueval@mail.fresnostate.edu

O. Al Sumeri
e-mail: omeralsum2018@mail.fresnostate.edu

D. R. Sheela
e-mail: drake3861@mail.fresnostate.edu

H. S. Sidhu
e-mail: hss08@mail.fresnostate.edu

J. Luscomb
e-mail: parchment@mail.fresnostate.edu

S. Tayeb
e-mail: tayeb@mail.fresnostate.edu

P. G. Bradford et al. (eds.), *Proceedings of IEMTRONICS 2024*, Lecture Notes in Electrical Engineering 1229, https://doi.org/10.1007/978-981-97-4780-1_4

Keywords Optimized k-nearest neighbor · Optimized color detection

1 Introduction

Several types of machine learning algorithms exist, popular two being supervised learning and unsupervised learning [1]. Supervised learning trains the system categorizing all the outputs based on the input. In unsupervised learning, the system will find patterns in data without categorizing from an external source. The algorithm KNN is considered a supervised learning algorithm due to it being given previously classified data. KNN is used to classify an unknown object by having pre-classified data points that can be compared to the input object [2]. Finding a specific amount of the closest number of neighboring values will determine what classification will be given to the new input.

The KNN algorithm has three main parts which make up the function of the algorithm [3]. First, the distances from the new input object to all other already preloaded classified objects are calculated by using a distance formula. The most common distance formula is the Euclidean distance formula. Equation 1 shows the Euclidean distance formula with n being the number of attributes [4]:

$$d(x, y) = \sqrt{\sum_{i=1}^{n} |x_i - y_i|^2} \tag{1}$$

The distances calculated are then stored inside a list and sorted from smallest to largest [3]. With the list now sorted, the closest data points to the input point are at the beginning of the list. Depending on the set k-value, a tally is done on which classification group is closest. Whichever group has more tallies, the new input point is added to that group due to the point being closer to multiple values of that classified group.

The k in KNN is a value that is used to select how many neighboring close points to compare the new point to. If a low k-value is selected then the result can be sensitive to noise points, while having a large k-value may include too many points from other classes.

For the k-value, a base standard has been set to $\sqrt{n}$, n being the size of the dataset [5]. The naïve implementation of KNN has a time complexity of $O(n)$, n being the training data size because n number of distances need to be calculated.

KNN is used in the industry for many applications such as making weather predictions for farming, health predictions for kidney failure and heart disease, and solar and wind energy production predictions [6–8]. Every algorithm has advantages and disadvantages; with KNN, the algorithm is known for a simple implementation while giving meaningful results. As seen in the KNN implementation used for health prediction article by Shahab Tayeb et al., the algorithm can result in a high yield of accuracy compared to other classifying methods [7]. The downfall of the algorithm comes when the given dataset is large, leading to longer computation times due to the large amount of data [9].

In this paper, KNN is used to classify color. A wide range of colors can be represented by a combination of three different colors: red, green, and blue (RGB). Together these colors can be combined to make new colors such as yellow or pink. RGB can be represented by 8-bit or 16-bit integers for each respective RGB value; the larger the bit size, the higher the precision [10].

An 8-bit integer value will be used to represent each color value in this paper's methodology, meaning the values will be limited from 0 to 255. These three colors can be represented on a three-axis graph. By implementing color on a three-axis graph with a range of 0–255, a possibility of 16,777,216 different colors can occur. By translating an RGB color to each of the three 8-bit values, it is now possible to feed the data into the KNN algorithm to classify what color best describes it [11].

2 Limitation of Existing Method

The limitations that exist for the naive KNN algorithm is the computation required to classify a new data point. The new data point must be compared to every pre-classified data point. A calculation for each data point using the Euclidean distance to determine the distance will take extra time even if it only needs to use k-data point distances, depicted in Fig. 1. Reference [12] shows how they used Euclidean distance to calculate distances for each data point of chili leaves using the hue, saturation, and brightness value of color. This means that the distance for each new leaf input must be calculated for each data point in the dataset.

The k-value the algorithm will use also increases the number of comparisons that will need to happen between the dataset to the new data point to determine its classification. Using a range of k-values can influence the accuracy of the algorithm. In the KNN implementation shown in Ref. [13], it is evident that with the k-value ranging there is also a range of accuracies that can occur. By generalizing the values, k can be reduced to determine the smallest distance and use the smallest distance as the choice of classification.

In a typical naive KNN algorithm, the k-value is chosen with the $\sqrt{n}$, n being the size of the data points [5]. So, as the size of the dataset increases so will the k-value, increasing the time and space complexity for the calculations as the dataset becomes larger.

Other algorithms have tried to either continually recalculate a k-value using certain methods for better accuracy and classification results or guess what the best k-value is for the given dataset. These methods of guess and check or continual recalculating adjustment of k lead to large amounts of time wasted finding the best k-value instead of classifying data [9]. This proposed improvement in KNN would release the time limitation of finding a k-value that best worked with the given data by allowing k to have the value of 1 or the closest data point to the unknown data point.

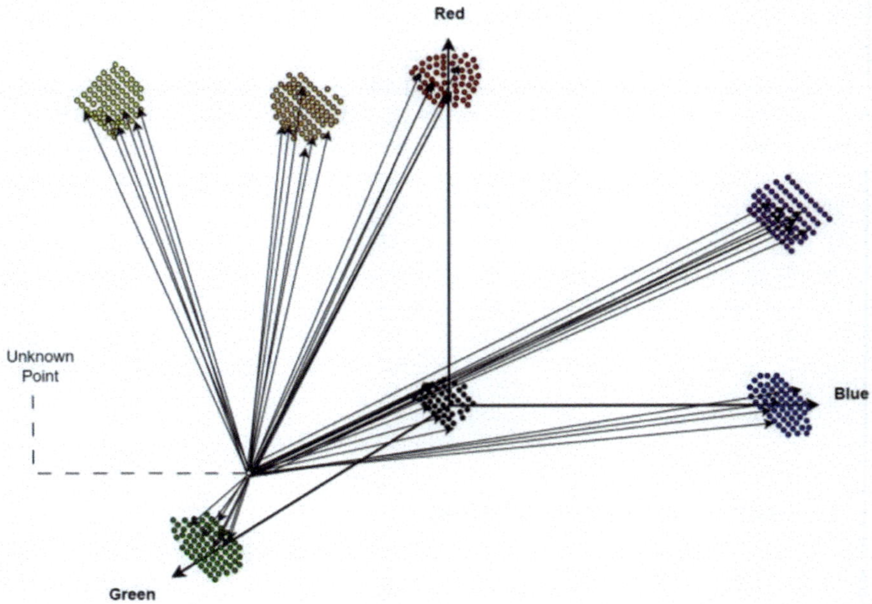

Fig. 1 Illustration of distances that need to be calculated for naive KNN

By overcoming the current limitations of high computational time to classify a point along with the limitation of finding the best *k*-value to use, this improved KNN algorithm could be used in applications where faster data computations could improve or save people's lives. One study found using the KNN algorithm to compute a given dataset to understand Hepatitis resulted in higher accuracy compared to other learning datasets [14]. The quicker and more accurate a classification can be, the quicker someone's life can be saved. Other examples where a faster KNN model can improve or save people's lives are faster Alzheimer's disease detection [15] and automobile accident prediction [16].

3 Proposed Method

The proposed method for improving KNN will be representing the clusters of data to a single point. The single representing point for each classification will be the average of all the data points in that classification. Every classification in the dataset will have one representing value; the representing values will be used to then measure the distance to the unknown point. The classification that has the least measured distance is then the classification of the unknown point.

If every classification is represented by a single data point, then the distance that needs to be calculated is only the number of classifications as shown in Fig. 2.

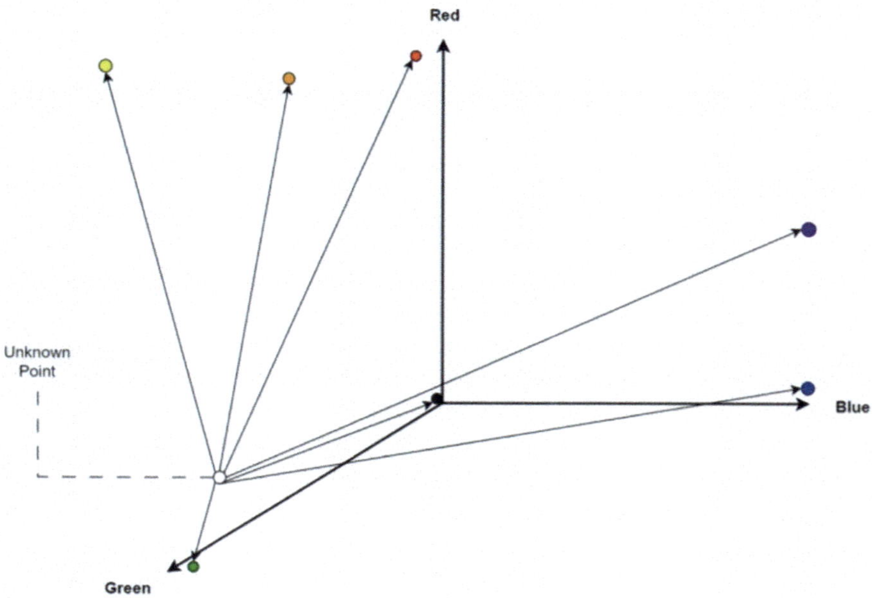

Fig. 2 Illustration of distances that need to be calculated for improved KNN

With the proposed method, the choice of k will go from $\sqrt{n}$, n being the size of the data points, to 1, the shortest distance calculated from the unknown point to the classifications. The proposed method reduces time complexity and space complexity. The time complexity is reduced because the number of distances calculated is reduced. The improved time complexity will be (C), C being the number of classifications in a dataset. The space complexity is reduced because the algorithm is not required to store the entire dataset, only the representation point for each classification. The improved space complexity will be (C) again.

In the context of color detection, each color will have a single representing point. The single representing RGB value chosen for the specific color will be a general representation of that color. The unknown color will have three values for it to be plotted on the RGB graph. The distance will be taken from the unknown color to all the known labeled colors on the graph. All the distances will be calculated and stored in a list. The main color that has the shortest distance to the unknown color will now be the color for the unknown color.

4 Methodology

The following implementation of the naive and proposed KNN algorithms will be tailored for the color detection application. The naive and improved programs will be able to detect 8 colors. The 8 colors are red, green, blue, yellow, green, orange, purple, black, and white.

For the naive and improved implementation, a color point class needs to be created. The attributes of the color point class are shown in Table 1. The three integers, r representing red, g representing green, and b representing blue, are the RGB attributes of the individual color point. The distance which is of data type double is used to store the distance from the individual color point to the unknown color point. The color given to the respective input will be held in the color type attribute.

For the naive implementation, color points need to be generated within a specific range. Table 2 contains the RGB value ranges for each color classification.

The process of the naive KNN is summarized as follows:

1. Find the distance from all the points in the dataset to the unknown point as shown in Algorithm 1.
2. Sort all the distances using a sorting algorithm, placing the values in an ascending-ordered list.
3. Classify the unknown color as shown in Algorithm 2:
   a. Take the square root of the size of the dataset to find the k-value
   b. Count the occurrence of each color up to k
   c. Find the color with the greatest occurrence

Table 1 Class and attributes created for a single-color point

Class name	ColorP
Attributes that are integers	r, g, b
Attributes that are doubles	distance
Attribute that is a string	colortype

Table 2 RGB value ranges for each color classification in the Naive implementation

Color	Red Min–Max	Green Min–Max	Blue Min–Max
Red	225–255	0–30	0–30
Blue	0–30	0–30	225–225
Yellow	225–255	225–255	0–30
Green	0–30	225–255	0–30
Orange	225–255	113–143	0–30
Purple	112–142	0–30	225–255
Black	0–30	0–30	0–30
White	225–255	225–255	225–255

d. Set the unknown color to the color that occurred the most.

Algorithm 1 Finding the distance from the unkonwn Color to all the other colors in the dataset

```
For i = 0 - > All_Colors_Datapoints.size()
ALD[i].distance = (UC.r-ALD[i].r)^2 + (UC.g-ALD[i].g)^2 + (UC.b-ALD[i].b)^2
```

Where: UC is UnkonwnColor, ALD is All_Colors_Datapoints.

Algorithm 2 Classifying the unknown point using the Naïve implementation of KNN

```
k=All_Colors.size()
if(k % 2==0)
     k = k + 1
for i=0 -> k
     f(All_Colors[i].colortype== "Red")
        Red_Frequency++
     else if(All_Colors[i].colortype== "Blue")
        Blue_Frequency++
     else if(All_Colors[i].colortype== "Orange")
        Orange_Frequency++
     else if(All_Colors[i].colortype == "Yellow")
        Yellow_Frequency++
     else if(All_Colors[i].colortype == "Purple")
        Purple_Frequency++
     else if(All_Colors[i].colortype == "Green")
        Green_Frequency++
     else if(All_Colors[i].colortype == "Black")
        Black_Frequency ++
     else if(All_Colors[i].colortype == "White")
        White_Frequency++
/* Put all the color frequencies into a
   Data structure called Freq_Color */
most_Frequent_Color = Freq_Color[0]
For i = 1 - > Freq_Color.size
     if(Freq_Color[i].count > max.count)
        most_Frequent_Color = Freq_Color[i]
     Unknown_Color.colortype = most_Frequent_Color
```

For the improved implementation, the representing point needs to be set for each classification. In the case of the color application, a representing RGB value will need to be created for each color. The representing values were chosen by selecting the general representation of each color. Table 3 contains each representing RGB value for each color.

The process of the improved algorithm is summarized as follows:

Table 3 Representing RGB value for each color classification

Color	Red value	Green value	Blue value
Red	255	0	0
Blue	0	0	255
Yellow	255	255	0
Green	0	255	0
Orange	255	128	0
Purple	127	0	255
Black	0	0	0
White	255	255	255

1. Find the distance from all the color classifications to the unknown point as shown in Algorithm 3.
2. Classify the unknown color as shown in Algorithm 4:
    a. Set the minimum distance to the first Color Classification
    b. Search through the whole Color Classification list to find the minimum distance calculated
    c. Set the unknown color to the color in the Color Classification list that has the least distance to the unknown point.

Algorithm 3 Calculating the distance from unknown color to all colors in the dataset for the improved implementation

1. For i = 0 - > All_Color_Classifications.size()
2. ACC[i].distance = (UC.r-ACC[i].r)2 + (UC.g-ACC[i].g)2 + (UC.b-ACC[i].b)2

Where: UC is UnkonwnColorACC, ACC is All_Color_Classifications.

Algorithm 4 Classifying unknown color using the improved implementation

1. minimum = All_Color_Classifications[0]
2. For i = 1 - > All_Color_Classifications.size()
3. if(minimum.distance > All_Color_Classifications[i].distance)
4. minimum = All_Color_Classifications[i]
5. Unknown_Color.colortype = minimum.colortype

5 Results/Analysis

To test the improved method from the naive method, three experiments were conducted. The experiments tested the space complexity, time complexity, and accuracy. The space complexity tests the amount of memory or predefined data the algorithm requires to execute. The time complexity tests the number of computations that the algorithm will process. The accuracy test assesses if the improved algorithm maintains accuracy compared to the Naive algorithm.

Both programs were implemented in the programming language C++. The testing was performed on a computer system with a macOS Mojave operating system, an Intel® Core™ i7-4980 HQ CPU @ 2.80 GHz, 16 GB DDR3 1600 MHz Micron Technology RAM, and an AMD Radeon R9 M370X 2 GB GPU. The development platform used was Code::Blocks 13.12 Mac version. For timing purposes, Chrono version 8.0.0 was used.

5.1 Time Complexity Experiment

For the time complexity experiment, two versions of the naive KNN algorithm were measured against the improved KNN algorithm. One version was used a nonoptimal sorting algorithm, bubble sort, which has a time complexity of (n^2). The other version of the naive KNN algorithm had a more optimal comparison-based sorting algorithm, with a time complexity of $(n \log n)$. The purpose of including two different sorting algorithms is to show there will be a huge delta no matter the sorting algorithm used in the naive KNN algorithm.

Figure 3 is a graph that displays the execution times for the two naive algorithms with the different sorting algorithms implemented and the improved KNN algorithm (Table of times in Appendix). The input number of unknown colors to be classified starts from 1000, with an increment of 1000, all the way up to 10,000. The naive algorithm with bubble sort takes much longer than the naive algorithm with $(n \log n)$ sort and the Improved KNN. The difference can be seen between the naive algorithm with $(n \log n)$ sorting implemented and the improved KNN, but it is hardly visible due to the naive algorithm with bubble sort implemented with high execution times.

Figure 4 is a graph that displays the execution times for the naive algorithm with an $(n \log n)$ sorting algorithm implemented and the improved KNN algorithm (Table of Times in Appendix Table 4). The input number of unknown colors to be classified is from 1000, with an increment of 1000, all the way to 10,000. The improved KNN is on average 300 times faster than the naive KNN algorithm with the most optimal comparison-based sorting algorithm, so much so that the growth of improved KNN is not visible due to delta between the two algorithms. The overall resultant time complexity of the improved algorithm is independent of the number of data points. This resulted in a constant $O(C)$ time complexity, with C being the number of predefined color classifications.

5.2 Space Complexity Experiment

The original naive algorithm needs to be fed with pre-classified data points to make a classification. The more pre-classified points given to the algorithm, the accuracy

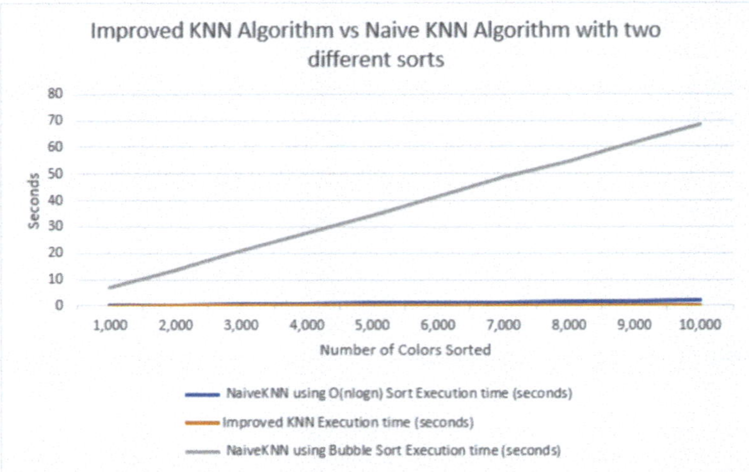

Fig. 3 Naive and improved algorithms using two different sorts

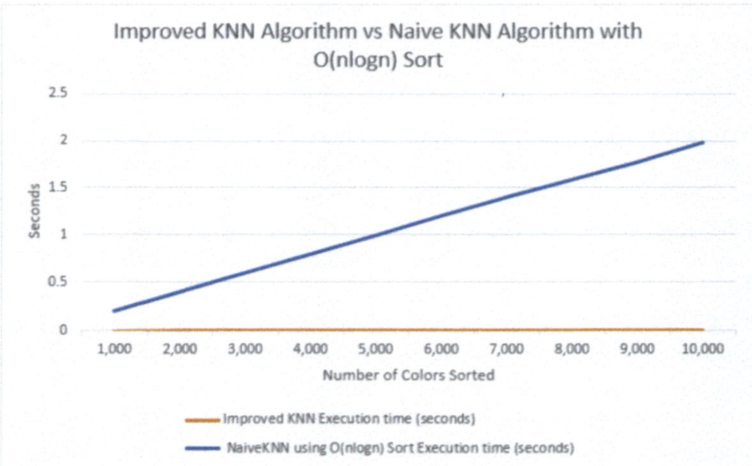

Fig. 4 Naive and improved algorithms using $O(n \log n)$ Sort

improves but comes at the cost of taking up more memory. The naive algorithm takes O(n * C), where n is the number of pre-classified points each group of colors will be populated with and C come to be fed with pre-classified data points to make a classification. The more pre-classified points given to the algorithm, the accuracy improves but comes at the cost of taking up more memory. The naive algorithm takes $O(n * C)$, where n is the number of pre-classified points each group of colors will be populated with and C comes from there being a constant amount of specified colors. There are eight classifications for colors, each of which requires a

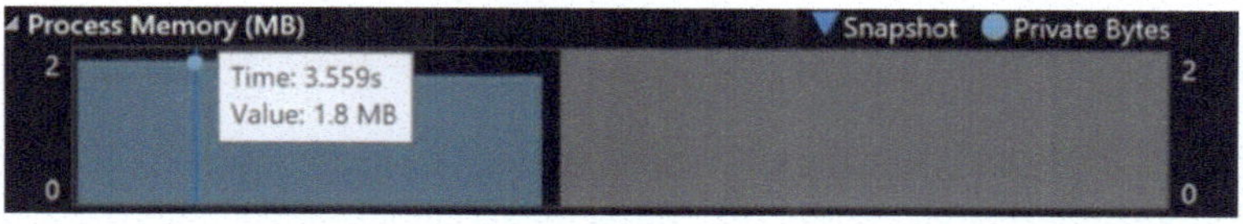

Fig. 5 Naive algorithm classifying 1000 points with 8 classification colors with 100 pre-classified points each

list of pre-classified color points making the space complexity $O(8*N)$. An example of classifying 1000 new points will have 100 pre-classified points exist for each color variety. This makes a total of 800 pre-classified points that will be saved in memory to be used for calculating closest neighbor distances. The result of this was of total memory size 1.8MB as seen in Fig. 5.

Increasing the naive algorithm's accuracy can be done by also increasing the number of pre-classified data points. By providing a large amount of pre-classified data points the algorithm has a larger dataset to compare itself to, having a better chance of finding a close match. This also increases the memory rapidly. Classifying 1000 new colors with 1 million pre-classified data points per color results in 8 million pre-classified data points which take 813 MB as seen in Fig. 6.

The improved algorithm only stores a list of eight predefined color points making the space complexity $O(8)$. The value will always be constant depending on how many colors the algorithm needs to detect which means the memory needed to run will always be very small and constant; by doing so, for the memory required to classify 1000 points with eight classification points only 1.6 MB are needed which is less than the naive algorithm MB as seen in Fig. 7. This value will stay constant no matter the amount of new data points.

Fig. 6 Classification of 1000 new data points, using 1 million pre-classified data points per color classification

Fig. 7 Improved algorithm classifying 1000 points with 8 classification color points

5.3 Accuracy Experiment

For the accuracy experiment, the naive and improved algorithms were tested with the same RGB values. For each color, five different variations of that color were tested. Test 1 is the lightest of the specific color and the color progressively gets darker until Test 5. The color of the tested RGB value is filled into the background of the cell (Table of tested values in Appendix Tables 5 and 6). Beside it is the color the algorithm detected. The total number of tests done on each algorithm is 40.

Figures 8 and 9 contain the results of the accuracy experiment using the naive algorithm. The RGB values that were detected incorrectly are (255, 102, 102), (102, 102, 255), and (153, 153, 0). The RGB value (255, 102, 102) is light red, but the naive algorithm detected it as Orange. The RGB value (102, 102, 255) is light blue, but the naive algorithm detected it as Purple. The RGB value (255, 102, 102) is dark yellow, but the naive algorithm detected it as Orange. 37/40 colors were detected correctly, concluding a 92.5% accuracy rate for the naive algorithm.

Figures 10 and 11 contain the results of the accuracy experiment using the improved algorithm. The RGB values that were detected incorrectly are (255, 102, 102), (102, 102, 255), and (153, 153, 0). All colors that were detected incorrectly by the naive algorithm were also detected incorrectly by the improved algorithm. 37/40 colors were detected correctly, concluding a 92.5% accuracy rate for the naive algorithm.

The naive algorithm has a 92.5% accuracy rate and the improved algorithm has a 92.5% accuracy rate. According to the experiment performed by testing five different variations of the eight colors that are being classified, there is no accuracy difference between the naive and improved algorithms.

Expected Value	Red		Green		Blue		Purple	
	Actual Color	Color Detected	Actual Color	Color Detected	Actual Color	Color Detected	Actual Color	Color Detected
Test 1		Orange		Green		Purple		Purple
Test 2		Red		Green		Blue		Purple
Test 3		Red		Green		Blue		Purple
Test 4		Red		Green		Blue		Purple
Test 5		Red		Green		Blue		Purple

Fig. 8 Accuracy test outcomes for red, green, blue, and purple using Naive implementation

Expected Value	Yellow		Orange		Black		White	
	Actual Color	Color Detected	Actual Color	Color Detected	Actual Color	Color Detected	Actual Color	Color Detected
Result 1		Yellow		Orange		Black		White
Result 2		Yellow		Orange		Black		White
Result 3		Yellow		Orange		Black		White
Result 4		Yellow		Orange		Black		White
Result 5		Orange		Orange		Black		White

Fig. 9 Accuracy test outcomes for yellow, orange, black, and white using the Naive implementation

Expected Value	Red		Green		Blue		Purple	
	Actual Color	Color Detected	Actual Color	Color Detected	Actual Color	Color Detected	Actual Color	Color Detected
Test 1		Orange		Green		Purple		Purple
Test 2		Red		Green		Blue		Purple
Test 3		Red		Green		Blue		Purple
Test 4		Red		Green		Blue		Purple
Test 5		Red		Green		Blue		Purple

Fig. 10 Accuracy test outcomes for red, green, blue, and purple using the improved implementation

Expected Value	Yellow		Orange		Black		White	
	Actual Color	Color Detected	Actual Color	Color Detected	Actual Color	Color Detected	Actual Color	Color Detected
Result 1		Yellow		Orange		Black		White
Result 2		Yellow		Orange		Black		White
Result 3		Yellow		Orange		Black		White
Result 4		Yellow		Orange		Black		White
Result 5		Orange		Orange		Black		White

Fig. 11 Accuracy test outcomes for yellow, orange, black, and white using the improved implementation

6 Conclusion

KNN is an algorithm that classifies a new data point by calculating the distance from the new data point to all the existing data points in the dataset. The k-number closest distances are then used to classify the data point. The standard method for calculating k is $\sqrt{n}$, n being the total number of data points in the dataset. The improved KNN algorithm optimizes the k-value to be 1 in datasets that have distinct clusters of condensed data points. Each of those clusters can be represented by a single point, reducing the time and space complexity consumption of the algorithm.

To test the improved algorithm, the application of classifying color was chosen. Three experiments were conducted, space, time, and accuracy to show that the improved algorithm outperforms the naive implementation of KNN. The improved algorithm has a final time complexity of $O(C)$, C being the number of color classifications we have in the program. The number of color classifications is 8, so the final time complexity is $O(8)$. The space complexity is also $O(C)$ since only the single-color classification data point needs to be stored. Again, the space complexity in this case is $O(8)$. The accuracy test demonstrated that the same level of accuracy is kept between the naïve and the improved implementation all while making significant development upon the space and time complexity.

The experiments hence show that the improved algorithm outperforms the naive algorithm in classifying new data points on datasets that have a structure of condensed data groups.

Table 4 Execution of times of the Naive and improved KNN algorithms

Number of unknown colors classified	Naive KNN using bubble sort execution time (s)	Naive KNN using $O(n\ n)$ sort execution time (s)	Improved KNN execution time (s)
1000	6.84933	0.195698	0.000652
2000	13.7052	0.398876	0.001299
3000	20.9106	0.596872	0.001926
4000	27.7479	0.788861	0.002422
5000	34.5055	0.985374	0.003288
6000	41.5026	1.19979	0.004632
7000	48.7438	1.39622	0.002874
8000	54.5085	1.58251	0.005347
9000	61.5478	1.77582	0.005891
10,000	68.5979	1.97525	0.007138

Table 5 RGB values that were tested for red, green, blue, and purple in the accuracy experiment

Tested color	Red	Green	Blue	Purple
Test 1	255, 102, 102	102, 255, 102	102, 102, 255	178, 102, 255
Test 2	255, 51, 51	51, 255, 51	51, 51, 255	153, 51, 255
Test 3	255, 0, 0	0, 255, 0	0, 0, 255	127, 0, 255
Test 4	204, 0, 0	0, 204, 0	0, 0, 204	102, 0, 204
Test 5	153, 0, 0	0, 153, 0	0, 0, 153	76, 0, 153

Table 6 RGB values that were tested for yellow, orange, black, and white in the accuracy experiment

Tested color	Yellow	Orange	Black	White
Test 1	255, 255, 102	255, 178, 102	51, 25, 0	226, 223, 235
Test 2	255, 255, 51	255, 153, 51	0, 51, 51	226, 217, 218
Test 3	255, 255, 0	255, 128, 0	25, 0, 51	255, 252, 218
Test 4	204, 204, 0	204, 102, 0	51, 0, 51	224, 224, 224
Test 5	153, 153, 0	153, 76, 0	0, 0, 0	255, 255, 255

Appendix

See Tables 4, 5 and 6.

References

1. Sindhu Meena K, Suriya S (2020) A survey on supervised and unsupervised learning techniques. In: Kumar L, Jayashree L, Manimegalai R (eds) Proceedings of international conference on artificial intelligence, smart grid and smart city applications. AISGSC 2019. Springer, Cham. https://doi.org/10.1007/978-3-030-24051-6_58
2. Zhang Z (2016) Introduction to machine learning: k-nearest neighbors. Ann Transl Med 4(11):218. https://doi.org/10.21037/atm.2016.03.37. PMID 27386492; PMCID PMC4916348
3. Wu X et al (2007) Top 10 algorithms in data mining. Knowl Inf Syst 14(1):1–37. https://doi.org/10.1007/s10115-007-0114-2
4. Abu Alfeilat HA et al (2019) Effects of distance measure choice on K-nearest neighbor classifier performance: a review. Big Data 7(4):221–248. https://doi.org/10.1089/big.2018.0175
5. Hassanat AB, Abbadi MA, Altarawneh GA, Alhasanat AA (2014) Solving the problem of the K parameter in the KNN classifier using an ensemble learning approach. arXiv (Cornell University)
6. Savitri P, Purwanto H, Wiharko T, Ramdon EJ, Ardiansyah E, Aulia BE (2023) Implementation of the K-nearest neighbors algorithm in the AgroWeather system for plant recommendations based on weather predictions. In: 2023 17th international conference on telecommunication systems, services, and applications (TSSA), Lombok, Indonesia, pp 1–5. https://doi.org/10.1109/TSSA59948.2023.10366995
7. Tayeb S et al (2017) Toward predicting medical conditions using k-nearest neighbors. In: 2017 IEEE international conference on big data (big data), Boston, MA, USA, pp 3897–3903. https://doi.org/10.1109/BigData.2017.8258395
8. Ahmad T, Manzoor S, Zhang D (2021) Forecasting high penetration of solar and wind power in the smart grid environment using robust ensemble learning approach for large-dimensional data. Sustain Cities Soc 75:103269. ISSN 2210-6707. https://doi.org/10.1016/j.scs.2021.103269
9. Jiang L, Cai Z, Wang D, Jiang S (2007) Survey of improving K-nearest-neighbor for classification. In: Fourth international conference on fuzzy systems and knowledge discovery (FSKD 2007), Haikou, China, pp 679–683. https://doi.org/10.1109/FSKD.2007.552
10. Süsstrunk S, Buckley R, Swen S (1999) Standard RGB color spaces. In: Proceedings of the IS&T 7th color and imaging conference, pp 127–134. https://doi.org/10.2352/CIC.1999.7.1.art00024
11. Abushmmala FF, Abughali HA (2020) Color harmony classification using machine learning algorithms: KNN and SVM. In: 2020 international conference on promising electronic technologies (ICPET), Jerusalem, Palestine, pp. 150–154. https://doi.org/10.1109/ICPET51420.2020.00037
12. Qur'ania A, Harsani P, Triastinurmiatiningsih, Por DP (2020) Hue saturation value (HSV) color detection to identify nutrient deficiencies in chili (*Capsium annuum* L.) using K-nearest neighbor. J Adv Res Dyn Control Syst 12:103–108. https://doi.org/10.5373/jardcs/v12sp6/sp20201012
13. Isnain AR, Supriyanto J, Kharisma M (2021) Implementation of K-nearest neighbor (K-NN) algorithm for public sentiment analysis of online learning. DOAJ (DOAJ: Directory of Open Access Journals). https://doi.org/10.22146/ijccs.65176
14. Alfyani R,Muljono (2020) Comparison of Naïve Bayes and KNN algorithms to understand hepatitis. In: 2020 international seminar on application for technology of information and communication (iSemantic), Semarang, Indonesia, pp 196–201. https://doi.org/10.1109/iSemantic50169.2020.9234299

15. Kaur M, Thacker C, Goswami L, Tamizhvani TR, Abdulrahman IS, Raj AS (2023) Alzheimer's disease detection using weighted KNN classifier in comparison with medium KNN classifier with improved accuracy. In: 2023 3rd international conference on advance computing and innovative technologies in engineering (ICACITE), Greater Noida, India, pp 715–718. https://doi.org/10.1109/ICACITE57410.2023.10183208
16. Akhiladevi M, Anitha K, Amrutha K, Amrutha M, Chandanashree KR (2022) Accident prediction using KNN algorithm. In: 2022 fourth international conference on emerging research in electronics, computer science and technology (ICERECT), Mandya, India, pp 1–5. https://doi.org/10.1109/ICERECT56837.2022.10059746

Development and Evaluation of LPWAN-Based Flood Monitoring and Warning System: Assessing Communication Range, Signal Quality, and Alert Mechanisms

Nur Adlah Mohamad Yusri, Sumendra Yogarayan, Siti Fatimah Abdul Razak, Mohd. Fikri Azli Abdullah, and Afizan Azman

Abstract This project aims to develop a flood monitoring and warning system using Low-Power Wide-Area Network (LPWAN) technology. The system employs ultrasonic sensors placed in sensor nodes to measure water levels in flood-prone areas. Data collected by the sensor nodes is wirelessly transmitted to a central gateway node using the Long Range (LoRa) protocol. The gateway node then forwards the data to a cloud-based server for storage and analysis. The system includes testing and optimisation to determine the optimal communication range and performance of the LoRa devices. Signal quality parameters such as Received Signal Strength Indicator (RSSI) and Signal-to-Noise Ratio (SNR) are evaluated to assess the system's performance. Additionally, a data visualisation and alert system is developed to provide users with real-time monitoring and timely alerts. The results demonstrate the successful implementation of the flood monitoring and warning system using LoRa technology, with accurate water level measurements and effective alert triggering. The system's communication range was observed to be limited to 400 m, but enhancements such as improved antennas and more powerful LoRa devices could overcome this limitation.

N. A. M. Yusri · S. Yogarayan (✉) · S. F. A. Razak · Mohd. F. A. Abdullah
Faculty of Information Science and Information (FIST), Multimedia University (MMU), Melaka, Malaysia
e-mail: sumendra@mmu.edu.my

N. A. M. Yusri
e-mail: nuradlahm@gmail.com

S. F. A. Razak
e-mail: fatimah.razak@mmu.edu.my

Mohd. F. A. Abdullah
e-mail: mfikriazli.abdullah@mmu.edu.my

A. Azman
School of Computing, Faculty of Information and Technology, Taylors University, Subang Jaya, Selangor, Malaysia
e-mail: afizanazman@taylors.edu.my

P. G. Bradford et al. (eds.), *Proceedings of IEMTRONICS 2024*, Lecture Notes in Electrical Engineering 1229, https://doi.org/10.1007/978-981-97-4780-1_5

The evaluation of signal quality parameters revealed variations in RSSI and SNR values, suggesting potential environmental factors and interference. Despite these challenges, the system shows potential for addressing flood-related issues and can be further improved for broader deployment in diverse areas.

Keywords LoRa · SX1278 · Flood · Warning · Monitoring

1 Introduction

Floods are the most frequent natural disaster, and their frequency and intensity are expected to increase due to climate change, making vulnerable populations living in flood-prone areas at higher risk. In Malaysia, floods are classified into flash floods and monsoon floods, with monsoon floods being more predictable based on monsoon winds. However, flash floods are harder to predict, requiring early warning systems. Recent floods in Malaysia have resulted in significant losses, estimated at RM6.1 billion, impacting livelihoods and the country's economy [1].

The main challenge in flood disaster management in Malaysia is ineffective communication and coordination, leading to delays and inefficiencies in rescue efforts. The lack of a centralised communication system among agencies and the absence of warning signs for villagers contributed to the confusion and inadequate preparation. To address this, improving the communication and warning system is essential. Additionally, the availability and accessibility of comprehensive flood data need to be enhanced to ensure reliable risk assessment and prediction for effective disaster management [2].

With the rise of the Internet of Things (IoT), sensor-enabled devices have become increasingly prevalent [3]. These devices gather and transmit data, and the choice of data transmission technology depends on factors like data size, speed, and range. While WLAN and Bluetooth are suitable for high-speed, large-scale data transmission, their range is limited. In contrast, ZigBee is ideal for low-power devices transmitting small amounts of data, but it has a restricted range due to its low power consumption. To overcome this challenge, the development of low-power wide-area network (LPWAN) technology has emerged. LPWAN enables the efficient transmission of low-bitrate data over long distances while consuming minimal power and offering cost-effective solutions [4]. Notable LPWAN technologies available in the market include LoRa, Sigfox, and NB-IoT.

To create an early flood monitoring system, the selection of suitable sensors is crucial. Pressure transducers offer high accuracy in water level measurement but require calibration and are sensitive to vertical displacement. Ultrasonic sensors are commonly used for water level monitoring due to their affordability, but they cannot be submerged in water and require calibration [5]. Soil moisture and carbon dioxide sensors have been suggested as indicators of flash flood, with increased carbon dioxide levels observed during wave run-up. Combined readings from these sensors can help predict the onset of flash floods [6, 7].

Various monitoring systems have been developed utilising different technologies. Ragnoli et al. [8] developed an Autonomous Flood Monitoring system using LoRaWAN, leveraging the advantages of LoRa technology such as its user-friendly nature, high energy efficiency, and exceptional performance in low signal-to-noise ratio (SNR) environments. Alternative LPWAN options like Sigfox and NB-IoT were considered but dismissed due to cost and energy consumption factors, respectively. Similarly, Leon et al. [9] implemented a flood early warning system using Twitter and LoRa, choosing LoRa over other LPWAN vendors due to its attractive range for accurate water distance measurements. Yassin et al. [10] proposed a flood detection and monitoring system based on LoRa technology, achieving accurate water level measurements and effective user alerts. Additionally, Lestari et al. [11] developed an IoT-based river water quality monitoring system utilising LoRa communication, demonstrating the feasibility of long-range communication for river water quality applications.

However, some authors prefer to use Sigfox over other LPWAN vendors. For instance, Soh et al. [12] developed a Cold Chain Management system to monitor the temperature of perishable products during transportation. The system utilised Sigfox UnaShield with Arduino UnoMaker as the prototype. Sigfox was chosen due to its wide coverage in Malaysia, facilitated by Xperanti. NB-IoT was not selected due to the difficulty in acquiring an IoT SIM Card, while LoRaWAN was considered costly due to the need for a gateway installation.

The rest of this article is organised as follows: Sect 2 describes the design of the prototype. Section 3 presents the testing and discussion. Finally, in Sect 4, the conclusion is summarised.

2 Methods

The workable prototype will utilise key hardware components such as the ultrasonic sensor for water level measurement and data transmission to the cloud. The system's blueprint will encompass a logical design section comprising flowcharts and breadboard diagrams. The hardware that will be used for the prototype includes two microcontroller boards (Arduino Uno) where one acts as a node, while the other acts as a gateway. An ultrasonic sensor (HC-SR04), water flow sensor, LoRa transceiver, and GPS will integrate with the node. On the other hand, the gateway is created with Arduino Uno based connected to a LoRa transceiver to receive the data from the node and transmit it to cloud over the Internet. A battery will act as a power supply once the node is launched. The components are laid out as shown in Table 1.

The design of the breadboard is divided into two parts. The first part represents the transmitting side, while the second part represents the receiving side. Figure 1 shows the sensing and receiving nodes. Figure 1a shows the transmitting side which is connected with LoRa module and ultrasonic sensor. The transmitter will collect and send the data that the sensor analysed. Figure 1b represents the receiving side

Table 1 Description of components

Model	Description
Arduino Uno	Three microcontrollers will be used in this project. One microcontroller will serve as the transmitter, another as the relay, and the third as the gateway
Ultrasonic Sensor (HC-SR04)	To measure the distance of the water
LoRa transceivers (SX1278)	The communication module to transmit and receive data

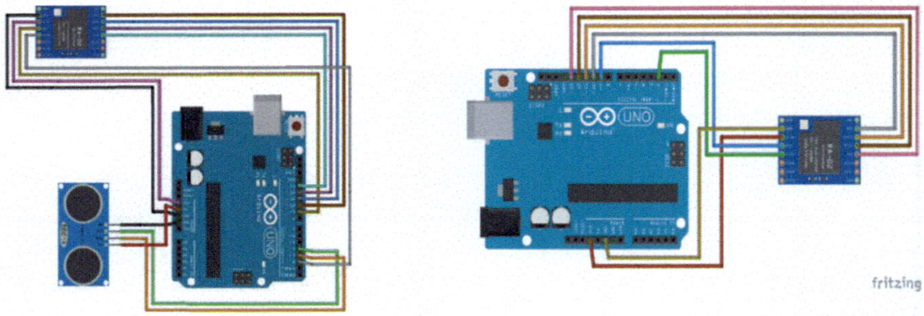

a) Sensing Node b) Receiving Node

Fig. 1 Sensing and receiving node connection

which is connected with a LoRa module with a Wi-Fi connectivity. The receiver here acts a gateway and will transmit the data to the user over the Internet.

The prototype will utilise the ThingSpeak platform as the destination for data transmission. ThingSpeak serves as a network server that facilitates the connectivity between the LoRa gateway and the application, enabling the uplink transmission of data from the sensor node to the ThingSpeak platform. This setup ensures seamless integration between the LoRa-based sensor network and the Internet, providing a reliable and efficient means of transmitting sensor data to the online platform for further analysis and visualisation of transmitting sensor data to the online platform for further analysis and visualisation.

3 Results and Discussion

The testing and discussion phase of the devices plays a crucial role in ensuring their functionality, reliability, and effectiveness in real-world scenarios. This phase involves rigorous testing, fine-tuning, and deploying the devices to the intended environment. An overview of the testing and implementation process consists of functional testing of the ultrasonic sensor and LoRa devices, data accuracy and visualisation, range test, and mobile notifications. The prototype consists of three

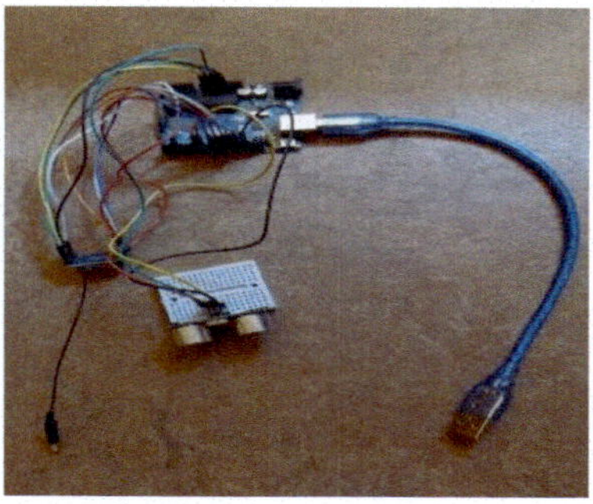

a) Sensor Node

b) Relay Node

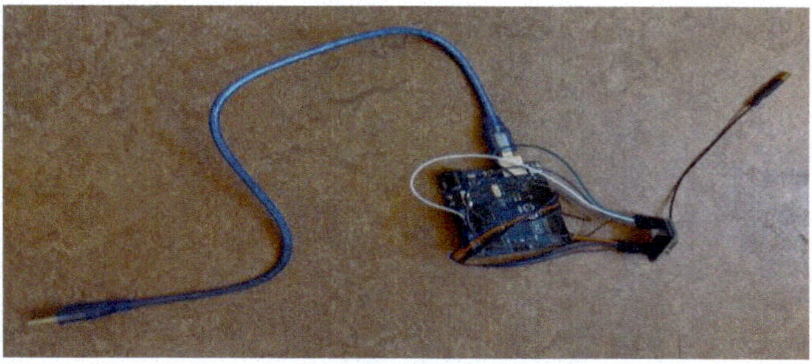

c) Gateway Node

Fig. 2 Prototype of three nodes

devices which are sensor node, the relay node, and the gateway. All the devices are connected to LoRa SX1278, and the gateway is connected to the laptop for Wi-Fi connectivity. The laptop will run a Python program which will upload the data to ThingSpeak (Fig. 2).

ThingSpeak was chosen as the preferred platform due to its cloud-based nature, seamless integration with MATLAB, data visualisation tools, support for various IoT connectivity options, alerting and triggering capabilities, and a supportive user community. These factors collectively make ThingSpeak an ideal choice for storing, analysing, and visualising sensor data, as well as leveraging MATLAB's advanced data analytics capabilities and setting up automated alerts and actions based on specific conditions or thresholds in the data. Figure 3 shows the assessing accuracy of the ultrasonic sensor measurements.

The test is conducted in a container filled with water and measure the distance from the surface of the water to the position of the ultrasonic sensor. Figure 4 shows how the test is conducted. The yellow arrow shows the distance of measuring from the top of the water surface all the way to below the ultrasonic sensor.

By utilising a measuring tape, the distance between the water and the ultrasonic sensor was determined to be 10 cm. This measurement was visually represented in Fig. 5, providing a clear depiction of the measured distance.

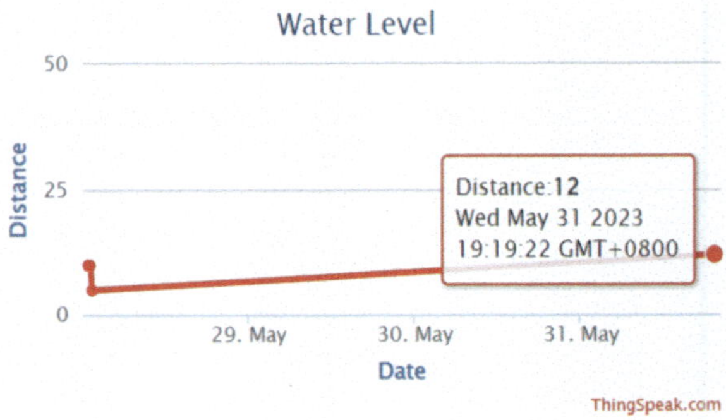

Fig. 3 Assessing accuracy and reliability of ultrasonic sensor measurements

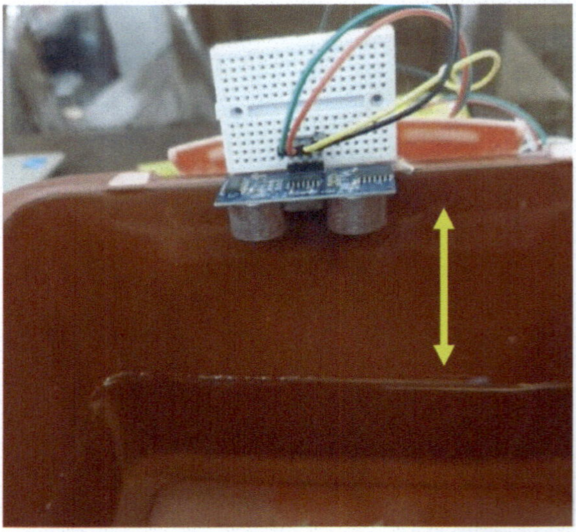

Fig. 4 Water level indicator

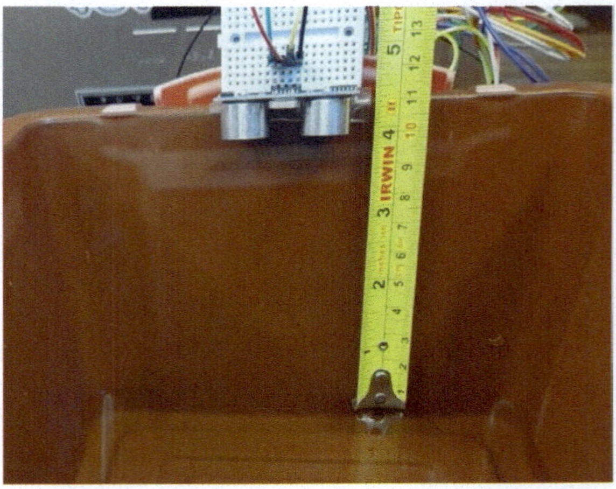

Fig. 5 Visual measurement

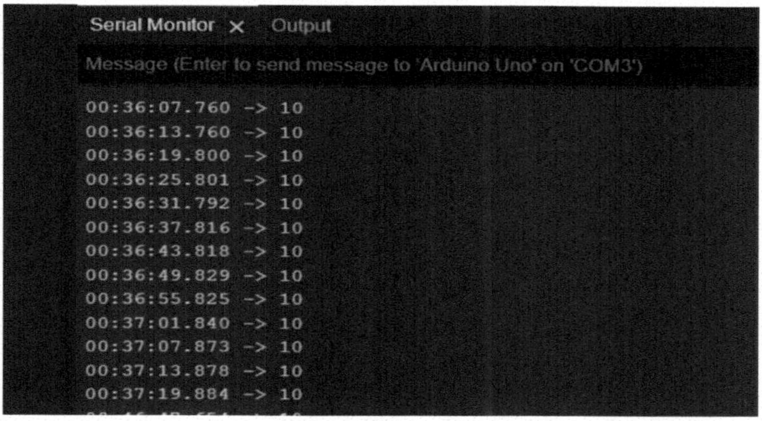

Fig. 6 Serial monitor of the sensor node

The distance measurement was accurately captured and displayed in the serial monitor of the Arduino IDE. The correct distance was detected and visually presented in Fig. 6, demonstrating the successful functionality of the distance measurement system.

3.1 Verifying Communication Between Devices

The setup of the three devices follows the configuration shown in Fig. 7. The sensor node has two possible routes for transmitting data: it can either send the data directly to the gateway node or first send it to the relay node, which then forwards it to the gateway node. Additionally, all these devices are connected to a laptop, allowing for data reading and monitoring via the serial monitor.

A water level of 10 cm was recorded during the measurement. Both the relay node and the gateway node successfully received the data transmitted by the sensor node. Moreover, the relay node was also able to transmit the data to the gateway node. The captured value is displayed in Fig. 8 and Fig. 9, showcasing the received data, respectively.

3.2 Data Visualisation and Alert System Test

The test was conducted on 28 May 2023 at 00:58 A.M. to verify if the data is uploaded to ThingSpeak every 5 min. Python code is used to read the serial port of the gateway and upload the data to ThingSpeak. The Python program is running and continuously uploads the data to ThingSpeak once every 5 min. In Fig. 10, it shows

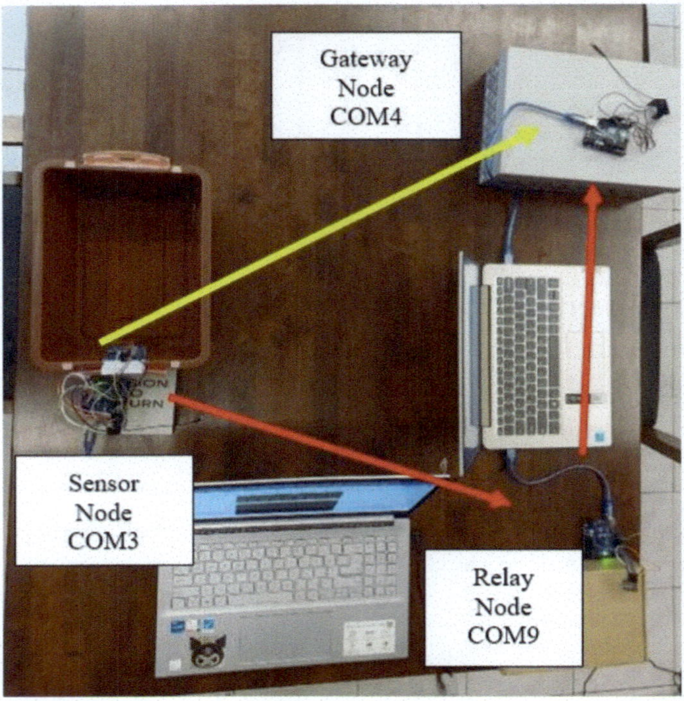

Fig. 7 Configuration of sensor, relay, and gateway nodes

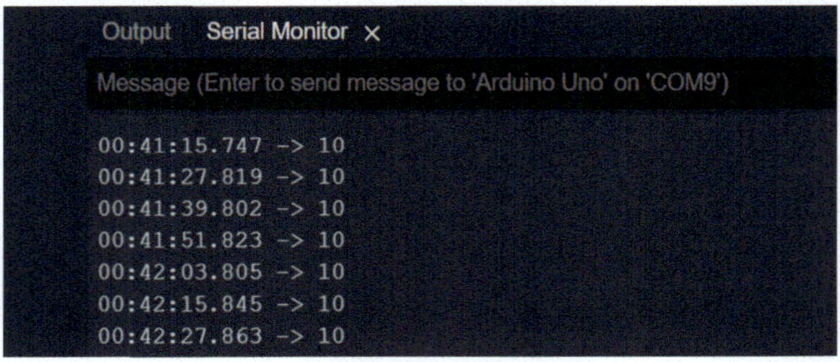

Fig. 8 Serial monitor of relay node

that the data is successfully uploaded to ThingSpeak. The data was successfully uploaded to ThingSpeak, and its visualisation can be observed in Fig. 11, depicting the graphical representation of the recorded data.

To assess the functionality of the alert notification system, a threshold of 5 cm was configured. Consequently, if the water level reaches 5 cm or falls below, an email alert is triggered. For testing, the water container was filled to a height of 5 cm.

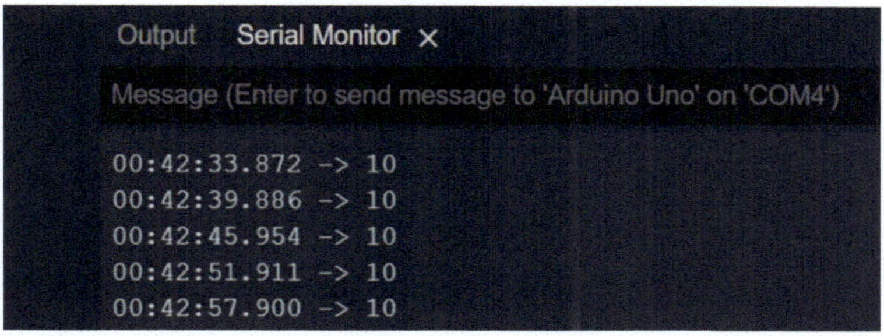

Fig. 9 Serial monitor gateway node

Fig. 10 Data successfully uploaded to ThingSpeak

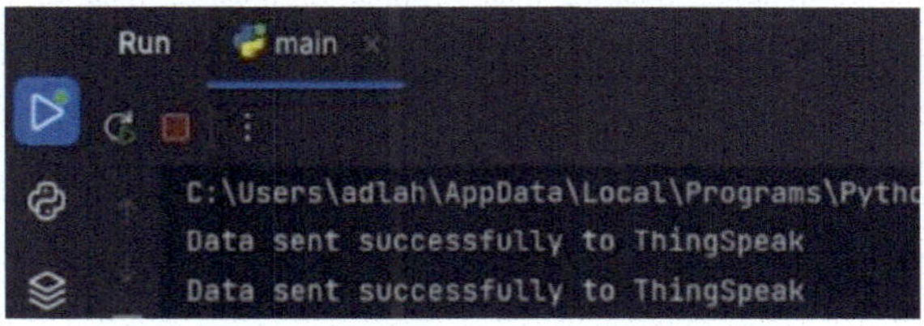

Fig. 11 Uploaded data in ThingSpeak

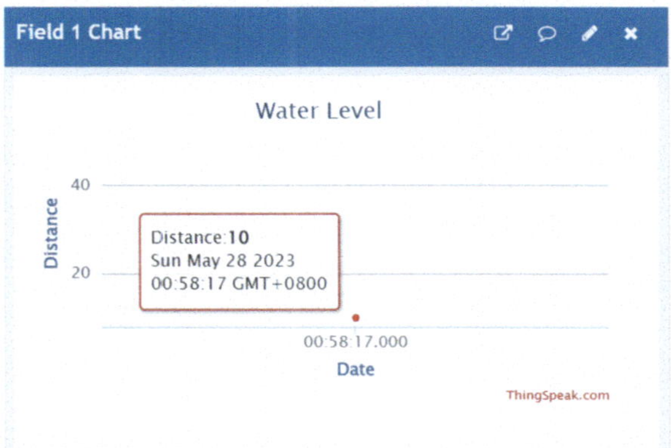

The updated water measurement in the container is depicted in Fig. 12. Figure 13 depicts the accurate measurement of water distance captured by the sensor node and visualises the corresponding data.

Upon successful data upload, the MATLAB analysis read the received data and promptly triggered an email notification to alert the user regarding the water level. The email alert functionality is demonstrated in Fig. 14.

Fig. 12 Sensor and visual measurement

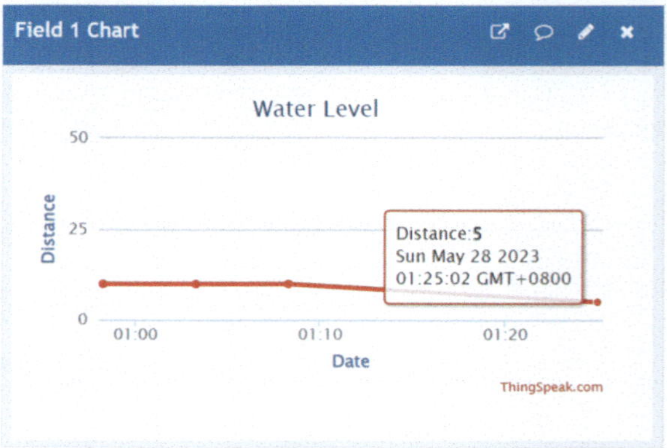

Fig. 13 Data uploaded in ThingSpeak

Fig. 14 Warning alert

3.3 *Packet Loss, RSSI, and SNR Test*

The first test was conducted to evaluate the performance and range of the communication between the two boards. The test focused on measuring packet loss between the boards, aiming to determine the maximum range at which reliable communication could be established. Additionally, the second test was conducted to assess the received signal strength indicator (RSSI) and signal-to-noise ratio (SNR) of the gateway board at regular intervals of 50 m. These tests aimed to assess the signal strength and quality of the received data as the distance increased. By conducting these tests, a better understanding of the system's range capabilities and signal characteristics was obtained, contributing to the overall evaluation of the communication performance. The test for evaluating the system's performance was carried out in two different locations: the MMU stadium and the faculty of Law, with a distance of 200 m between them. Figure 15 depicts a map indicating the calculated distance between the two test locations using Google Maps.

Through this test, it was identified that the packet loss occurred at 200 m. This finding indicated that the two devices, the sensor node, and the gateway node, could effectively communicate within a range of up to 200 m without experiencing packet

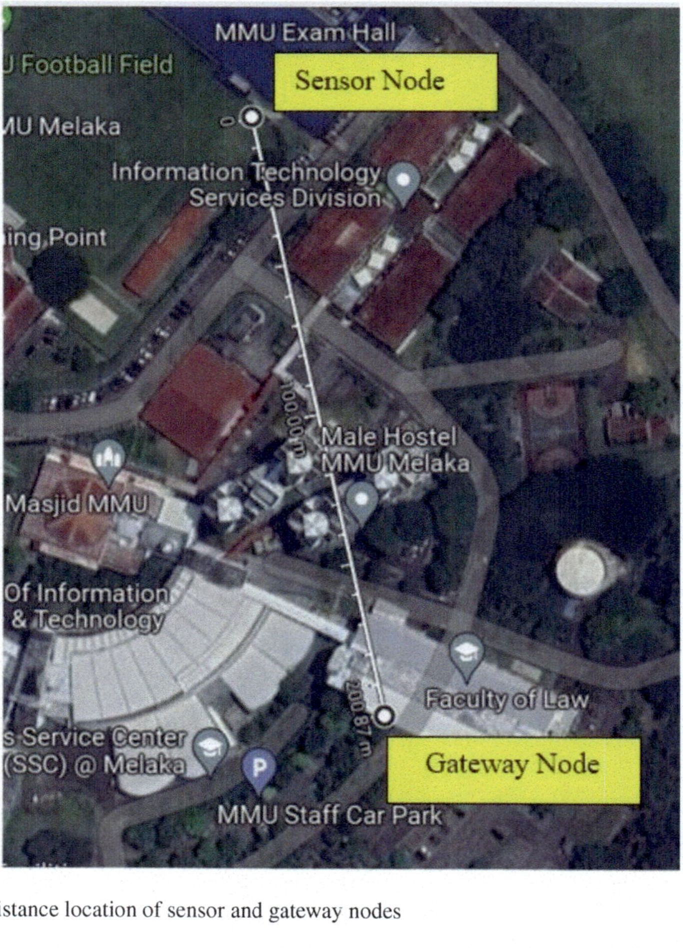

Fig. 15 Distance location of sensor and gateway nodes

loss. The sensor node would now send data to the relay node, which would then forward it to the gateway node. This setup allowed for a more robust and reliable data transfer between the sensor node and the gateway node communication distance beyond 200 m. However, the maximum distance is now 400 m as the packet loss is observed in that mark. Figure 16 shows the location of all the three devices.

Table 2 shows the result of the testing which includes the distance in meters, the RSSI readings, and SNR values. The RSSI values show a decreasing trend as the distance increases. This is expected as the signal strength typically weakens with distance. Firstly, as the distance between the devices increases, there is a consistent decrease in the RSSI values. This indicates a gradual reduction in signal strength as the devices move further apart. The largest drop in RSSI is observed

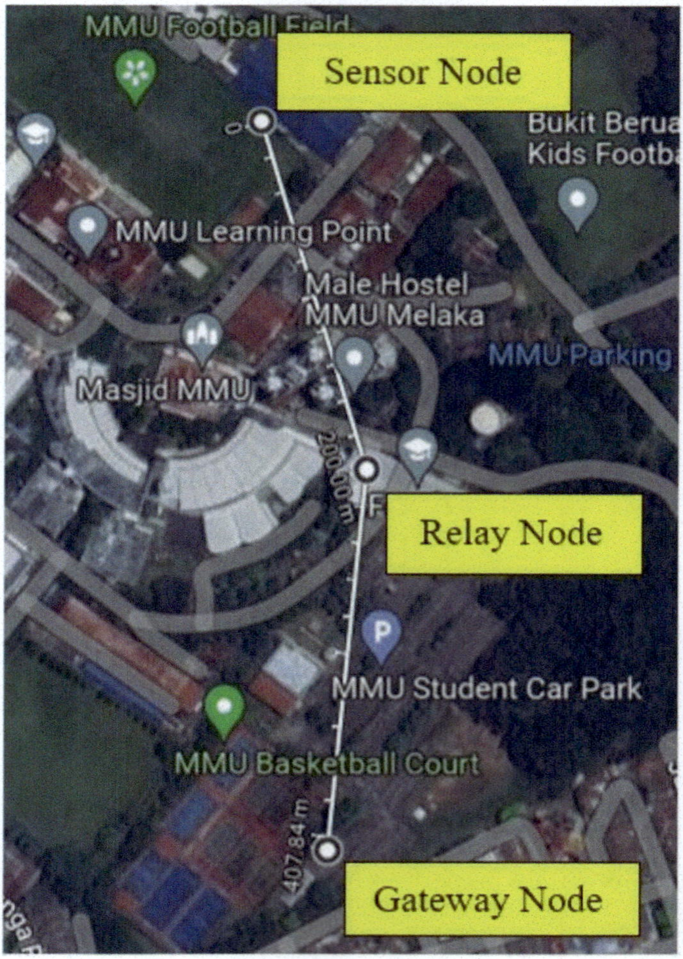

Fig. 16 Distance location of sensor, relay, and gateway nodes

Table 2 Testing result of RSSI and SNR

Distance (m)	RSSI (dBm)	SNR
50	− 70	8.75
100	− 103	8.25
150	− 106	− 5.00
200	− 112	− 6.50
250	− 111	− 6.00
300	− 105	− 10.50
350	− 112	− 5.50
400	− 114	− 7.25

between 50 and 100 m, suggesting a significant decrease in signal strength within this range. Secondly, the SNR values show some variation, with fluctuations observed at different distances. However, it is worth noting that the SNR values generally remain within a moderate range, indicating a reasonable signal-to-noise ratio for communication. The lowest SNR value is observed at 300 m, indicating a relatively weaker signal quality at that distance. Overall, these measurements suggest that the signal strength diminishes with distance, with the largest decrease occurring within the first 100 m. The SNR values show some sensitivity to distance, but generally remain within an acceptable range for reliable communication within the tested distance range of 50–400 m.

4 Conclusion and Future Works

In conclusion, the project successfully designed and configured LPWAN IoT devices for flood monitoring and warning, establishing a communication network for accurate data collection and transmission. Tests confirmed the accuracy of the sensor node in measuring water levels and verified the effectiveness of the data visualisation and notification system. However, limitations such as communication range, scalability, and location testing were identified. To enhance the system, future works should focus on improving antenna designs, exploring advanced LoRa devices, conducting testing in diverse locations and environments, and implementing prediction and modelling techniques. By addressing these recommendations, the flood monitoring and warning system can overcome existing limitations and improve its performance, reliability, and practicality, providing a more comprehensive solution for addressing flood-related challenges. Continued research, development, and innovation will contribute to the advancement of this critical system.

Acknowledgements This project was supported by Telekom Malaysia Research and Development (Grant No. MMUE/220020). The authors would like to thank all the anonymous reviewers for their constructive comments. The authors would also like to thank all the volunteers for their participation in this work.

References

1. Cea L, Costabile P (2022) Flood risk in urban areas: modelling, management and adaptation to Climate change. A review. Hydrology 9(3):50. https://doi.org/10.3390/hydrology9030050
2. Abdul Munir Z, Kaliani Sundram VP, Adham KA (2022) Inter-organisation practices for humanitarian supply chain: a case study of flood disaster in Malaysia. Glob Bus Manage Res Int J. http://www.gbmrjournal.com/pdf/v14n1/V14N1-18.pdf
3. Di Gennaro P, Lofù D, Daniele V, Tedeschi P, Boccadoro P (2018) WaterS: a Sigfox-compliant prototype for water monitoring. Internet Technol Lett 2(1). https://doi.org/10.1002/itl2.74
4. Razak SFA, Yogarayan S, Kamis NH, Abdullah MFA, Yusof I (2023) Application and growth of long-range communications technology in vehicular communications. Int J Electr Comput Eng (IJECE) 13(3):3484–3497
5. Arshad B, Ogie RI, Barthélemy J, Pradhan B, Verstaevel N, Perez P (2019) Computer vision and IoT-based sensors in flood monitoring and mapping: a systematic review. Sensors 19(22):5012. https://doi.org/10.3390/s19225012
6. Khan TA, Alam MM, Shahid Z, Ahmed SF, Mazliham (2018) Artificial intelligence based multi-modal sensing for flash flood investigation. In: 2018 IEEE 5th international conference on engineering technologies and applied sciences, ICETAS 2018. https://doi.org/10.1109/icetas.2018.8629147
7. Yogarayan S, Razak SFA, Abdullah MFA, Kwong FOW (2021) Connected vehicle communication concept for flood level warning using low cost microcontroller. In: 2021 9th international conference on information and communication technology (ICoICT), pp 423–428. IEEE
8. Ragnoli M, Barile G, Leoni A, Ferri G, Stornelli V (2020) An autonomous low-power LORA-based flood-monitoring system. J Low Power Electron Appl 10(2):15. https://doi.org/10.3390/jlpea10020015
9. Leon E, Alberoni C, Wister MA, Hernández-Nolasco JA (2018) Flood early warning system by Twitter using LoRa. In: 12th international conference on ubiquitous computing and ambient intelligence (UCAmI 2018). https://doi.org/10.3390/proceedings2191213
10. Yassin MIHM, Wan J, Newaz SHS, Omar S (2021) LORA based real-time flood detection and monitoring system: a Brunei Darussalam based study. In: 2021 international conference on electronics, communications and information technology (ICECIT). https://doi.org/10.1109/icecit54077.2021.9641097
11. Lestari RD, Rusdinar A, Murti MA, Tawaqal G, Lee D (2019) Design of IoT-based river water monitoring robot data transmission model using low power wide area network (LPWAN) communication technology. In: Proceedings—2019 IEEE international conference on internet of things and intelligence system, IoTaIS 2019. https://doi.org/10.1109/iotais47347.2019.8980377
12. Soh ZHC, Zamri LBM, Sulaiman SN, CheAbdullah SA, Yahaya SZ, Ibrahim MH (2021) Cold chain management system via low power wide area network and IoT. In: 2021 6th IEEE international conference on recent advances and innovations in engineering, ICRAIE 2021. https://doi.org/10.1109/icraie52900.2021.9704002

Heart Rate Measurement Using Face Recognition Techniques and Sensors

Tan Hao Ze, Sumendra Yogarayan, Siti Fatimah Abdul Razak, Mohd Fikri Azli Abdullah, and Afizan Azman

Abstract The human heart rate serves as a critical metric of physiological condition and a significant indicator of an individual's circulatory system. To measure heart rate, a non-intrusive facial recognition method has been proposed. The framework's architecture comprises both software and hardware integration, enabling real-time heart rate measurement and validation. The system is developed with a series of steps including face detection, region of interest (ROI) extraction, green channel value extraction, signal processing, peak signal extraction, and heart rate calculation. The accuracy of the system's heart rate measurements is validated against the heart rate results obtained from the IoT sensor, MAX30100, integrated with Arduino Mega. Testing involved 10 participants assessing the developed system in an indoor environment. Mean square error (MAE) and root mean squared error (RMSE) were calculated between the heart rate (BPM) measured by the system and the sensor to evaluate the difference and accuracy.

Keywords Heart rate · Face detection · Arduino Mega · Pulse sensor · max30100

T. H. Ze · S. Yogarayan (✉) · S. F. A. Razak · M. F. A. Abdullah
Faculty of Information Science and Information (FIST), Multimedia University (MMU), Melaka, Malaysia
e-mail: sumendra@mmu.edu.my

T. H. Ze
e-mail: joshua966@gmail.com

S. F. A. Razak
e-mail: fatimah.razak@mmu.edu.my

M. F. A. Abdullah
e-mail: mfikriazli.abdullah@mmu.edu.my

A. Azman
School of Computing, Faculty of Information and Technology, Taylors University, Subang Jaya, Selangor, Malaysia
e-mail: afizanazman@taylors.edu.my

P. G. Bradford et al. (eds.), *Proceedings of IEMTRONICS 2024*, Lecture Notes in Electrical Engineering 1229, https://doi.org/10.1007/978-981-97-4780-1_6

1 Introduction

Heart rate serves as a vital indicator for evaluating physical health, identifying potential ailments, and monitoring recovery [1]. It can aid in detecting symptoms of various diseases, including fatigue, drowsiness, fainting, chest discomfort, shortness of breath, palpitations, and a racing heart [2]. Monitoring heart rate during activities like driving is crucial for ensuring driver safety and may even save lives. According to [3], heart disease stands as the leading cause of mortality in Malaysia, accounting for 17.0% of medically certified deaths in 2020, hence, the proposal of heart rate measurement using face recognition techniques. The primary objective of this project is to design and develop a tool for detecting heart rate using face recognition, aligning with the problem statement concerning inconsistent factors in environments and subjects that affect heart rate measurement results. The secondary objective is to validate the accuracy of the developed tool, corresponding to the problem statement that heart rate measurement may be influenced by numerous factors. Recent research indicates that variations in skin color can be captured by video cameras. However, this presents challenges due to fluctuations caused by environmental and subject conditions, such as mobility. Moreover, while continuous heart rate monitoring may offer additional insights into heart health, there are limited techniques available for real-time measurement.

The Remote Photoplethysmography (rPPG) technique enables contactless measurement of heart rate [4]. rPPG calculates variations in diffused and specular reflection and the variance of red, green, and blue light reflection from the skin. Diffused reflection, influenced by blood volume fluctuations, serves as the primary indicator [5]. For instance, in [6], rPPG was utilized to monitor the driver's heart rate under various outdoor driving conditions. Challenges such as light intensity affecting accuracy were addressed by employing adaptive neural network model selection (ANNMS) to forecast the driver's heart rate, resulting in improved accuracy in outdoor driving scenarios. Similarly, in [7], a video analytics-based approach was proposed to estimate remote heart rate from face pictures. This method involved identifying and cropping the region of the driver's face, recovering relevant components from mixed noise components, and smoothing the collected signal using temporal filtering and frequency-domain analysis. Bland–Altman plots demonstrated superior performance in driving situations [8].

In other studies, convolutional neural networks (CNN) in facial recognition were combined with pulse rate monitoring to create embedded systems for real-time pulse measurement [9]. These systems utilized webcams to record photographs and imported pretrained models into Caffe using OpenCV and KCF. Additionally, research by [10] employed a laptop's webcam to measure real-time heart rate, incorporating a face-based HR measuring method leveraging LAB color. The variation in skin color caused by blood flow was utilized to compute the heart rate. Signal processing methods such as FFT, ICA, PCA, and the Viola-Jones algorithm were used to transform the video from RGB to LAB color space and determine heart rate from the face area using peak detection methods. Similarly, [11] proposed a remote sensing technique for heart rate detection using near-infrared face video data. This method involved tracking face photos to create time-series signals, which were subjected to empirical mode decomposition, bandpass filtering, and fast Fourier transformation. Field tests demonstrated 95% measurement accuracy, outperforming existing commercial devices and webcam-based HR monitoring methodologies.

In this project, Python is employed for system development, utilizing OpenCV and Dlib for face detection and the Dlib 68 Points Face Landmark Detection Library for facial landmark detection. Upon detecting the face, the ROIs (forehead and cheeks) are extracted for green channel value extraction. Signal processing techniques such as detrending, demeaning, and bandpass filtering are applied. Subsequently, Fast Fourier Transform and peak signal extraction are performed to calculate the heart rate. The hardware includes a laptop webcam with 720 p, 16:9, and 30 fps, an Arduino Mega 2560 Rev3, and a MAX30100 pulse sensor, integrated to validate system accuracy.

2 Literature Review

This section provides an overview of the implemented face detection and heart rate measurement methods. It outlines the system's process and includes relevant diagrams for illustration.

2.1 *Face Detection*

The Dlib 68 Points Face Landmark Detection library was utilized to detect the face. This computer vision algorithm examines facial characteristics and geometry to identify 68 facial landmarks, including the eyes, nose, and mouth [12]. Figure 1 illustrates the Dlib Facial Landmark of the face.

The system has been restricted to detecting only one face due to various factors, including isolating the area of interest, minimizing signal contamination, and simplifying processing. Focusing on a single face reduces the likelihood of signal contamination from other individuals. The system will not initiate heart rate measurement if multiple faces are detected. Figure 1 demonstrates the system detecting more than one face.

2.2 *Region of Interest (ROI) Extraction*

The Region of Interest (ROI) of the face is selected as the forehead and cheeks because of their proximity to vasculature, accessibility, stability, and reduced interference from facial emotions, movements, or speech. These regions are crucial for obtaining accurate heart rate readings as they exhibit pulsatile blood flow associated with the

Fig. 1 Illustration of vehicle collision prevention assistance

heart rate. Figure 2 illustrates the system detecting the face and ROIs, initiating the heart rate calculation process.

After obtaining the ROI corresponding to the forehead and cheeks, the green channel values within each ROI are extracted and averaged. The choice of extracting green channel values is due to their sensitivity to changes in blood volume. The green channel of an image is commonly used for Photoplethysmography (PPG) measurements because the amount of green light absorbed and reflected changes as blood pulses through the capillaries near the skin's surface. Tracking these changes in the green channel over time enables the determination of an individual's heart rate:

$$\begin{aligned}&\text{Average Green Channel Value}\\&\quad = (\text{Sum of average Green Channel Pixel Values})/2.0\end{aligned} \tag{2.1}$$

The equation provided above describes the process of averaging the green channel values. It involves taking the average of the green channel values within each ROI, and then dividing the sum by 2.0 to obtain the average of the two regions.

2.3 *Signal Processing*

After obtaining the signal of average green channel values from the ROIs, signal processing is applied. Initially, the signal is checked to ensure it does not contain

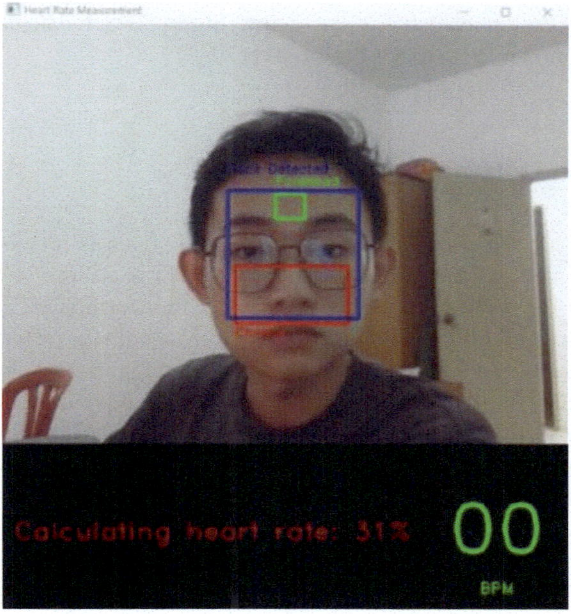

Fig. 2 Face and ROI detected

any infinite or Not a Number (NaN) values. Next, detrending and demeaning are applied to the signal. Detrending involves removing any linear trend or drift that may have developed in the data. A linear trend refers to an incremental rise or fall in the signal values over time. Detrending is commonly performed to eliminate the long-term trend, making it easier to observe and understand the short-term behavior of the signal. The 'signal' module is utilized in the program to perform detrending, employing the 'signal.detrend' function to implement linear detrending.

Following detrending, the signal is demeaned. Demeaning involves subtracting the mean value of the signal from each data point, resulting in a new signal with a mean value of zero. This process centers the signal around zero, eliminating the average or baseline value. Demeaning is often used as a preprocessing step to focus on the relative changes or fluctuations in the signal while ignoring the impact of the mean. In the program, the signal and the desired number of sliding windows are input into the demean function, which creates windows in the signal, determines the mean value of each window, and then subtracts that value from the contents of each window's components, resulting in a demeaned signal where each window is centered at zero.

Once the signal is demeaned, a Butterworth bandpass filter is applied. In signal processing, a Butterworth filter, a type of infinite impulse response (IIR) filter, is commonly used to attenuate frequencies outside a certain range while preserving those within the range. This type of filter is also known as a bandpass filter because it allows a specific frequency band to pass through while attenuating frequencies outside that band. The Butterworth filter is implemented using the 'signal.lfilter' function available in the ‘scipy’ library, which calculates the filter coefficients and applies them to the data to filter it.

2.4 Heart Rate Calculation

After filtering the signal, the filtered signal is prepared for heart rate calculation. Initially, Fast Fourier Transform (FFT) is utilized to identify peaks in the signals. FFT computes the discrete Fourier transform (DFT) of a sequence or its inverse, transforming a signal from its original domain (usually time or space) to a representation in the frequency domain, and vice versa. The ‘np.fft.rfft()’ function from the NumPy library is employed to apply a one-dimensional real-valued FFT on the signal. This function returns the non-redundant, positive half of the complex spectrum, specifically utilized for real-valued input signals. Additionally, the ‘np.abs()’ function computes the absolute values of the complex spectrum, yielding an array of real-valued magnitudes from ‘np.fft.rfft()’.

Next, to filter out unwanted frequency components or identify peaks in the frequency spectrum obtained through FFT, the program generates a list of frequencies corresponding to the FFT values, using the frame per second (fps) and the buffer size. Subsequently, the result of the FFT undergoes further processing with peak frequency extraction. The ‘argmax()’ function is employed to return the index of the highest value in the FFT result. This process enables the detection of the highest peaks in the signal, ultimately determining the heart rate.

To filter out abnormal peak values in the signal, a conditional statement checks the frequency. If any peak values fall outside the allowed heart rate range in frequency (Hz)—between 50 bpm (0.83 Hz), the minimum allowed heart rate, and 200 bpm (3.33 Hz), the maximum allowed heart rate—they are removed, and a diagnostic message is issued. The frequency corresponding to the abnormal peak value is also set to zero. If the frequency falls within the defined range, it is multiplied by 60 to convert from beats per second (bps) to beats per minute (bpm). Since heart rate per minute is the standard measure, converting the frequency from seconds to minutes yields the BPM value. Adjustments are necessary to obtain a heart rate value at a specific unit of BPM:

$$\mathrm{BPM} = \mathrm{BPS} \times 60 \tag{2.2}$$

The equation provided above shows how the BPM is calculated: by multiplying the BPS by 60. Therefore, this formula allows us to measure the heart rate.

Table 1 Pins connection between MAX30100 and Arduino

MAX30100	Arduino Mega 2560 Rev3
Vin	5 V or 3.3 V
GND	GND
SDA	SDA
SCL	SCL

Fig. 3 Setup pins with MAX30100 and Arduino

2.5 *IoT Setup*

To validate the accuracy of the heart rate measurement, a pulse oximeter and heart rate sensor are used alongside the program measuring the heart rate using the MAX30100 sensor. To facilitate the functionality of the sensor, an Arduino Mega 2560 Rev3 is employed for integration with the MAX30100 sensor. Initially, we need to connect the MAX30100 pins to the Arduino pins. Table 1 illustrates the pin connections, and Fig. 3 displays the Arduino Mega and MAX30100 connected.

Subsequently, a program written in the C programming language was prepared to connect the MAX30100 sensor with Arduino using the Arduino IDE. The program utilized a MAX30100 library and the 'Wire.h' library to facilitate communication between the two devices and to read the values of heart rate (BPM). Figure 4 illustrates the output of the MAX30100 in Arduino Mega.

2.6 *Participants*

After completing all the setups, 10 participants gathered to undergo the testing procedure. Each participant was required to fill out a Google Form to provide the necessary data, ensuring privacy and security of the information. The gathered data indicated

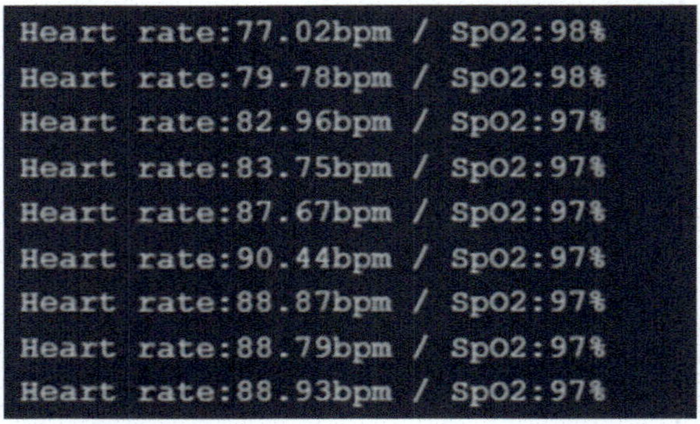

Fig. 4 Output of MAX30100 in Arduino Mega

that all 10 participants were Chinese, aged between 20 and 25 years old, and engaged in sports activities regularly, ranging from 2–3 times to 4–5 times a week.

2.7 *Testing*

In the testing phase, participants were instructed to sit in front of a laptop, facing the webcam steadily without making any head movements for 6 min to measure their heart rate. The recorded data was stored in a CSV file. If participants made any heavy movements or turned their faces away from the webcam, causing the program to fail in detecting their faces, the program would reset the buffer and stop measuring the heart rate. Additionally, participants were asked to place their finger on the MAX30100 sensor to obtain pulse values simultaneously. The testing environment was indoors. Figures 5 and 6 depict how participants held the sensor while measuring heart rate with their faces.

After measuring heart rate for 6 min, the participants were asked to perform a 2-min exercise, which involved jumping jacks. Since jumping jacks is a cardio exercise,

Fig. 5 Participant holding sensor

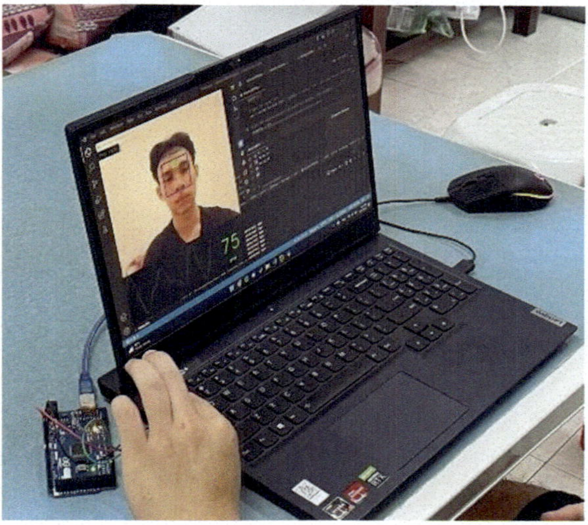

Fig. 6 Participant holding the sensor while measuring heart rate with face

it helps in increasing the heart rate of the participants while performing it. Figure 7 shows the participants doing jumping jacks.

After the 2-min exercise, the participants were asked to sit in front of the webcam again for 6 min to observe the increase in heart rate.

3 Results and Discussion

After testing, the BPM data of 10 participants were obtained before and after exercise. Analysis of the data resulted in the creation of graphs and tables to illustrate the results. Mean Square Error (MAE) and Root Mean Squared Error (RMSE) were calculated to demonstrate the accuracy of the results. MAE measures the average absolute difference between expected and actual values. It provides a simple and understandable way to quantify the average size of prediction mistakes. Both projected and actual values are given in the same units as MAE. RMSE, on the other hand, is the square root of the average of the squared discrepancies between projected values and actual values. Due to the squaring procedure, RMSE gives greater weight to higher errors, making it more susceptible to outliers. Like MAE, RMSE is represented in the same units as the anticipated and actual values:

$$\text{MAE} = \frac{1}{n}\sum_{i=1}^{n}|x_i - x| \tag{2.3}$$

The equation above shows the equation of MAE. The steps can be break down into the following:

Fig. 7 Participant with jumping jacks

1. Calculate the absolute difference between the average BPM values of face and IoT measurements.
2. Absolute difference = |Average BPM (Face)—Average BPM (IoT)|.
3. MAE = Absolute difference.

$$\mathrm{RMSE}_{F0} = \left[\sum_{i=1}^{n} (z_{f_i} - Z_{o_i})^2 / N \right]^{1/2} \tag{2.4}$$

The Equation above shows the equation of RMSE. The steps can be break down into the following:

1. Calculate the squared difference:
2. Squared difference = (Average BPM (Face)—Average BPM (IoT))^2
3. Calculate the RMSE:
4. RMSE = sqrt(Squared difference).

Table 2 shows the average heart rate (BPM) in 6 min before exercise for 10 participants. FR represents the heart rate measured by face recognition, and IoT represents the heart rate measured by the sensor. The MAE column represents the Mean Absolute Error (MAE), and the RMSE column represents the Root Mean Squared Error (RMSE). From the result table, the tester that has the highest values of MAE and RMSE is tester 5. With the higher values of MAE and RMSE, it indicates that the FR BPM value and IoT BPM value have a wider difference between them. This means that the FR BPM is less accurate and has a larger average difference from the IoT values. However, there were also the lowest MAE and RMSE values from the table, which are from tester 10. It has the value of 0 for both MAE and RMSE, which means that the FR BPM is as accurate as the IoT BPM.

Table 3 shows the average heart rate (BPM) in 6 min after exercise for 10 participants. After exercise, the FR BPM and IoT BPM values have increased. From the result table, the highest MAE and RMSE values are from tester 4, which is less accurate. However, the lowest MAE and RMSE values are from tester 1, which is more accurate with a value of 1 for both MAE and RMSE.

Figure 8 shows the line chart for tester 10. Figure 8a depicts the line chart of tester 10's BPM 6 min before, and Fig. 8b after the exercise. The blue lines indicate the FR BPM values, and the red lines indicate the IoT BPM values. From the charts, it shows us that the BPM values between FR BPM and IoT BPM have much overlapping, which means that the values of the FR BPM and IoT BPM are quite similar. Hence, it is quite accurate.

Figure 9 shows the line chart of tester 1. Figure 9a depicts the tester 1's BPM in 6 min before, and Fig. 9b after the exercise. The charts show us that the BPM values between FR BPM and IoT BPM in Fig. 9a, which is before exercise, do not have much overlapping, which means that the values of the FR BPM and IoT BPM are not quite similar; hence it is not quite accurate. But for the chart after exercise, it shows that the lines are overlapping, with similar BPM, then more accurate. Hence, the BPM after exercise is more accurate than before exercise. In summary, with implementation, this project has developed a real-time system which monitors heart

Table 2 Average 6 min of BPM before exercise

Tester	FR (BPM)	IoT (BPM)	MAE	RMSE
1	76	96	20	20
2	85	78	7	7
3	86	79	7	7
4	84	75	9	9
5	73	94	21	21
6	83	93	9	9
7	79	88	10	10
8	82	72	10	10
9	81	79	2	2
10	82	82	0	0
		Average	9	

Table 3 Average 6 min of BPM after exercise

Tester	FR (BPM)	IoT (BPM)	MAE	RMSE
1	99	100	1	1
2	102	83	19	19
3	109	86	23	23
4	116	81	35	35
5	82	98	16	16
6	92	94	3	3
7	98	105	7	7
8	99	95	4	4
9	108	89	19	19
10	101	97	4	4
		Average	13	

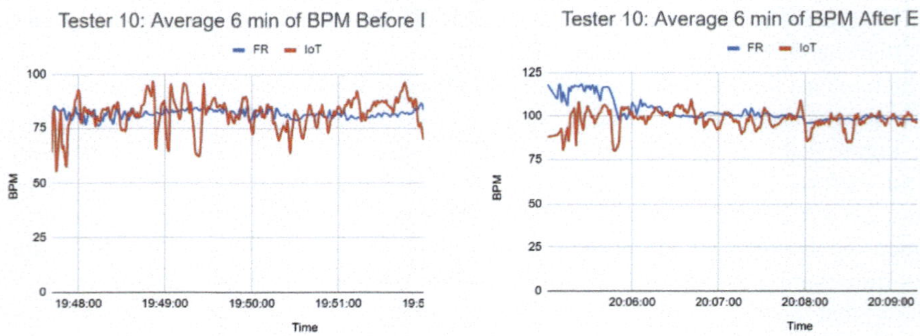

a) Before Exercise b) After Exercise

Fig. 8 Line chart for tester 10

rate using face recognition techniques. Testing was conducted to validate the accuracy of the system with MAX30100 sensor and Arduino Mega. Results were extracted and analyzed. However, the accuracy of the heart rate may vary, due to the factors of the environment, light intensity, skin conditions, motion, and facial expression.

4 Conclusion and Future Works

In conclusion, our endeavor to design and develop a tool for detecting heart rate using face recognition has been met with considerable success, marked by several significant achievements. First and foremost, our approach involved the utilization of the Dlib 68 Points Face Landmark Detection algorithm to accurately detect facial

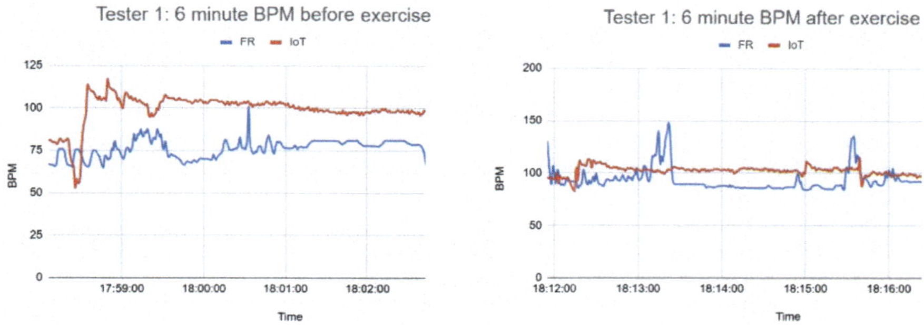

a) Before Exercise b) After Exercise

Fig. 9 Line chart for tester 1

landmarks, enabling precise localization of regions of interest (ROIs) on the forehead and cheeks. Subsequently, the green channel signal was meticulously extracted from these ROIs, laying the foundation for signal processing techniques, including detrending, demeaning, and bandpass filtering. Leveraging the power of Fast Fourier Transform (FFT) and peak extraction algorithms, we successfully computed the heart rate, thus fulfilling the primary objective of our project. The resulting system is not only adept at detecting faces but also excels in extracting heart rate information in real-time, providing users with instantaneous feedback. Furthermore, our secondary objective centered on validating the accuracy of the developed tool, a task accomplished through meticulous testing and analysis. By employing the MAX30100 pulse sensor integrated with an Arduino Mega IoT device, we conducted comprehensive data collection exercises involving 10 participants, capturing heart rate readings before and after exercise routines. By meticulously analyzing the collected data and computing metrics such as Mean Square Error (MAE) and Root Mean Squared Error (RMSE), we effectively assessed the accuracy of our tool. Our findings revealed varying degrees of accuracy, with the system exhibiting remarkable precision in certain scenarios while encountering limitations in others.

Acknowledgements This project was supported by Telekom Malaysia Research and Development (Grant No. MMUE/220020). The authors would like to thank all the anonymous reviewers for their constructive comments. The authors would also like to thank all the volunteers for their participation in this work.

References

1. Darzi E, Soltanian-Zadeh H (2019) Using image-extracted features to determine heart rate and blink duration for driver sleepiness detection. arXiv (Cornell University). [Online]. Available: https://arxiv.org/pdf/1911.01333
2. Shmerling RH (2020) How's your heart rate and why it matters? Harvard Health, 25 March 2020. https://www.health.harvard.edu/heart-health/hows-your-heart-rate-and-why-it-matters

3. Khan AYF et al (2022) The Malaysian HEalth and WellBeing AssessmenT (MYHEBAT) study protocol: an initiation of a national registry for extended cardiovascular risk evaluation in the community. Int J Environ Res Public Health 19(18):11789. https://doi.org/10.3390/ijerph191811789
4. Karthick R, Dawood MS, Meenalochini P (2023) Analysis of vital signs using remote photoplethysmography (RPPG). J Ambient Intell Humaniz Comput 14(12):16729–16736. https://doi.org/10.1007/s12652-023-04683-w
5. Rohmetra H, Raghunath N, Narang P, Chamola V, Guizani M, Lakkaniga NR (2021) AI-enabled remote monitoring of vital signs for COVID-19: methods, prospects and challenges. Computing. https://doi.org/10.1007/s00607-021-00937-7
6. Wu B-F, Chu YW, Huang PC, Chung ML (2019) Neural network based luminance variation resistant remote-photoplethysmography for driver's heart rate monitoring. IEEE Access 7:57210–57225. https://doi.org/10.1109/access.2019.2913664
7. Lee K, Han DK, Ko H (2018) Video analytic based health monitoring for driver in moving vehicle by extracting effective heart rate inducing features. J Adv Transp 2018:1–9. https://doi.org/10.1155/2018/8513487
8. Panigrahi A, Sharma A, Sharma H, Mukherjee A (2023) Contactless HR measurement from facial videos using alternative color spaces with CEEMDAN. In: 2023 9th international conference on signal processing and communication (ICSC). https://doi.org/10.1109/icsc60394.2023.10441625
9. Lin J, Lu M-H, Lin Y-H (2019) A contactless healthcare system with face recognition. In: 2019 4th international conference on intelligent green building and smart grid (IGBSG). https://doi.org/10.1109/igbsg.2019.8886237
10. Oviyaa M, Renvitha P, Swathika R (2020) Real time tracking of heart rate from facial video using webcam. In: 2020 second international conference on inventive research in computing applications (ICIRCA). https://doi.org/10.1109/icirca48905.2020.9183124
11. Zhang Q, Zhou Y, Song S, Liang G, Ni H (2018) Heart rate extraction based on near-infrared camera: towards driver state monitoring. IEEE Access 6:33076–33087. https://doi.org/10.1109/access.2018.2845390
12. Shekharpandey (2023) Dlib 68 points Face landmark Detection with OpenCV and Python. Studytonight, 22 May 2023. https://www.studytonight.com/post/dlib-68-points-face-landmark-detection-with-opencv-and-python#:~:text=You%20can%20use%20Dlib%2068

Analysis of Fatigue Characteristics in Muscle Groups Using the FREEEMG 1000

Bakhtiyor Makhkamov, Kamoliddin Shukurov, A'lokhan Kakhkharov, Shokhrukhmirzo Kholdorov, and Dilshodbek Mamajonov

Abstract This study explored the relationship between muscle fatigue and changes in sEMG amplitude during isometric exercises. Using the FREEEMG 1000 hardware and software system, we monitored muscle activity biosignals in a cohort of healthy schoolchildren. Our findings suggest a correlation between fluctuations in sEMG amplitude and indications of muscle fatigue. However, the potential of fatigue indices rooted in sEMG amplitude as an objective metric for evaluating the efficacy of endurance training requires investigation that is more comprehensive.

Keywords sEMG amplitude · Muscle fatigue · Isometric exercises · FREEEMG 1000 · Muscle activity biosignals · Endurance training · Objective metrics · Athletes with disabilities · Training methodologies · Biosignal analysis · Biosignal processing · Biomechanical sensors · Athlete performance · Sports science · Biomechanics laboratory · Muscle activity monitoring

1 Introduction

Muscle fatigue that arises during exercise is evaluated using a variety of assessment methods, with surface electromyography (sEMG) being the most commonly used. Studying muscle fatigue across different exercises, muscle groups, and conditions enhances our understanding in this domain.

Static muscle tension is a natural component of movement activity and is incorporated at all stages of a child's growth and development. Additionally, it is integrated into the curriculum of physical education activities to promote the harmonious physical development of children. In school-aged children, the balance between dynamic

B. Makhkamov · K. Shukurov · A. Kakhkharov · S. Kholdorov
Tashkent University of Information Technologies Named After Muhammad Al-Khwarizmi, Amir Temur 108, Tashkent, Uzbekistan

D. Mamajonov (✉)
Andijan State University, Andijan, Uzbekistan
e-mail: olimpia-1992@mail.ru

P. G. Bradford et al. (eds.), *Proceedings of IEMTRONICS 2024*, Lecture Notes in Electrical Engineering 1229, https://doi.org/10.1007/978-981-97-4780-1_7

and static components of muscle activity shifts, leading to an increase in static tension [1–3].

When engaging in static exercises, it is crucial to understand their unique characteristics. Specifically, as muscle tension increases, blood flow through the muscle decreases. The capillaries within the muscles attained maximum compression, leading to a cessation of blood flow. Consequently, vital nutrients like oxygen and glucose cannot enter the muscles, and byproducts of muscle activity, such as lactic acid, are not expelled [2–4].

Given this perspective, it becomes imperative to adopt a methodologically sound approach and ensure moderation in the load when incorporating static exercises into the physical education curriculum for students.

In our country school settings, it is evident that a majority of teachers abstain from incorporating static exercises into their instructional routines. Among those who do, many encounter challenges related to the application methodology and determining appropriate load moderation. A common obstacle is the imprecise determination of exercise duration, often done without proper guidance. One potential solution is a deeper exploration of the intricacies of muscle fatigue during static exercises. The prevailing method, adopted by many, gauges fatigue based on observable external signs. However, our pedagogical observations have discernibly shown that this approach falls short in efficacy when applied to static exercises.

In research by J. Finsterer, muscle fatigue assessment revolves around analyzing ATP metabolism (e.g., lactate, ammonia, and oxpurines), oxidative stress markers (e.g., lipid and protein peroxidation biomarkers, among others), and inflammatory biomarkers (e.g., leukocytes). However, identifying the most representative parameters for fatigue in these studies remains a matter of debate. This is because the relevance of specific parameters can vary significantly based on experimental conditions and the populations studied [5–7].

Another challenge is the need for biological samples, which makes continuous monitoring during exercise problematic. Several non-invasive methods, such as mechanomyography (MMG), surface electromyography (sEMG) [8, 9], near-infrared spectroscopy (NIRS) for both isometric and isotonic contractions, and sonomyography (SMG) [6–9], offer potential solutions to address this challenge.

In this research, as a natural progression and supplement to prior experiments utilizing numerous non-invasive methods, we observed specific characteristics of fatigue emergence during test exercises designed to enhance endurance. These observations focused on particular muscle groups and employed the surface electromyography (sEMG) technique.

To date, the impacts of muscle strength and muscle geometry on mean frequency (MF) have proven to be subject-dependent, and findings across various studies have not always aligned. Nevertheless, a review of existing literature suggests that MF is widely recognized as a benchmark variable for pinpointing muscle fatigue, especially in the context of isometric muscle contractions [7–9].

sEMG stands out as a versatile and non-invasive method, facilitating the extraction of myoelectric properties linked to neuromuscular activity during muscle contractions. These properties, along with their variations, provide insights into the biochemical and physiological changes occurring in the subject muscle. Furthermore, this method holds potential for real-time fatigue monitoring [7, 9].

Surface EMG and intramuscular EMG signals are captured using non-invasive and invasive electrodes, respectively. Presently, surface-detected signals are preferred for gathering information regarding the timing or intensity of superficial muscle activation [10, 11]. Electromyography (EMG) signals are invaluable in both medical and engineering fields as electrophysiological indicators. The recording of EMG signals provides a fundamental method for comprehending human body behavior under both normal and pathological conditions. However, the analysis and classification of EMG signals can be challenging due to contamination by various types of noise, particularly during motion [12].

EMG signals serve a variety of purposes, including generating control commands for rehabilitation equipment such as robotic prostheses and for generic man–machine interfaces in Human–Computer Interaction (HCI). They find application in numerous clinical and industrial settings [13]. Processing and classifying EMG signals typically involve using the Electromyography Control technique. Control systems based on the classification of EMG signals are commonly referred to as Myoelectric Control Systems (MCSs), with powered upper-limb prostheses and electric-powered wheelchairs being two main potential applications of MCSs [14]. However, for these applications to be effective, accurate EMG signal acquisition is essential. During EMG signal acquisition, various background noises are inevitably picked up due to the presence of electronic equipment and physiological factors.

The characteristics of the amplifying and displaying systems are paramount for accurately monitoring the various voltage changes originating from the muscles [15]. Over the years, only a limited number of specific processing techniques have been recommended for extracting information from the EMG signal [16–19]. Few researchers have fully capitalized on existing computer capabilities to design automated EMG systems applicable in both research and rehabilitation [20, 21]. The integration of digital computers into EMG signal analysis represents a significant advancement, enabling the rapid and objective execution of complex data analysis procedures. Moreover, in certain scenarios, it may offer analysis techniques that are otherwise unavailable or impractical.

2 Subjects and Devices

The experiment utilized a set of four test exercises specifically designed to measure endurance, targeting distinct muscle groups. Exercises 1 and 2 were designed to observe the abdominal muscles (rectus abdominis) and the front portion of the thigh (vastus lateralis). In contrast, exercises 3 and 4 were adapted to study the front portion of the thigh (vastus lateralis) and the calf muscles (gastrocnemius).

Fig. 1 The process of performing the first exercise

First Exercise: Maintaining a Tilted Body Position While Sitting.

- Positioning: The examinee begins by lying on their back. Legs are bent at the knees, forming a 90° angle. Both the thigh and the sole of the footrest flat on the floor. The upper body is elevated to a 40° angle from the floor. Hands are placed behind the head with fingers interlocked.
- Support: A partner holds the examinee's legs to keep them in position.
- Objective: The examinee attempts to maintain this position for as long as possible.
- Measurement: The duration for which the examinee can hold this position is recorded in seconds.
- Muscles Observed: The abdominal muscle (rectus abdominis) and the front thigh muscle (vastus lateralis) are the primary muscles under observation during this exercise (Fig. 1).

Second Exercise: Holding Straight Legs at an Angle:

- Positioning: The examinee begins by sitting on the floor. They lean back, supporting themselves with their hands. Legs are kept extended, elevated to a 40° angle from the floor.
- Objective: The examinee attempts to maintain this leg position for as long as possible.
- Muscles Observed: The primary muscles under observation during this exercise are the abdominal muscle (rectus abdominis) and the front thigh muscle (vastus lateralis) (Fig. 2).

Third Exercise Process: Standing and Raising One Straight Leg Forward:

- Preparation: Ensure a clear area with ample space to stand and lift the leg. Have a timer or stopwatch nearby to measure endurance time. A protractor or angle measuring tool could be handy to check deviations in leg angle, though this is optional.
- Starting Position: The subject stands straight with feet together, arms by the side, and ensures a good balance.

Fig. 2 The process of performing the second exercise

- Performing the Exercise: Without bending the knee, the subject raises the right leg straight out in front to its maximum height. The objective is to maintain the leg's position for as long as possible.
- Observation and Measurement: Monitor the muscles of the front part of the thigh (vactus lateralis) to check for signs of fatigue or strain. Using the timer or stopwatch, record how long the subject can maintain the raised leg position. It is essential to watch for any drop in the leg's position. If the leg deviates and drops more than 10° from its initial raised position, stop the timer. The time noted at this point is the subject's static endurance for this exercise (Fig. 3).

Fourth Exercise Process: Semi-Sitting on Tiptoes with Bent Knees:

- Preparation: Ensure a clear area with ample space for the examinee to perform the exercise safely. Have a comfortable, non-slip floor surface to avoid slipping or injury.
- Starting Position: The examinee begins by standing upright.
- Transition to Exercise Position: From the standing position, the examinee moves into a semi-sitting posture. Knees are bent outward, with thighs and calves forming

Fig. 3 The process of doing the third exercise

Fig. 4 The process of doing the fourth exercise

a 90° angle. The examinee stands on the tips of their feet, ensuring the heels touch each other. The body should remain upright, held in a vertical position.

- Performing the Exercise: Once in position, the examinee attempts to maintain this posture for as long as possible, using the strength of the anterior thigh (vastus lateralis) and calf (gastrocnemius) muscles.
- Observation: The focus is on monitoring the resistance and potential fatigue of the vactus lateralis and gastrocnemius muscles (Fig. 4).

In our study, we utilized the FREEEMG device from BTS Bioengineering to conduct our surface electromyography (sEMG) analyses. This equipment allowed us to obtain precise readings and insights into muscle activities during the exercises.

The BTS FREEEMG 1000 sensor offers real-time visualization of acquired signals, making it an invaluable tool for functional assessments, biofeedback, and monitoring applications. Its versatility extends to a range of fields including sports, rehabilitation, ergonomics, neurology, and orthopedics [22–24].

Designed for user-friendliness, the BTS FREEEMG 1000 sensor adheres easily to the human body and transmits information rapidly. It eliminates the need for excessive adhesive and wetting tapes, ensuring a comfortable user experience. Moreover, its design, free from excess cables, ensures that it doesn't interfere with natural human behavior or compromise the integrity of the incoming signal [22–24] (Fig. 5).

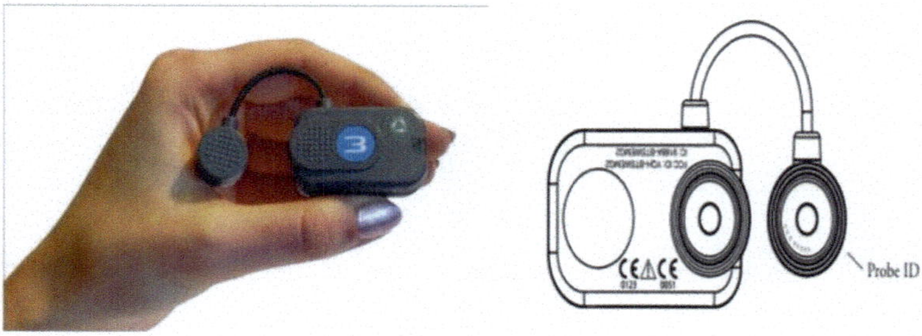

Fig. 5 View of the BTSFREEEMG 1000 sensor

3 Results and Discussion

The Politecnico di Milano's Ethics Committee approved the study, and it adhered to the guidelines of the Declaration of Helsinki. Prior to the study, every participant and his or her respective parents were briefed about the research methodology. Written informed consent was subsequently procured from each. It was ensured that all participants were free from any neurological or orthopedic conditions that might affect their performance. Moreover, none had pre-existing training related to the specific test exercises chosen for the experiment.

Mean Absolute Value (MAV) is a parameter used to ascertain the degree of muscle contraction [25–27]. It is extensively employed in myoelectric control (1, Fig. 6).

$$\mathrm{MAV}_k = \frac{1}{N}\sum_{i=1}^{N}|x_i| \tag{1}$$

Here, N is the number of values in the segment, k is the sequence number of the segment, i is the value index, and x is the signal value.

Simple Square Integral (SSI) represents the energy of muscle activity biosignals. It is extensively utilized in myoelectric control and classification (2, Fig. 7) [25–27]:

$$\mathrm{SSI}_k = \sum_{i=1}^{N}\left|x_i^2\right| \tag{2}$$

Here, N is the number of values in the segment, k is the sequence number of the segment, i is the value index, and x is the signal value.

Signal Dispersion (VAR—Variance of EMG) denotes the distribution or fluctuation in the intensity of muscle activity biosignals across both time and frequency (3, Fig. 8) [25–27]:

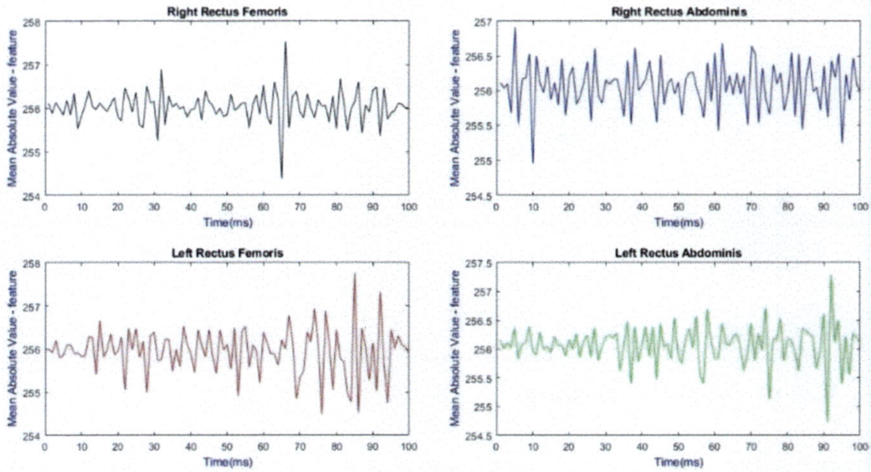

Fig. 6 Mean absolute value results

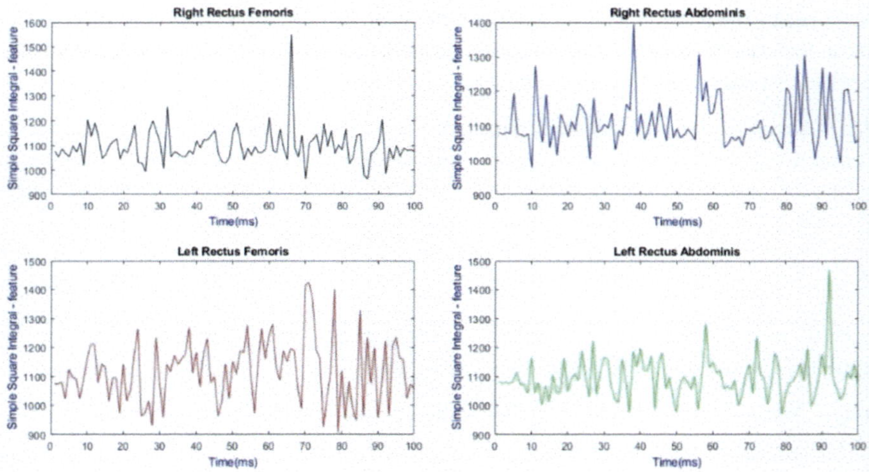

Fig. 7 Simple square integral results

$$\mathrm{VAR}_k = \frac{1}{N}\sum_{i=1}^{N}(x_i - \overline{x})^2 \tag{3}$$

Here N is the number of values in the segment, k is the sequence number of the segment, i is the value index, x is the signal value, and $\overline{x}$ is the threshold value.

Root Mean Square (RMS)—this parameter captures the fluctuations, indicative [25–27] of muscle fatigue, in the contraction–expansion dynamics of muscle activity

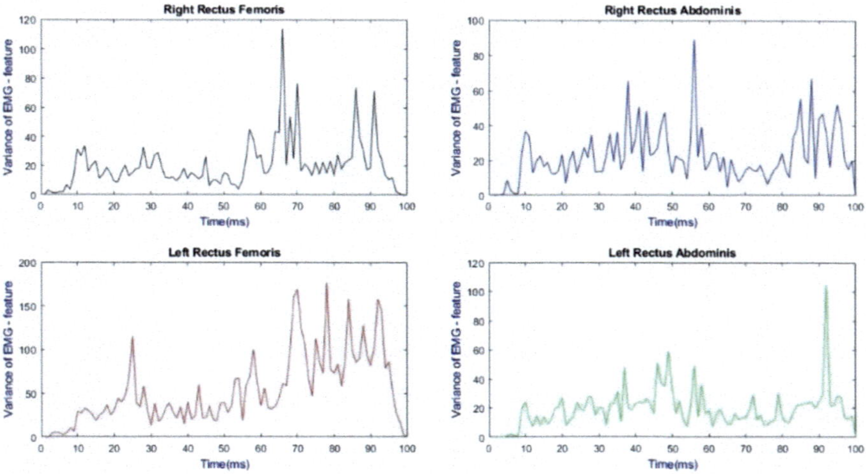

Fig. 8 Signal dispersion results

biosignals (4, Fig. 9):

$$\mathrm{RMS}_k = \sqrt{\frac{1}{N}\sum_{i=1}^{N} x_i^2} \tag{4}$$

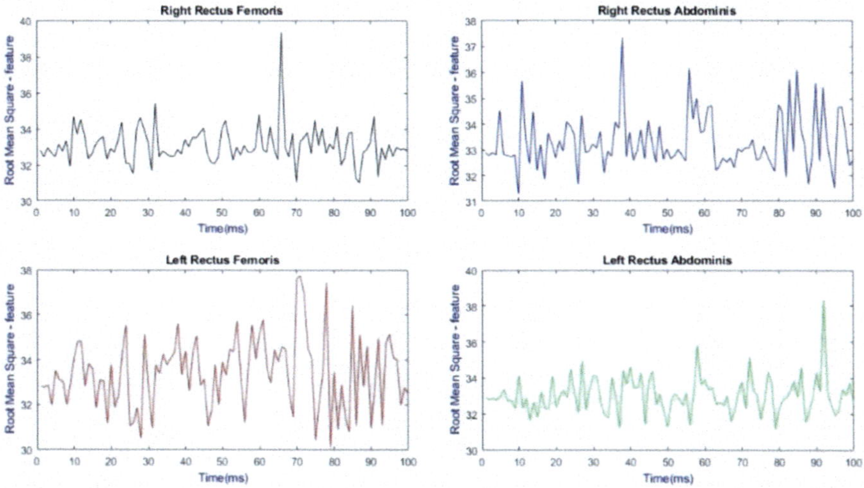

Fig. 9 Root mean square results

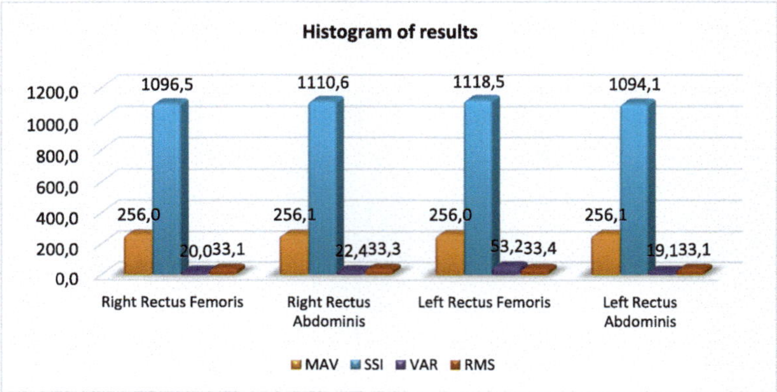

Fig. 10 Fatigue characteristics histogram of muscle groups

Table 1 Feature values of biosignals

	Right rectus femoris	Right rectus abdominis	Left rectus femoris	Left rectus abdominis
MAV	256.0	256.1	256.0	256.1
SSI	1096.5	1110.6	1118.5	1094.1
VAR	20.0	22.4	53.2	19.1
RMS	33.1	33.3	33.4	33.1

Here, N is the number of values in the segment, k is the sequence number of the segment, i is the value index, and x is the signal value.

In the experiment, 14 healthy schoolchildren, averaging 12.1 ± 0.5 years of age, participated. All subjects had a body mass index (BMI) ranging from 20 to 25, which is within the normal weight category.

Throughout the assessment, the subjects undertook four predetermined test exercises. The objective for each exercise was to sustain the contraction until they reached exhaustion, ensuring that they consistently maintained the appropriate posture (Fig. 10) (Table 1).

4 Conclusion

Through the analysis of muscle activity biosignals using the FREEEMG 1000 hardware and software system, we observed that fluctuations in sEMG amplitude during isometric exercises could correlate with muscle fatigue. However, the application of fatigue indices grounded in sEMG amplitude as an objective measure for gauging the effectiveness of endurance training in students necessitates further investigation.

This research opens a window to the broader application of such technology, potentially revolutionizing training strategies, especially for athletes with disabilities. The significance of these findings underscores the need for continued exploration in this domain, aiming to optimize training methodologies across various sports disciplines.

Looking ahead, a deeper exploration of this technology promises transformative breakthroughs. It has the potential to revolutionize training methodologies and enhance performance for athletes with disabilities across a wide spectrum of sports.

Acknowledgements While this study has provided valuable insights into the relationship between muscle fatigue and sEMG amplitude during isometric exercises, investigating a more diverse range of populations, including different age groups, fitness levels, and health conditions, would contribute to a comprehensive understanding of how muscle fatigue manifests across various demographics. Further validating and refining fatigue indices rooted in sEMG amplitude as objective metrics for evaluating the efficacy of endurance training will be essential for practical applications in both educational and clinical settings.

This research was carried out within the framework of the project AM-292306L1314—"Development of a selection system according to kinematic and psycho-physiological characteristics for preparing reserve athletes for Paralympic sports competitions" funded by the Ministry of Higher Education, Science and Innovation of the Republic of Uzbekistan.

References

1. Pilosyan NA (1999) Fiziologicheskoe obosnovanie primeneniya staticheskix uprajneniy na urokax fizicheskoy kulturы u mladshix shkolnikov: dissertasiya... kandidata biologicheskix nauk: 03.00.13. - Krasnodar, - 209 s [In Russian]
2. Drabkin AS (2012) Unikalnaya sistema izometricheskix uprajneniy Jeleznogo Samsona / A.S. Drabkin, Yu.V. Shaposhnikov. – Moskva: Izdatelskiy sentr AST, – S. 41 [In Russian]
3. Moalla W, Merzouk A, Costes F, Tabka Z, Ahmaidi S (2006) Muscle oxygenation and EMG activity during isometric exercise in children. J Sports Sci 24(11):1195–1201. https://doi.org/10.1080/02640410500457893
4. Allen DG, Lamb GD, Westerblad H (2008) Skeletal muscle fatigue: cellular mechanisms. Physiol Rev 88(1):287–332
5. Georgakis A, Stergioulas LK, Giakas G (2003) Fatigue analysis of the surface EMG signal in isometric constant force contractions using the averaged instantaneous frequency. IEEE Trans Biomed Eng 50(2):262–265
6. Finsterer J (2012) Biomarkers of peripheral muscle fatigue during exercise. BMC Musculoskelet Disord 13(1):218
7. De Luca CJ (1997) The use of surface electromyography in biomechanics. J Appl Biomech 13:135–163
8. Al-Mulla MR, Sepulveda F, Colley M (2011) A review of non-invasive techniques to detect and predict localised muscle fatigue. Sensors 11(4):3545–3594
9. Scano A, Pirovano I, Manunza ME, Spinelli L, Contini D, Torricelli A, Re R (2020) Sustained fatigue assessment during isometric exercises with time-domain near infrared spectroscopy and surface electromyography signals. Biomed Opt Express 11:7357. https://doi.org/10.1364/boe.403976
10. Chowdhury RH, Reaz MB, Ali MA, Bakar AA, Chellappan K, Chang TG (2013) Surface electromyography signal processing and classification techniques. Sensors (Basel) 13(9):12431–12466. https://doi.org/10.3390/s130912431. PMID 24048337; PMCID PMC3821366

11. Merletti R, de Luca CJ (1989) New techniques in surface electromyography. In: Desmedt JE (ed) Computer aided electromyography and expert systems, vol 2, chap 9 (sect 3). Elsevier, Amsterdam, The Netherlands; New York, NY, USA; Oxford, UK, pp 115–124
12. Alkan A, Günay M (2012) Identification of EMG signals using discriminant analysis and SVM classifier. Expert Syst Appl 39:44–47
13. De la Rosa R, Alonso A, Carrera A, Durán R, Fernández P (2010) Man-machine interface system for neuromuscular training and evaluation based on EMG and MMG signals. Sensors 10:11100–11125
14. Phinyomark A, Hu H, Phukpattaranont P, Limsakul C (2012) Application of linear discriminant analysis in dimensionality reduction for hand motion classification. Meas Sci Rev 12:82–89
15. Strzecha K, Krakós M, Więcek B, Chudzik P, Tatar K, Lisowski G, Mosorov V, Sankowski D (2021) Processing of EMG signals with high impact of power line and cardiac interferences. Appl Sci 11:4625. https://doi.org/10.3390/app11104625
16. Basmajian JV (1973) Electromyography. In: Bourne GH (ed) The structure and function of muscle, vol II, 2nd edn. Academic Press, New York, pp 273–283
17. Agarwal GG (1975) An analysis of the EMG by Fourier, simulation and experimental techniques. IEEE Trans Biomed Eng 22(3):225–229
18. Broman H (1974) Frequency domain methods applied to the analysis of gross muscle function in humans. Technical Report No. 45, School of Electrical Engng, Chalmers Univ. of Technology, Sweden
19. Bigland-Ritchie B (1981) EMG force relations and fatigue of human voluntary contractions. Exercise Sports Sci Rev 9:75–117
20. Grieve DW, Cavanagh PR (1973) The quantitative analysis of phasic electromyograms. In: Desmedt JE (ed) New developments in electromyography and clinical neurophysiology, vol 2. Karger, Basel, pp 489–496
21. Jackson KM (1982) Digital analysis of electromyograms: a Fortran package. Electromyogr Clin Neurophysiol 22:65–87
22. BTS Bioengineering (2016) User manual of BTS G-WALK. Tech. Eng. version 8.1.0., BTS SpA., 149, 15–35
23. BTS Bioengineering. https://www.btsbioengineering.com/products/bts-sportlab-motion-capture-sport/
24. BTS Bioengineering. https://www.btsbioengineering.com/products/g-walk-inertial-motion-system/
25. Qakhkharov A, Kholdorov S, Atadjanova N, Davletova S, Khayitov N (2021) Analysis of methods and algorithms for feature extraction of biosignals of muscle activity. In: 2021 international conference on information science and communications technologies (ICISCT), pp 1–5. https://doi.org/10.1109/ICISCT52966.2021.9670012
26. Shukurov K, Berdanov U, Khasanov U, Kholdorov S, Turaev B (2021) The role of adaptive filters in the recognition of speech commands. In: 2021 international conference on information science and communications technologies (ICISCT), Tashkent, Uzbekistan, pp 1–4. https://doi.org/10.1109/ICISCT52966.2021.9670084
27. Kamoliddin Elbobo ugli S, Shokhrukhmirzo Imomali ugli K, Umidjon Komiljon ugli K (2020) Uzbek speech commands recognition and implementation based on HMM. In: 2020 IEEE 14th international conference on application of information and communication technologies (AICT), Tashkent, Uzbekistan, pp 1–6. https://doi.org/10.1109/AICT50176.2020.9368591

Optimizing Task Handling in Mobile Edge Computing: A Dynamic Load Balancing Model with Intermediate Coordination Node and Nearest Neighbor Integration

Hoa Doan Nguyen Thanh, Phu Nguyen Ngoc Thien, Nghia Phan Duc, Anh Nguyen Phan Tuan, and Lap Hoang Van

Abstract This paper delves into the realm of Mobile Edge Computing, addressing the imperative need for efficient task handling. Our study introduces an innovative load balancing model enhanced by an intermediary node tasked with coordination functions. The core methodology incorporates dynamic load balancing coupled with the integration of the nearest neighbor model, providing a comprehensive solution to the complexities of task distribution in Mobile Edge Computing environments. To validate the effectiveness of our proposed approach, we conducted experiments utilizing a dataset comprising telephone data from the Shanghai area. The results of our implementation showcased substantial improvements in processing completion times compared to traditional techniques such as least connection, least time, and round-robin approaches. This substantiates the efficacy of our dynamic load balancing model with intermediate coordination node and nearest neighbor integration in optimizing task handling within the Mobile Edge Computing paradigm. Index Terms—Edge Computing, Load Balancing, Mobile Edge Computing, Dynamic Load Balancing, Nearest Neighbor.

Keywords Edge computing · Load balancing · Mobile edge computing · Dynamic load balancing · Nearest neighbor

---

D. N. T. Hoa (✉) · N. N. T. Phu · P. D. Nghia · N. P. T. Anh · H. V. Lap
FPT University, Ho Chi Minh, Vietnam
e-mail: hoadnt@fe.edu.vn

N. N. T. Phu
e-mail: phunntse182426@fpt.edu.vn

P. D. Nghia
e-mail: nghiapdse182439@fpt.edu.vn

N. P. T. Anh
e-mail: anhnptse182494@fpt.edu.vn

H. V. Lap
e-mail: laphvse180503@fpt.edu.vn

P. G. Bradford et al. (eds.), *Proceedings of IEMTRONICS 2024*, Lecture Notes in Electrical Engineering 1229, https://doi.org/10.1007/978-981-97-4780-1_8

1 Introduction

In the dynamic landscape of modern computing, Mobile Edge Computing (MEC) has emerged as a transformative paradigm, bringing computational resources closer to the edge of the network [1]. However, with this proximity comes a host of challenges, necessitating innovative solutions to ensure the optimal performance of tasks in MEC environments. Mobile Edge Computing faces challenges stemming from the dynamic and resource-constrained nature of edge devices. The need for efficient task handling becomes paramount as MEC strives to meet the demands of real-time applications, such as augmented reality, IoT analytics, and responsive mobile services [2]. In this context, the traditional load balancing techniques commonly employed in centralized computing systems prove insufficient, calling for tailored approaches to address the unique challenges posed by MEC [3]. The focal point of this paper is the imperative need to enhance task handling efficiency within the Mobile Edge Computing paradigm. The challenges of load balancing in MEC environments prompt the introduction of a novel approach a dynamic load balancing model with an intermediate coordination node and nearest neighbor integration. This model seeks to overcome the limitations of existing methodologies by introducing an intermediary node dedicated to coordination functions, providing adaptability to the dynamic workloads inherent in MEC. The statement of the problem revolves around the inadequacy of conventional load balancing techniques in MEC scenarios, leading to suboptimal task distribution and system performance. The motivation for the proposed model lies in the quest for a comprehensive solution that not only addresses the challenges of load balancing in MEC but also optimizes task handling efficiency. This paper presents an in-depth exploration of the proposed dynamic load balancing model, elucidating the role of the intermediate coordination node and the integration of the nearest neighbor model. The model's adaptive nature and its ability to respond to the dynamic characteristics of MEC environments are highlighted, setting the stage for a robust and efficient approach to task distribution. In Section I-The Introduction provides an overview of MEC challenges. Section 2—The literature review critically surveys existing work in load balancing for MEC, identifying gaps that the proposed model aims to fill. The methodology in Sect. 3 provides a detailed exposition of the novel model. The experiment and discussion are in Sect. 4. Finally, the discussion in Sect. 5 interprets the results, compares them with theoretical expectations, addresses limitations, and suggests avenues for further refinement.

2 Literature Overview

Current approaches to load balancing in MEC exhibit notable challenges and shortcomings. Traditional techniques, including Round Robin and Least Connection, lack the adaptability required for the dynamic nature of tasks in MEC environments. These methods often struggle to efficiently distribute workloads, leading to suboptimal resource utilization and increased latency. Additionally, the limited computational capabilities of edge devices pose a significant challenge, as traditional load balancing algorithms may not consider the unique constraints of these devices, resulting in uneven task distribution and potential bottlenecks. The existing literature on load balancing in MEC reflects a dynamic field grappling with the unique challenges posed by the distributed and edge-centric nature of MEC environments. Numerous studies have explored traditional load balancing techniques and their applicability in MEC scenarios. Early works, such as [4], investigated approaches like Round Robin and Least Connection to distribute tasks among edge servers. However, as the demand for real-time applications surged, shortcomings in these conventional methods became apparent, prompting researchers to delve into more adaptive strategies. Recent literature has witnessed a shift toward dynamic load balancing techniques tailored for MEC. Studies, such as Dasoriya [5], have explored the effectiveness of dynamic load balancing algorithms that can adapt to the fluctuating workloads inherent in edge computing. The introduction of intermediate coordination nodes has gained prominence, with [5] proposing models that strategically manage task distribution and enhance overall system performance. Nearest Neighbor models have been integrated, as seen in [6], to further optimize load balancing decisions based on proximity and real-time context. Moreover, researchers have recognized the significance of considering the unique characteristics of MEC, such as limited resources and varying network conditions. In [7], the literature highlights the importance of load balancing algorithms that factor in the constraints of edge devices and dynamically allocate tasks to ensure efficient resource utilization. Additionally, the integration of machine learning and artificial intelligence techniques for load prediction and decision-making has gained traction, as evidenced by [8]. Despite these advancements, gaps in the literature persist, necessitating further exploration. Comprehensive surveys, like [9], emphasize the need for standardized benchmarks and metrics to evaluate the performance of load balancing algorithms in diverse MEC scenarios. The ongoing evolution of edge computing architectures and the emergence of new technologies continue to drive the quest for innovative load balancing strategies. Overall, the literature survey underscores the dynamic nature of load balancing research in MEC, reflecting a continuous pursuit of adaptive and efficient solutions to address the evolving challenges posed by edge computing environments. Within the current body of literature on load balancing in MEC, discernible gaps persist, presenting opportunities for innovative contributions. Existing studies often concentrate on specific facets of load balancing in MEC, lacking a unified approach that comprehensively addresses the challenges posed by dynamic and resource-constrained environments. A key gap lies in the absence of models that seamlessly integrate coordination nodes and nearest neighbor strategies to optimize task distribution. Moreover, the lack of standardized benchmarks and metrics for evaluating load balancing algorithms in diverse MEC scenarios poses a hindrance, impeding effective comparisons and limiting the generalizability of findings. Another critical gap arises from the need for research explicitly tailored to the dynamic and resource-constrained nature of edge devices, ensuring that load balancing strategies can adapt to varying workloads and constraints. The proposed dynamic load balancing model with intermediate coordination node and nearest neighbor integration seeks to fill these identified gaps by offering a comprehensive solution. This model aspires to provide a unified approach that combines coordination nodes and nearest neighbor integration, addressing the specific challenges posed by MEC. By contributing to the establishment of benchmarks and metrics, it aims to enhance the standardized evaluation of load balancing algorithms, ultimately advancing the field and fostering the development of more effective and adaptable strategies for task distribution at the edge of the network. The proposed model aims to bridge these gaps by providing a holistic solution that combines coordination nodes and nearest neighbor integration, offering adaptability to the dynamic nature of MEC environments. This model seeks to contribute to the establishment of benchmarks and metrics, fostering a more standardized evaluation of load balancing algorithms in MEC. By addressing these identified gaps, the proposed model aspires to advance the current state of load balancing in MEC and contribute to the development of more effective and adaptable strategies for task distribution at the edge of the network.

3 A New Model for Dynamic Load Balancing in Edge Computing

The load balancing system architecture depicted in Fig. 1 comprises several key components orchestrating the efficient distribution of tasks in MEC environments. At its core lies the Queue Pool, serving as the centralized repository for all incoming user requests. Utilizing the First In First Out (FIFO) mechanism, the Queue Pool ensures fair and orderly task processing. Adjacent to the Queue Pool is the intermediate coordination node, a pivotal element responsible for executing computations based on the dynamic load balancing mechanism. Functioning as a multi-objective optimization processing function, the intermediate coordination node considers various factors, including the distance between edge devices and user requests, the processing capabilities of available resources, and the current system workload. This node acts as the brain of the load balancing system, making informed decisions regarding task assignment and resource allocation to optimize system performance. By leveraging real-time data and sophisticated algorithms, the intermediate coordination node dynamically adjusts task assignments to minimize latency, enhance resource utilization, and ensure equitable distribution of workload across available resources. Through its strategic positioning and intelligent decision-making capabilities, the intermediate coordination node plays a pivotal role in orchestrating the seamless operation of the load balancing system, facilitating efficient task handling in MEC environments. Multi-objective optimization processing functions play a crucial role in modern distributed computing environments, where the efficient allocation of computational tasks across diverse resources is paramount. In this context, the intermediate coordination node emerges as a pivotal component, tasked with orchestrating task assignments while considering a multitude of factors to optimize system performance. This introduction explores the significance of multi-objective optimization processing functions and elucidates how the intermediate coordination node integrates various factors to achieve optimal task allocation in distributed systems. In the landscape of distributed computing, optimizing processing functions involves balancing multiple objectives simultaneously, such as minimizing latency, maximizing throughput, and optimizing resource utilization. This multidimensional optimization task poses a complex challenge, particularly in MEC environments, where resources are geographically dispersed, and computational tasks exhibit diverse characteristics. The intermediate coordination node serves as the linchpin in addressing this challenge, leveraging a multi-objective optimization framework to make informed task allocation decisions. By considering various factors, including the distance between edge devices and user requests, the processing capabilities of available resources, and the current system workload, the coordination node endeavors to optimize multiple objectives concurrently. Mathematically, the multi-objective optimization problem can be formulated as Eq. 1:

$$\text{Minimize } F(x) = \{f_1(x), f_2(x), \ldots, f_m(x)\} \tag{1}$$

subject to $g_j(\mathrm{x})\ 0, j = 1, 2, \ldots, p$ and $h_k(x) = 0, k = 1, 2, \ldots, q$.

Where x represents the decision variables governing task allocation. $F(\mathrm{x})$ denotes the vector of objective functions to be minimized, including m objectives $f_1(\mathrm{x}), f_2(x), \ldots, f_m(x)$. $g_j(x)$ and $h_k(x)$ represent inequality and equality constraints, respectively, on the decision variables. p and q denote the number of inequality and equality constraints, respectively. In the context of MEC, decision variables may include task assignment to specific resources, resource utilization levels, and communication overhead. The objective functions $f_i(x)$ aim to optimize various performance metrics, such as response time, energy consumption, and resource utilization, while the constraints $g_j(x)$ and $h_k(x)$ capture system requirements and limitations. By formulating the task allocation problem as a multi-objective optimization problem, the intermediate coordination node can explore trade-offs between conflicting objectives and identify Pareto-optimal solutions that offer significant improvements in system performance. This approach enables the efficient utilization of computational resources, minimizes latency, and enhances the overall user experience in MEC environments.

4 Experimental Results

The experiment conducted aimed to evaluate the performance of the proposed dynamic load balancing model in a simulated MEC environment. We edited the dataset provided by Shanghai Telecom containing more than 7.2 million records of

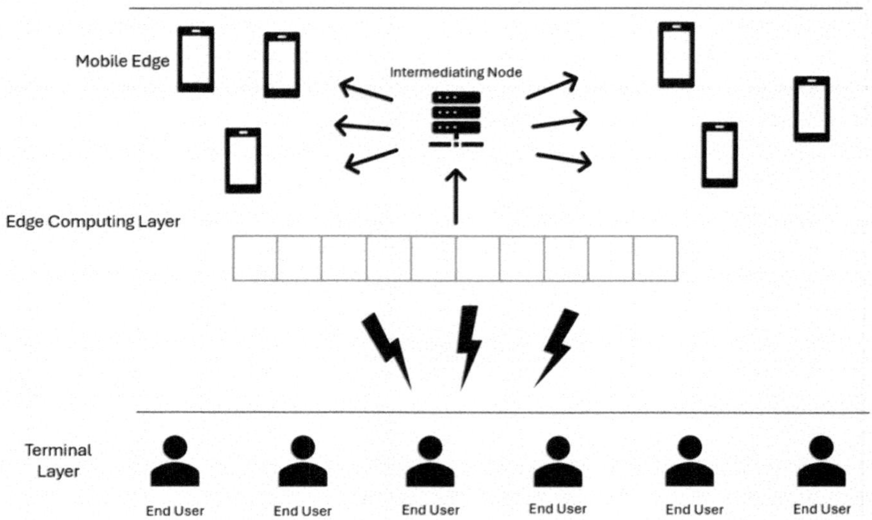

Fig. 1 The load balance system architecture

accessing the Internet through 3233 base stations from 9481 mobile phones [10]. The experiment utilized a simulation dataset consisting of 36,000 edge devices, each characterized by their geographical coordinates and processing capabilities. These edge devices represent the computational resources available within the MEC network. To simulate real-world scenarios, the experiment generated a set of n simultaneous requests from random locations within the network. Each request was defined by its location and the number of tasks requiring processing. These tasks represent computational workloads that need to be distributed and executed across the edge devices. The experiment focused on comparing the performance of the proposed dynamic load balancing model with three traditional approaches: least connection, random, and round-robin approaches. These approaches served as benchmarks for evaluating the effectiveness of the proposed model in optimizing task handling within the MEC environment. During the simulations, the performance of each approach was assessed based on the processing completion times metric: the time taken for tasks to be completed and responses to be delivered to the requesting entities. To calculate the time to complete a task sent from node A to node B in a network environment, we need to consider several factors, including the processing ability at node B, the processing status or free time of node B, and the distance between nodes A and B. We can formulate this calculation using a combination of these parameters. Equation 2 is a basic formula:

$$\text{Complete Time} = \text{Transmission Time} + \text{Processing Time} \quad (2)$$

where Transmission Time is the time taken for the task to travel from node A to node B. This can be calculated based on the distance between the nodes and the transmission speed of the network. Processing Time is the time taken for node B to process the task. This depends on the processing ability of node B and its current processing status. Transmission time can be calculated using the following Eq. 3:

$$\text{Transmission Time} = \frac{\text{Distance}}{\text{Transmission Speed}} \tag{3}$$

where Distance is the physical distance between nodes A and B. Transmission Speed is the speed at which data can be transmitted between nodes, usually measured in meters per second or a similar unit. Processing Time can be calculated based on the processing ability of node B and its current processing status. A simplified formula to estimate processing time is shown in Eq. 4:

$$\text{Processing Time} = \frac{\text{Task Size}}{\text{Ability B}} \tag{4}$$

where Task Size is the size or complexity of the task being processed, typically measured in bytes or a similar unit. Ability B is the processing capacity of node B, usually measured in operations per second (OPS) or a similar metric. Status Factor is a factor that accounts for the current processing status or workload of node B. This factor can be adjusted based on factors such as CPU utilization or the number of tasks currently being processed. In every demonstration, we generate 25,000 requests, each comprising a set of mathematical operations requiring execution. This demonstration is repeated 15,000 times, and statistical outcomes are illustrated, as depicted in Fig. 2. The results reveal that the processing time of our proposed model may not be optimal across all scenarios; however, for requests entailing a substantial number of calculations, the completion time exhibits superior convergence compared to alternative methods.

5 Conclusion and Future Work

This paper has successfully navigated the challenges of Mobile Edge Computing task handling, presenting a dynamic load balancing model with intermediate coordination node and nearest neighbor integration. As we recapitulate the key findings, it becomes evident that our proposed model has demonstrated significant improvements in optimizing task handling efficiency within the dynamic MEC environment. The empirical results, derived from experiments with telephone data from Shanghai Telecom, affirm the model's effectiveness in surpassing traditional load balancing techniques, such as least connection, random, and round-robin approaches. Our model not only addresses the challenges associated with load balancing but also contributes to the advancement of responsive and efficient computing at the edge of the network. By

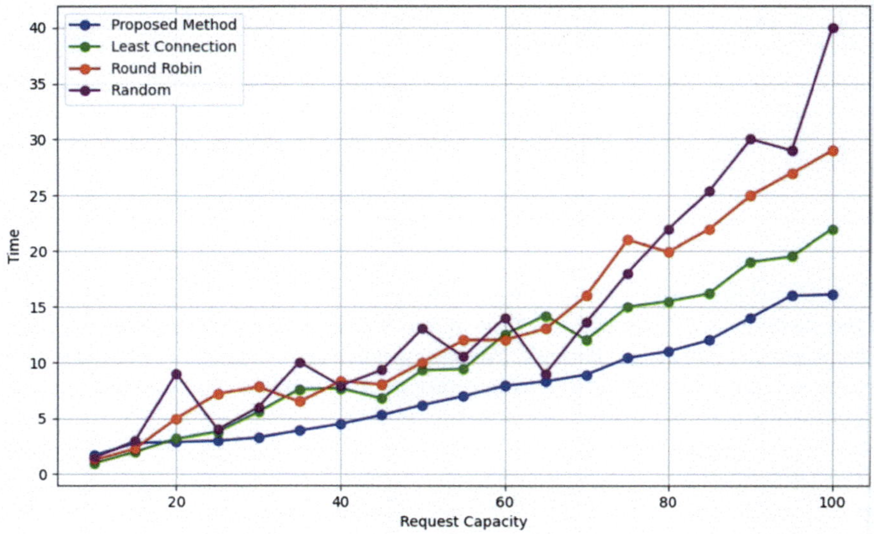

Fig. 2 The experience results

emphasizing the adaptability of our model to the dynamic nature of tasks, we envision its potential application in diverse MEC scenarios, from IoT analytics to augmented reality applications. This research opens avenues for future exploration and innovation in MEC. As the demand for responsive edge computing solutions continues to grow, the insights gained from this study can serve as a foundation for future advancements, ultimately shaping the landscape of Mobile Edge Computing and its role in the broader realm of distributed systems.

References

1. Abbas N, Zhang Y, Taherkordi A, Skeie T (2018) Mobile edge computing: a survey. IEEE Internet Things J
2. Ahmed A; Ahmed E (2016) A survey on mobile edge computing. In: 2016 10th international conference on intelligent systems and control (ISCO)
3. Wilson M, Nunoo-Mensah H, Boateng KO (2023) A review of computational load-balancing for mobile edge computing. Lecture notes in networks and systems book series (LNNS, vol 711)
4. Malik A, Shukla G, Sharma D, Singh S, Kumar S (2023) Enhancement of edge security using dynamic load-balancing algorithm for 5g cloud computing network. Lecture notes in networks and systems book series (LNNS, vol 831)
5. Dasoriya R, Kotadiya P, Arya G, Nayak P, Mistry K (2017) Dynamic load balancing in cloud a data-centric approach. In: 2017 international conference on networks and advances in computational technologies (NetACT)
6. Zhang W, Chen X, Liu Y, Xi Q (2020) a distributed storage and computation k-nearest neighbor algorithm based cloud-edge computing for cyber-physical-social systems. IEEE Access 8

7. Shao S, Liu S, Li K, You S, Qiu H, Yao X, Ji Y (2023) LBA-EC: load balancing algorithm based on weighted bipartite graph for edge computing. Chin J Electron 32(2)
8. Chen J, Ran X (2019) Deep learning with edge computing: a review. IEEE Open Access
9. Malazi HT, Chaudhry SR, Kazmi A, Palade A, Cabrera C, White G, Clarke S (2022) Reinforcement learning for communication load balancing: approaches and challenges. IEEE Open Access
10. Xu J, Wang S, Bhargava BK, Yang F (2019) A blockchain-enabled trustless crowd-intelligence ecosystem on mobile edge computing. IEEE Trans Ind Inf 15(6):3538–3547

Review Paper on the Use Cases and Types of Virtual Reality

Pratik Chakraborty, Debabrata Ghosh, Shivanshu Srivastav, Sayan Hait, and Srijita Chakraborty

Abstract This comprehensive review delves into the multifaceted applications of Virtual Reality (VR) and explores the contemporary technologies supporting its evolution. Examining advanced features like Eye Tracking, Facial Expression Tracking, and Heartbeat Tracking, the review highlights the dynamic capabilities that enhance user experiences within virtual environments. In addition to technology exploration, the review thoroughly categorizes VR into three distinct types: Immersive, Non-Immersive, and Semi-Immersive, providing an insightful overview of their characteristics and applications. By uncovering the diverse use cases and emerging technologies, this abstract goes beyond the surface, emphasizing VR's potential to redefine industries, from revolutionizing gaming and education to transforming healthcare practices. The nuanced exploration underscores VR's role as a transformative force in shaping the future of immersive digital experiences.

Keywords Virtual reality · Eye tracking · Facial expression tracking · Immersive · Non-immersive · Semi-immersive

1 Introduction

In current times the substantial transformation in the utilization of Virtual Reality (VR) across various disciplines has been a revelation. Notably, modern technology, particularly in the field of eye tracking, has significantly reduced rendering workloads. When it comes to the different types of VR experiences, Immersive VR provides a full range of sensory engagement, Non-Immersive VR offers minimal sensory interaction, and Semi-Immersive VR strikes a balance between reality and immersion. Virtual Reality (VR) technology has seen significant advancements, with eye tracking being a prominent use case. Next-gen VR devices employ eye tracking, introducing Foveated rendering. This method optimizes rendering by focusing high

---

P. Chakraborty (✉) · D. Ghosh · S. Srivastav · S. Hait · S. Chakraborty
Institute of Engineering and Management Kolkata, University of Engineering and Management, Newtown, India
e-mail: Pratik.Chakraborty2022@iem.edu.in

P. G. Bradford et al. (eds.), *Proceedings of IEMTRONICS 2024*, Lecture Notes in Electrical Engineering 1229, https://doi.org/10.1007/978-981-97-4780-1_9

resolution on the central vision, enhancing performance and visual quality while reducing GPU workload. Facial expression tracking further enhances user experience by capturing emotions, aiding in immersive interactions and avatar realism. Heartbeat tracking in VR monitors users' health in real time, crucial for scenarios like training simulations. Immersive VR, characterized by head-mounted displays (HMDs) and motion tracking sensors, offers heightened sensory engagement. Non-immersive VR utilizes 2D displays, while semi-immersive VR blends elements of both, catering to varied user preferences and needs. Eye tracking methods such as electro-oculography (EOG) and video oculography (VOG) enable precise tracking within HMDs. Semi-immersive VR systems like VRDD and CAVE offer immersive experiences while balancing cost and functionality. Heart rate monitoring in VR aids medical training and patient education. Educationally, VR applications like VRCT combine cognitive training with physical activity, enhancing cognitive function in older adults. 360° content creation offers immersive experiences, especially when combined with VR echnology, revolutionizing fields like education and construction. VR education with 360° videos provides an engaging alternative to conventional lectures, fostering active participation and comprehension. In summary, VR technology continues to evolve, offering diverse applications ranging from healthcare to education. These advancements promise to revolutionize various industries, providing immersive and engaging experiences for users worldwide.

2 Review on Use Cases and Types of VR

The first use case of VR, where implementation is evident, lies in the domain of eye tracking technology. Next-generation VR devices feature eye tracking capabilities that introduce a groundbreaking rendering method known as Foveated rendering. It's important to note that the fovea, a small region on the retina of our eyes, contains tightly packed sensors capable of processing high-resolution images. Beyond this area, resolution significantly declines. This underscores the inefficiency of the traditional rendering approach, where pixels maintain even resolution throughout the image. Foveated rendering, in contrast, renders peripheral vision with low resolution while enhancing the pixels in the central focus with high resolution. This dramatically improves system performance, visual quality, and frame rates by alleviating the GPU workload.

Moving on to the second use case, the analysis of facial expression tracking technologies is pivotal. Next-generation VR devices are equipped with built-in facial expression sensors capable of capturing users' facial expressions as they interact within the virtual environment. This feedback provides insights into the user's mental and emotional state, including factors like concentration and enjoyment during training sessions. This technology also enables the creation of expressive and dynamic 3D avatars in collaborative enterprise-Metaverses, elevating the level of realism.

Another use case in the realm of VR involves Heartbeat Tracking. Integrated heart rate trackers in next-gen VR devices monitor the user's real-time health status. They can track and record the user's stress levels during interactions within the VR environment, particularly in situations involving critical and life-threatening scenarios like plant malfunctions and firefighting.

Shifting our focus to the various types of VR, we begin with Immersive VR. To immerse themselves in the virtual environment, users don head-mounted displays (HMDs) such as VR goggles or headsets, often equipped with high-resolution displays and motion tracking sensors. Immersive VR systems detect the user's head and body movements in real time through sensors and cameras, enabling the virtual environment to react accordingly. Surround sound or 3D audio, adapting to the user's head movements, enhances the sense of presence. Some immersive VR setups include haptic feedback devices, such as gloves or suits, allowing users to experience sensations and interact with virtual objects.

During the 1970s, filmmakers established groundbreaking practices to capture the locomotion and movement of different actors using 3D optical technology. Computers were assigned to capture and calculate the movements of characters with markers and tracked with a camera. This technique was very creatively assigned to the famous Star Wars Character known as Jar Jar Binks which was seen in the movie "Star Wars: The Phantom Menace". After its success in the industry, the motion capture alongside the movement was blended into camera-based capturing to generate movement data, leading to the development of new motion capture techniques [1].

The HTC VIVE is a virtual reality head-mounted display equipped with two controllers. In addition to the costs and experimental requirements, three additional trackers have been incorporated [2].

Further motion capture options for immersive VR include the Quest 2, which features a lighter weight, updated internal specifications, a display with a higher refresh rate and per-eye resolution, and updated Oculus Touch controllers with improved battery life. Similar to its predecessor, the Quest 2 can operate as either a standalone headset with an internal Androidbased operating system or with Oculus Rift-compatible VR software running on a desktop computer.

Non-Immersive VR technology primarily utilizes 2D displays, such as computer monitors or smartphones, to present digital content to users. Unlike Immersive VR, which aims to engage multiple senses, Non-Immersive VR primarily focuses on visual and, at times, auditory engagement, omitting elements like haptic feedback or full-body tracking. Users typically interact with Non-Immersive VR using standard input devices like a keyboard, mouse, or touchscreen, rather than specialized controllers.

Semi-Immersive VR blends the capabilities of both immersive and non-immersive VR. Semi- Immersive VR typically employs larger screens or projectors compared to Non-Immersive VR, providing a wider field of view (FOV) and a more immersive visual experience. While Semi- Immersive VR primarily focuses on visual engagement, it may incorporate some sensory feedback, such as spatial audio, to enhance immersion. Users may have access to basic tracking systems, such as head tracking or hand tracking, allowing them to interact with the virtual environment more naturally.

In Eye tracking, the following procedures are standard in the modern VR products and research.

Three methods have been used to track eye movements in HMDs: (1) electro-oculography (EOG), (2) scleral search coils, and the most common, (3) video oculography (VOG) [3].

EOG primarily works in such a way that it detects the position of the eye based on the resting potential of the eye when some type of electrode is spaced out on the skin surrounding the eye. Electrodes are easily integrated into the HMD that contacts the face of the user. This solution is feasible because of the eye being dipole which means it is positively attracted in the direction of the cornea and negatively attracted to the retina. The orientation of the eye clearly suggests that there is a definitive voltage difference between the electrodes which are placed directly on the contradicting sides of the eye. The drawback that comes with EOG is that it is easily prone to error in predicting the correct position of the eye but it is the only solution in which the eyes are tracked even when they are closed [3].

The scleral search coil method is responsible for keeping track of the wire embedded in the user's contact lens. The user puts their head in between the space of the Helmholtz coils to produce a uniform magnetic field. As the eye moves within a uniform, known magnetic field, electrical currents are induced in the scleral coil, indicating the horizontal, vertical, and torsional orientation of the eye. This method is reported to be highly accurate and precise, with spatial resolution of less than 0.1° and temporal resolution of greater than 1 kHz. However, implementing this method in his HMD is difficult because it not only requires a wired contact lens, but also requires a Helmholtz coil to be attached to the head, which makes the volume of the resulting magnetic field significantly smaller. Nevertheless, his HMD with a scleral coil system was developed to be used in cases where there's a need in precise tracking is required in a HMD, such as the validation of alternative tracking systems [3].

Time and cost savings were key success factors for the rapid development of the semi-immersive VR system. Reproducing the entire real-world environment surrounding a user in a virtual environment in a complete rebuilding process would be extremely time-consuming in practice. For example, recreating a user's hand in a virtual environment requires high computing power due to the huge number of polygons and kinematics limitations. Additionally, multi-user functionality is one of the most important features of these systems. In particular, the project validation step, and even final project approval, requires a presentation device that can allow users to move in and around the photorealistic product being approved. The Workbench and CAVE are the most famous Fake Space products and the best examples of semi-immersive technology [4].

Based on the workbench concept, VRDD (Virtual Reality Design Desk) is the first VR device developed at the V-Lab (v-lab.ingfo.unibo.it) of the Faculty of Aerospace Engineering of the University of Bologna Forli. A translucent screen approximately 2 m × 1.5 m wide is rear-projected using a 120 Hz compatible video projector. Using a powerful computing machine (PC workstation) and a stereo-enabled system, he can run four buffer images in real time. In addition, users are equipped with an integrated head tracking system and a set of clamp gloves combined with a pen device for interacting with virtual objects and scenery. The state-of-the-art flight, ship, and vehicle simulator is semi-immersive. The cockpit, bridge, or seat of a driver is a type of physical model, but other outside world features are generally computer-generated. This is one of the most important features of semi-immersive environments [4].

CAVE systems are usually considered to be the best VR solution for ambient environments due to their multi-screen configuration. The extremely wide field of view (more than 230° in virtual CAVE space) allows users to immerse themselves in a complete VR experience. Recently, a new generation of his CAVE was considered and manufactured at V-Lab, based on the key concepts of flexibility and scalability. (Immersive Reconfigurable Space) (Figs. 10 and 11). This device is based on three 2.5×1.9 screens with relative rotation functionality. In other words, the side displays rotate relative to the central unit to fit any configuration between a flat wall (all three screens aligned) and a CAVE setup (side screens positioned 90 ° from center) You can do as you like. The rear projection of each unit also makes this system ideal for object manipulation and interaction-based simulations. Additionally, a PC graphics cluster was chosen as the image generator with each node driving a projector, and C++ code was written to manage data exchange and synchronization of all graphics servers using the Open Inventor library [4].

In terms of heart rate monitoring in VR, there has been steady growth.

The Stanford Virtual Heart Extend is working with Lighthaus, Inc. to utilize immersive VR headsets for instructive purposes. This extend has a few distinctive objectives. The primary is pointed at quiet and family instruction, making a difference in families superior get the life systems of their child's heart, but is right now restricted to drawings and plastic models. This moved forward understanding is expected to empower guardians to superior take an interest in their child's complex therapeutic care. This application has been expanded to Stanford restorative understudies and inhabitants who can visualize typical and irregular life systems and get how birth surrenders influence physiology. Completely immersive VR headsets permit understudies to investigate, associate with, and walk through models, giving them a more comprehensive understanding of life structures and physiology. A library of roughly 20 common intrinsic injuries is accessible to inhabitants. The objective of these encounters is to supply a more profound anatomical understanding of these injuries and move forward the understanding and learning rate of these complex irregular physiology and hemodynamic impacts [5].

The goal of REAtouch is to create a visual environment that allows the therapist to make informed decisions about how to design the intervention. It also focuses on the implementation of motor skill training principles: Intensity (high dose and repetitive practice) Target shaping and shaping (properly targeted and appropriate training with gradually increasing the intensity of the activity level training), hands-of (self-generated movements, driving by adapting the therapeutic environment's offerings), motor motivation, and feedback skills task performance.

It consists of 45 inches. The screen is mounted on a frame that can be adjusted. The therapist can adjust the height of the screen (from 55 to 121 cm) as well as the tilt angle of the reactive surface (from 0° to 85°) [6].

The environment was a side room in a hospital ward, where a patient was in a bed, and an avatar was being controlled by a student. The VR simulation was developed by Daden Ltd, a virtual reality company, using Unity 3D. The scenario was inspired by a worsening patient scenario developed by nursing colleagues. Students portrayed a nurse avatar in a room adjacent to the ward. The simulation began with the charge nurse handing over to the student avatar. The handoff (a form of SBAR communication) provided nursing students with knowledge and recorded the patient's status, current prescriptions, and observations from the previous night. The handover included the patient's medical history, including type 2 diabetes, chest infection, and hyperglycemia at admission. The patient subsequently developed hypoglycemia, which educators anticipated kids would identify and treat. The nurse in charge instructed the student to review the patient's observation sheet and make observations. The room's amenities mirrored nursing facilities at a nearby hospital. The student assumed responsibility for the patient's care and consulted with the charge nurse as needed [7].

The hypothesis of VR in instructive applications does not appear to have progressed sufficiently to permit reliable utilization of related terms such as inundation and authenticity. To decrease the restraining impacts of equivocalness and the need for clarity, encouraged work is required to contribute to a common understanding. A common understanding can be built on articles like our own that summarize classification systems from hypothesis and endeavor to classify other works based on sound criteria. Finally, a future research challenge is to propose learning theories and other framework elements for educational VR applications such as taxonomy [8].

VRCT (Virtual Reality Training) is a semi-immersive virtual reality (VR) program that combines cognitive training with movement activities. It utilizes a three-way projection that encircles the user and allows them to complete cognitive tasks by touching the screen. It can be executed. The program on this device handles various cognitive components, including B. Introduces users to visual and spatial awareness, memory, learning, attention, and decision-making, as well as challenging game tasks. Users perform cognitive gaming tasks projected on three screens while dynamically moving their bodies. Most of the virtual reality (VR) programs that pair physical activities with his games focus on specific motor learning exercises. However, there are rare VR programs that include complex tasks that require cognitive and motor skills in older adults. The goal of this study was to look at how semi-immersion VRCT in combination with exercise activity improves cognitive function, equilibrium, and walking performance in community living older adults [9].

360° content is produced by capturing and recording images from every angle simultaneously using an omnidirectional camera or multiple cameras. Unlike 2D videos that have a restricted and fixed perspective, 360° videos provide the viewer with a distinct advantage of being able to specifically capture different aspects of his 3D space along with all information about his surroundings. 360° video represents a novel approach to video streaming and potentially stands as one of the most auspicious manifestations of immersive video technology. 360° video scans can be viewed on a PC or digital device. Nevertheless, ongoing research is currently exploring the potential of HMD-based VR, which offers a captivating audiovisual experience within a three-dimensional environment. In terms of immersion, VRHMDs surpass smartphones, cardboard VR, and even VR HMDs with built-in displays, showcasing the most significant positive impact. The combination of 360° video and VR has led to the creation of content that is being explored for its potential use in different fields including education, experimentation, and construction. This innovative approach enables the recording of lectures and presentations in a way that captures the presenter, audience, and environment in one immersive experience, thus facilitating the sharing of participants' interactions within a VR environment. Also, viewers can experience the realism of the scene and recognize the emotions of the people around them during the lecture, such as joy, engagement, empathy, concern, recognition, and knowledge. VR environments that utilize 360° videos are highly efficient and cost-effective in producing educational content and allow students to experience the level of active participation in real-time instructional sessions that is unattainable through conventional online education. VR education with 360° videos seems to provide an adequate substitute for conventional lectures, enabling students to comprehend the material and enhancing their level of immersion [10] (Figs. 1, 2, 3, 4, 5, 6, 7, 8, 9, 10, 11, 12, 13, 14 and 15; Tables 1 and 2).

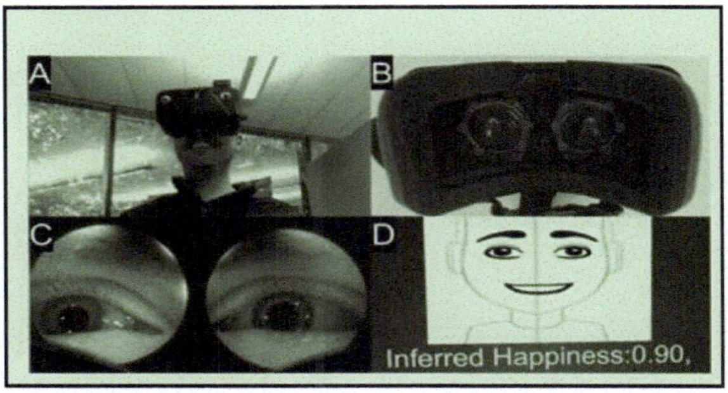

Fig. 1 Different types of eye and face tracking parameters [11]

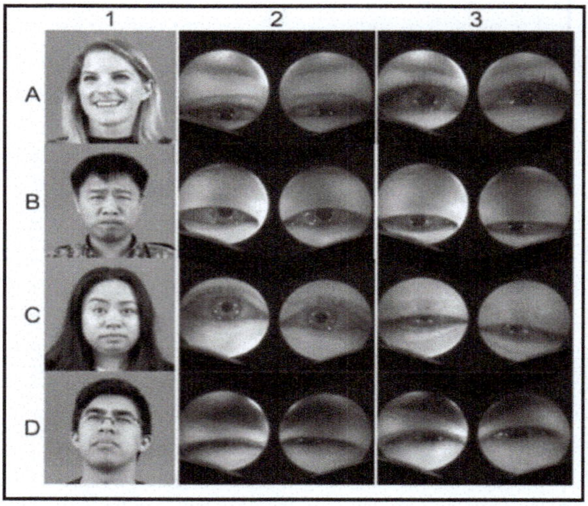

Fig. 2 Different types of expressions and their effects on eye orientation [11]

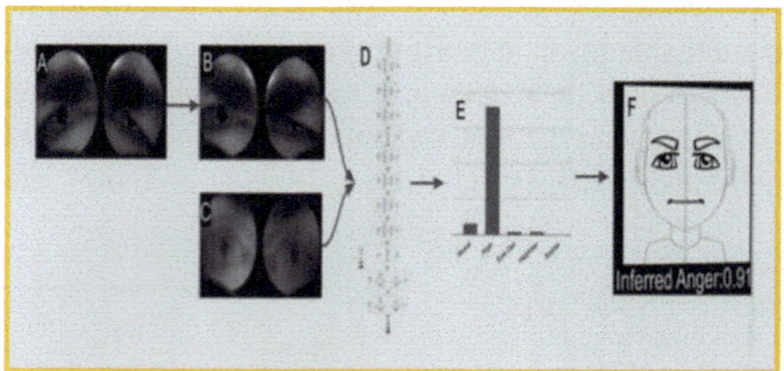

Fig. 3 Input eye tracking -> output face expression [11]

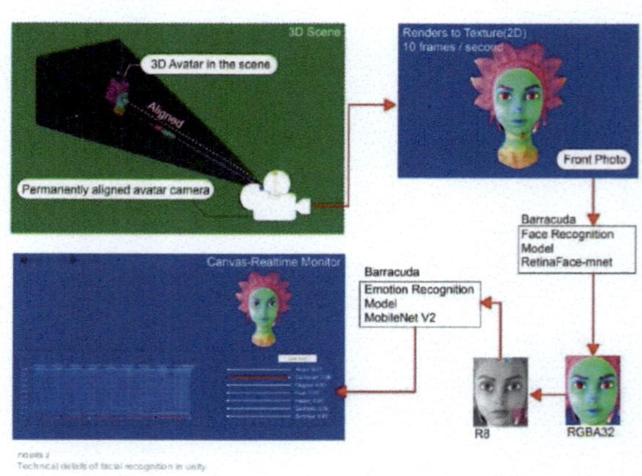

Fig. 4 Facial recognition tracking [12]

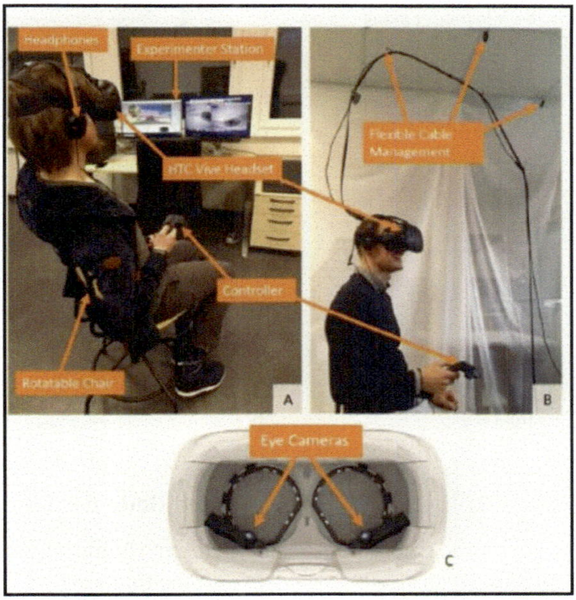

Fig. 5 Eye tracking demo [13]

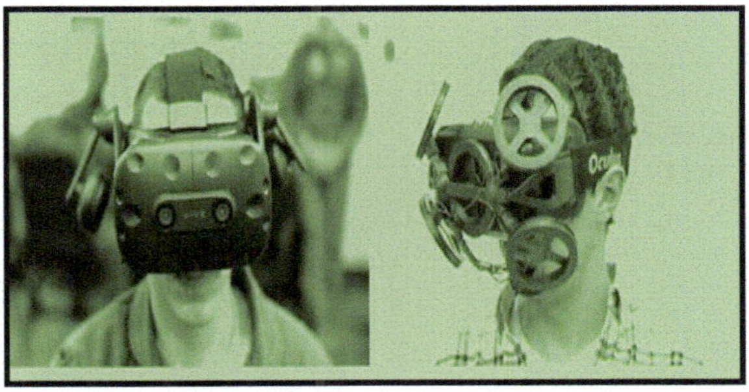

Fig. 6 Oculus quest [3]

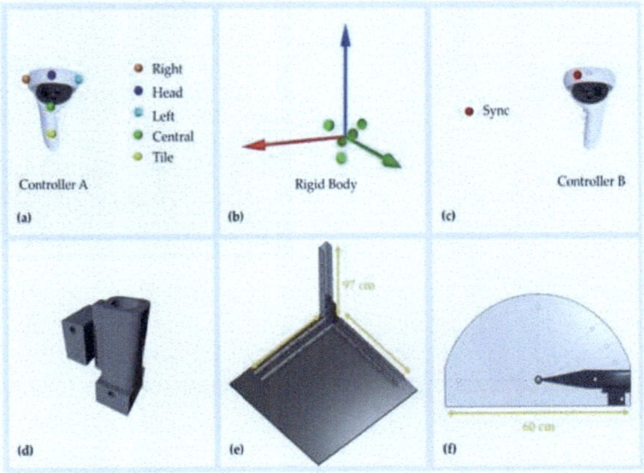

Fig. 7 Oculus quest controllers for rotational & translational movement [14]

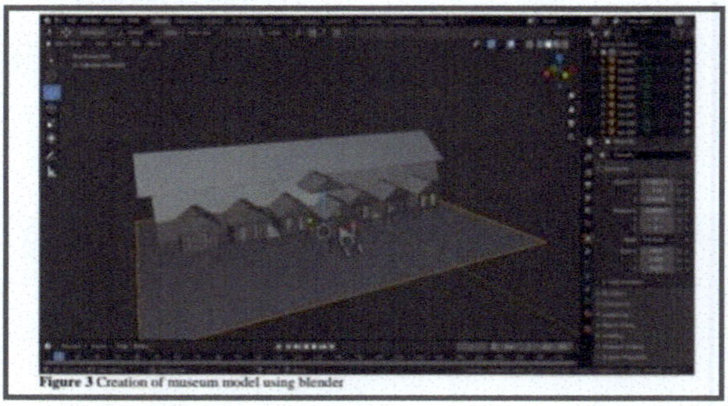

Fig. 8 VR model using Blender [15]

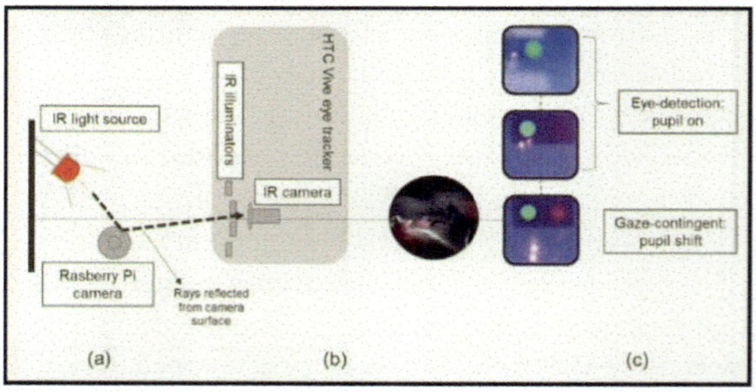

Fig. 9 HTC VIVE eye tracking [16]

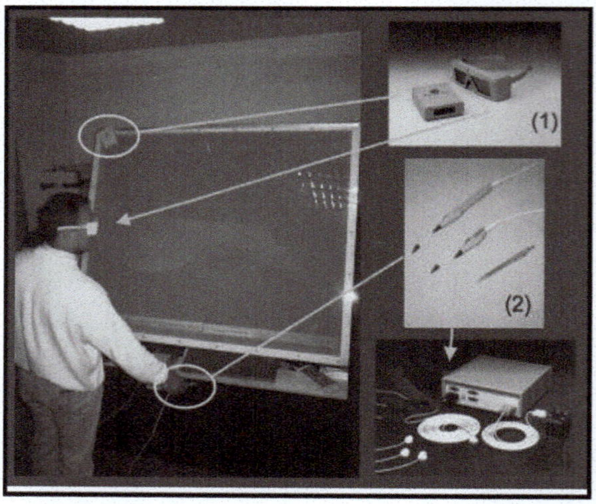

Fig. 10 VRDD at V-Lab, Forlì [4]

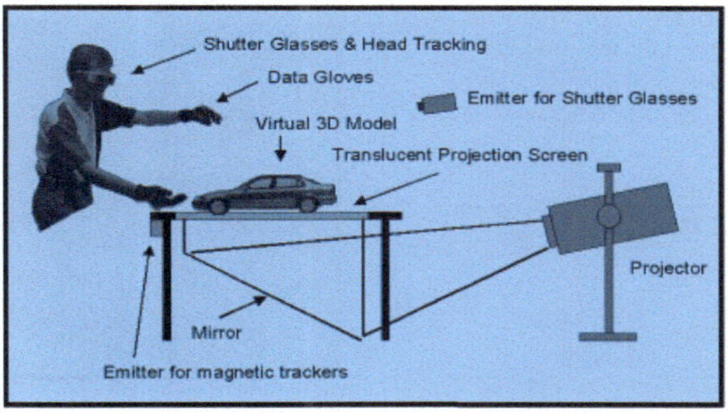

Fig.11 VRDD rear-projection layout [4]

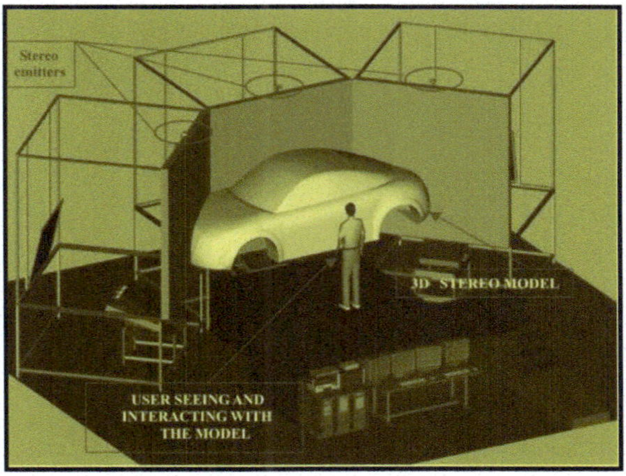

Fig. 12 I.R.R layout [4]

Fig. 13 I.R.R at a V-lab [4]

Fig. 14 HTC vive with the pupil labs add-on [17]

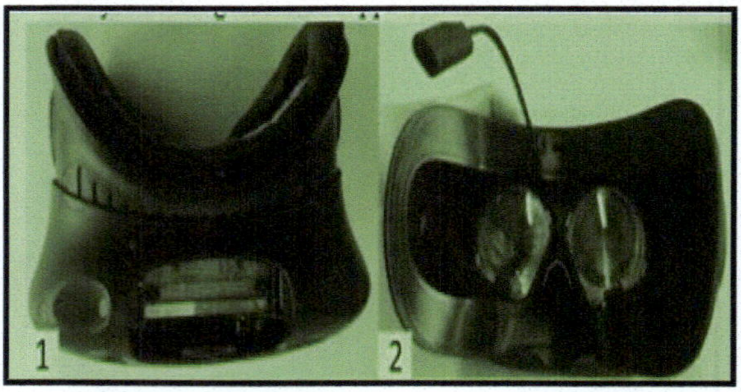

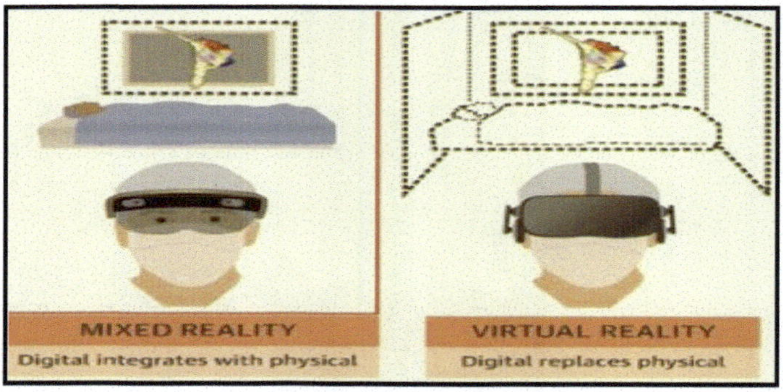

Fig. 15 In cardiovascular system [5]

Table 1 Comparison between previous gen. and next-gen VR

Name	Positional tracking	Display	Max resolution	Aspect ratio	Refresh rate	Field of view
Prev. Gen VR	Hardware and software	LCD	276 × 372	16:9	N/A	60°–70°
Oculus quest 2	Inside-out	LCD	1832 × 1920	N/A	120 Hz, 90 Hz, 72 Hz, 60 Hz	89°
HTC Vive Focus 3	Inside-out	LCD	2448 × 2448	1:1	90 Hz	120°

Table 2 Comparison between different non-immersive VR devices

Component/feature	PlayStation 4(PS4)	Xbox One
Processor	Custom AMD "Jaguar" 8- core CPU	Custom AMD "Jaguar" 8-core CPU
GPU	Custom AMD Radeon GCN, 1152 cores	Custom AMD Radeon GCN, 853 cores
RAM	8 GB DDR5	8 GB DDR3
Storage	500 GB/1 TB HDD (various models)	500 GB/1 TB HDD (various models)
Optical Drive	Blu-ray/DVD	Blu-ray/DVD

3 Conclusion

This investigation has delivered a comprehensive overview of Virtual Reality (VR), traced its historical evolution, and explored its diverse applications. Immersive VR, known for its complete sensory engagement, has emerged as a transformative force in fields encompassing gaming, training, and therapy. In contrast, Non-Immersive VR, characterized by 2D displays, offers accessibility and affordability, while Semi-Immersive VR strikes a harmonious balance between immersion and practicality. The future of VR holds tremendous promise, positioned to redefine education, industry, and entertainment. With the continual evolution of technology, VR remains at the forefront of human-computer interaction, promising innovative experiences in the uncharted digital landscape.

References

1. Ha E, Byeon G, Yu S (2022) Full-body motion capture-based virtual reality multi-remote collaborative system, Department of Culture Technology, Changwon National University, Changwon 51140, Korea
2. Cai S, Deng D, Li J, Wen J, Chen C, Ming Z (2019) Research on virtual reality arm motion capture and recognition. In: 2019 IEEE 4th international conference on advanced robotics and mechatronics (ICARM)
3. Adhanom IB, MacNellage P, Folmer E (2023) Eye tracking in virtual reality: a broad review of applications and challenges, Published online
4. Lorusso ML (2020) Semi-immersive virtual reality as a tool to improve cognitive and social abilities in preschool children. Scientific Institute IRCCS, E. Medea, Unit of Neuropsychology of Development Disorders, Bosisio Parini, 23842 Lecco, Italy
5. Josupeit J (2022) Let's get it started: eye tracking in VR with the pupil labs eye tracking add-on for the HTC Vive. Technische Universitat Dresden, Germany
6. Saussez G, Bailly R, Araneda R, Paradis J, Ebner-Karestinos D, Klöcker A, Sogbossi ES, Riquelme I, Brochard S, Bleyenheuft Y (2023) Efficacy of integrating a semi-immersive virtual device in the HABIT-ILE intervention for children with unilateral cerebral palsy: a non-inferiority randomized controlled trial. UCLouvain, Institute of Neuroscience, COSY Pole, MSL-IN Lab, Brussels, Belgium

7. Singleton H, James J, Falconer L, Holley D, Priego-Hernandez J, Beavis J, Burden D, Penfold S (2022) Effect of non-immersive virtual reality simulation on type 2 diabetes education for nursing students: a randomised controlled trial. Department of Nursing Science, Bournemouth University, Bournemouth, BH8 8AJ, UK
8. Radianti J, Majchrzak TA, Fromm J, Wohlgenannt I (2020) Asystematic review of immersive virtual reality applications for higher education: design elements, lessons learned, and research agenda. University of Agder, Kristiansand, Norway
9. Hwang N-K, Choi J-B, Choi D-K, Park J-M, Hong C-W, Park J-S, Yoon T-H (2021) Effects of semi-immersive virtual reality-based cognitive training combined with locomotor activity on cognitive function and gait ability in community-dwelling older adults. Department of Occupational Therapy, Seoul North Municipal Hospital, 38 Yangwonyeokro, Seoul 02062, Korea
10. Kim J, Kim K, Kim W (2022) Impact of immersive virtual reality content using 360-degree videos in undergraduate education. IEEE
11. Hickson S, Dufour N, Sud A, Kwatra V, Essa I (2019) Eyemotion: classifying facial expressions in VR using eye-tracking cameras. In: 2019 IEEE Winter conference on applications of computer vision (WACV).
12. Zhang Z, Fort JM, Mateu LG (2023) Facial expression recognition in virtual reality environments: challenges and opportunities. Universität Politecnica de Catalunya, Barcelona, Spain (2023)
13. Clay V, König P, König S (2019) Eye tracking in virtual reality. University of Osnabruck, Germany
14. Carnevale A, Mannocchi I, Sassi MSH, Carli M, De Luca G, Longo UG, Schena VDE (2022) Virtual reality for shoulder rehabilitation: accuracy evaluation of oculus quest 2. Research Unit of Orthopaedic and Trauma Surgery, Fondazione Policlinico Universitario Campus Bio-Medico, Via Alvaro del Portillo, 200, 00128 Roma, Italy
15. Samah KAFA, Nasaruddin NIS, Afandi MAR, Rahim NZA, Rum SFM, Saman FI (2021) Non-immersive virtual reality for Malay and Islamic world museum Melaka: effects from covid-19 pandemic. Faculty of Computer and Mathematical Sciences, University Teknologi MARA Cawangan melaka Kampus Jasin, Melaka, Malaysia
16. Sipatchin A, Rifai K, Wahl S (2021) Eye-tracking for clinical ophthalmology with virtual reality (VR): a case study of the HTC vive pro eye's usability. Institute for Ophthalmic Research, University of Tubingen, 72076 Tubingen, Germany
17. Silva JNA, Southworth M, Raptis C, Silva J (2018) Emerging applications of virtual reality in cardiovascular medicine. Published by Elsevier on behalf of the American College of Cardiology Foundation

Fault-Tolerant Routing in IoT Based on WBAN

Sakina Oussane, Haroun Benkaouha, and Amir Djouama

Abstract The Internet of Things (IoT) is a technology that connects physical objects to each other using a variety of digital systems. This technology has many applications, particularly in the field of medical monitoring. For this, a specific network called Wireless Body Area Network (WBAN) can be integrated into the IoT network. These networks are made up of sensors that measure patients' health parameters. The growing use of wireless body area networks (WBANs) in the medical field offers many advantages in terms of monitoring and healthcare. However, like any distributed system, WBANs also face design challenges such as fault-tolerance, quality of service, and data loss, while ensuring patient comfort. In this article, we present a new tree-based routing protocol, called Fault-tolerance Routing-Based Tree (FTRBT). The aim of this protocol is to guarantee availability, fault-tolerance, and quality of service in terms of delay and storage. To evaluate the effectiveness of this protocol, we compared its performance to that of a random routing protocol, without resorting to the use of a pre-established virtual architecture. This evaluation was carried out using the Castalia simulator based on OM-Net++. The results of our simulations and comparison demonstrate the effectiveness of the FTRBT protocol in detecting and tolerating faults within WBANs.

Keywords Fault-tolerance · Architecture · WBAN · Sensor · Sink · Failure · Recovery · Redundancy · Routing protocol · Quality of service

---

S. Oussane (✉) · H. Benkaouha
LSI Laboratory, USTHB, Algiers, Algeria
e-mail: s.oussane@usthb.dz

H. Benkaouha
e-mail: haroun.benkaouha@usthb.edu.dz

A. Djouama
National School of Artificial Intelligence, Algiers, Algeria
e-mail: amir.djouama@ensia.edu.dz

P. G. Bradford et al. (eds.), *Proceedings of IEMTRONICS 2024*, Lecture Notes in Electrical Engineering 1229, https://doi.org/10.1007/978-981-97-4780-1_10

1 Introduction

The Internet of Things (IoT), a major technological advance in the field of communications, encompasses intelligent devices integrating physical components such as sensors and actuators. These devices are capable of detecting the internal state of objects and their external environment, processing the data collected, and exchanging information via the Internet. This interconnection facilitates remote control and encourages closer integration between the physical world and IT systems [1].

Sensors communicate with each other to form wireless sensor networks (WSNs), and these networks are used in a variety of fields including healthcare, where a specific type of WSN called Wireless Body Sensor Network (WBAN) has been developed. This network offers many advantages, such as the ability to provide remote healthcare to distant patients, personalized health services for the elderly, and the monitoring of groups of people for the detection and prevention of epidemics. It enables discrete and continuous monitoring of the body's health parameters, while preserving the individual's daily routine without disrupting normal activities [2].

In a WBAN, several medical sensors are used to collect physiological signs such as heart rate, body temperature, and blood pressure. Each sensor, considered as a node, is placed on the human body in the form of a wearable device or implanted inside the body [3]. The data collected is then transmitted to a collection node, often called a sink, which forwards it to a monitoring center or remote processing platform via gateways and wireless technologies such as Bluetooth, Zigbee, or Wi-Fi [4].

However, healthcare systems exploiting the Internet of Things (IoT) face major challenges linked to the sensitivity and criticality of the data manipulated by these networks. In the context of wireless body area networks (WBANs), the implementation of fault-tolerant routing is of crucial importance. Existing routing protocols mentioned in the literature have shortcomings such as the inability to ensure reliable data transmission, high end-to-end delays, high energy consumption, and reduced network lifetime. It is imperative to deliver data quickly and reliably to care centers, while ensuring data availability despite potential system failures, such as loss of sensor nodes, broken communication links, or malfunctioning of certain WBAN network components.

To overcome these challenges, the deployment of a Quality of Service (QoS) routing protocol is essential for efficient data management and fault resilience. This article examines various existing studies on fault-tolerant routing protocols in WBANs, leading to the creation of a new protocol, FTRBT (Fault-tolerance Routing-Based Tree). FTRBT organizes WBANs in the form of a virtual tree, guaranteeing increased data availability.

The aim is to guarantee the reliability and availability of remotely transmitted healthcare data, particularly in the context of the Internet of Things (IoT) applied to healthcare. The authors address the challenges of data management, communication between the various system components, and fault-tolerance and communication errors.

The rest of the article is organized as follows: Sect. 2 explores previous research on routing protocols and fault-tolerance techniques in healthcare systems. Sect. 3 sets out the context of the working environment. Section 4 presents and details the design of the proposed fault-tolerant routing protocol. In Sect. 5, the performance of the solution is evaluated and compared using the OMNet simulator against a random solution. Section 6 concludes the article.

2 Related Works

Assessing resilience to failures in the healthcare sector is essential for improving the quality of care. In this section, we explore routing protocols specifically developed for WBANs, focusing on fault management and QoS optimization within WBANs.

In 2010, Ben-Othman and Yahya [5] introduced the EQRS protocol (Energy-efficient, QoS-based routing protocol for wireless sensor networks) to meet quality of service (QoS) requirements while optimizing the use of energy resources in wireless sensor networks. This multipath protocol aims to minimize energy consumption, guarantee reliable data transmission, and maximize network lifetime. To achieve this, EQRS balances energy consumption between nodes, differentiates between real-time and delayed traffic, and uses various criteria such as remaining energy, buffer size, and signal-to-noise ratio to discover and maintain multiple paths between source and destination nodes. It divides data packets into sub-packets with error correction codes to improve transmission reliability and uses an early error correction technique to ensure data recovery in the event of node failure. However, implementing this protocol can be complex and require additional resources in terms of computing power and storage, and managing multiple paths can introduce a control and maintenance overhead that can increase energy consumption.

In 2016, Zhuoming et al. [6] developed a Dynamic Routing Algorithm (DRA) to balance energy consumption between various sensor types when aggregating and transmitting vital signals within a WBAN. The aim of DRA is to optimize data transmission paths while extending the overall network lifetime. An improved version of the DRA incorporates a multipath selection mechanism using Dijkstra's algorithm to minimize energy consumption. However, using Dijkstra's algorithm can lead to significant computational complexity, especially in a dynamic environment such as a WBAN, potentially impacting network performance in terms of latency and responsiveness.

3 Environment and Assumptions

In intelligent healthcare systems, WBANs play a crucial role in facilitating remote, real-time monitoring of patient health. Communication between these WBANs requires the implementation of specific routing protocols to overcome various challenges.

Our architecture represents WBANs as nodes, each uniquely identified by "*IdWBAN*". Each WBAN follows a defined architecture, consisting of sensor nodes positioned on the human body, establishing wireless communication with a data collection device called a "*sink*". The sink collaborates with sinks from other WBANs to ensure system fault-tolerance, as illustrated in Fig. 1. The collected data can then be transmitted to a health monitoring system or healthcare professional, enabling continuous real-time monitoring of patient health.

In addition, the following assumptions are necessary for the proper functioning of our system presented below:

- Links are bidirectional.
- Each WBAN has a unique *IdWBAN*.
- An intra-WBAN routing protocol is already in place, and medical data is available at the sink level.
- Each WBAN has K blocks of size T to store data from K other WBANs.
- The data from each WBAN is treated as a single entity.

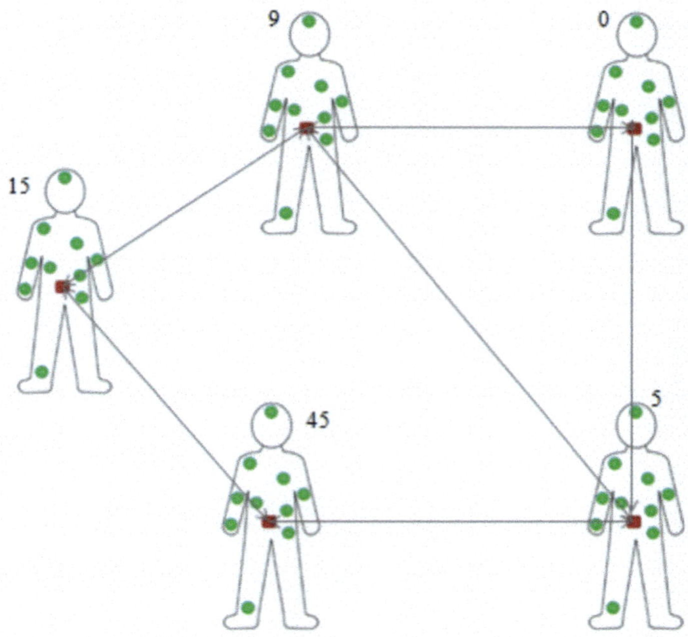

Fig. 1 WBANs communication architecture

- A specific block *B0* is reserved for storing information relating to a given WBAN.
- The number of data copies is set so that it is less than the storage capacity *K* and is the same for all nodes.

4 Proposed Approach

In this section, we present the design of our proposed Fault-tolerance Routing-Based Tree (FTRBT) protocol. This protocol relies on replication as a fault-tolerance strategy to manage potential failures. Its approach is based on mechanisms for efficiently routing data to the cloud.

4.1 Protocol Description

In this design approach, we have chosen an efficient and well-organized data replication method, structuring the WBANs in the form of a virtual tree.

In our protocol, we employ a tree-like structure for organizing WBAN sinks, aiming to optimize the distribution of replicas across nodes and ensure maximum availability of copies in the event of failure. Furthermore, we have implemented a prioritization hierarchy for messages, giving precedence to alert messages over best-effort messages. When a connected node is accessible, data is directed to it for cloud backup. Replication of data generated by WBANs occurs periodically, led by the associated sink node. If the data designated for replication is deemed sensitive, it is treated as an alert message; otherwise, it is processed as a best-effort message.

The binary tree is built in levels, starting with the sink affiliated to the WBAN with the highest identifier, taking advantage of the physical links available (Fig. 2). Periodically, the sink of each WBAN performs a backup of the data received, before replicating it. The Cloud is called upon to back up data when one of the sinks is connected, thus freeing up storage space.

The protocol utilizes various message types to establish and maintain connectivity in the WBAN network, ensuring efficient transmission and communication of health data. During the recognition phase, *Hello* messages are exchanged to create and update the routing table for each WBAN, crucial for detecting link breaks used in tree formation. Another type is the *Share* message, sent by the initiator node to share the created tree with all other nodes. *Data* messages carry sensor measurements from each WBAN, transmitted to the sink during replication or cloud storage (in alert or best-effort mode). *ACK* messages confirm data storage in the queue, while *Free* messages are release notifications sent by a storage node connected to the cloud to relevant nodes, including those whose data has been backed up by this sink and those who have backed up data from this sink.

The flowchart shown in Fig. 3 describes the overall operation of the protocol.

Fig. 2 Tree creation flowchart

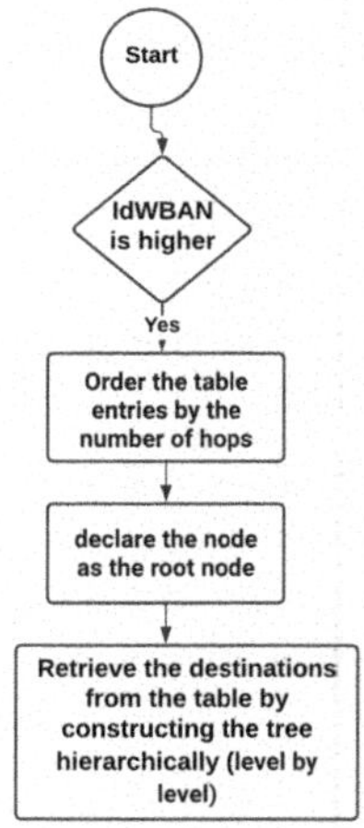

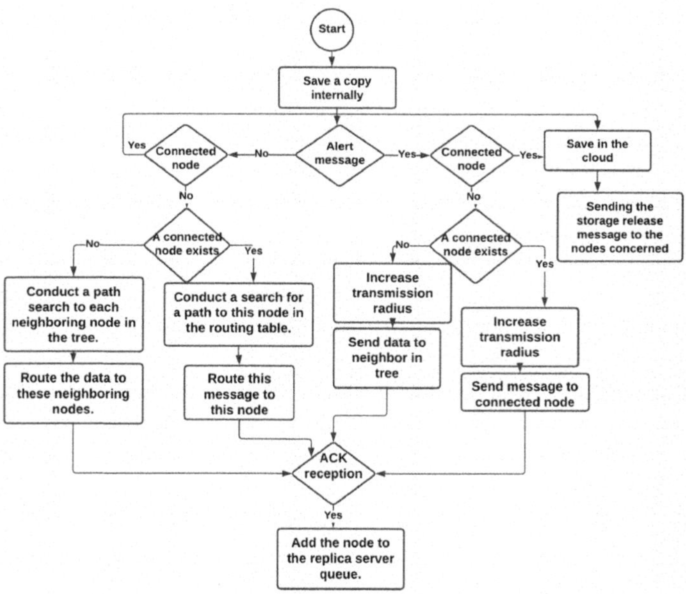

Fig.3 FTRBT protocol flowchart

Reconnaissance phase: At initialization, each sink initiates a Hello message to its neighbors, facilitating the construction of the routing table (*TR*). This data structure is instrumental in recording paths to different destinations, aiding nodes in determining the next hop for efficient message routing.

Upon receiving a Hello message, each node updates its internal sequence number if necessary. If the destination in the Hello message is absent in *TR*, a new entry is added. If the message is more recent or has fewer hops than the entry in *TR* for that destination, an update occurs, including the next hop and hop count. The updated message is then propagated to neighbors to update their routing tables. After a set number of Hello messages (*Hello N*), each receiving node checks if it possesses the highest IdWBAN. If so, it assumes responsibility for tree creation, sending a Share message to inform all nodes of the tree's establishment. Regular exchanges of Hello messages maintain the routing table and tree. Additionally, nodes remove entries from the routing table if no Hello message has been received for three periods.

Routing phase: At regular intervals, each WBAN sink node receives data and processes it according to its nature (Alert or Best effort), as well as its current connection status (connected or disconnected).

- **If the data received is sensitive and the node is not connected**: Upon saving a data copy, an alert message is dispatched, expanding the transmission radius to a connected node when feasible. If a connected node is accessible, it stores the data in the Cloud, acknowledging the save with an ACK message, enabling the source node to enqueue it in the replica server. In the absence of a connected node, the sender node extends the transmission radius, transmitting the data to direct neighbors in the tree and specifying the replication copy count. If neighbors possess ample storage capacity, they execute the backup and reciprocate with an ACK message.
- **If the data is not sensitive and the node is not connected**: For best-effort messages, the node attempts to find a path to a connected node using its routing table established via the DSDV protocol. In the absence of a connected node in this table, it directs the data to be replicated to its tree neighbor by determining the next hop. After saving a copy of the data, a Best-Effort message is sent to a connected node, if available. The connected node records the data and responds with an ACK message to confirm the save. If no connected node is available, the message is routed to the node's neighbors in the tree, specifying the replication copy count. If the neighbors have sufficient storage capacity, they perform the backup and send an ACK message.

If the node is online, the data is stored directly in the Cloud. When a node reconnects after a disconnection, the data is sent to the Cloud and a Free message is sent to the nodes where the node has made replicas, so that they can free up their occupied storage space as well as the space of the nodes they have replicated in turn.

5 Performance Analysis

To evaluate the proposed solution, we used the Omnet++ simulator, a modular, object-oriented discrete-event simulator [7]. To meet the specific requirements of WBAN simulation, we also used the Castalia framework [8], designed specifically for WBAN networks and based on the OMNeT++ platform. In order to analyze the performance of our proposed FTRBT protocol, we compared simulation results with a random protocol called RRP (Random Routing Protocol).

5.1 Simulation Parameters

To evaluate the effectiveness of our proposed FTRBT protocol, we defined a simulation area of 5 m by 5 m, with a total simulation time of 1000 s. The transmission power range was set between – 7 and 1 dBm, and each node was configured to send a maximum of 1000 data packets.

5.2 Metrics

To assess the performance of the two proposed protocols, we used the following metrics as evaluation criteria:

- **Packet Delivery Rate (PDR)**: It represents the proportion between the number of packets received by a node connected or involved in the backup copy and the number of packets sent.
- **Backup Time**: It refers to the time required to save a package.

5.3 Results Interpretation

In this section, we will undertake a series of simulations and tests to evaluate and compare the performance of the proposed FTRBT protocol against a randomized RRP protocol. These simulations will be carried out by exploring various scenarios and analyzing the performance indicators mentioned above.

- **Packet Delivery Rate (PDR)**: Fig. 3 shows the packet delivery rate (PDR) of the transmitting nodes. In this case, we see that the FTRBT protocol has a higher packet delivery rate (PDR) than the RRP protocol due to the high node density, which leads to stronger interference between packets transmitted between neighboring nodes in the case of the RRP protocol. The FTRBT protocol uses a structured approach in which the path to the destination is predetermined in its routing table, thus avoiding interference and maintaining a higher RRP. In summary, the FTRBT protocol has a better PDR thanks to its structured approach and reduced traffic between nodes.
- **Backup Time**: Fig. 4 shows the rate of packets saved as a function of time. Compared with the RRP protocol, the FTRBT protocol has significantly longer delays, exceeding 500 ms, while those of the RRP protocol remain below 100 ms. This disparity is mainly due to the exchange of maintenance control messages

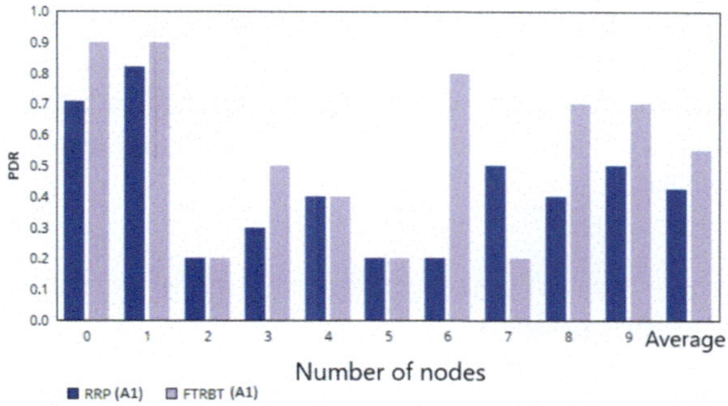

Fig. 4 Packet delivery rate (no node connected)

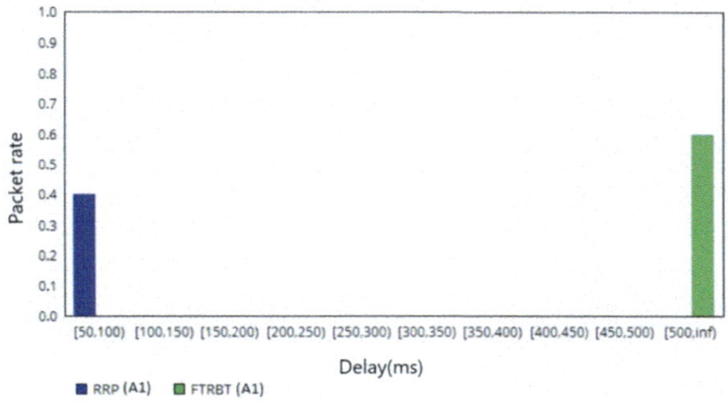

Fig. 5 Backup time

in the FTRBT protocol, introducing additional delays into the communication process. These messages are crucial for the maintenance and updating of the broadcast tree used by the FTRBT protocol, resulting in significant delays. In addition, access to the routing table needed to find the optimal path for packets in the FTRBT protocol can be time-consuming, especially in complex networks with many possible nodes and routes, contributing to further delays. Finally, network convergence in the FTRBT protocol can also generate extended delays, requiring adjustments and updates to the broadcast tree in the event of a topology change or failure, prolonging the time needed to re-establish optimal communication. In short, the increased delays in the FTRBT protocol compared with the RRP protocol are attributable to maintenance control messages, routing table access, and network convergence, contributing to an overall increase in delays in the FTRBT protocol (Fig. 5).

5.4 Discussion

The performance comparison between the proposed FTRBT protocol and the randomized RRP protocol reveals that the RRP protocol has faster transmission times. This is because the RRP protocol sends packets directly to its neighbors, whereas the FTRBT protocol requires access to the routing table, thus prolonging the process. However, the FTRBT protocol has a better packet delivery rate due to its high node density, reducing interference and data loss compared with the RRP protocol, which is particularly susceptible to these problems due to its lower node density.

6 Conclusions

This article proposes a resilient solution for Internet of Things (IoT) networks, with a particular focus on Wireless Body Area Networks (WBANs), aimed at guaranteeing data preservation, backup, and recovery. The aim is to minimize losses and ensure continuous system operation, even in the event of sensor failure or transmission problems. The FTRBT protocol, specially designed for fault-tolerant tree routing in WBAN networks, has significant implications in the field of real healthcare, particularly with regard to remote medical monitoring., to evaluate the effectiveness of this protocol and compare its performance with that of a random routing protocol (RRP).

In the future, we plan to enhance our solution by developing a fault-tolerant inter-WBAN routing protocol with Quality of Service (QoS) integration in the medical domain to provide seamless and lossless data services.

References

1. Patel KK, Patel SM (2016) Internet of things-IOT : definition, characteristics, architecture, enabling technologies, application and future challenges. Int J Eng Sci Comput. https://doi.org/10.4010/2016.1482
2. Gokhale P et al (2018) Introduction to IOT. https://doi.org/10.17148/IAR-JSET.2018.517
3. Rathee D, Rangi S et al (2014) Recent trends wireless body area network (WBAN) research and cognition based adaptive WBAN architecture for healthcare. Health Technol
4. Panhwar MA, Kamran A et al (2019) Wireless body area networks : architecture, standards, challenges and application. Int J Comput Sci Netw Secur 19(12)
5. Ben-Othman J, Yahya B (2010) Energy efficient and QoS based routing protocol for wireless sensor networks. J Parallel Distrib Comput 70(8):849–857
6. Li Z et al (2016) Weighted energy-balanced efficient routing algorithm for wireless body area network. Int J Distrib Sens Netw 2016, Article ID 7364910, 7 p
7. Varga A, Hornig R (2010) An overview of the OMNeT++ simulation environment. In: 1st international ICST conference on simulation tools and techniques for communications, networks and systems
8. Waheed T, Shaikh FK, Khan IU et al (2019) WBAN performance evaluation at PHY/MAC/network layer using Castalia simulator. In 2019 13th international conference on mathematics, actuarial science, computer science and statistics (MACS). IEEE, pp 1–6

Review on Ultra-wideband Multi-input and Multi-output Antennas with Notched Band Characteristics

Arunima Samanta, Koushiki Ghosh, Pritam Aich, Shatavisha Dasgupta, Srijita Chakraborty, and Mrinmoy Chakraborty

Abstract This paper presents a comprehensive review of Ultra-Wideband (UWB) Multiple Input Multiple Output (MIMO) antennas with notch characteristics. UWB communication systems have emerged as promising solutions for high-speed data transmission, positioning, and radar applications due to their wide bandwidth and low power consumption. However, coexistence with narrowband wireless technologies poses challenges, necessitating the integration of notch characteristics into UWB MIMO antennas to mitigate interference. The review is conducted based on antenna parameters such as antenna size, operating bandwidth, gain and number of notch bands, envelope correlation coefficient (ECC), and isolation. A comparative analysis has been provided based on the characteristics parameters. Various techniques to minimize the mutual coupling within the UWB MIMO antenna have been explored. Several methods have been employed to attain the notch characteristics in the ultra-wideband MIMO antenna. The practical application of the UWB MIMO antenna has been extensively discussed for the operational frequency band. Overall, this review provides valuable insights into the design and implementation of UWB MIMO antennas with notch characteristics, offering a foundation for further advancements in ultra-wideband communication systems.

Keywords UWB MIMO antenna · Mutual coupling · Band notch characteristics

---

A. Samanta (✉) · K. Ghosh · P. Aich · S. Dasgupta · S. Chakraborty
Institute of Engineering and Management Kolkata, University of Engineering and Management Kolkata, Kolkata, India
e-mail: ARUNIMA.SAMANTA2022@iem.edu.in

M. Chakraborty
Dr. B. C. Roy Engineering College, Durgapur, India

P. G. Bradford et al. (eds.), *Proceedings of IEMTRONICS 2024*, Lecture Notes in Electrical Engineering 1229, https://doi.org/10.1007/978-981-97-4780-1_11

1 Introduction

Over the recent years, research about UWB MIMO antenna has increased to make it most suitable in wireless communication. This antenna also has several applications in other areas such as in PAN wireless connections, wide-range, low data rate communication, radar and imaging systems, and wireless applications. The MIMO antenna is designed such that the envelope correlation coefficient as well as mutual coupling is reduced. Two elements or more are used to design the MIMO antenna. Decoupling technique is used, where the elements of the antenna are placed near the periphery which is useful for increasing isolation.

In this proposition, a review on ultra-wideband multi-input and multi-output antennas is presented, where the UWB range is operating typically in a range of frequency 3–10.6 GHz. Different methods are used for the UWB MIMO antenna to obtain band-notched characteristics with high isolation and are employed in several wireless applications.

2 Review on UWB MIMO Antenna

A square patch is present in the MIMO antenna which has a microstrip line and a radiator (cup-shaped) which is shortened along the edges leading to an increase in the current path and wider bandwidth. The designed MIMO antenna has elements that are placed perpendicular to each other. The isolation of the antenna is increased using a ground plane. Reduced coupling of the antenna also helps to attain better isolation. It consists of two ports where port 1 is used as an input and port 2 as output.

Lesser transmission coefficient means the antenna has better port-isolation and low mutual coupling. This parameter determines the quality of the antenna and its performance. The radiation pattern of the MIMO antenna is also used to determine its quality; in order to be suitable for different communication systems, the antenna should have stable gain, low mutual coupling (lower than—16 dB), and bandwidth in the range of 2.9–11.6 GHz [1].

The antenna also has a low envelope correlation coefficient (ECC) and good radiation pattern which will make it convenient in UWB communication systems. The geometry of the MIMO antenna contains two radiators (arrow-shaped) which are fabricated on a substrate of a permittivity 4.4 The MIMO antenna operates in a range of frequency 2–12 Hz [2]. The substrate material has a thickness of 1.6 mm and the dimension of the antenna is $43 \times 34.9 \times 1.6\ mm^3$ [3].

The band notch characteristics are made by keeping slot-1 inverted (*L*-shaped), which forms ant-2.In the same way, Ant-3 is formed by keeping another inverted (*L*-shaped) Slot-3. Ant-4 is formed by placing slot 3 (*U*-shaped) on the feed line of Ant-3. If the parasitic element is placed along the ground plane, − 19.7 dB isolation can be obtained [4].

If the parasitic element is placed along the ground plane, − 19.7 dB isolation can be the antenna that is designed such that—18 dB of isolation is obtained [5]. The designed MIMO antenna has a *T* shape. Antenna elements are placed near each other to produce good isolation and make the antenna compact in nature. The mutual coupling of the antenna is reduced by cutting a rectangular slot over a ground plane [6].

A slot which is *T*-shaped is cut on the ground plane which increases the isolation of elements of the antenna further. Another advantage of this slot (*T*-shaped) is it decreases surface current in the middle of two ports causing less mutual coupling [7]. The decrease in mutual coupling within radiating elements is achieved by introducing stubs (*L*-shaped) on the ground surface. Isolation is further increased by placing a slot on the ground surface [8].

Two customized printed monopole antenna elements in a triangular shape, as well as a mixture of varactor and PIN diodes, make up the antenna construction. There are three operational modes for the proposed antenna: a frequency in UWB mode for spectrum sensing. The UWB MIMO antenna comprises four circular monopoles with a periodic electromagnetic band gap (EBG) construction and modified defective ground plane [9].

The suggested EBG structures differ from conventional mushroom-shaped ones in that the design has several vias connecting the top and bottom planes in addition to grid patterns on the top patch and metallic ground plane. The center of the dielectric is optimized to reduce electromagnetic coupling between parallel parts. A crisp band-notched feature is additionally rendered attainable by the radiators' four crescent-shaped resonant slots.

The inverted *L*-shaped slits on the tapered microstrip-fed slot antenna operate as a single radiating element to integrate notched bands at WLAN and Super-Extended C-bands [10]. Relative coupling over the operational band is less than − 22 dB which is attained in between 2.93 and 20 GHz. For UWB applications, a dual notched 4-element MIMO antenna with gap sleeves and an *H*-slot was proposed and constructed [11].

The proposed antenna has four well-isolated, orthogonal elements and is CPW (Co-Planar Waveguide) fed. With notches in the 3.3–4.1 GHz and 8.2–8.6 GHz frequency bands, the suggested MIMO antenna obtained an impedance bandwidth ($S11 < -10$ dB) from 2.1–20 GHz. WiMAX interference (3.3–3.7 GHz) and military/radar applications spectrum (8.2–8.6 GHz) can be filtered by these acquired notches. A 4 by 4 MIMO antenna array with 5.5 GHz WLAN spectrum rejection properties was presented for ultra-wideband (UWB) applications [12].

A modified patch antenna with a portion of a ground plane makes up the single UWB antenna element. The antenna element has a *U*-shaped slot that has been etched to reject the WLAN band. Each antenna element is orthogonal to its neighboring element in the planar geometry of the antenna arrangement. Among the antenna elements is a parasitic decouple in the form of a fan that consists of four strip lines joined by a center conductor with a square cross section. A four-element MIMO antenna with quasi-self-complementary (QSC) elements is suggested [13]. It has a notch band in a higher WLAN frequency range (5.15–5.85 GHz) and ultra-wideband (UWB) features (3.1–10.6 GHz). The UWB components are monopoles with a trapezoidal shape and complementary ground plane cuts. The IEEE 802.11ac frequency spectrum from 5.1 to 5.95 GHz is rejected by a dual-polarized ultra-wideband (UWB) multiple-input multiple-output (MIMO) antenna that uses microstrip line feeding [14].

By utilizing strategies like orthogonal polarization, defective ground structure, and materials, the unavoidable mutual coupling is suppressed. In order to lessen the interference between the antenna elements, a split-ring is integrated. An UWB MIMO antenna design for a dual band notched, quad-element UWB with multiple inputs and outputs (MIMO) is provided [15]. Narrowband interference (NBI) is capable of being suppressed by the suggested design. The geometry of the proposed antenna design with the corresponding results is illustrated in Figs. 1, 2, 3, 4, 5, 6, 7, 8, 9, 10, 11, 12, 13, 14, 15, 16, 17, 18, 19, 20, 21, 22, 23, 24, 25, 26, 27, 28, 29 and 30.

Thus, the UWB MIMO antenna is designed to make it convenient and useful for a wide range of communication systems including UWB communication MIMO communication systems and also has many other applications. A comparison of various parameters such as antenna size, operating bandwidth, gain, number of notch bands, envelope correlation coefficient (ECC), and isolation is presented in Table 1.

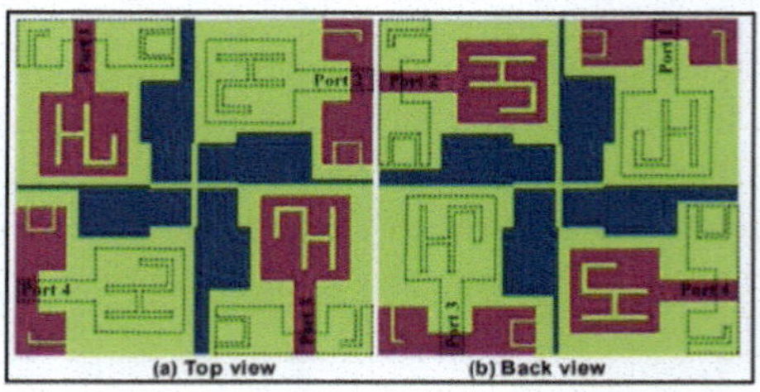

Fig. 1 Geometry of the MIMO antenna [1]

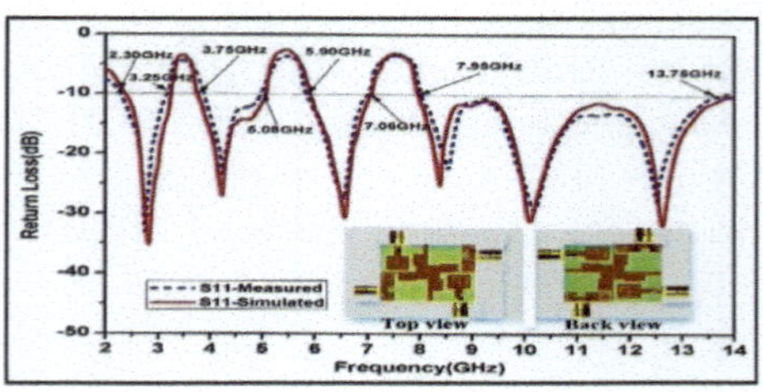

Fig. 2 Results of return loss parameter of the antenna based on simulation and measurement [1]

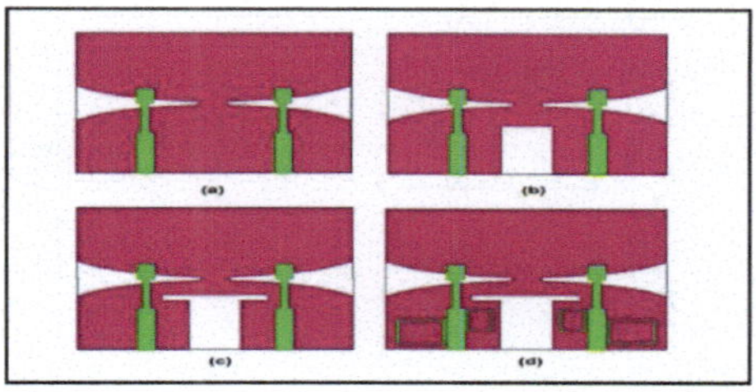

Fig. 3 Geometry of UWB antenna. **a** Antenna 1. **b** Antenna 2. **c** Antenna 3. **d** Antenna 4 [2]

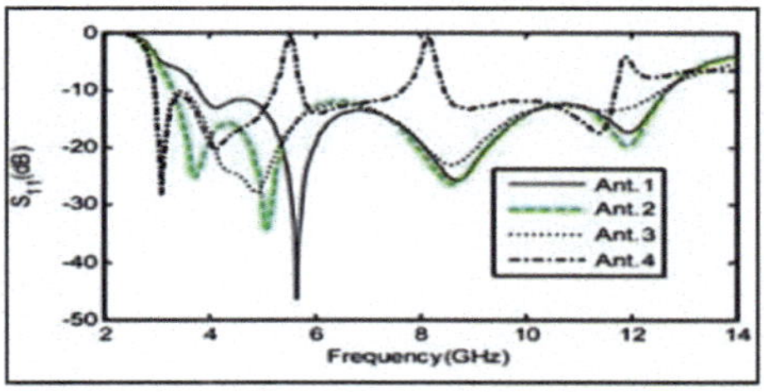

Fig. 4 Results of return loss parameter of the antenna based on simulation [2]

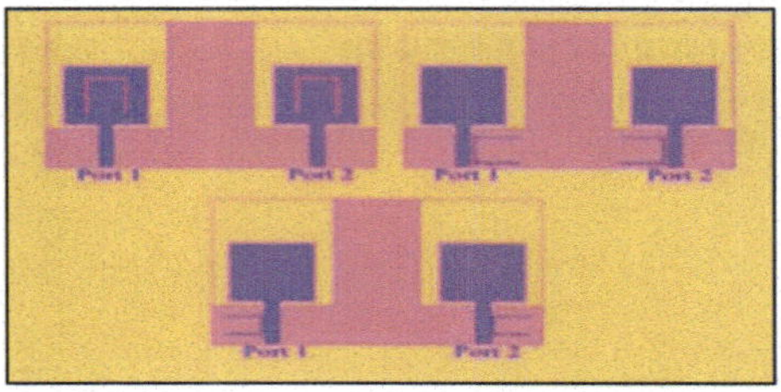

Fig. 5 Geometry of UWB antenna [3]

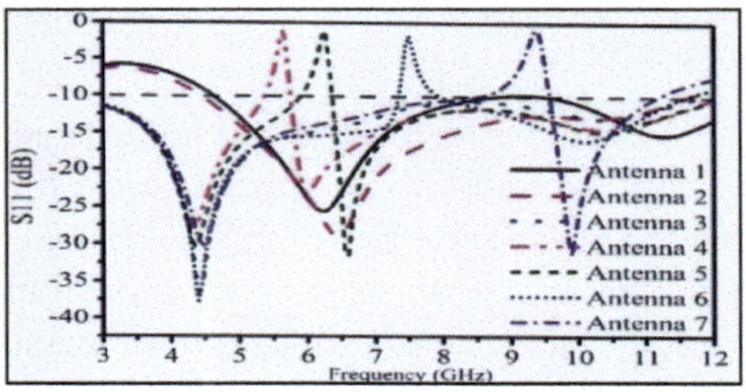

Fig. 6 Results of return loss parameter of the Antenna based on simulation [3]

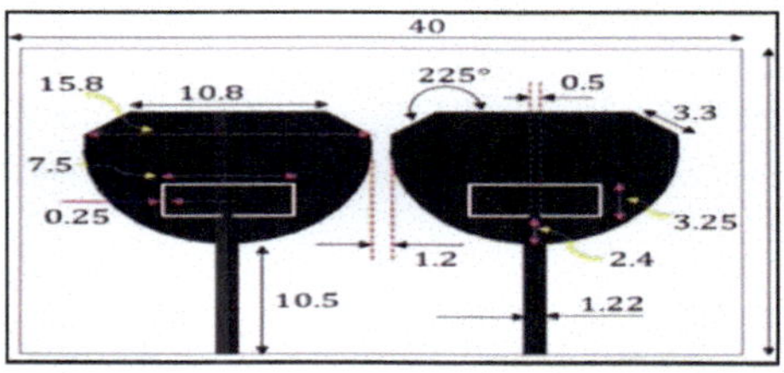

Fig. 7 Geometry of UWB antenna given in millimeter [4]

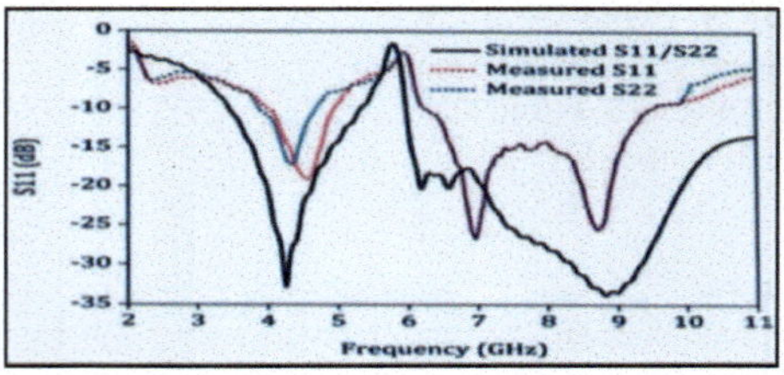

Fig. 8 Results of return loss parameter of the antenna based on simulation and measurement

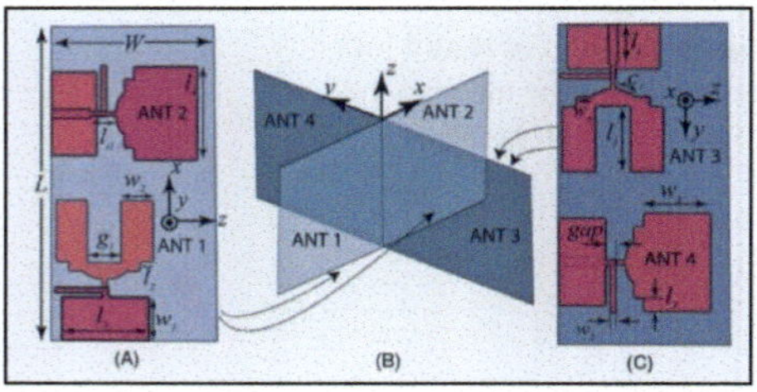

Fig. 9 Geometry of UWB antenna [5]

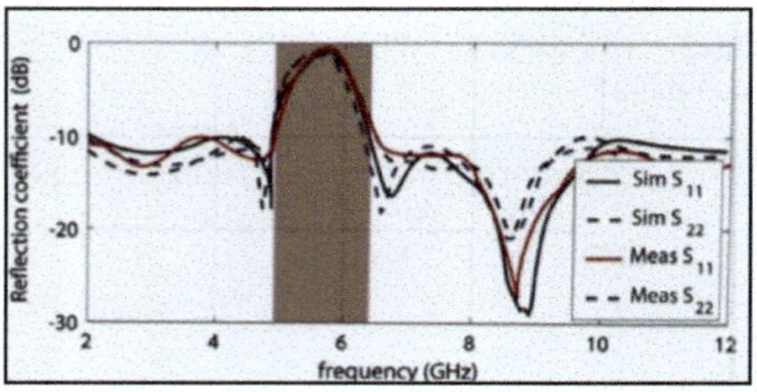

Fig. 10 Results of return loss parameter of the antenna based on simulation and measurement [5]

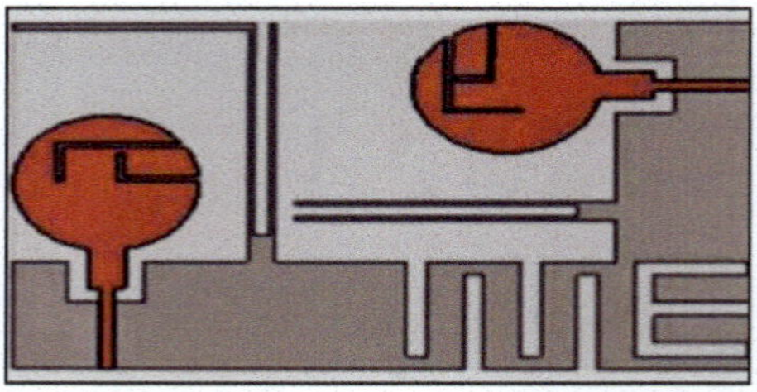

Fig. 11 Geometry of UWB antenna [6]

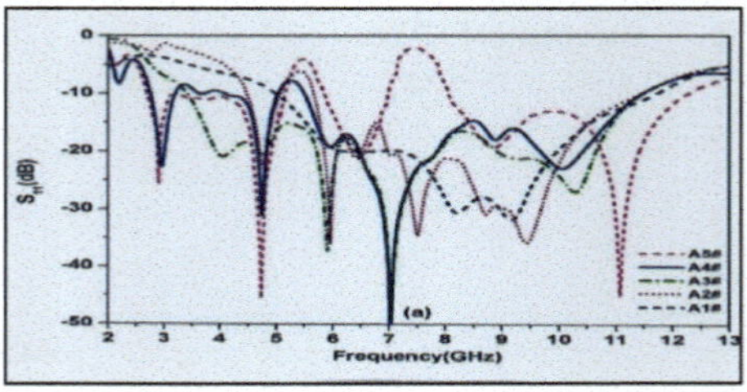

Fig. 12 Results of return loss parameter of the antenna based on simulation [6]

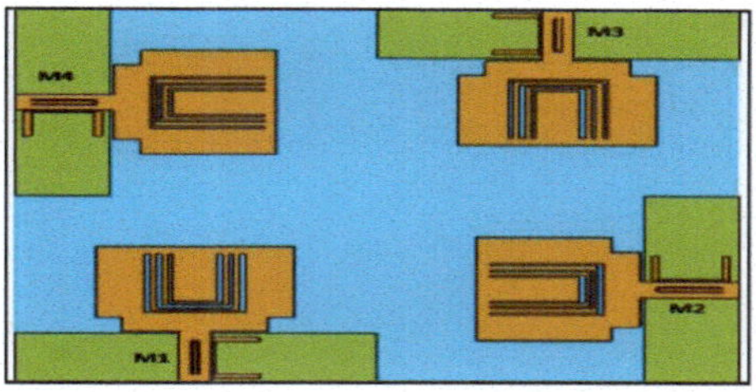

Fig. 13 Geometry of UWB antenna [7]

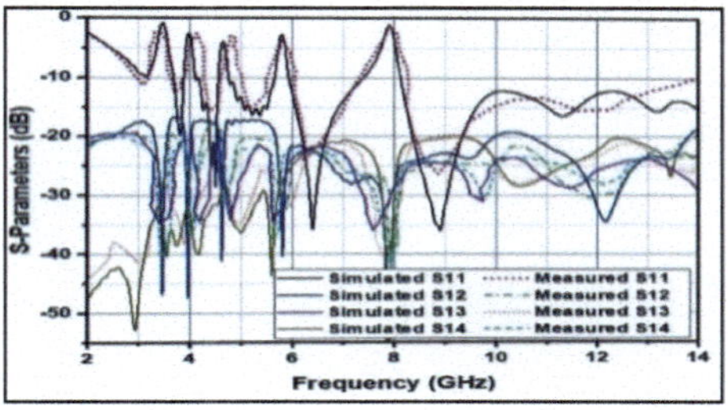

Fig. 14 Results of return loss parameter of the antenna based on simulation and measurement

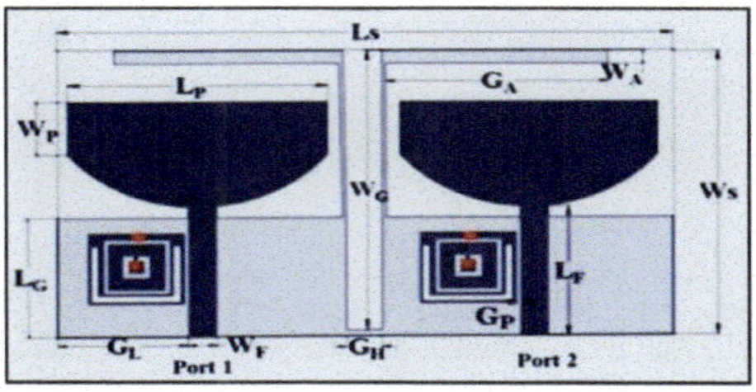

Fig. 15 Geometry of UWB antenna [8]

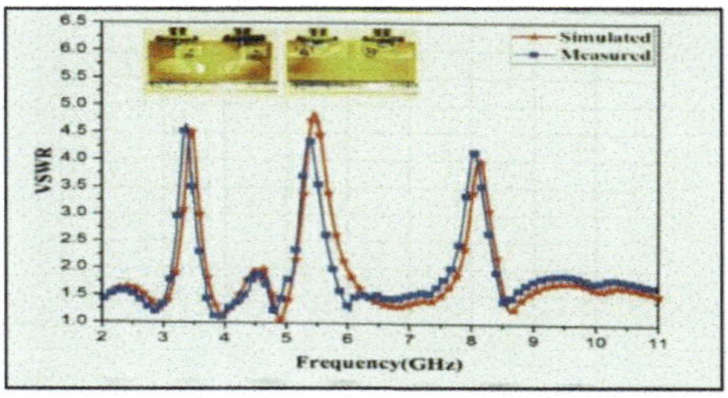

Fig. 16 Results of VSWR of the antenna based on simulation and measurement [8]

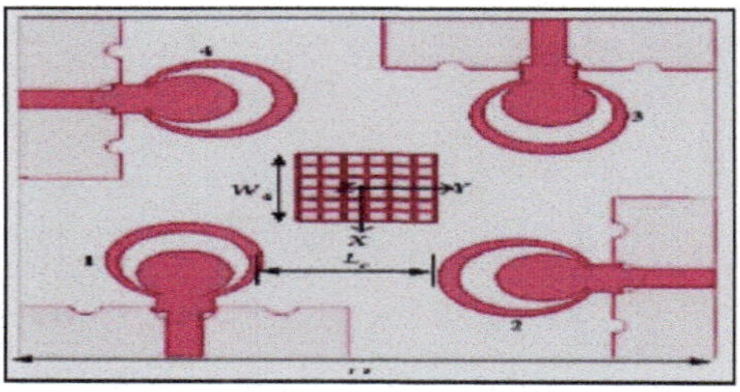

Fig. 17 Geometry of UWB antenna [9]

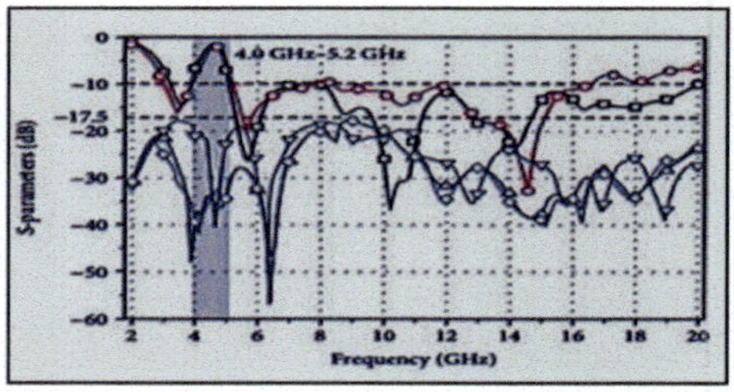

Fig. 18 Results of Scatter parameter of the antenna based on simulation and measurements [9]

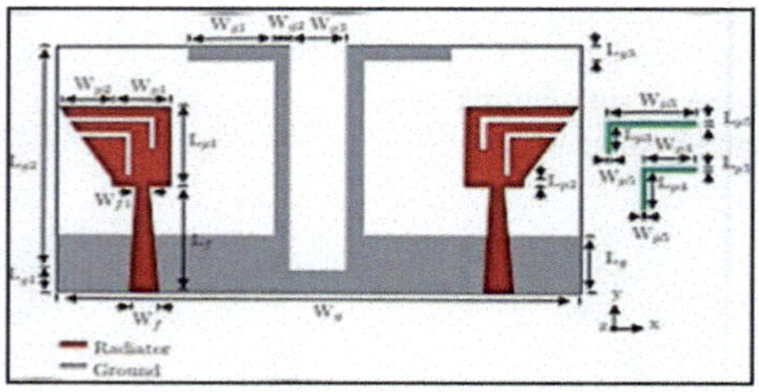

Fig. 19 Geometry of UWB antenna [10]

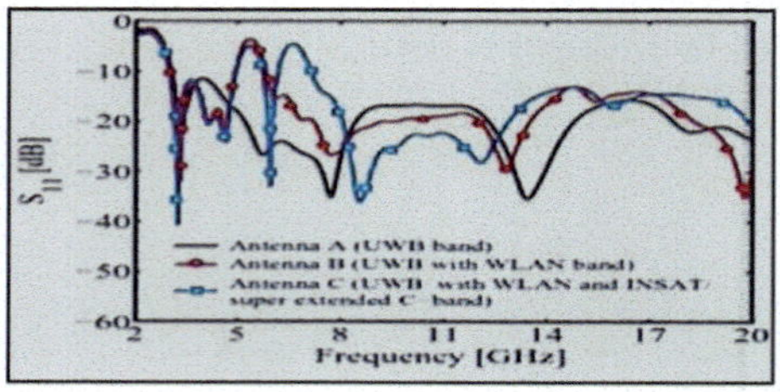

Fig. 20 Scatter parameter of the antenna based on simulation [10]

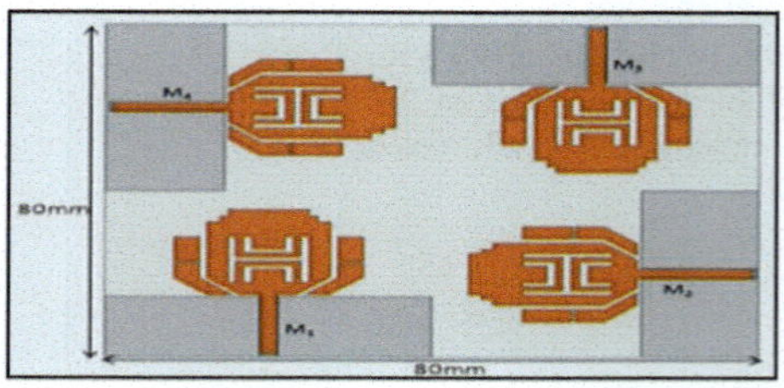

Fig. 21 Geometry of UWB antenna [11]

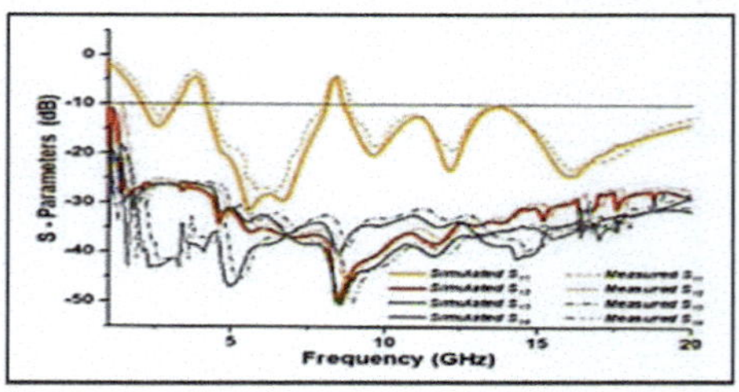

Fig. 22 Scatter parameter of the antenna based on simulation [11]

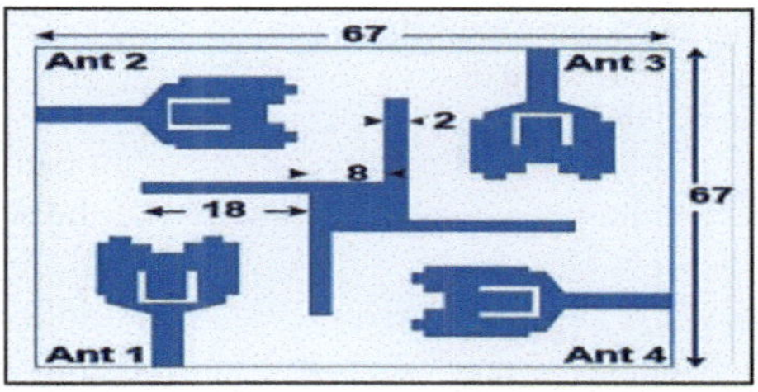

Fig. 23 Geometry of UWB antenna [12]

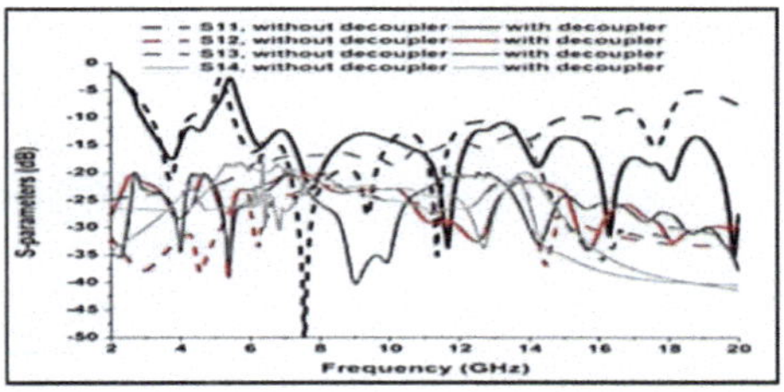

Fig. 24 Scatter parameter of the antenna based on simulation [12]

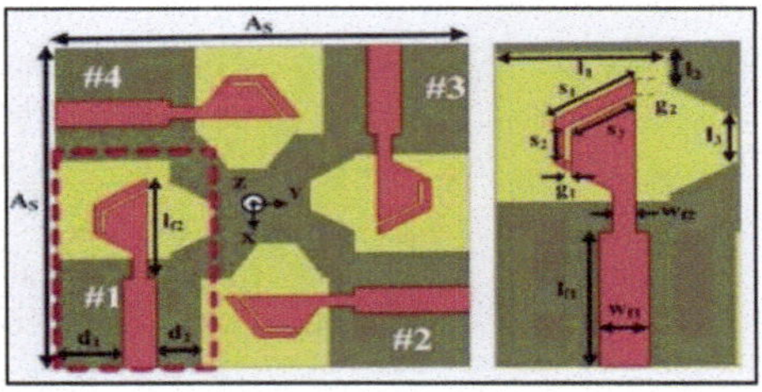

Fig. 25 Geometry of UWB antenna [13]

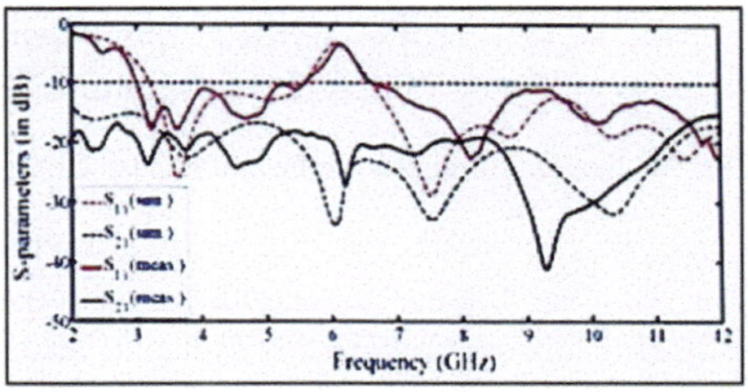

Fig. 26 Scatter parameter of the antenna based on simulation [13]

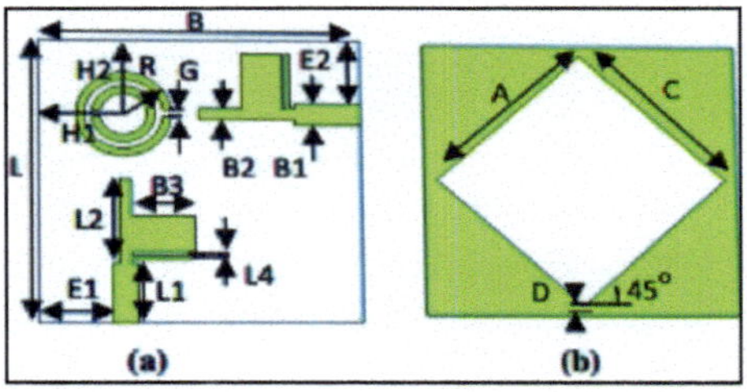

Fig. 27 Geometry of UWB antenna [14]

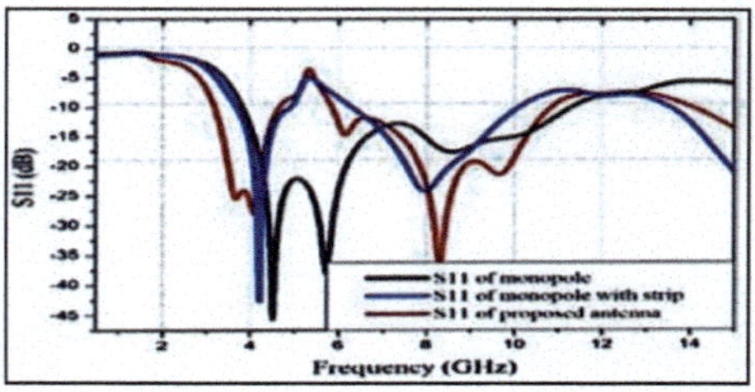

Fig. 28 Scatter parameter of the antenna based on simulation [14]

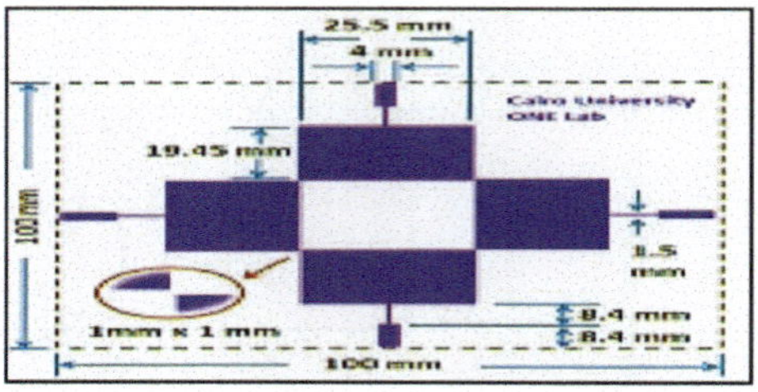

Fig. 29 Geometry of UWB antenna [15]

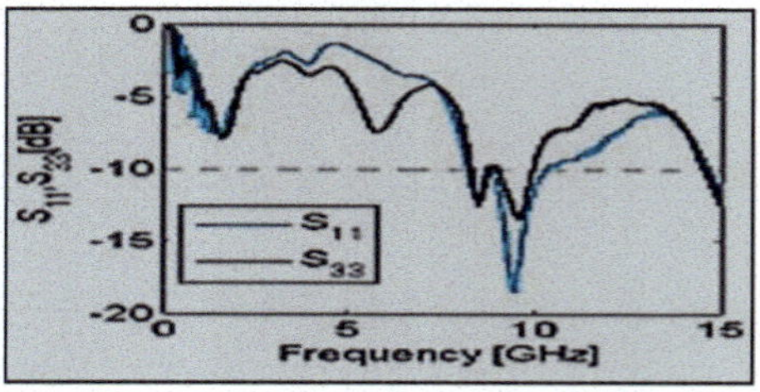

Fig. 30 Scatter parameter of the antenna based on simulation [15]

Table 1 Comparison of performance of designed MIMO antenna with other antenna

References	Antenna size	Operating	Gain (dB)	No. of	ECC	Isolation (in dB)
[1]	29 × 40 × 0.608 mm^3	2.4–12	3.7	1	< 0.28	> 20
[2]	23 × 29 × 0.9 mm^3	3.3–10.9	4	1	< 0.01	> 20
[3]	50 × 20 × 0.7 mm^3	2.6–11.3	6	3	< 0.02	> 15
[4]	33 × 42 × 1.7 mm^3	3.6–10.5	5.3	1	< 0.02	> 17
[5]	14 × 24 × 1.8 mm^3	2.8–12	5.6	1	< 0.03	> 17
[6]	19 × 23 × 1.7 mm^3	3.0–11.0	3.5	1	< 0.02	> 20
[7]	16 × 22 × 60 mm^3	3.1–11	2.5	3	< 0.01	> 20
[8]	23 × 29 × 0.9 mm^3	2.5–11.0	3	1	< 0.04	> 15
[9]	0 6λ × 0 6λ × 0 016λ	3.0–16.2	8.4	1	< 0.3	17.5
[10]	18 × 34 × 1.6 mm^3	2.93–20	0–7	2	< 0.01	22
[11]	80 × 80 × 1.6 mm^3	2.1–20	5.8	2	< 0.02	20
[12]	67 × 67 × 1.6 mm^3	3.5–20	8.1	1	< 0.01	20
[13]	36 × 36 × 1.6 mm^3	3.1–10.6	4.5	1	< 0.5	20
[14]	40.5 × 40.5 × 1.6 mm^3	3.1–10.6	6	1	< 0.001	20
[15]	25.5 × 19.45 × 1.6 mm^3	3.1–10.6	–	2	< 0.1	20

3 Conclusion

A review on ultra-wideband antenna with multi input/multi-output with notched characteristics is presented with its applications. The quality and performance of the mimo antenna are determined by parameters like radiation pattern, gain, isolation, and envelope correlation coefficient (ECC).The antenna operates within a certain frequency range and is found convenient for UWB MIMO and many other communication systems.

Acknowledgements The researchers intend like to thank and express our gratitude toward faculty members of IEM for their relentless support and who have willingly helped us with their abilities to complete the work.

References

1. Tang Z, Zhan XJ, Hu S, Xi Z, Liu Y, Compact UWB-MIMO antenna with high isolation and triple band-notched characteristics. IEEE Access. https://doi.org/10.1109/ACCESS.2019.2897
2. Li Z, Yin C, Zhu X, Compact UWB-MIMO Vivaldi antenna with dual band-notched characteristics. IEEE. https://doi.org/10.1109/ACESS.2019.2906338
3. Luo H, Wang D, Chen Y, Li E (2021) Compact dual-port UWB-MIMO antenna with quadruple band—notched characteristics. Int J Electron Commun (AEU) 136:153770
4. Ishteyaq I, Masoodi IS, Muzaffar K, Magray MI, Compact band-notched antennas with high isolation for UWB MIMO applications. Int J Microw Wirel Technol 1–7. https://doi.org/10.1017/S1759078720001427
5. Khan MS, Naqvi SA, Iftikhar A, Asif SM, Fida A, AWLAN band-notched compact four element UWB MIMO antenna. Islamabad. https://doi.org/10.1002/mmce.22282
6. Bhattacharjee A, Karmakar A, Saha A, Bhattacharya D, Design of a compact UWB MIMO-diversity antenna incorporating fractal inspired isolation structure with band notch characteristics. AEIE Department, Netaji Subhash Engineering College. https://doi.org/10.1002/mop.32927
7. Sadineni RB, Gowda DP (2022) Design of penta-band notched UWB MIMO antenna for diverse wireless applications. Prog Electromagn Res M 107:35–49
8. Thakur E, Jaglan N, Devgupta S, Design of compact triple band-notched UWB MIMO antenna with TVC-EBG structure. J Electromagn Waves Appl. ISSN: 0920-5071. https://doi.org/10.1080/09205071.2020.1775136
9. Wu W, Yuan B, Wu A (2018) A quad-element UWB-MIMO antenna with band-notch and reduced mutual coupling based on EBG structures. Hindawi Int J Antennas Propag 2018:10. Article ID 8490740. https://doi.org/10.1155/2018/8490740
10. Chandel R, Gautam AK, Rambabu K, Tapered fed compact UWB MIMO-diversity antenna with dual band-notched characteristics. IEEE Trans Antennas Propag. https://doi.org/10.1109/TAP.2018.2803134
11. Rekha VSD, Pardhasaradhi P, Madhav BTP, Ushadevi Y, Dual band notched orthogonal 4-element MIMO antenna with isolation for UWB applications. IEEE Access. https://doi.org/10.1109/ACCESS.2020.3015020
12. Hassan MM, Rasool M, Asghar MU, Zahid Z, Khan AA, Rashid I, Rauf A, Bhatti FA, A novel UWB MIMO antenna array with band notch characteristics using parasitic decoupler. J Electromagn Waves Appl. https://doi.org/10.1080/09205071.2019.1682063
13. Aquil J, Sarkar D, Srivastava KV (2017) A quasi self-complementary UWB MIMO antenna having WLAN- band notched characteristics. In: 2017 IEEE applied electromagnetics conference (AEMC), Aurangabad, India, pp 1–2. https://doi.org/10.1109/AEMC.2017.8325722
14. Irene G, Rajesh A, A dual-polarized UWB-MIMO antenna with IEEE 802.11ac band-notched characteristics using split-ring resonator. J Comput Electron. https://doi.org/10.1007/s10825-018-1213-x
15. Shehata M, Said MS, Mostafa H (2018) Dual notched band quad-element MIMO antenna with multitone interference suppression for IR-UWB wireless applications. IEEE Trans Antennas Propag 66(11):5737–5746. https://doi.org/10.1109/TAP.2018.2868725

A Transformer-Based Model for Network Intrusion Detection: Architecture, Classification Heads, and Transformer Blocks

Nourhan Ibrahim, Sarama Shehmir, Aman Yadav, and Rasha Kashef

Abstract This paper introduces a transformer-based Network Intrusion Detection System (NIDS). Addressing a common oversight in current NIDSs that often neglect networks' long-term behavior and characteristics, our proposed model capitalizes on transformer models to effectively identify these features. Our proposed model enables adaptability across diverse flow-based network datasets by offering various transformer components, including the classification head, transformer, and data preprocessing. We leveraged our proposed model with different transformer architectures, such as shallow encoder transformer, shallow decoder transformer, and GPT 2.0, on three widely used NIDS benchmark datasets, including those specific to IoT environments. The evaluation covers key metrics like accuracy, $F1$ score, precision, and recall, with a notable finding emphasizing the pivotal role of classification head selection in determining model performance. Our proposed model provides crucial insights into optimizing transformer architectures for enhanced accuracy, efficiency, and applicability in the domain of network intrusion detection.

Keywords Cyber-attacks · Network intrusion detection systems (NIDS) · Transformers · Deep learning (DL) · Internet of things (IoT)

N. Ibrahim · S. Shehmir · A. Yadav (✉) · R. Kashef
Electrical, Computer, and Biomedical Engineering, Toronto Metropolitan University, Toronto, ON M5B 2K3, Canada
e-mail: aman.yadav@torontomu.ca

N. Ibrahim
e-mail: nourhan.ibrahim@torontomu.ca

S. Shehmir
e-mail: sarama.shehmir@torontomu.ca

R. Kashef
e-mail: rkashef@torontomu.ca

N. Ibrahim
Faulty of Engineering, Alexandria University, Alexandria, Egypt

P. G. Bradford et al. (eds.), *Proceedings of IEMTRONICS 2024*, Lecture Notes in Electrical Engineering 1229, https://doi.org/10.1007/978-981-97-4780-1_12

1 Introduction

The demand for robust intrusion detection systems (IDSs) is paramount in IoT cybersecurity. Traditional methods struggle to adapt to the evolving threat landscape and the overwhelming volume of data generated by IoT devices. The importance of effective IDSs in IoT's dynamic and interconnected world cannot be overstated. In selecting network traffic data as the input for evaluating transformer abilities in Intrusion Detection Systems (NIDS), we strategically chose flow records for several compelling reasons. Flow records serve as a condensed representation of network communication between two endpoints, aggregating essential information that captures the dynamics of data exchanges [1]. This choice is pivotal in optimizing the efficiency and scalability of NIDS, as flow records significantly reduce the volume of data requiring analysis. The aggregation of network data into flow records offers a comprehensive snapshot of communication patterns and streamlines the analytical process. This data volume reduction enhances NIDS's scalability, enabling the system to process and analyze large amounts of network traffic effectively. By focusing on flow data, there is a balance between capturing crucial information for intrusion detection and maintaining the computational efficiency necessary for real-time or near-realtime processing. Recently, transformers have become powerful tools for various language tasks, like talking and understanding [2]. They excel at recognizing complex patterns without needing specialized knowledge. One notable example is ChatGPT [3], created by OpenAI, which impressively engages in conversations using a GPT [4] (Generative Pre-trained Transformer) setup. This showcases how transformers can intelligently understand and generate language, transforming our interactions with computers. This paper explores the potential of transformers, redirecting their capabilities toward bolstering cybersecurity for Internet of Things (IoT) networks. Transformers, known for their adaptability to diverse data types beyond language, offer a promising solution to the unique challenges posed by cybersecurity in the IoT landscape. By harnessing the transformative power of transformers, this paper aims to enhance intrusion detection capabilities within IoT networks. While transformers have traditionally excelled in language-related tasks, their inherent ability to capture intricate relationships positions them as ideal candidates for addressing the complexities of cybersecurity in the IoT domain. Beyond their success in natural language processing, transformers harbor the potential to revolutionize the approach to securing interconnected devices and systems. With their adaptability and pattern recognition capabilities, transformer-based models present an innovative solution to the challenges faced in maintaining the security of IoT networks. This paper delves into the potential impact of applying transformers in the cybersecurity landscape, aiming to fortify practices within the IoT domain and contribute to the evolving field of transformer applications.

In this paper, we developed a transformer-based model for NetFlow data processing. This model entails the utilization of a transformer to examine sequences of network data, presenting a concise summary of the ongoing network activities. This summary is used for identifying intrusion patterns in the network. To validate the performance of our model, the model has been rigorously evaluated on three distinct Netflow datasets NF-ToN-IoT, NF-BoT-IoT, and NF-UNSWB-NB15 datasets, with a diverse range of network scenarios [5]. The model showcased remarkable results across various IoT datasets with a balanced accuracy above 99%. The rest of the paper can be structured as follows: Sect. 2 reviews related work, offering insights into the existing literature and highlighting key findings in the transformer models and NIDS. In Sect. 3, we outline the methodology adopted in our study, detailing the implementation of our transformer-based model and its components for evaluating transformer models on various datasets. Section 4 presents the evaluation results, emphasizing key performance metrics. Section 5 summarizes our conclusions with future directions.

2 Related Work

This section covers related studies that concentrate on transformer-based approaches to NIDS that handle sequences of flows, whereas typical ML-based NIDS function on a single flow. Abdallah Ghourabi [6] proposed a comprehensive security framework for medical systems to safeguard healthcare environments from various attacks. The hybrid security system includes an intrusion detection system for monitoring IoMT networks and a malware detection system protecting medical staff computers. Employing a Light Gradient Boosting Machine (lightGBM) and a BERT-based transformer, the method excels in processing big data with parallel, distributed, and GPU learning capabilities. Their proposed method follows three main stages: extracting network flow from captured activities, preprocessing the data, and utilizing lightGBM and BERT-based transformers for classifying network activities as benign or attacks. The evaluation of four datasets, including ECU-IoHT, TF-ToN-IoT, Edge-IIoTset, and EMBER, demonstrated the model's effectiveness using accuracy, precision, recall, $F1$ score, ROCAUC, and MCC metrics. However, a potential limitation of their work is the deployment complexity, requiring a correlation operation for result aggregation. The authors in [7] developed a modified transformer neural network (MTNN) that was proposed for robust intrusion detection in IoT networks. Traditional cybersecurity mechanisms prove inadequate for IoT attacks, making the MTNN model an innovative solution. Experimental results on the TF-ToN-IoT dataset showed significant accuracy, precision, recall, and $F1$ score enhancements. The MTNN model, with lower parameters and information gain for feature selection, is deemed suitable for deployment in distributed IoT networks. The study emphasizes transformers for intrusion detection in IoT networks, discussing the potential use of generative adversarial networks for false data injection. For further performance improvement, hyperparameter optimization using grid search or Bayesian Optimization is recommended. In [8], the Robust Transformer-based Intrusion Detection System (RTIDS) introduced a transformative approach to intrusion detection systems. RTIDS outperformed contemporary detection algorithms, achieving higher accuracy than SVMIDS, RNN IDS, LSTMIDS, and FNN-IDS. Evaluations on CICIDS2017 and CICDDoS2019 datasets demonstrated impressive accuracy, precision, recall, and $F1$ score for both datasets. RTIDS reported an accuracy of 98.45% for CICIDS2017 and 98.58% for CICDDoS2019, highlighting its efficacy in intrusion detection. The intrinsic power of RTIDS is found in its capacity to identify network anomalies and traffic infractions with greater skill than either deep learning-based or classical intrusion detection methods. Examining the system's architecture in greater detail, the authors emphasize the critical function of data preparation strategies and self-attention processes. With a dataset that contains more than 30 million records, the thorough analysis highlights how useful RTIDS may be in practical applications. Future research could focus on improving the effectiveness of the intrusion detection system's transformer algorithm to speed up processing and lessen the effects of unusual occurrences. Furthermore, the authors anticipate that incorporating meta-learning will be a viable remedy for the difficulties presented by few-shot classification scenarios. An intrusion detection system for the Internet of Things networks utilizing an attention mechanism and a bidirectional gated recurrent unit (BiGRU) was presented by Yalong Song and colleagues [9]. In order to address the issues with imbalanced datasets and insufficient feature information learning in the Deep Learning models currently in use for IoT network intrusion detection, this work proposes SEW-MBiGD, an intrusion detection model based on the SEW model using balanced datasets and the attention mechanism with BiGRU fusion neural network. The SEW model can learn the dataset's minority class features, and model balancing improves the data quality. According to the experimental results on NSL-KDD, the SEW method successfully balances the data, which makes it easier for the detection model to learn minority class samples. With the addition of multi-head self-attention (MHSA) to BiGRU, the MBiGD model can learn features more efficiently by allowing attention to temporal class features and enhancing the model's assessment of the relationships between various features. The outputs of the comparative and ablation experiments show that the SEW-MBiGD model can extract data characteristics more thoroughly and produce better results for each evaluation indicator. SVM's binary classification accuracy rose from 77.7 to 81.2%, while Decision Tree and CNN had improvements of about 1%. The suggested SEW-MBiGD model fared better than the other models. Overall improvement was seen in the MBiGD model ablation studies, with BiGRU's incorporation of the MHSA layer improving accuracy by 4.7% for multi-classification and 5.3% for binary classification. Accuracy was increased even further by training on a balanced dataset. In [10], a novel intrusion detection model using a combination of multi-head attention and Bidirectional Long Short-Term Memory (BiLSTM) was introduced. This contemporary multi-classification approach, tested on the modern Network Intrusion Detection Systems (NIDS) dataset, uses an embedding layer to convert the features of intrusion data into vector format, enhancing data representation. The embedding process transforms original vectors into two-dimensional vectors. The multi-head attention mechanism, on the other hand, allows the model to concentrate selectively on essential features within the vector, enhancing its interpretability. This mechanism collaborates seamlessly with BiLSTM, which, although not designed for time series data, can differentiate relationships between distant features, associating different features for predictions. The study employed datasets such as KDD-CUP99, NSL-KDD, and CICIDS2017 as part of the NIDS standard for training and testing. Data processing techniques, including normalization and one-hot encoding, were applied to ensure optimal model performance. SMOTE (synthetic minority oversampling technique) addressed class distribution imbalances. The suggested model exhibited superior performance concerning accuracy and $F1$ score when benchmarked against other models. On the KDD-CUP99, NSL-KDD, and CICIDS2017 datasets, the suggested model demonstrated accuracies of 98.29%, 95.19%, and 99.08%, respectively. The researchers point out certain limitations, such as the model's inability to accurately identify or report novel intrusion types. However, these intrusions can still be classified for additional scrutiny. The authors in [11] report that a hierarchical attention model for IDS achieved a detection accuracy of over 98.76% and a false alarm rate of 1.49% when the timestep was set to 10. Previous studies have proposed ML techniques for IDS, including feature selection and the use of DNN. The attention mechanism used in the model helps capture relevant features and can be further developed for feature selection and parallel computing. The new model has a satisfactory performance on the UNSW-NB15 dataset, with an accuracy of over 98.76% and a false alarm rate lower than 1.2%. Compared to the BiLSTM model, the proposed model showed an improvement of 3.05%. Table 1 summarizes the related work surveyed in this section.

Table 1 A summary of related models

Model	Class type	Dataset	Evaluation metrics	Limitations
[6] LightGBM, transformer	Binary	Edge-IIoTset, ECU-IoHT, ToN-IoT, EM-BER	MCC, ROC-AUC, $F1$ score, recall, accuracy, and precision	The model is complex to deploy
[7] MTNN	Multi-class	TF-ToN-IoT	Recall, precision, accuracy, and $F1$ scores	False data injection using GANs
[8] RTIDS	Multi-class	CICIDS2017, CIC-DDoS2019	Precision, recall, accuracy, and $F1$ scores	Integration of meta-learning to overcome few-shot occurrences
[9] SEW-MBiGD	Binary	NSL-KDD	$F1$ score, accuracy, precision, and recall	Cannot identify misclassified attacks
[10] Multi-attention head and BiLSTM	Multi-class	KDD-CUP99, NSL-KDD, CI-CIDS2017	Accuracy, $F1$ score	Cannot identify novel intrusion types
[11] Hierarchical Attention mechanism	binary	UNSW-NB15	Accuracy, false Alarm rate	Cannot identify misclassified attacks

3 The Proposed Model

As highlighted earlier, a noticeable gap exists in the exploration of NIDS capable of effectively managing sequences of traffic flows within the IoT environment. Transformers have become a compelling option for researchers, harnessing their inherent ability to process sequences and uncover complex relationships among elements in a sequence. In this sense, we thoroughly discuss the key components essential to our proposed transformer-based model designed for NIDS.

Figure 1 is a conceptual framework that adopts a transformer-based approach for NetFlow data processing. This methodology entails the utilization of a transformer to examine sequences of network data. To be more specific, the flow records that a NetFlow exporter gathers are fed into the transformer model, which makes it easier to find classifications for each record. The NetFlow exporter is a network device or software component responsible for collecting and consolidating flow records. These flow records contain essential details about the communication dynamics between various devices on the network, presenting a concise summary of the ongoing network activities [12].

Figure 2 illustrates the main components of our proposed transformer-based model for NIDS in IoT networks. Next, will explore our transformer model's diverse components in depth. This comprehensive analysis encompasses preprocessing techniques, transformer blocks, and classification heads.

3.1 Dataset Ingestion

Our model is built to handle different types of flow data formats commonly found in networks. When working with datasets, we carefully inspect them to identify their unique features. We create a detailed list that includes both numerical and categorical features. This list becomes crucial during preprocessing, where we refine the dataset to help the model better understand the information. Additionally, we specifically identify the label column, which plays a crucial role in distinguishing between the "Attack" and "Benign" classes. This intentional identification ensures that our model is well equipped to accurately classify instances of both attacks and benign activities during the intrusion detection process.

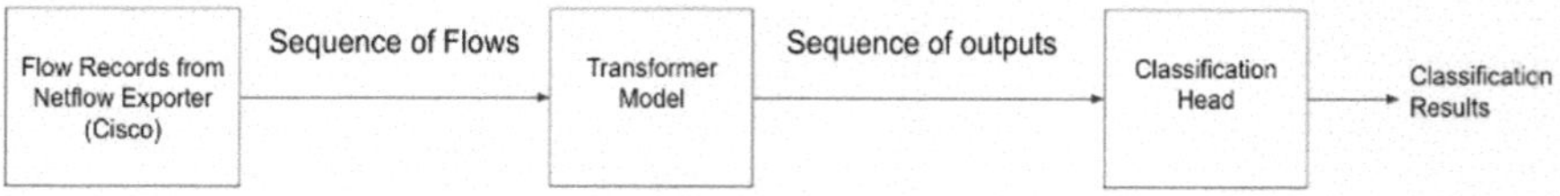

Fig. 1 A simple framework for NetFlow data processing using transformers

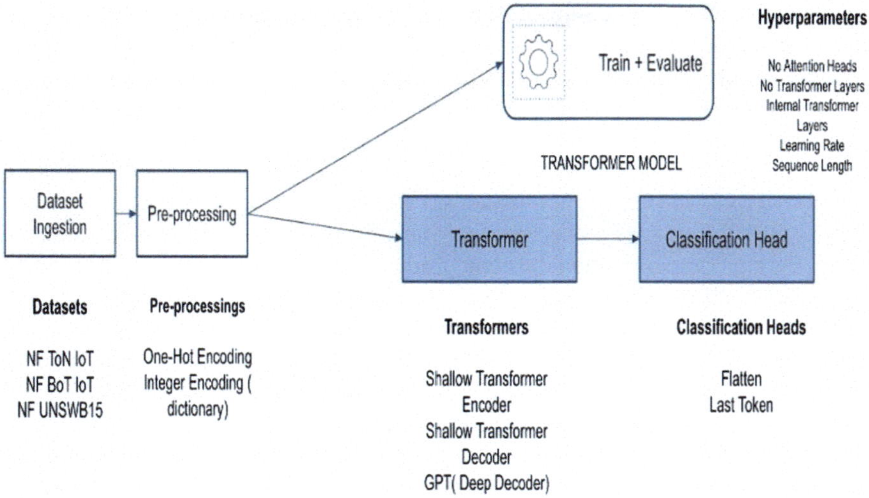

Fig. 2 The proposed model architecture

3.2 *Preprocessing*

In the preprocessing stage of our model, a crucial decision is made regarding the encoding of categorical variables, where the preprocessing layer is adept at either one-hot encoding or integer encoding. Additionally, numerical features undergo min–max scaling to ensure standardized processing. Notably, the expected categorical format, whether one-hot or integer encoded, is a parameter during the fit and transform methods for the categorical fields.

3.3 *Transformer Model*

Our proposed model investigates three transformer architectures: shallow encoder-based transformer (2-layered encoder-based transformer), shallow decoder-based transformer (2-layered decoder-based transformer), and deep decoder-based transformer (GPT 2.0). The difference between shallow and deep models lies primarily in their depth, number of attention heads, and internal size, while their core transformer block structure is the same. The transformer-based deep learning models consist of blocks, each designed to execute a specific transformation on input sequences. These blocks can either be encoder or decoder blocks. In the context of transformers, the role of an encoder block is to take in an input and convert it into a fixed-length feature representation. This representation encapsulates the semantic meaning of the input, considering its contextual relationship with other inputs in the sequence [13]. In Network Intrusion Detection Systems (NIDS), the encoder's task would involve transforming each network flow into a fixed-length feature vector. On the other hand, decoder blocks, the counter-part of encoder blocks, are commonly employed in generational tasks. Decoders receive a sequence of feature representations and generate an output sequence. This translates to taking a flow's feature vector and generating a reconstructed raw flow record in the NIDS domain. This transformative architecture allows transformers to process and understand sequential data efficiently, making them potent tools for various applications, including intrusion detection in network security. GPT-2.0 utilizes a deep decoder transformer architecture [4]. It adopts an innovative approach by treating the input prompt as part of a sequence and generating output with each generated word serving as context for predicting the subsequent word. Notably, GPT-2.0 exclusively employs transformer decoder blocks, departing from the traditional encoder-decoder structure. The model's internal block structure is iteratively replicated, with smaller GPT-2.0 variants featuring 12 decoder blocks. In the decoding process, the input sequence traverses these blocks sequentially, where the output of each block becomes the input for the subsequent one. This stacking of decoder blocks enables GPT-2.0 to model the distribution of natural language effectively, enhancing its language generation capabilities. Unlike traditional transformer models, GPT's autoregressive nature considers tokens solely to the left of the token it generates, progressing through a sequence one token at a time. This unique architectural choice empowers GPT-2.0 to generate contextually rich and coherent language predictions.

3.4 Classification Head

The classification head in our transformer-based NIDS model serves as a pivotal component, wielding significant importance in the overall performance and efficacy of the system. Positioned at the end of the transformer architecture, the classification head is responsible for interpreting the intricate representations learned by the transformer blocks and making informed decisions regarding the nature of network traffic—whether it indicates a potential intrusion or benign behavior. The design and selection of the classification head play a crucial role in determining the model's ability to distinguish between normal and malicious network activities accurately. An adeptly chosen classification head can enhance the model's sensitivity to subtle patterns and anomalies within the data, thereby improving detection accuracy [14]. Figure 3 illustrates the classification head approaches adopted in our model.

In our proposed transformer-based NIDS model, we explore two distinct classification head approaches: "flatten" and "last token." The "flatten" approach transforms the output sequence from the transformer blocks into a flattened representation, effectively collapsing the sequence into a one-dimensional array. This flattened representation is fed into the classification head for intrusion detection decisions. On the other hand, the "last token" approach focuses specifically on the final token generated by the transformer blocks. In this technique, the classification head receives only the

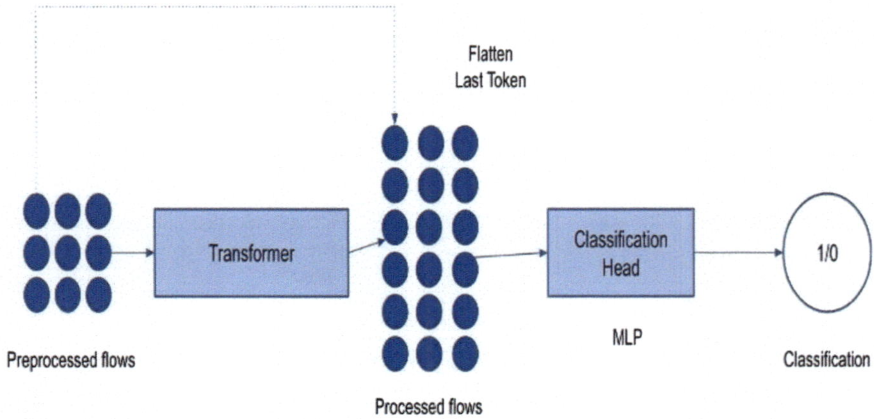

Fig. 3 Classification head approaches

output corresponding to the last token in the sequence, considering it as the representative feature for the entire input. By investigating both the "flatten" and "last token" approaches, we aim to compare their efficacy in capturing relevant information from the transformer-generated representations, providing insights into the optimal choice for the classification head in the context of NIDS.

4 Experimental Setup

Our proposed model uses the Python programming language within the Google Colab environment. The Google Colab environment, a cloud-based platform, provides a convenient and collaborative space for coding, experimentation, and model training. Its integration with popular libraries and frameworks for machine learning, such as TensorFlow and PyTorch, facilitates seamless implementation and efficient utilization of computational resources. We used already built libraries for transformer-based models in [14].

4.1 Datasets

The performance of our model has undergone comprehensive evaluation across three distinct datasets, each offering unique insights into the model's capabilities. The first dataset, NF-ToN-IoT [15], focuses on network traffic within the Internet of Things (IoT) domain. This dataset is specifically curated to capture the complexities and nuances associated with IoT environments, providing a valuable testbed for evaluating the model's adaptability to modern network architectures. The second dataset,

Table 2 Datasets

Dataset	Features	Records	Types of attacks
NF-ToN-IoT [15]	8	1,379,274	Backdoor, DoS, DDOS, injection, MITM, password, ransomware, scanning, XSS
NF-BoT-IoT [16]	8	600,100	Reconnaissance, DDOS, DOS, theft
NF-UNSWB-NB15 [17]	43	2,390,275	Fuzzers, backdoor, DoS, exploits, generic, reconnaissance, shellcode, worms

NF-BoT-IoT [16], delves into Botnet traffic within IoT networks. This dataset emphasizes detecting malicious activities associated with Botnets in IoT scenarios, challenging the model to discern subtle patterns indicative of such threats. Lastly, the algorithm is evaluated on NF-UNSWB-NB15 [17], a traditional dataset representative of more conventional network settings. This dataset enables a broader assessment of the model's performance across various network behaviors and threats commonly encountered in standard network environments. The diversity of these datasets ensures a comprehensive evaluation, highlighting the model's robustness and effectiveness in handling various network scenarios and security challenges. Table 2 contains a detailed description of the three above-mentioned datasets.

4.2 Model Hyperparameters

We investigated different hyperparameter settings for our proposed model. We conducted a comprehensive grid search in our model development process to optimize the hyperparameters and achieve the best possible results. Table 3 summarizes the values of the hyperparameters used.

Table 3 Hyperparameters

Hyperparameter	Value
Transformer block	Encoder, decoder
Layers	2, 4, 6, 8
Feed forward (FF) dimensions	128, 256, 512
Attention heads	2, 4, 6, 8, 12
Learning rate	0.01, 0.001, 0.0005, 0.0001, 0.00001

4.3 Model Training and Testing

In our model development, we have adhered to use 80% of the dataset allocated for training the model, and the remaining 20% is reserved for testing. This ratio is a commonly employed practice in machine learning to balance providing the model with sufficient data for learning and retaining an independent subset for evaluating its generalization capabilities. For the model training, we used early stopping and an epoch limit. The early stopping was set to a patience of 5 epochs, and the maximum number of epochs was limited to 20. We chose 20 epochs as in our initial experimentation the majority of models had converged to within 1% of their final performance by the 20th epoch. We used the Adam optimizer for training, with specific learning rates (as a part of the grid search).

4.4 Evaluation Metrics

We employed standard performance metrics to evaluate the effectiveness of various transformer models, including the $F1$ score and accuracy. These metrics are calculated based on the combination of True Positives (TP), True Negatives (TN), False Positives (FP), and False Negatives (FN).

5 Experimental Results

Evaluation is conducted on separate datasets, including TF-ToN-IoT, TF-BoT-IoT, and UNSW NB15. The evaluation results include metrics such as balanced accuracy, true positives, false positives, false negatives, and the $F1$ score. For the TF-ToN-IoT dataset, the model achieved outstanding performance with a balanced accuracy of 100%, indicating that it correctly classified both benign and attack instances. The $F1$ score is also very high at 0.999958. For the TF-BoT-IoT dataset, the model demonstrated a balanced accuracy of 91.14%, indicating good overall performance. The $F1$ score is 0.906167, a good balance between precision and recall. The model exhibited exceptional performance for the UNSW NB15 dataset with a balanced accuracy of 99.17%. The model achieved a high number of true positives and true negatives, resulting in an $F1$ score of 0.945971. Table 4 summarizes the results achieved from the three proposed transformer models with different classification head approaches. Figure 4a shows the $F1$ score for the three benchmark datasets across the proposed transformer models integrating different classification head techniques. While both transformer encoder and decoder blocks proved effective, the smaller size and broader applicability of encoder blocks suggest their preference for achieving optimal results.

Table 4 Datasets, transformer configuration, and performance metrics

Dataset	Transformer	Layers	Heads	FF dim.	Classification head	$F1$ score (%)	Detection rate (%)
Ton-IoT	Shallow encoder	2	2	128	Flatten	99.99	100
	Shallow decoder	2	2	128	Last token	99.99	100
	Shallow decoder	2	2	128	Flatten	99.99	100
	GPT model	12	12	768	Flatten	84.49	50
BoT-IoT	Shallow encoder	2	2	128	Flatten	90.62	91.14
	Shallow decoder	2	2	128	Last token	89.84	90.73
	Shallow decoder	2	2	128	Flatten	90.57	91.21
	GPT model	12	12	768	Flatten	0	50
UNSW-NB15	Shallow encoder	2	2	128	Flatten	94.60	99.17
	Shallow decoder	2	2	128	Last token	99.24	99.25
	Shallow decoder	2	2	128	Flatten	94.67	99.35
	GPT model	12	12	768	Flatten	0	50

6 Conclusion and Future Directions

In this paper, we introduced a transformer-based Network Intrusion Detection System (NIDS). We have shown the paramount importance of the classification head in determining the performance of our transformer-based Network Intrusion Detection System (NIDS) model. Notably, the "LastToken" classification head emerged as the most crucial factor for optimizing model accuracy, demonstrating its superiority over alternative choices. Additionally, we propose the exploration of Deep Encoder Transformers like BERT for enhanced performance in future iterations. Furthermore, future work will incorporate input encoders post-preprocessing to strategically reduce the number of parameters. This reduction aims to enhance computational efficiency, reducing the overall complexity of the model while maintaining or improving performance. Integrating Input Encoders presents a promising avenue for achieving a more streamlined and resource-efficient NIDS framework. Overall, our findings contribute to the evolution of transformer-based models in network security and intrusion detection. Future directions include adopting distributed computing [18–22] and blockchain strategy [23–25] for scalable and privacy-aware transformer models. We also plan to investigate the comparison of transformer-based models

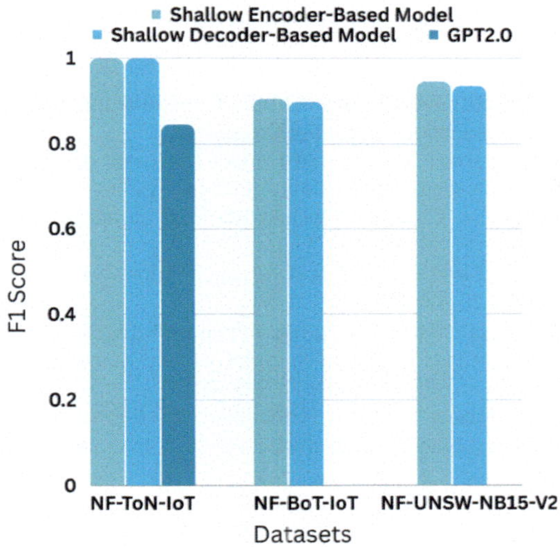

(a) F1 Score of the three transformer models with last token classification head across the three benchmark datasets.

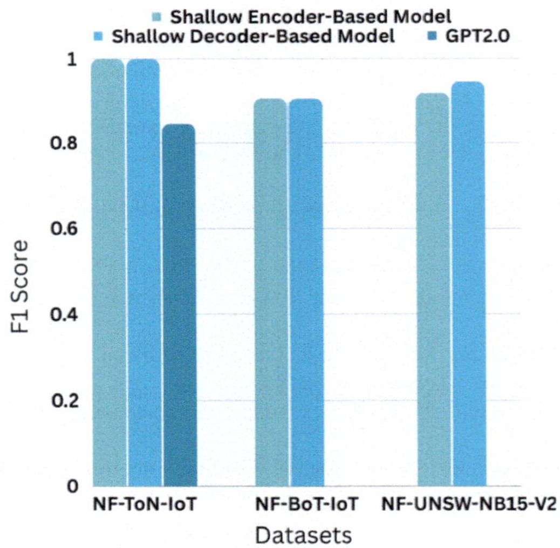

(b) F1 Score of the three transformer models with flattened classification head across the three benchmark datasets.

Fig. 4 Comparison of $F1$ scores

against GANs-based models [26–30]. In future research, integrating anomaly detection algorithms [31–35] with intrusion detection systems (IDS) presents a promising direction for enhancing cybersecurity measures.

References

1. Sarhan M, Layeghy S, Moustafa N, Portmann M (2020) NetFlow datasets for machine learning-based network intrusion detection systems. arXiv, cs.NI
2. Vaswani A, Shazeer N, Parmar N, Uszkoreit J, Jones L, Gomez AN, Kaiser, Ł, Polosukhin I (2017) Attention is all you need. Adv Neural Inf Process Syst 30
3. OpenAI (2022) Introducing ChatGPT. OpenAI Blog. https://openai.com/blog/chatgpt, Nov 2022
4. OpenAI (2019) GPT-2 1.5B release. OpenAI Research. https://openai.com/research/gpt-2-1-5b-release/. Accessed on 20 Apr 2023
5. Netflow datasets (2020). http://staff.itee.uq.edu.au/marius/NIDS_datasets/
6. Ghourabi A (2022) A security model based on LightGBM and transformer to protect healthcare systems from cyberattacks. IEEE Access 10:48890–48903
7. Ahmed SW, Kientz F, Kashef R (2023) A modified transformer neural network (MTNN) for robust intrusion detection in IoT networks. In: 2023 international telecommunications conference (ITC-Egypt)
8. Wu Z, Zhang H, Wang P, Sun Z (2022) RTIDS: a robust transformer-based approach for intrusion detection system. IEEE Access 10:64375–64387
9. Song Y, Zhang D, Li Y, Shi S, Duan P, Wei J (2023) Intrusion detection for Internet of things networks using attention mechanism and BiGRU. In: 2023 5th international conference on electronic engineering and informatics (EEI)
10. Zhang J, Zhang X, Liu Z, Fu F, Jiao Y, Xu F (2023) A network intrusion detection model based on BiLSTM with multi-head attention mechanism. Electronics 12(19):4170
11. Liu C, Liu Y, Yan Y, Wang J (2020) An intrusion detection model with hierarchical attention mechanism. IEEE Access 8:67542–67554
12. Cisco Systems (2011) Cisco IOS NetFlow version 9 flow-record format—white paper. https://www.cisco.com/en/US/technologies/tk648/tk362/technologies_white_paper09186a00800a3db9.pdf
13. Islam S et al (2024) A comprehensive survey on applications of transformers for deep learning tasks. Expert Syst Appl 241(122666):122666
14. Manocchio LD, Layeghy S, Lo WW, Kulatilleke GK, Sarhan M, Portmann M (2023) Flow-Transformer: a transformer framework for flow-based network intrusion detection systems. arXiv cs.CR
15. NF-ToN-IoT (2024) Edu.au [Online]. Available: https://rdm.uq.edu.au/files/2ad93cd0-ef9c-11ed-827d-e762de186848. Accessed: 06 Feb 2024
16. NF-BoT-IoT, Edu.au [Online]. Available: https://rdm.uq.edu.au/files/48fa1900-ef9c-11ed-827d-e762de186848
17. NF-UNSW-NB15-v2, Edu.au [Online]. Available: https://rdm.uq.edu.au/files/8c6e2a00-ef9c-11ed-827d-e762de186848
18. Manjunath YSK, Kashef RF (2021) Distributed clustering using multi-tier hierarchical overlay super-peer peer-to-peer network architecture for efficient customer segmentation. Electron Commer Res Appl 47:101040
19. Kashef R, Kamel M (2006) Distributed cooperative hard-fuzzy document clustering. In: Proceedings of the annual scientific conference of the LORNET research network
20. Kashef R, Warraich M (2020) Homogeneous vs. heterogeneous distributed data clustering: a taxonomy. In: Data management and analysis: case studies in education, healthcare and beyond, pp 51–66
21. Kashef R, Kamel M (2008) Distributed peer-to-peer cooperative partitional-divisive clustering for gene expression datasets. In: 2008 IEEE symposium on computational intelligence in bioinformatics and computational biology. IEEE, pp 143–150
22. Kashef R, Niranjan A (2017) Handling large-scale data using two-tier hierarchical super-peer P2P network. In: Proceedings of the international conference on big data and Internet of thing, pp 52–56

23. Yeh, T.-Y., Kashef, R.: Trust-based collaborative filtering recommendation systems on the blockchain. Adv Int Things 10(4):37–56
24. Schmid P, Schaffhäuser A, Kashef R (2023) IoTBChain: adopting blockchain technology to increase PLC resilience in an IoT environment. Information 14(8):437
25. Saleminezhadl A, Manuel R, Ravikumar C, Kashef R (2021) IoT analytics and Blockchain. arXiv preprint arXiv:2112.13430
26. Soleymanzadeh R, Kashef R (2023) Efficient intrusion detection using multi-player generative adversarial networks (GANs): an ensemble-based deep learning architecture. Neural Comput Appl 35(17):12545–12563
27. Soleymanzadeh R, Kashef R (2022) A stable generative adversarial network architecture for network intrusion detection. In: 2022 IEEE international conference on cyber security and resilience (CSR), Rhodes, Greece, pp 9–15. https://doi.org/10.1109/CSR54599.2022.9850286
28. Nguyen H, Kashef R (2023) TS-IDS: traffic-aware self-supervised learning for IoT network intrusion detection. Knowl-Based Syst 279:110966
29. Ahmed SW, Kientz F, Kashef R (2023) A modified transformer neural network (MTNN) for robust intrusion detection in IoT networks. In: 2023 international telecommunications conference (ITC-Egypt). IEEE, pp 663–668
30. Soleymanzadeh R, Kashef R (2022) The analysis of the generator architectures and loss functions in improving the stability of GANs training towards efficient intrusion detection. In: 2022 9th international conference on soft computing and machine intelligence (ISCMI), 26 Nov 2022. IEEE, pp 246–252
31. Kashef RF (2018) Ensemble-based anomaly detetction using cooperative learning. In: KDD 2017 workshop on anomaly detection in finance. PMLR, pp 43–55
32. Achiluzzi E, Li M, Georgy MFA, Kashef R (2022) Exploring the use of data-driven approaches for anomaly detection in the Internet of things (IoT) environment. arXiv preprint arXiv:2301.00134
33. Li M, Kashef R, Ibrahim A (2020) Multi-level clustering-based outlier's detection (MCOD) using self-organizing maps. Big Data Cogn Comput 4(4):24
34. Kashef R, Gencarelli M, Ibrahim A (2020) Classification of outlier's detection methods based on quantitative or semantic learning. In: Combating security challenges in the age of big data: powered by state-of-the-art artificial intelligence techniques, pp 45–59
35. Mohammed K, Jebamikyous AH, Nawara HD, Kashef R (2021) IoT cyber-attack detection: a comparative analysis. In: International conference on data science, E-learning and information systems 2021, pp 117–123

Parameter Identification in Magnetorheological Dampers via Physics-Informed Neural Networks

Yuandi Wu, Brett Sicard, Patrick Kosierb, and S. Andrew Gadsden

Abstract This paper presents an investigation into the utilization of physics-informed neural networks for parameter identification in the domain of magnetorheological dampers. MR dampers are known for their controllable rheological properties, making them integral components in various engineering applications such as vibration control and structural dynamics. Efficient utilization of MR dampers relies on accurate characterization of their material properties, necessitating robust parameter identification techniques. The proposed methodology integrates physics-informed neural networks, a class of neural networks that embed physical principles into their architecture, enabling the incorporation of governing equations and boundary conditions during the training process. This fusion of physics-based constraints with machine learning facilitates the extraction of meaningful parameters from experimental data, enhancing the accuracy of the identification process. Through a series of simulations and experiments, this study assesses the efficacy of physics-informed neural networks in capturing the complex nonlinear behaviour exhibited by MR dampers. The neural network is trained on a dataset comprising experimental observations of the damper's response under varying conditions. The results demonstrate the capability of physics-informed neural networks to discern and infer key material parameters. The findings presented herein contribute to the growing body of research on the application of machine learning techniques in structural dynamics and control. The demonstrated results of physics-informed neural networks in parameter identification for MR dampers showcase their potential as a valuable tool for engineers and researchers seeking to optimize the design and control of these adaptive devices in real-world engineering applications.

Y. Wu (✉) · B. Sicard · P. Kosierb · S. A. Gadsden
McMaster University, Hamilton, ON L8S 4L8, Canada
e-mail: wuy187@mcmaster.ca

B. Sicard
e-mail: sicardb@mcmaster.ca

P. Kosierb
e-mail: kosierbp@mcmaster.ca

S. A. Gadsden
e-mail: gadsdesa@mcmaster.ca

P. G. Bradford et al. (eds.), *Proceedings of IEMTRONICS 2024*, Lecture Notes in Electrical Engineering 1229, https://doi.org/10.1007/978-981-97-4780-1_13

Keywords Physics-informed machine learning · Magnetorheological dampers · Neural networks · Parameter estimation · Deep learning

1 Introduction

Magnetorheological (MR) dampers are pivotal in engineering systems requiring dynamic response control due to their adaptive and tunable properties. However, accurately determining the material characteristics of MR dampers poses challenges due to their inherent complexity and nonlinearity [33]. Traditional methods for identifying parameters often struggle to manage these complexities, resorting to iterative techniques or local search algorithms that may necessitate substantial prior assumptions or system simplifications [7].

In addressing these challenges, this study investigates the integration of Physics-Informed Neural Networks (PINNs) as a potential approach to improve parameter identification accuracy in the context of MR dampers. PINNs utilize deep neural networks as solvers for differential equations, enabling relatively precise predictions for unknown terms even with limited data [25]. Their capacity as universal function approximators allows them to handle nonlinear problems without the need for predefined assumptions or system simplifications [11, 22, 26]. Additionally, PINNs leverage automatic differentiation to effectively explore parameter spaces and enhance model performance [2, 5, 21].

For MR dampers, characterized by significant nonlinearity in their rheological processes and the simultaneous estimation of multiple parameters, PINNs offer several compatible characteristics for the estimation process. The commonly used modified Bouc-Wen model [27], describing MR damper behaviour with a set of differential equations and multiple parameters, often poses challenges for traditional approaches in identifying satisfactory parameters due to initial value setting difficulties and high-dimensional parameter spaces [14, 32, 36]. PINNs present a promising solution by integrating physical principles and experimental data into the learning process, enabling accurate parameter estimation for MR dampers [26]. Results indicate strong alignment between PINN predictions and experimental data, although inherent noise in the data introduces some discrepancies. Factors contributing to this noise include the composite loss function used during training, the complexity of the model, and variations in voltage inputs. Further refinement of network architecture, loss weighting schemes, and consideration of system dynamics are crucial for improving the prediction accuracy of parameters, especially in capturing hysteresis behaviours and responses to low voltage inputs.

The paper is organized as follows: Sect. 2 provides an overview of current methodologies used for mathematically modelling MR dampers. Following this, Sect. 3 explores background information related to PINNs, which have shown promise in solving complex physics-based problems. The proposed parameter estimation scheme is then explained in Sect. 4, detailing how the model parameters are estimated using the PINNs framework. Subsequently, Sect. 5 presents the results of the parameter estimation scheme and discusses these findings. Finally, the paper is concluded in Sect. 6, summarizing the key points discussed and suggesting directions for future research in this area.

2 Parametric Modelling for Magnetorheological Dampers

Magnetorheological dampers are devices designed to provide variable damping in response to changes in an applied magnetic field, relying on the unique rheological properties exhibited by MR fluids. Beyond their fundamental contributions to the understanding of controllable damping systems, MR dampers have found widespread applications in civil engineering for structural vibration control [23, 44], seismic mitigation [6, 8], and adaptive suspension systems in automotive engineering [9, 37, 39]. Moreover, MR dampers have found their way into robotics, offering precise control over the damping characteristics in robotic joints and limbs [3, 34, 40]. This adaptability contributes to improved stability and maneuverability in various robotic applications. Key to their operation is the MR fluid itself, which is composed of a suspension of micron-sized ferrous particles within a liquid carrier. The dynamic behaviour may be actively controlled by adjusting the alignment of the aforementioned ferrous particles through the application of an external magnetic field, typically induced through an externally applied voltage [10]. The responsiveness of MR dampers to changes in voltage stems from the magnetic flux-induced alignment of ferrous particles along the field lines. As the applied voltage increases or decreases, corresponding adjustments in the magnetic field strength occur, leading to alterations in the alignment of particles and, consequently, changes in the rheological properties of the MR fluid [8]. Through this property, the damper may be adjusted to provide varying levels of resistance to motion, thereby influencing the damping characteristics of the overall system.

The mathematical modelling of MR dampers poses considerable challenges, primarily attributed to the intricate dynamics arising from force–velocity hysteresis and history dependency [37, 41, 46]. The nonlinearity inherent in these dampers results from complex interactions involving magnetic field strength, particle distribution, and rheological properties of the fluid. Force–velocity hysteresis, a key characteristic of MR dampers, brings in a complicated connection between the loading and unloading phases, making it challenging to develop a precise mathematical representation. Moreover, the history-dependent nature of these dampers implies that their response is influenced not only by the current state but also by preceding loading conditions [28]. To address these complexities, researchers have turned to phenomenological approaches, employing parametric models to capture the intricate nonlinear dynamics. Phenomenological models are empirical or semi-empirical models that are developed based on observed phenomena and experimental data, rather than being derived from underlying physical principles. These models aim to capture the essential features of a system's behaviour without necessarily delving into the detailed internal mechanisms or physics governing the phenomena. Various authors have proposed models based on empirical observations and experimental data, aiming to characterize the system's behaviour using parameters that encompass the interplay of magnetic, fluid, and structural elements [15, 27, 43]. However, the identification of these parameters remains challenging, involving navigation through a high-dimensional parameter space and necessitating comprehensive experimental data to ensure the accuracy of model predictions.

Prominent among the various mathematical models proposed for describing the behaviour of MR dampers is the modified Bouc-Wen model introduced by Spencer and colleagues [27]. This model has gained recognition for its effectiveness in capturing the intricate dynamics, especially in addressing the roll-off region, an aspect that posed challenges in the original Bouc-Wen model [12, 27]. The modified Bouc-Wen model is a phenomenological mathematical model extensively utilized to describe the behaviour of MR dampers. An illustration of the model is provided in Fig. 1 [27]. Spencer's modification has demonstrated improved accuracy in predicting the damping force over a wide range of velocities. However, it is essential to acknowledge that the modified Bouc-Wen model comes with a notable degree of complexity. The model involves a system of differential equations, necessitating a comprehensive understanding of the underlying physics and intricate interactions within the damper system [12]. Furthermore, the model demands the identification and tuning of 14 parameters, adding a layer of complexity to the calibration process. While the modified Bouc-Wen model has proven its efficacy in capturing the nonlinearity of MR dampers, its complexity raises challenges regarding computational implementation and the requirement for extensive experimental data to accurately identify and finetune the numerous model parameters [28, 35]. The modified Bouc-Wen model is characterized by the following equation:

$$F = \alpha z + c_0(\dot{x} - \dot{y}) + k_0(x - y) + k_1(x - x_0). \tag{1}$$

Equation (1) may also be simplified as

$$F = c_1\dot{y} + k_1(x - x_0), \tag{2}$$

where the evolutionary variable z (also denoted as the hysteretic displacement in certain publications), and the intermediary displacement y is represented in the following differential equations:

$$\dot{z} = -\gamma|\dot{x} - \dot{y}|z|z|^{n-1} - \beta(\dot{x} - \dot{y})|z|^n + A(\dot{x} - \dot{y}), \tag{3}$$

$$\dot{y} = \frac{1}{c_0 + c_1}(\alpha z + k_0(x - y) + c_0\dot{x}). \tag{4}$$

whereas the model outlined above is static for a set magnetic field strength within the damper, the model may be extended to incorporate the changes in magnetic field strength through varying applied voltage. Parameters α, c_0, and c_1 were shown to be

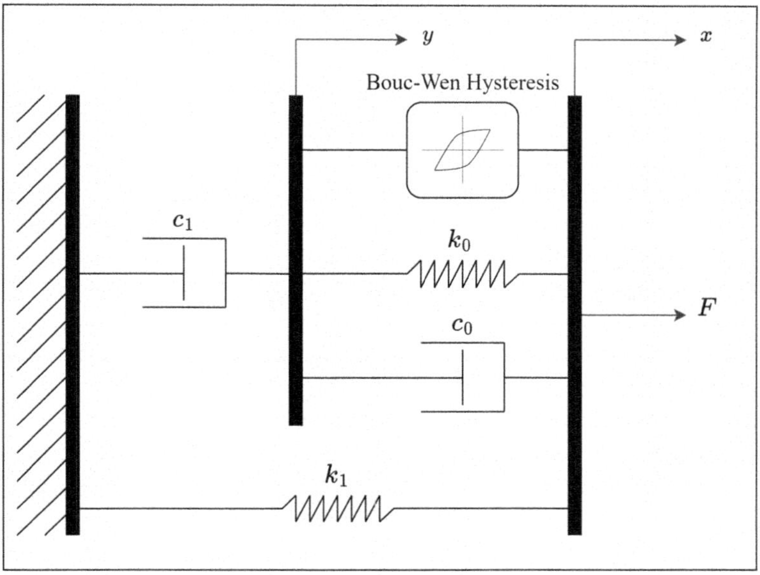

Fig. 1 The modified Bouc-Wen model. Adapted from [27]

changing linearly with an efficient voltage u, whereby u represents the filtered input voltage v [1, 16, 42]. The relations are given as follows:

$$\alpha = \alpha_a + \alpha_b u, \tag{5}$$

$$c_0 = c_{0,a} + c_{0,b}\, u, \tag{6}$$

$$c_1 = c_{1,a} + c_{1,b} u, \tag{7}$$

$$\dot{u} = \eta(u - v), \tag{8}$$

whereby parameters $\left[\alpha_a, \alpha_b, \beta, \gamma, \eta, A, c_{0a}, c_{0b}, c_{1a}, c_{1b}, k_0, k_1, n, x_0\right]$ are system parameters to be identified prior to simulations with the modified Bouc-Wen model.

3 Physics-Informed Neural Network for Parameter Estimation

In recent years, PINNs have emerged as a powerful paradigm for solving partial differential equations and elucidating complex physical phenomena. This section delves into the application of PINNs within the context of identifying solutions to partial differential equations, building upon the foundational work conducted by Raissi [26]. PINNs seamlessly integrate neural network architectures with the governing physics of a system, offering a data-driven approach to characterize latent solutions. Specifically, we explore the methodology employed in prior works of various authors, which involves utilizing PINNs to discern optimal parameters λ that effectively describe observed data [4, 13, 25, 26].

In the initial study conducted, the investigation primarily centred on the utilization of data-driven methodologies for the identification of the solution $u(t, x)$ to partial differential equations in a generalized form [26]. The considered equations are expressed concerning spatial and temporal variables, denoted as $x \in \Omega$ and $t \in [0, T]$, respectively.

$$0 = \frac{\delta u}{\delta t} - \mathcal{N}[u; \lambda]. \tag{9}$$

Here, the variable u symbolizes the latent solution of the differential equation, while $\mathcal{N}$ signifies the nonlinear operator parameterized by λ. In the context of a system with a hidden state $u(t, x)$, characterized by a sparse and potentially noisy set of observations, the authors employed the PINN to discern the optimal parameters λ that effectively characterize the observed data. This endeavour involved determining the parameters λ through the PINN methodology, which is designed to efficiently handle scattered and potentially noisy observations of the hidden state $u(t, x)$.

In the subsequent section, the framework established by Raissi, which was originally designed for the solution of partial differential equations, undergoes adaptation to address a system of ordinary differential equations (ODEs). The adaptability of the methodology becomes apparent as we extend its application from partial to ordinary differential equations, catering to a broader range of dynamical systems.

4 Parameter Estimation Framework Utilizing PINNs

To address the challenges associated with the complexity of the modified Bouc-Wen model, we propose to integrate this model into a PINN framework for parameter estimation. PINNs leverage the power of neural networks to learn the underlying physics of a system while simultaneously incorporating physical principles in the form of partial differential equations. By integrating the modified Bouc-Wen model into a PINN, we aim to harness the modelling accuracy of the former while benefiting from the data-driven capabilities of neural networks for parameter identification. It is worth noting that various authors have recognized the potential of PINNs for parameter estimation in diverse engineering applications [24, 29, 31, 38, 45]. The capability of PINNs to seamlessly integrate physical laws with data-driven approaches has been harnessed to identify and tune parameters in complex dynamical systems efficiently [26]. In the context of MR dampers, the application of PINNs for parameter estimation has gained traction due to the inherent challenges associated with the complexity of models such as the modified Bouc-Wen model. Leveraging the strengths of PINNs, we aim to contribute to this growing body of work by employing the network architecture to accurately estimate and fine-tune the 14 parameters of the modified Bouc-Wen model.

The challenge of parameter estimation is commonly formulated as an inverse problem, wherein the objective is to infer the parameters of a given model based on observed data [30]. In the context of parameter estimation, the utilization of NNs is particularly advantageous due to their inherent ability to be configured as inverse models [20]. In many real-world scenarios, the physical parameters of a system are often challenging to directly measure or quantify [45]. However, by casting these parameters as neural network parameters, the neural network can efficiently learn and approximate their values through the optimization of weights and biases. This approach aligns with the inherent capability of neural networks to adapt and generalize complex patterns from data. The neural network, equipped with the task of minimizing discrepancies between its predictions and the training dataset, naturally extends its ability to handle additional parameters. By incorporating the physical parameters into the architecture, the neural network explores various combinations to achieve the best fit with observed data.

Neural networks have emerged as powerful tools for approximating complex functions in various fields, owing to their universal approximation capabilities. The universal approximation theorem states that a neural network may approximate any continuous function to arbitrary precision, given a sufficiently large number of neurons in its hidden layers [22]. In the context of dynamical systems, ordinary differential equations (ODEs) represent a ubiquitous framework for describing the evolution of physical systems over time. The ability of neural networks to approximate functions renders them suitable candidates for approximating the solutions to systems of ODEs, as illustrated in the work of Lagrais and colleagues, who initially conceptualized the idea [17]. A specific implementation considered in this investigation involves employing a PINN to tackle the parameter estimation task, integrating ODEs governing the behaviour of the modified Bouc-Wen model, as illustrated in Eqs. (1), (2), (3), (4), (5), (6), (7) and (8). A visualization of the workflow adhered to herein is presented in Fig. 2.

For a neural network denoted as $N\left(x, \frac{dx}{dt}, v, t\right)$, where $x, \frac{dx}{dt}, v, t$ represents the input variables. Leveraging the universal approximation property, it is possible to express the neural network function as a suitable approximation for the solution to a system of ODEs. In this study, we focus on a specific scenario where the universal approximative capabilities of a neural network are employed to obtain the solution

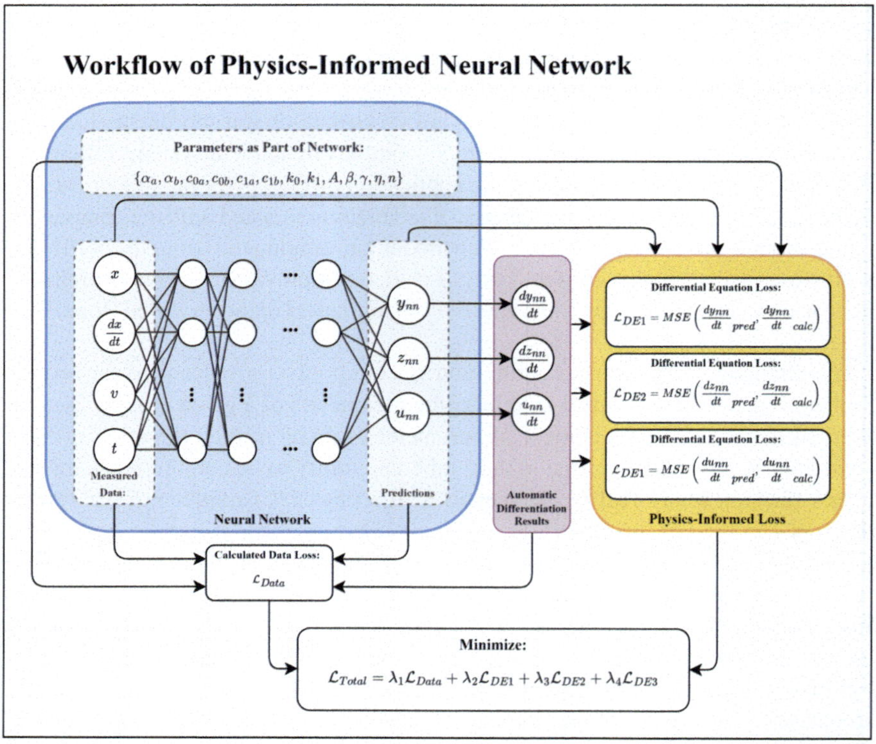

Fig. 2 Workflow of proposed physics-informed neural network for estimation of parameters

vector $\left[u_{NN}, y_{NN}, z_{NN}\right]^T$, representing the dependent variables of the system:

$$N\left(x, \frac{\mathrm{d}x}{\mathrm{d}t}, v, t\right) \approx [u, y, z]^T \tag{10}$$

where the subscript *NN* denotes predictions via the neural network. To facilitate the analysis of the dynamical system, it becomes necessary to determine the derivatives of the neural network function with respect to the independent variable *t*, such that variables $\frac{\mathrm{d}u_{NN}}{\mathrm{d}t}$, $\frac{\mathrm{d}y_{NN}}{\mathrm{d}t}$, and $\frac{\mathrm{d}z_{NN}}{\mathrm{d}t}$ are available for computation of Eqs. (3), (4), and (8). This differentiation process is accomplished through automatic differentiation, a technique that efficiently computes the derivatives of a function with respect to its input variable.

Thus, Eqs. (3), (4), and (8) may be reformulated in this context as

$$0 = \frac{\mathrm{d}u_{NN}}{\mathrm{d}t} - (\eta(u_{NN} - v)) \tag{11}$$

$$0 = \frac{\mathrm{d}y_{NN}}{\mathrm{d}t} - \left(\frac{1}{c_0 + c_1}\left(\alpha z_{NN} + k_0(x - y_{NN}) + c_0\frac{\mathrm{d}x}{\mathrm{d}t}\right)\right) \tag{12}$$

$$0 = \frac{\mathrm{d}z_{NN}}{\mathrm{d}t} - \left(-\gamma\left|\frac{\mathrm{d}x}{\mathrm{d}t} + \frac{\mathrm{d}y_{NN}}{\mathrm{d}t}\right| z_{NN}|z_{NN}|^{n-1} - \beta\left(\frac{\mathrm{d}x}{\mathrm{d}t} + \frac{\mathrm{d}y_{NN}}{\mathrm{d}t}\right)|z_{NN}|^n + A\left(\frac{\mathrm{d}x}{\mathrm{d}t} - \frac{\mathrm{d}y_{NN}}{\mathrm{d}t}\right)\right) \tag{13}$$

Thus, for a set of parameters to be identified:

$$\lambda = \left[\alpha_a, \alpha_b, \beta, \gamma, \eta, A, c_{0a}, c_{0b}, c_{1a}, c_{1b}, k_0, k_1, n\right], \tag{14}$$

the above equations outlined are in alignment with the original formulation by Raissi, whereby the network is trained to minimize discrepancies between predicted time derivatives of solution space, with the calculated value of time derivatives utilizing governing differential equations:

$$0 = \frac{\delta N}{\delta t} - \mathcal{N}[N; \lambda] \tag{15}$$

From the above, the physics-based mean squared error (MSE) loss functions are defined, and subsequently minimized by the network. The loss function is defined for each sample point i of total samples taken N for the sequence parsed:

$$\mathcal{L}_u = \frac{1}{N}\sum_{i=1}^{N}\left[\left(\frac{\mathrm{d}u_{NN}^{(i)}}{\mathrm{d}t}\right) - \left(\eta\left(u_{NN}^{(i)} - v^{(i)}\right)\right)\right]^2 \tag{16}$$

$$\mathcal{L}_y = \frac{1}{N}\sum_{i=1}^{N}\left[\left(\frac{\mathrm{d}y_{NN}^{(i)}}{\mathrm{d}t}\right) - \left(\frac{1}{c_0^{(i)} + c_1^{(i)}}\left(\alpha^{(i)} z_{NN}^{(i)} + k_0\left(x^{(i)} - y_{NN}^{(i)}\right) + c_0^{(i)}\frac{\mathrm{d}x^{(i)}}{\mathrm{d}t}\right)\right)\right]^2 \tag{17}$$

$$\mathcal{L}_z = \frac{1}{N}\sum_{i=1}^{N}\left[\frac{\mathrm{d}z_{NN}}{\mathrm{d}t} + \gamma\left|\frac{\mathrm{d}x}{\mathrm{d}t} - \frac{\mathrm{d}y_{NN}}{\mathrm{d}t}\right| z_{NN}|z_{NN}|^{n-1} + \beta\left(\frac{\mathrm{d}x}{\mathrm{d}t} - \frac{\mathrm{d}y_{NN}}{\mathrm{d}t}\right)|z_{NN}|^n - A\left(\frac{\mathrm{d}x}{\mathrm{d}t}\frac{\mathrm{d}y_{NN}}{\mathrm{d}t}\right)\right]^2 \tag{18}$$

As direct observations of y and z are difficult, the data-driven section of the network will be reformulated such that the object of comparison is the force instead. From Eqs. (1) and (2), force may be represented as a function of variables from both the prediction and input space. The data-driven loss may then be derived as the MSE between observed force at each time point $F^{(i)}$, and the function of predicted and input variables. The new physics-informed data-driven loss is given as follows:

$$\mathcal{L}_{\text{data},1} = \frac{1}{N}\sum_{i=1}^{N}\Big[\Big(F^{(i)}\Big) - \Big(\alpha^{(i)} z_{NN}^{(i)} + c_0^{(i)}\Big(\frac{\mathrm{d}x^{(i)}}{\mathrm{d}t} - \frac{\mathrm{d}y_{NN}^{(i)}}{\mathrm{d}t}\Big) + k_0\Big(x^{(i)} - y_{NN}^{(i)}\Big) + k_1\Big(x^{(i)} - x_0\Big)\Big)\Big]^2 \quad (19)$$

$$\mathcal{L}_{\text{data},2} = \frac{1}{N}\sum_{i=1}^{N}\Big[\Big(F^{(i)}\Big) - \Big(c_1^{(i)} \dot{y}_{NN}^{(i)} + k\big(x^{(i)} - x_0\big)\Big)\Big]^2 \quad (20)$$

The total loss by which the network is trained is thus a composite loss comprising the aforementioned component losses. The loss is given as

$$\mathcal{L} = \phi_{\text{data}}\mathcal{L}_{\text{data},1} + \phi_{\text{data}}\mathcal{L}_{\text{data},2} + \phi_{\text{phys}}\mathcal{L}_u + \phi_{\text{phys}}\mathcal{L}_y + \phi_{\text{phys}}\mathcal{L}_z \quad (21)$$

The learning process incorporates weights, denoted as $\phi = [\phi_{\text{data}}, \phi_{\text{phys}}]$, to achieve a balanced optimization between adherence to physically derived differential equations and alignment with observed measurements, specifically pertaining to force. The training process for the PINN involves the minimization of the overall loss function, which encompasses both the physical loss and data loss components. To ensure the model's robustness, a training and validation process was conducted with an 80–20 data split for training and validation data. Subsequently, the model was evaluated on a novel dataset that had not been seen during training. A Bayesian optimization strategy was employed to tune the hyperparameters of the PINN developed. Bayesian optimization was chosen over its counterparts such as grid and random search for its pragmatic utility in validation and hyperparameter optimization. This approach systematically explores the parameter space by leveraging probabilistic models, efficiently balancing the trade-off between exploration and exploitation. This approach allows for the navigation of high-dimensional parameter space efficiently, facilitating the convergence of our neural network model.

5 Results and Discussion

In this section, we present and discuss the results obtained from employing the PINN, as discussed in Sect. 4 for the estimation of key parameters within a modified Bouc-Wen model for MR dampers. Employing a Python environment with PyTorch, we constructed and tested the neural network framework to ascertain key parameters. Table 1 presents the parameters determined through the aforementioned processes.

Force–time, force–velocity, and force–displacement curves were plotted to compare the estimated values obtained from the neural network against the measured data from physical experimental setups. Notably, the curves generated by parameters identified by the neural network were found to be in agreement with the observed data. A sample of the predicted results, with varying voltage applied to the MR

Table 1 List of parameters and their corresponding values, as determined by the PINN parameter estimation algorithm outlined in Sect. 4

Parameter	Value	Units
α_a	1.92114100e+03	$\frac{\mathrm{N}}{\mathrm{m}}$
α_b	5.88251000e+03	$\frac{\mathrm{N}}{\mathrm{m}}\,\mathrm{V}$
β	3.63320700e+04	m^{-2}
γ	3.63320700e+04	m^{-2}
η	6.00043108e+01	s^{-1}
A	1.55320000e+02	–
c_{0a}	1.65073317e+05	$\frac{\mathrm{N\,s}}{\mathrm{m}}$
c_{0b}	− 3.33399584e+05	$\frac{\mathrm{N\,s}}{\mathrm{m}}\,\mathrm{V}$
c_{1a}	7.73465300e+01	$\frac{\mathrm{N\,s}}{\mathrm{m}}$
c_{1b}	2.40504070e+04	$\frac{\mathrm{N\,s}}{\mathrm{m}}\,\mathrm{V}$
k_0	2.60786039e+04	$\frac{\mathrm{N}}{\mathrm{m}}$
k_1	1.72270067e+02	$\frac{\mathrm{N}}{\mathrm{m}}$
n	1.99999659e+00	–
x_0	0	m

damper, are illustrated in Figs. 3, 4 and 5 for force–time, force–displacement, and force–velocity plots, respectively.

Upon examining the outcomes, it is apparent that the PINN has shown promising performance in estimating the parameters of the modified Bouc-Wen model. Figures 3, 4 and 5 illustrating the results reveal a significant correspondence between

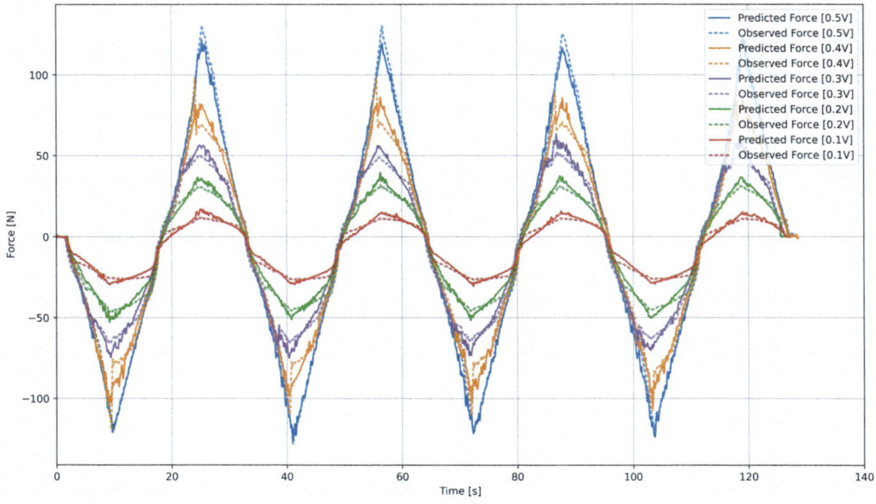

Fig. 3 Sample of predicted and observed force over time, with varying voltages applied to the MR damper

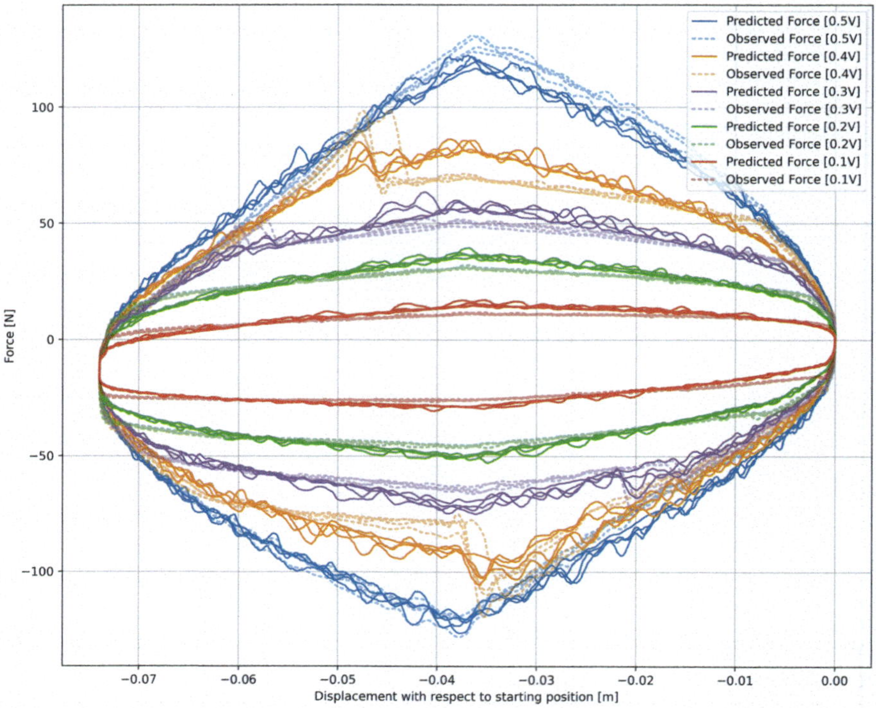

Fig. 4 Sample of predicted and observed force plotted with damper displacement, with varying voltages applied to the MR damper

the PINN predictions and experimental data, indicating the effectiveness of the physics-informed data-driven approach. However, it is important to note the presence of inherent noise in the experimental data used to train the PINN, which contributes to some level of noise in the obtained results.

The predictions obtained from the PINN exhibit some level of noise, which may be attributed to several factors. One potential reason for the noise in the predictions could be the composite loss function utilized during the training process. Since the PINN aims to simultaneously satisfy the governing equations of the physical system and match the observed data, tuning the weights of the loss function becomes crucial. However, achieving an optimal balance between these objectives is inherently challenging. The composite loss function's weighting scheme may inadvertently prioritize one aspect over the other, leading to discrepancies between the predicted and observed outputs. Moreover, the inherent complexity of the modified Bouc-Wen model for MR dampers introduces nonlinearities and uncertainties that further contribute to the noise in the predictions. Despite these challenges, efforts to fine-tune the network architecture and training parameters could potentially mitigate the noise and improve the accuracy of the predictions. Further investigation into the network's sensitivity to different loss weighting schemes and regularization techniques may

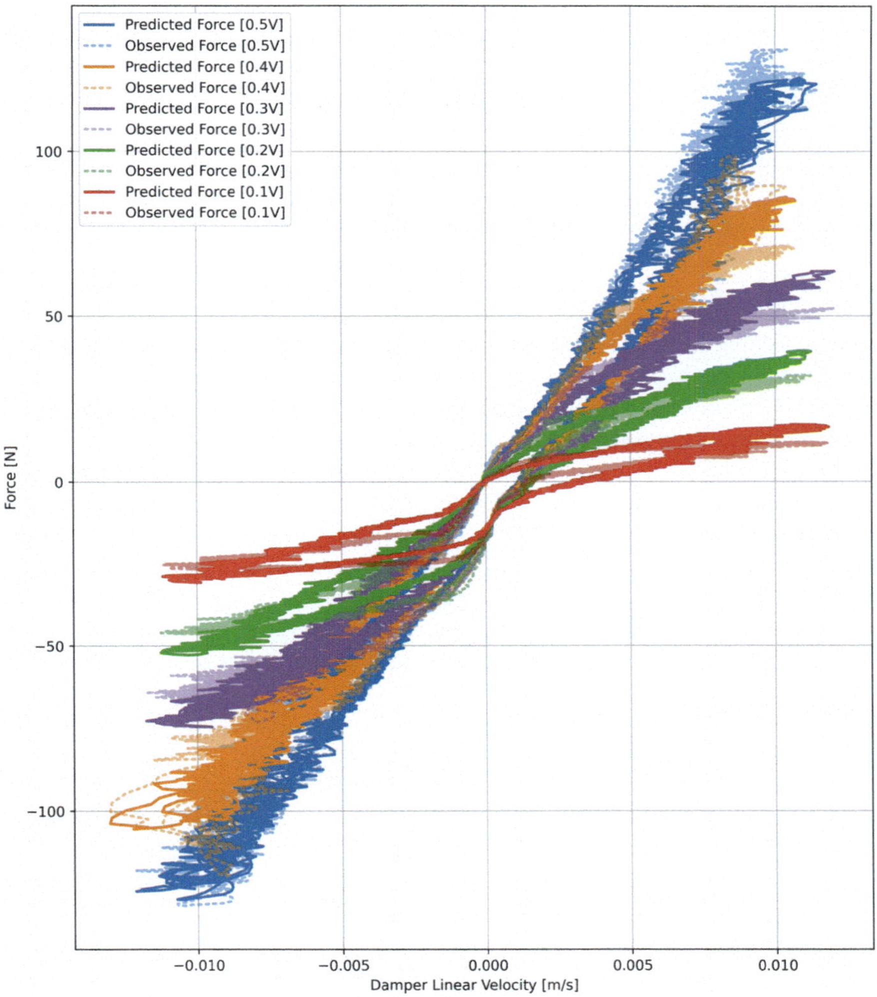

Fig. 5 Sample of predicted and observed force over time, with varying voltages applied to the MR damper

also offer insights into enhancing the predictive capabilities of the PINN in complex physical systems.

The observed discrepancy in the accuracy of predictions at different voltage inputs, particularly the notable deviation in capturing aspects of hysteresis at low voltage inputs, can be attributed to several factors inherent to both the physics of the system and the limitations of the machine learning approach employed. Root mean Squared Error (RMSE) of force predictions for a set sampling period of 100 s may be seen in Table 2. At higher voltage inputs, where the predictions demonstrate greater accuracy, the increased voltage likely induces a more pronounced response within the MR

Table 2 Root mean squared error obtained at various voltages

Applied voltage (V)	RMSE of force prediction
0	18.7936
0.1	17.7621
0.2	13.8159
0.3	11.8298
0.4	16.8806
0.5	10.0988

damper, leading to clearer and more distinguishable patterns in the force–displacement, force–velocity, and force–time curves. This heightened response facilitates the learning process of the PINN, resulting in more accurate parameter estimations. Conversely, at lower voltage inputs, the response of the MR damper is relatively weaker, potentially leading to smaller signal-to-noise ratios and increased susceptibility to measurement inaccuracies. One potential explanation for these discrepancies could be attributed to the effect of the accumulator within the MR damper. Additionally, considering the low velocities and extended stroke lengths examined in our study, it is important to acknowledge that the force exerted by the diaphragm and compressed nitrogen gas becomes non-negligible, potentially influencing the observed discrepancies in predictions, as explored in other publications on the subject [18, 19].

These observations underscore the importance of considering various factors, such as voltage inputs and the physical characteristics of the MR damper components when interpreting and refining the performance of machine learning models for parameter estimation in complex systems. Further investigation into these nuances is warranted to enhance the accuracy and robustness of future predictions. Additionally, it is essential to consider the inherent limitations of the machine learning approach itself. Despite its capabilities in learning complex relationships from data, the PINN relies on the quality and representativeness of the training dataset. If the dataset does not encompass a diverse range of operating conditions, including scenarios with low voltage inputs and intricate hysteresis behaviours, the network may struggle to generalize effectively to such conditions during inference. As a result, the discrepancies observed in the predictions at low voltage inputs may reflect the inherent challenges in training machine learning models to accurately capture the full range of dynamics exhibited by MR dampers, particularly under conditions of low excitation.

6 Conclusion

This study explored the application of PINNs for parameter identification in MR dampers, crucial components in engineering applications like vibration control and structural dynamics. Through the integration of physical principles into the neural network architecture, PINNs enabled the incorporation of governing equations during the hybrid physics-informed data-driven training process, enhancing the accuracy of parameter identification. The study assessed the efficacy of PINNs in capturing the complex nonlinear behaviour exhibited by MR dampers. The results illustrated the ability of PINNs to discern and infer key physical parameters from experimental data, providing insights into the underlying physics governing MR damper behaviour. Notably, the PINN framework demonstrated promising performance in estimating parameters within the modified Bouc-Wen model, as evidenced by the alignment between PINN predictions and experimental observations. In addition, results reveal that PINNs exhibit promising performance in discerning and inferring key material parameters from experimental data, despite encountering challenges such as the representation of hysteresis at low voltage inputs. Notably, there was greater accuracy in predictions at higher voltage inputs, indicating the network's proficiency in capturing pronounced system responses. However, the limitations in capturing subtle nuances of hysteresis underscore the need for further refinement in both network architecture and dataset representation.

References

1. Azar BF, Rahbari NM, Talatahari S (2011) Seismic mitigation of tall buildings using magnetorheological dampers. Asian J Civ Eng (Build Hous) 12(5):637–649
2. Baydin AG, Pearlmutter BA, Radul AA, Siskind JM (2018) Automatic differentiation in machine learning: a survey. J Mar Learn Res 18:1–43
3. Case D, Taheri B, Richer E (2011) Design and characterization of a small-scale magnetorheological damper for tremor suppression. IEEE/ASME Trans Mechatron 18(1):96–103
4. Chen Y, Lu L, Karniadakis GE, Dal Negro L (2020) Physics-informed neural networks for inverse problems in nano-optics and metamaterials. Opt Express 28(8):11618–11633
5. Chiu PH, Wong JC, Ooi C, Dao MH, Ong YS (2022) CAN-PINN: a fast physics-informed neural network based on coupled-automatic–numerical differentiation method. Comput Methods Appl Mech Eng 395:114909
6. Cruze D, Hemalatha G, Jebadurai SVS, Sarala L, Tensing D, Christy SJE (2018) A review on the magnetorheological fluid, damper and its applications for seismic mitigation. Civ Eng J 4(12):3058–3074
7. Dréo J, Pétrowski A, Siarry P, Taillard E (2006) Metaheuristics for hard optimization: methods and case studies. Springer Science & Business Media
8. Dyke S, Spencer B Jr, Sain M, Carlson J (1998) An experimental study of MR dampers for seismic protection. Smart Mater Struct 7(5):693
9. El Majdoub K, Giri F, Chaoui FZ (2020) Adaptive backstepping control design for semi-active suspension of half-vehicle with magnetorheological damper. IEEE/CAA J Automat Sin 8(3):582–596
10. Goncalves FD, Koo JH, Ahmadian M (2006) A review of the state of the art in magnetorheological fluid technologies–part i: MR fluid and MR fluid models. Shock Vibr Dig 38(3):203–220
11. Hornik K, Stinchcombe M, White H (1990) Universal approximation of an unknown mapping and its derivatives using multilayer feedforward networks. Neural Netw 3(5):551–560
12. Ikhouane F, Rodellar J (2007) Systems with hysteresis: analysis, identification and control using the Bouc-Wen model. Wiley

13. Jagtap AD, Mao Z, Adams N, Karniadakis GE (2022) Physics-informed neural networks for inverse problems in supersonic flows. J Comput Phys 466:111402
14. Jiang M, Rui X, Zhu W, Yang F, Zhu H, Jiang R (2021) Parameter sensitivity analysis and optimum model of the magnetorheological damper's Bouc–Wen model. J Vib Control 27(19–20):2291–2302
15. Jiang R, Rui X, Wei M, Yang F, Zhu H, Gu L (2023) A phenomenological model of magnetorheological damper considering fluid deficiency. J Sound Vibr 117851
16. Jung HJ, Spencer BF Jr, Lee IW (2003) Control of seismically excited cable-stayed bridge employing magnetorheological fluid dampers. J Struct Eng 129(7):873–883
17. Lagaris IE, Likas A, Fotiadis DI (1998) Artificial neural networks for solving ordinary and partial differential equations. IEEE Trans Neural Netw 9(5):987–1000
18. Lee AS (2021) Sliding innovation filtering: theory and applications. Ph.D. thesis. University of Guelph
19. Lee AS, Wu Y, Gadsden SA, AlShabi M (2023) Interacting multiple model estimators for fault detection in a magnetorheological damper. Sensors 24(1):251
20. Li H, Schwab J, Antholzer S, Haltmeier M (2020) Nett: solving inverse problems with deep neural networks. Inverse Prob 36(6):065005
21. Li Z, Zheng H, Kovachki N, Jin D, Chen H, Liu B, Azizzadenesheli K, Anandkumar A (2021) Physics-informed neural operator for learning partial differential equations. arXiv preprint arXiv:2111.03794
22. Lu L, Jin P, Pang G, Zhang Z, Karniadakis GE (2021) Learning nonlinear operators via DeepONet based on the universal approximation theorem of operators. Nature Mach Intell 3(3):218–229
23. Or SW, Duan Y, Ni Y, Chen Z, Lam KH (2008) Development of magnetorheological dampers with embedded piezoelectric force sensors for structural vibration control. J Intell Mater Syst Struct 19(11):1327–1338
24. Pagnier L, Chertkov M (2021) Physics-informed graphical neural network for parameter & state estimations in power systems. arXiv preprint arXiv:2102.06349
25. Raissi M, Perdikaris P, Karniadakis GE (2019) Physics-informed neural networks: a deep learning framework for solving forward and inverse problems involving nonlinear partial differential equations. J Comput Phys 378:686–707
26. Raissi M, Perdikaris P, Karniadakis GE (2017) Physics informed deep learning (part i): data-driven solutions of nonlinear partial differential equations. arXiv preprint arXiv:1711.10561
27. Spencer B Jr, Dyke S, Sain M, Carlson J (1997) Phenomenological model for magnetorheological dampers. J Eng Mech 123(3):230–238
28. Talatahari S, Kaveh A, Mohajer Rahbari N (2012) Parameter identification of Bouc-Wen model for MR fluid dampers using adaptive charged system search optimization. J Mech Sci Technol 26:2523–2534
29. Tanyu DN, Ning J, Freudenberg T, Heilenkötter N, Rademacher A, Iben U, Maass P (2023) Deep learning methods for partial differential equations and related parameter identification problems. Inverse Prob 39(10):103001
30. Tarantola A (2005) Inverse problem theory and methods for model parameter estimation. SIAM
31. Tartakovsky AM, Marrero CO, Perdikaris P, Tartakovsky GD, Barajas-Solano D (2020) Physics-informed deep neural networks for learning parameters and constitutive relationships in subsurface flow problems. Water Resourc Res 56(5):e2019WR026731
32. Tsang H, Su R, Chandler A (2006) Simplified inverse dynamics models for MR fluid dampers. Eng Struct 28(3):327–341
33. Xiaomin X, Qing S, Ling Z, Bin Z (2009) Parameter estimation and its sensitivity analysis of the MR damper hysteresis model using a modified genetic algorithm. J Intell Mater Syst Struct 20(17):2089–2100
34. Xu J, Xu, L, Li, Y, Liu, J, Li, X, Lu, L (2018) Design of lower extremity rehabilitation robots with magnetorheological dampers and wire-driven system. In: 2018 IEEE international conference on information and automation (ICIA). IEEE, pp 395–400

35. Yamin, AHM, Ab Talib MH, Zaurah I, Darus M, Nor NSM, Magneto-rheological (MR) damper–parametric modelling and experimental validation for lord RD 8040-1
36. Yang G, Spencer B Jr, Carlson J, Sain M (2002) Large-scale MR fluid dampers: modeling and dynamic performance considerations. Eng Struct 24(3):309–323
37. Yang J, Ning D, Sun S, Zheng J, Lu H, Nakano M, Zhang S, Du H, Li W (2021) A semi-active suspension using a magnetorheological damper with nonlinear negative-stiffness component. Mech Syst Signal Process 147:107071
38. Yang L, Meng X, Karniadakis GE (2021) B-PINNs: Bayesian physics-informed neural networks for forward and inverse pde problems with noisy data. J Comput Phys 425:109913
39. Yao G, Yap F, Chen G, Li W, Yeo S (2002) MR damper and its application for semi-active control of vehicle suspension system. Mechatronics 12(7):963–973
40. Yun D, Park C, Park D, Kim HS (2022) Magnetorheological damper for vibration reduction in a robot arm. Intell Serv Robot 15(5):671–678
41. Zambare H, Khoje A, Patil S, Razban A (2021) MR damper modeling performance comparison including hysteresis and damper optimization. IEEE Access 9:24560–24569
42. Zapateiro M, Karimi HR, Luo N, Phillips BM, Spencer BF Jr (2009) Semiactive backstepping control for vibration reduction in a structure with magnetorheological damper subject to seismic motions. J Intell Mater Syst Struct 20(17):2037–2053
43. Zeinali M, Mazlan SA, Abd Fatah AY, Zamzuri H (2013) A phenomenological dynamic model of a magnetorheological damper using a neuro-fuzzy system. Smart Mater Struct 22(12):125013
44. Zemp R, de la Llera JC, Almazan JL (2011) Tall building vibration control using a TM-MR damper assembly. Earthq Eng Struct Dynam 40(3):339–354
45. Zhai W, Tao D, Bao Y (2023) Parameter estimation and modeling of nonlinear dynamical systems based on Runge–Kutta physics-informed neural network. Nonlinear Dyn 111(22):21117–21130
46. Zhang S, Shi W, Chen Z (2021) Modeling and parameter identification of MR damper considering excitation characteristics and current. Shock Vib 2021:1–17

Understanding Energy Consumption in Real-World E-IoT Systems with CoAP, RPL, and 6LoWPAN

Jiawei Lu, Kirk Martinez, and Alex Weddell

Abstract This paper investigates the energy consumption of Environmental Internet of Things (E-IoT) systems using CoAP, RPL, and 6LoWPAN. We focus on Sleepy Networks, a method to save energy in battery-powered nodes where nodes spend most of their time sleeping and have regular communications windows. Our experiments used the Microchip SAM R30 Xplained Pro platform and measured the current and time duration of system activities. This included reading sensors, storing data, and the radio activities controlled by the network protocols. After measuring the power consumption of the various states in detail, we explored the effect of network maintenance on scheduling and energy consumption. The results indicate that reducing sleep energy consumption is fundamentally important. While considering the user requirements and the volume of produced data, the results show that optimising the communication window can reduce energy consumption further. Therefore, we suggest to further reduce the energy consumption of the system, and it is necessary to take a balance between network reliability and energy consumption.

Keywords E-IoT · Energy consumption · RPL · CoAP · 6LoWPAN

1 Introduction

Wireless sensor networks (WSN) can be used to monitor remote/hazardous environments such as unpopulated or geohazard-prone areas where the resources are limited, and system maintenance is challenging. Martinez et al. [1] proposed the notion of Environmental Sensor Networks (ESNs) which are WSNs specifically

J. Lu (✉) · K. Martinez · A. Weddell
University of Southampton, University Road, Southampton, UK
e-mail: Jiawei.Lu@soton.ac.uk

K. Martinez
e-mail: km@ecs.soton.ac.uk

A. Weddell
e-mail: asw@ecs.soton.ac.uk

P. G. Bradford et al. (eds.), *Proceedings of IEMTRONICS 2024*, Lecture Notes in Electrical Engineering 1229, https://doi.org/10.1007/978-981-97-4780-1_14

designed to carry out continuous sensing of the natural environment. The development of low-power protocols standardised for sensor networks and IP-based standards [2] can enable full Internet connectivity and enhance the interoperability of ESNs, enabling them to become more open and heterogeneous networks. The notion of Environmental IoT (E-IoT) [3] emerges to allow bidirectional access between users and sensor nodes which emphasises device connectivity across vendors and integrates with the Internet. The system design should consider the collaboration of different protocols and technologies to withstand various real-world environmental challenges. Power consumption is one of the most critical elements that will decide the system's lifetime and reliability [1]. Thus, Low-Power protocols and efficient power management of the system are critical when developing remote E-IoT systems.

From the application layer to the physical layer, various low-power protocols can be applied to the low-power network stack, such as the Constrained Application Protocol (CoAP), Low-power Wireless Personal Area Networks based on IPv6 (6LoWPAN), and the IPv6 Routing Protocol for Low-Power and Lossy-Networks (RPL). We concentrate on an IP-based E-IoT system so non-IP-based protocols such as Long Range Wide Area Network (LoraWAN) will not be considered here. 6LoWPAN was designed for the requirements of low-power networks and brings uniformity to IoT networks. CoAP [4] is an application protocol designed for these resource-limited systems [4]. CoAP has low complexity as it uses UDP, not connection-based TCP protocol. While UDP lacks built-in mechanisms for error recovery or retransmission of lost packets CoAP provides these features. To enhance packet reliability, a CoAP transaction can optionally wait for an acknowledgement (ACK) from the receiver and re-transmit the packet after a timeout. The inclusion of a 2-byte Message ID in the CoAP header allows the sender to identify ACKs to avoid duplicate transmissions. RPL can effectively extend the range of the network through packet forwarding. Previous studies have focused on improving the efficiency, applicability, and operation of 6LoWPAN, RPL, and CoAP by simulation [5] and have concluded that limited power is the main challenge of WSNs based on these protocols [6]. We are interested here in the real-world energy efficiency of E-IoT systems using these low-power protocols. Improving system energy efficiency to extend system lifetime is a common challenge in designing a real-world E-IoT system. It is critical for remote and hazardous environment monitoring systems as it fundamentally decides the feasibility and maintenance issues of deployment. IEEE 802.15.4 MAC is the de facto underlying MAC specification for devices to run in low-power mode to save power [7]. The mechanism of IEEE 802.15.4 MAC that has been defined to avoid collisions is Carrier Sense Multiple Access with Collision Avoidance (CSMA-CA). MAC protocols such as ContikiMAC [8] which build on top of IEEE 802.15.4 are designed to optimise the power consumption further. ContikiMAC uses a Radio Duty Cycle (RDC) and CSMA [8]. It applies a duty cycle mechanism (8 Hz by default), waking up nodes for transmission after two successive Clear Channel Assessments (CCAs) are identified [8]. The Mountain Sensing [9] project used ContikiMAC as it saves power while making the network appear to be always alive. The Sleepy network implemented in this study schedules the nodes to intermittently communicate and run in a low-power sleep state after completing planned tasks. Nodes are designed to wake-up periodically to execute required tasks such as sensing and data transmission on a regular schedule. We focus on a sleepy network rather than a MAC-based solution, as it has the potential to reduce unnecessary transitions and potentially optimise overall energy consumption.

In previous research, we explored the practical implementation of energy-efficient E-IoT systems in some real-world scenarios. The Mountain Sensing project [9] implemented an E-IoT system using ContikiMAC which proved that 868 MHz 6LoWPAN with Contiki can support the ESNs more efficiently compared with 2.4 GHz-based ESNs. It also showed that the multi-hop 6LoWPAN network could cover 3.5 km over a mountainous environment. A system based on a low-cost system-on-chip (ATSAMR30) was tested in a forest environment, which provides an in-forest range of typically 150-200 m depending on the environment [10]. This SoC has an 868 MHz low-power radio (13mW) and runs the RIOT operation system (RIOT-OS) [11].

In this experiment, our objective is to analyse the energy consumption of a sleepy E-IoT system using CoAP, RPL, and 6LoWPAN based on RIOT-OS. This aims to explore the impact of different design choices on the system's lifetime. The system features can be split into three aspects: (1) user-configurable system, (2) energy-efficient sleep schedule, and (3) adaptive energy-efficient RPL configuration. This E-IoT system should be able to allow the user to configure and change the features of the system, such as communication and sensing frequency, depending on their requirements. Meanwhile, the system is adaptive to dynamic changes which means that the system should be able to intelligently modify the configurations such as schedule and routing parameters, to adapt to the configuration change. When designing such an efficient sleepy network, the fundamental task is to ensure the system is energy-efficient while providing the users with data when they require it. With efficient collaboration between system software and hardware, all the components should be able to reliably work with the sleepy network schedule. The IoT network stack requires a certain level of system activity for maintenance, which could potentially clash with our use of minimal communications windows. The analysis of the system's current consumption for different activities during sleep, sensing, and communication states will be introduced in Sect. 3.

This paper is an experimental investigation into the energy consumption of the sleepy E-IoT network. By measuring and analysing current consumption under various scenarios, we not only contribute experimental data to the field but also reveal potential areas for optimisation. This research is particularly relevant in the context of advancing the understanding of low-power IoT device behaviour, offering valuable benchmarks for future studies. Moreover, our approach to system design, including the utilisation of low-power protocols and sleep scheduling, demonstrates practical experience in improving energy efficiencies.

2 System Architecture and Experiment Setup

2.1 Hardware and Software Configuration

In this experiment, Microchip SAM R30 Xplained Pro (SAMR30 Xpro) [12] was selected as this hardware platform. It has an ultra-low power ATSAMR30G18A SoC which supports sub-GHz (868 MHz band) radio (AT86RF212B integrated transceiver). This MCU only has 256 kB flash, which is used for code, so an I/O1 Xplained Pro SD card extension board was used to provide sensor data storage/buffering. A DS18B20 Digital Temperature Sensor was used with its default 12-bit resolution supported by RIOT-OS (the conversion time is constrained by the driver leading to typically 750 ms sensing time). The embedded crystal of SAMR30 Xpro provides the functional RTC with an accuracy of ± 20 ppm. In our experiment, the accuracy of RTC can affect the synchronisation and schedule timekeeping. An accurate RTC, DS3231, was used to ensure the accuracy of synchronisation of wake-up and scheduling (±2 ppm from 0 to + 40 °C, and (± 3.5 ppm from − 40 to + 85 °C). By applying the DS3231, assuming that the working temperature is between 0 and + 40 °C, the maximum time drift per week is about 1.2 s. Compared with using the SAMR30 Xpro's built-in oscillator (about 10% as accurate as the DS3231from 0 to + 40 °C), using DS3231 can enhance the synchronisation reliability, as the DS3231 provides enhanced accuracy across a wide temperature. Wireshark and a RIOT-OS Sniffer [13] were used to capture the wireless packet exchanges.

2.2 Protocol Stack and Topology

6LoWPAN, RPL, and CoAP are used in the system. These low-power protocols allow devices to form a star/tree topology, reduce energy consumption, and ensure robust communications. The RPL standard [14] extends the range of the network and supports scalability, making it well-suited for constrained environment deployments. CoAP provides a simple RESTful request/response interaction model and supports features to make it possible for devices to reply to data requests as well as respond to configuration changes. This IoT network protocol stack is shown above in Fig. 1.

RIOT-OS default network stack, Generic (GNRC) network stack, fully sup- ports IPv6 with 6LoWPAN and implements RPL [15]. The Destination-Oriented Directed Acyclic Graph (DODAG) is the fundamental structure of RPL; the parent node and child node have different Rank values within a single DODAG. RPL allows the node to communicate with surrounding nodes by exchanging ICMPv6 control messages: DODAG Advertisement Object (DAO), DODAG Information Objective (DIO), and DODAG Information Solicitation (DIS) are used to select a root and join a DODAG [16]. Four RPL instances are (a) Single DODAG with one root; (b) Multiple DODAG; (c) Signal DODAG that uses a virtual root on the backbone network; and (d) a combination of (a), (b), and (c) dependent on the scenario [17]. Figure 2 shows

Fig. 1 Network protocol stack

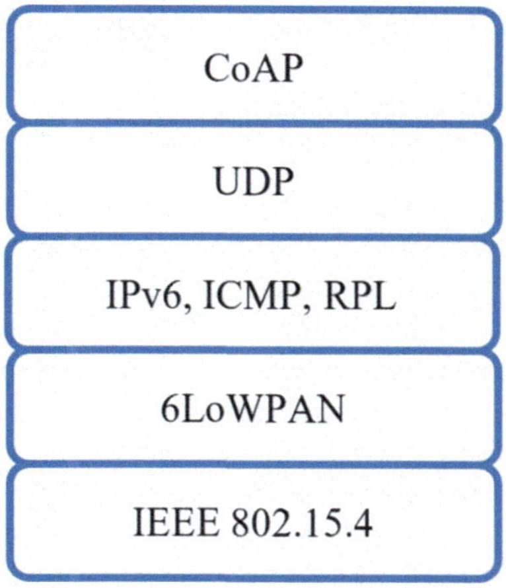

an example of an RPL star topology DODAG. Our experiment currently only deploys nodes in a star topology with one border router (BR) as the parent.

The exchange of the RPL control message is periodic and controlled by the RPL "Trickle Timer" which has four parameters:

(1) $I_{\min}$: the minimal interval of DIO ($I_{\min} = 3$ in this experiment which equals 8 ms).
(2) $I_{\max}$: the maximum time interval between DIOs ($I_{\min} = 20$ in this experiment, approximately 2.33 h).
(3) K: the constant determines doubling the interval between transmissions ($K = 10$ in this experiment).
(4) DIO timer: Interval for DIO sending is 0, $I_{\min}, \ldots, I_{\max}$.

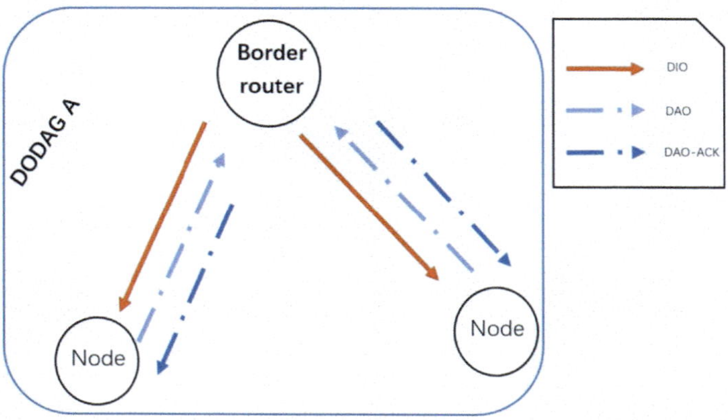

Fig. 2 RPL node star topology-DIO and DAO exchange and DODAG

2.3 Experiment Setup

In our experiment shown in Fig. 3, GNRC network stack, 6LoWPAN, RPL, and CoAP are all provided by RIOT-OS. RTC3231, Atmel I/O1 Xplained Pro, and a DS18 sensor were connected to the SAM R30 Xplained Pro board. Radio packets were captured by a RIOT-OS sniffer to help correlate activity timings with current consumption changes, simultaneously measured using a Nordic Semiconductor Power Profiler Kit II and Agilent 34411A multimeters. The initial phase involved investigating the current consumption of the fundamental networking activities controlled by GNRC network stack. Subsequently, the experiment enabled RPL to explore the characteristics and current consumption of the various RPL control messages.

Two scenarios were considered: (1) a predetermined sensing rate with one daily communication window, and (2) a predetermined sensing rate with regular communication windows distributed daily. The Border Router (BR) was powered by an always-on Linux PC. Thus, the measurement focused on the energy consumption analysis of the individual nodes. In this experiment, the Linux PC sent CoAP GET requests to the nodes to retrieve their data. These are routed through the BR which converts IPv6 packets to and from 6loWPAN. Several activities were measured in terms of the current and time duration. Table 1 provides an overview of the expected activities for the sensor node.

Energy consumption was measured across three states: Sleep, Sensing, and Communication. Specifically, during the sensing and data storage state, the node takes readings from the DS18B20 and stores the data on the SD card. During the communication state, the focus of network activities is ICMPv6 control messages (Router/

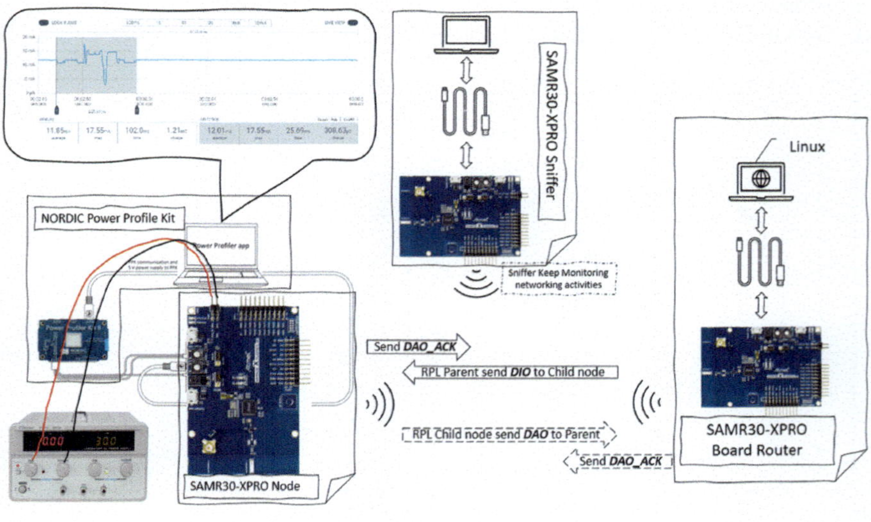

Fig. 3 Example of system current measurement experiment

Table 1 Node's activities overview

State	Activity	Description
Sensing	Set system to sensing state	Node wake-up from sleep mode
	Read sensor and store data	DS18 temperature data read and stored on the IO1 Xplained Pro SD card
	Sensing state	Node stays in IDLE mode and radio is off
	Set system to sleep state	Node enters sleep mode
Communication	Turn on the radio	Set transceiver to IDLE mode
	6LoWPAN activity and router advertisement	Node sends 6LoWPAN and router advertisement packets to BR
	Neighbour solicitation	Node sends neighbour solicitation to determine neighbours
	Neighbour advertisement	In response of receiving neighbour solicitation
	DIO	DIO control message activities of the node
	DAO-Tx and DAO-ACK-Rx	Node sends DAO message to BR and receives DAO-ACK from BR
	CoAP response communication	Receive CoAP request and respond node stays in IDLE mode with radio on
	Set system to sleep state	Node turns radio off and enters sleep state

Neighbour Solicitation/Advertisement), 6LoWPAN packets, DIO, and DAO/DAO ACK managed by the GNRC network and RPL routing protocol.

The aim is to identify the current consumption, time duration, and hence energy consumption of these network activities.

3 Results Analysis and Discussion

3.1 Sleepy Network SoC Activity Current Measurement

Network activities for maintaining GNRC network and RPL network are managed by the border router. The periodicity of these activities which construct GNRC and RPL network was observed every 60 s. Figure 4 is a sleepy network system SoC current measurement of 60 s communication activities. It has been tested that periodic sleep and wake-up schedules can work well with the 60 s communication window, and the SoC current consumption of the sleep state is sensible low, which also proves that the fundamental system can support an available sleepy E-IoT system. To begin with, a simple sleep and wake-up schedule was used with a periodic 1-min

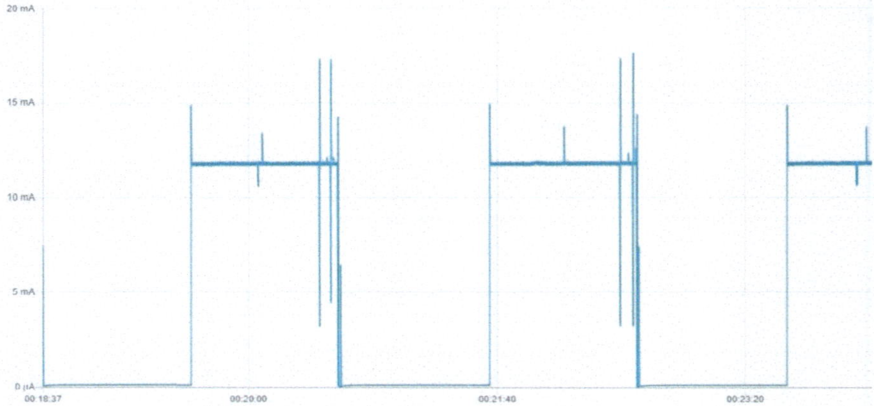

Fig. 4 SoC current measurement at 3.3 V of all the activities of the sleepy network

Table 2 Average SoC current consumption measurements at 3.3 V for different activities

Activity information	Average current
DIO	11.2 mA
DAO-Tx and DAO-ACK-Rx	12.5 mA
Neighbour solicitation	11.4 mA
Neighbour advertisement	11.4 mA
CoAP response	12.4 mA
Sleep state–standby mode	2.56 μA

communication and sleep state to capture the measurements. In summary, (1) the SoC sleep state current drops to a low level of 2.56 μA, (2) the average current of the RPL control message is over 11 mA, and (3) the average current of the board in the IDLE state reaches 11.8 mA. Activities such as Router/Neighbour Solicitation/ Acknowledgement DIO, DAO, DAO ACK, and response to CoAP requests were then identified. Table 2 below summarises the average SoC current for different activities.

Theoretically, more network maintenance and control messages can be expected with the increment of communication windows which increases the system's energy consumption. To comprehensively understand the impact of energy consumption with an increased number of communication windows, we measured the current and time duration of the sleep and wake-up activities in a practical system.

3.2 *Sleep E-IoT System Schedule Design*

A general sleep E-IoT system schedule example is shown in Fig. 5. As illustrated, devices are scheduled to switch between the sensing state and sleep state. Also, the communication window will appear at the beginning of one period, followed by a sensing window, if the communication frequency is once a day. If the required communication frequency is more than once per day, the communication window will be evenly distributed in one day, followed by a sensing state. For example, if the communication rate configured is 4 times daily, the communication window will occur at the beginning of every 6 h. Similarly, the communication window will open at the beginning of every hour if the communication is configured to happen every hour, then turn off the radio to go to the sensing state subsequently. As Fig. 5 shows above, the system will start to sleep after all the activities are finished. This design can avoid unnecessary energy consumption by switching the device on from the sleep state. System configurations such as the sensing rate and communication rate are changeable by sending a CoAP PUT to the nodes from the Linux PC. To investigate the energy consumption, all activities' current consumption and time duration were measured and listed in Table 3. These results can be used to calculate the system's daily energy consumption. We note that the sleep state of this development board has an increased sleep current consumption compared with the findings illustrated in Table 2 due to the peripherals such as the SD card extension board.

The nodes are synchronised by sending a CoAP PUT from the Linux PC to set the RTC of each node. This initial configuration allows users to accurately set the RTC timer for all nodes. Once the RTCs are configured, all nodes seamlessly enter the synchronised schedule at the commencement of the next hour. This synchronisation strategy supported by DS3231, compared with using the board's built-in oscillator, improves the synchronisation accuracy of every node to operate within the schedule.

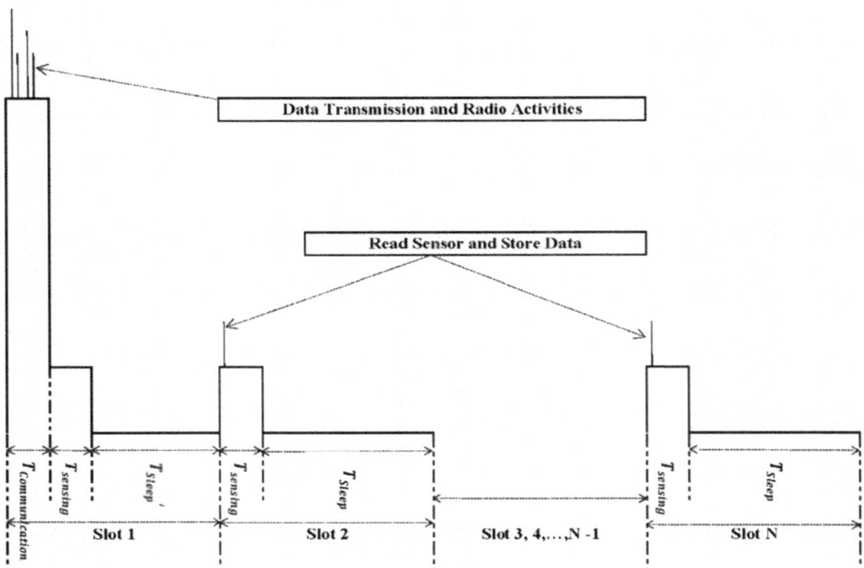

Fig. 5 Sleepy E-IoT system schedule overview

Table 3 Average current and time duration of different sleepy E-IoT system activities

Activity information	Time duration (s)	Average current (mA)
6LoWPAN and router advertisement	0.188	14.915
Neighbour solicitation	0.144	13.848
Neighbour advertisement	0.142	13.788
DIO	0.093	12.524
DAO-Tx and DAO-ACK-Rx	0.1395	14.702
CoAP response	0.13	14.832
Set system to sensing state	0.049	1.476
Set system to sleep state	0.048	1.475
Turn on the radio	0.048	7.679
Set system to communication state	0.095	6.523
Read sensor and store data	0.798	2.963
Communication state	NA	12.85
Sensing state	NA	2.85
Sleep state–standby mode	NA	0.132

This synchronisation method ensures that the entire system functions operate cohesively and reduces potential risks in timing and activity. Currently, all the nodes are synchronised to wake-up sequentially, where they take turns to communicate. Thus, in this experiment, the energy consumption caused by overhearing is minimised in the laboratory environment. In future work, synchronisation process optimisations will be explored to further facilitate the initial setup, fundamentally supporting the overall efficiency of the system.

3.3 *Overall System Energy Consumption*

The time for the BR to periodically maintain the network was observed to be 60 s including all the GNRC and RPL activities that cyclically construct the network. To begin with, the communication window defaults to 60 s to allow all the network communication setups. Figure 6 shows the daily energy stack of different schedule examples. This result indicates that daily communication energy consumption has increased as the number of communication windows increases. Significantly, the dominant factor in energy use is the sleep state so any deployable system should minimise this. As a result, in the case of one 60 s communication window per day with a sensing rate of every 10 min, the daily energy consumption is 38.6 J and the estimated system lifetime is 712 days (estimated using 2600 mAh 18,650 Lithium-Ion rechargeable battery). Obviously, in a star topology, that 60 s is longer than the necessary time overhead as the communication window can be shut after all the data is transmitted.

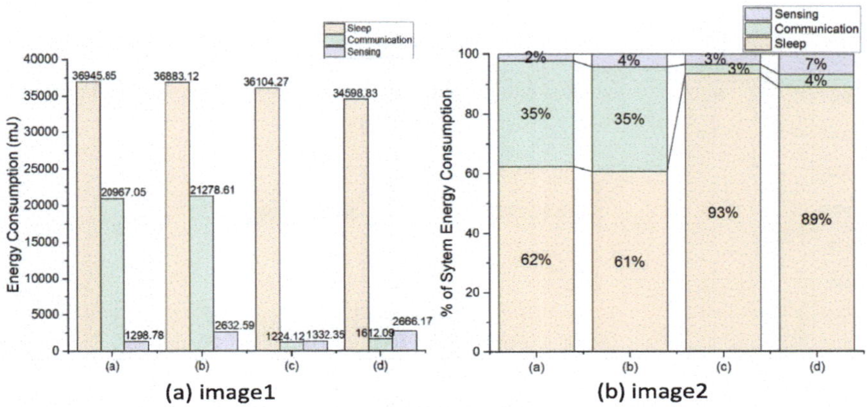

Fig. 6 Daily energy consumption of different states and energy stack (**a** and **b** communicate every hour, **c** and **d** communicate once a day). For both panels, from left to right, they are: **a** sensing every 10 min, 60 s communication every hour; **b** sensing every 5 min, 60 s communication every hour; **c** sensing every 10 min, 60 s communication every day; and **d** sensing every 5 min, 60 s communication every day

A potential solution to further optimise the system energy consumption is to reduce the unnecessary communication window length. The following assumption is made to focus on evaluating the energy consumption related to data transmission through CoAP. After the initial network construction phase, subsequent communication windows can be considered to exclude network maintenance activities, such as ICMPv6, RPL, and 6LoWPAN control messages. Each sensing event will create 18 Bytes of data, and each CoAP packet can maximumly contain 97 Bytes of payload [18]. Thus, 5 groups of readings can be transmitted with one CoAP packet. Assuming N sensor readings will be taken before each communication window, the minimum time required for one communication window can be summarised in Function 1. T_{IDLE} includes the time cost to wake-up the node and turn the radio on. T_{CoAP} is the time response cost for a single CoAP request. $T_{\text{Radio ST ANDBY}}$ is the time cost for turning off radio after completing data transmission. Adding up the minimum communication activity and sensing activity (sensing every 5/10 min and communication every hour) implies a 2 s communications window could safely be used.

$$T_{\text{communication}} = T_{\text{IDLE}} + \text{ROUNDUP}\left(N * \frac{1}{5}\right) * T_{\text{CoAP}} + T_{\text{Radio\_STANDBY}} \quad (1)$$

We can make similar assumptions for different communication frequencies. Referring to Fig. 5, if the communication is scheduled once per day, the value of ROUNDUP ($T_{\text{communication}} + T_{\text{sensing}}$) is equal to 5 s (sensing every 10 min) and 9 s (sensing every 5 min). Estimated results illustrated in Fig. 7 show that total energy consumption increased by 6.4% compared to (a) and (c), and a nearly 9.9% increase

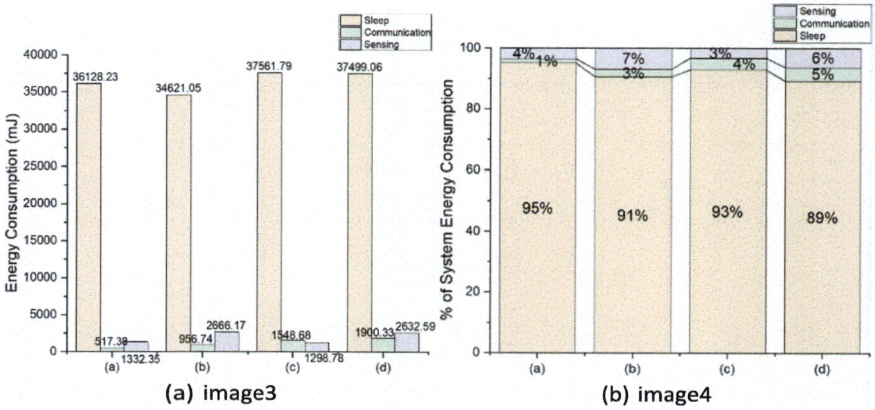

Fig. 7 Sleep network daily energy stack with optimised prediction. For both panels, from left to right, they are: **a** sensing every 10 min, communication every day and **b** sensing every 5 min, communication every day; **c** sensing every 10 min, communication every hour; and **d** sensing every 5 min, communication every hour

compared with (b) and (d). Theoretically, optimising the system energy consumption by minimising the time overhead of network maintenance such as RPL control message is potentially valuable. In future work, it is potentially worth reducing the communication time overhead by focusing on minimising unnecessary RPL control messages while ensuring basic RPL reconstruction requirements.

4 Conclusion

Our investigation into the star topology sleepy E-IoT system has provided our understanding of the relationship between communication frequency, energy consumption, and network maintenance. This Sleepy E-IoT system design focuses on using CoAP, RPL, and 6LoWPAN. Also, the system can enter the sleep state and wake-up state as scheduled to finish the planned tasks with reliable communication features. Furthermore, CoAP is applied to ensure the ability of the user to update and reconfigure the system presets. To comprehensively understand the impact of energy consumption with an increased number of communication windows, we have measured the current and time duration of system activities such as read sensor, store data, and radio activities controlled by the standard IoT protocols. The observed increases in daily communication energy consumption with additional windows emphasise the importance of a balanced approach that considers network reliability and communication time. The estimation result shows that optimising the communication time overhead can reduce energy consumption further. According to Figs. 6 and 7, when the system's sensing frequency remains the same, optimising the communication window results in a reduction of system energy consumption by more than 1.6%

(communication once a day) and 30.86% (communication every hour). We suggest optimising communication windows and network activities to control the communication energy consumption, and this can further guide similar systems to take a balance between network reliability and energy consumption. In the upcoming stages, our attention will be directed towards optimising energy consumption. This will be achieved through optimising the system in routing configuration and communication patterns, aiming to further reduce overall system power requirements.

References

1. Martinez K, Hart JK, Ong R (2004) Environmental sensor networks. Computer 37:50–56
2. Nikoukar A, Raza S, Poole A, Güneş M, Dezfouli B (2018) Low-power wireless for the Internet of things: standards and applications. IEEE Access 6:67893–67926
3. Hart JK, Martinez K (2015) Toward an environmental Internet of things. Earth Space Sci 2:194–200
4. Shelby Z, Hartke K, Bormann C, Frank B (2014) RFC 7252: the constrained application protocol (CoAP). Internet Engineering Task Force
5. Vasudha M, Giri A (2023) Analysing the performance of 6LoWPAN-CoAP and RPL-CoAP on LoRaWAN in constrained environment. In: 2023 7th international conference on computation system and information technology for sustainable solutions (CSITSS), pp 1–6
6. Ashrif FF, Sundararajan EA, Ahmad R, Hasan MK, Yadegaridehkordi E (2024) Survey on the authentication and key agreement of 6LoWPAN: Open issues and future direction. J Netw Comput Appl 221:103759
7. Khanafer M, Guennoun M, Mouftah HT (2014) A survey of beacon-enabled IEEE 15.4 MAC protocols in wireless sensor networks. IEEE Commun Surv Tutor 16:856–876
8. Dunkels A (2011) The ContikiMAC radio duty cycling protocol. Swedish Institute of Computer Science
9. Bragg GM, Martinez K, Basford PJ, Hart JK (2016) 868MHz 6LoWPAN with ContikiMAC for an Internet of things environmental sensor network. In: 2016 SAI computing conference (SAI), pp 1273–1277
10. Martinez K, Lu J, Kendall S, Hart JK (2020) Testing low power IP-based sensor networks an a forest environment. In: AGU fall meeting abstracts, pp IN025-003
11. Baccelli E, Gündoğan C, Hahm O, Kietzmann P, Lenders MS, Petersen H, Schleiser K, Schmidt TC, Wählisch M (2018) RIOT: an open source operating system for low-end embedded devices in the IoT. IEEE Internet Things J 5:4428–4440
12. Microchip Technology (2017) SAM R30 IEEE 802.15.4 sub-GHz system-in-package datasheet
13. RIOT: RIOT sniffer application. https://github.com/RIOTOS/applications/blob/master/sniffer/tools/README.md
14. Brandt A, Hui J, Kelsey R, Levis P, Pister K, Struik R, Vasseur J, Alexander R (2012) RPL: IPv6 routing protocol for low-power and lossy networks. In: RFC 6550
15. Lenders M (2016) Analysis and comparison of embedded network stacks. Master's thesis. Freie Universitat Berlin
16. Chen Y, Chanet JP, Hou KM (2012) RPL routing protocol a case study: precision agriculture. In: First China-France workshop on future computing technology (CF-WoFUCT 2012), p 6
17. Winter T, Thubert P, Brandt A, Hui JW, Kelsey R, Levis P, Pister K, Struik R, Vasseur J-P, Alexander RK (2012) RFC 6550: RPL: IPv6 routing protocol for low-power and Lossy networks. In: Internet requests for comments, pp 1–157
18. Jeong Y, Son S, Lee S, Lee B (2018) A total crop-diagnosis platform based on deep learning models in a natural nutrient environment. Appl Sci 8:1992

Analysis of Step Up or Boost Converter Design for Charging Station of Electric Vehicles

Aung Kyaw Htay, Hla Myo Tun, Lei Lei Yin Win, Ei Ei Khin, and Devasis Pradhan

Abstract The paper presents the analysis of step up or boost converter design for charging station for electric vehicles. The charging station for electric vehicles is important for generating stabilized outputs such as voltage and current. The research challenge in this study is to confirm the stabilize converter design for charging station of electric vehicles. The mathematical modelling for boost converter design is also evaluated. The specific objective of this study is to model the optimized design for boost converter used in charging station for electric vehicles. The SIMULINK model for boost converter design is implemented. The achievement or contribution of this study is to estimate the desired output voltage and current level based on the designed boost converter for this purpose. The simulation results from the implemented SIMULINK model were confirmed with the recent works. The main compared parameters are optimized level and amount of boosting in this study. The current study meets the high level in optimization level and two times boosting than other recent works.

Keywords Step up converter · Charging station · Electric vehicles · Mathematical modelling · Net zero energy

A. K. Htay · L. L. Y. Win · E. E. Khin
Department of Electronic Engineering, Yangon Technological University, Yangon, Myanmar

H. M. Tun (✉)
Research Department, Yangon Technological University, Yangon, Myanmar
e-mail: hmt.ytu@gmail.com

D. Pradhan
Department of ECE, Acharya Institute of Technology, Bengaluru, India

P. G. Bradford et al. (eds.), *Proceedings of IEMTRONICS 2024*, Lecture Notes in Electrical Engineering 1229, https://doi.org/10.1007/978-981-97-4780-1_15

1 Introduction

Any altered types of vehicles moving around us cause air pollution, global warming, and depletion of the earth's resources. In order to solve these challenges, a key is to substitute conventional vehicles with Electric vehicles, Hybrid vehicles, and fuel-cell electric vehicles. Electric and hybrid vehicles use two sorts of energy storage devices. The first one and only is MES (Multi-function Energy Storage), used to store high energy, despite the fact that the second individual is a rechargeable energy storage system (RESS) which affords high power capability and reversibility. MES affords a high driving range and RESS has a fine regenerative braking and acceleration system. Energy storage device deviations output voltage based on load and the high voltage of the DC link creates complications for vehicle engineers during the formation of energy storage mechanisms with traction drive. DC-to-DC converters can be utilized for interfacing components in electrical powertrains. Power electronic converters in automobile must be light, small, reliable, effective, and generate low electromagnetic interference [1–3]. The DC-to-DC converter is an electromechanical device or power electronic circuitry used to convert a fixed or variable DC voltage from one level to another based on electronic circuit requirements. According to the power electronics converter group, the DC-to-DC converter could be operated for small amounts of voltage applications such as batteries, or high amounts of voltage applications like High Voltage power transmission systems. When there were no semiconductors, a general scheme for converting DC voltage into higher voltage for low-power utilization was to modify it into AC voltage by means of vibrator circuits. After that, a step up transformer could be used to increase the output voltage level, applied by a rectifier circuit for accomplishing the DC conversion process. A combination of both motor and generator was utilized for several high-power applications. The motor could operate the generator, providing the necessary load voltage. These schemes were not cheap and less amount of efficient but utilized since no alternative approach occurred at that time [4, 5]. The innovation of power semiconductor devices and integrated circuits led to different very cost-effective solutions, such as DC power supply, where high-frequency AC is input to a transformer. That device is very cost-effective and easy to control. It could vary the voltage that again converts into DC through high-quality rectifiers.

2 Mathematical Model of Boost Converter Design

The DC-to-DC converter is an electromechanical device or power electronic circuitry used to convert a fixed or variable DC voltage from one level to another based on electronic circuit requirements. According to the power electronics converter group, the DC-to-DC converter could be operated for small amount of voltage applications such as batteries, or high amount of voltage applications like High Voltage power transmission system. The use of renewable energy sources has attracted significant

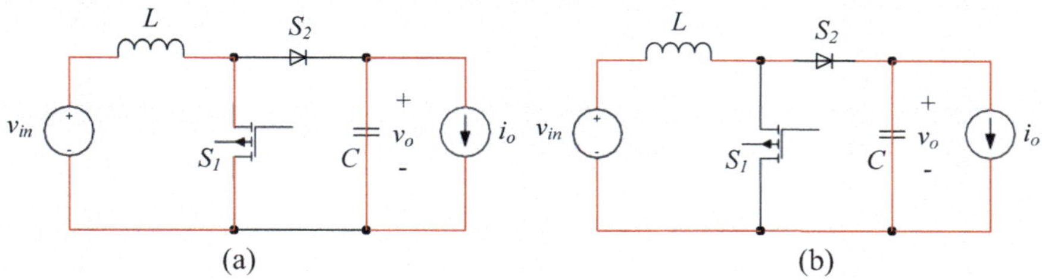

Fig. 1 DC-to-DC boost converter design. **a** S1 ON, **b** S2 ON

interest due to their abundant availability and clean nature, as conventional generating stations are putting pressure on existing fossil fuel reserves. DC-to-DC boost converters are mostly used in solar photovoltaic systems. The output of PV is not fixed, it varies with temperature and irradiance value. When using a conventional boost converter, we get a high ripple in voltage. Interleaved Boost Converter is one type of power electronics converter with the help of which we get less ripple in output voltage, power, and input current as compared to the conventional boost converter (Fig. 1).

The DC–DC converter is the one type of power converter that converts from one level of voltage to another level of voltage. The DC–DC converter is categorized as isolates and non-isolates converter topologies. The isolated topologies are generally used SMPS. Buck converters, boost converters, and Buck-Boost converters are examples of non-isolated topologies. The output voltage is stepped down using the Buck converter. The boost converter gives the input voltage to the output by boosting it, so that the output voltage is always greater than the input voltage. Mostly DC–DC boost converters are used in solar photovoltaic systems so the overall system is more efficient [6–8].

For On-time equations,

$$v_{L1} = V_{in} \tag{1}$$

$$i_{C1} = -I_o \tag{2}$$

$$V_o = V_C \tag{3}$$

For Off-time equations (Fig. 2),

$$v_{L2} = V_{in} - V_o \tag{4}$$

$$i_{C2} = I_L - I_o \tag{5}$$

$$V_o = V_c \tag{6}$$

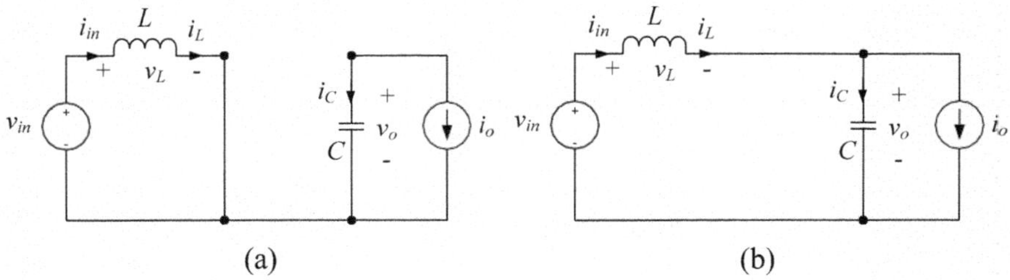

Fig. 2 Operating condition of DC-to-DC boost converter design. **a** S1 ON, **b** S2 ON

For averaging,

$$D \cdot v_{L_1} + D' \cdot v_{L_2} = 0 \quad \text{and} \quad D \cdot i_{C_1} + D' \cdot i_{C_2} = 0 \tag{7}$$

$$\begin{aligned} & D \cdot V_{in} + D' \cdot (V_{in} - V_o) = 0 \\ & \rightarrow 1)\left(D + D'\right) V_{in} - D' V_o = 0 \\ & D \cdot (-I_o) + D' \cdot (I_L - I_o) = 0 \\ & \rightarrow 2)\, D' I_L - \left(D + D'\right) I_o = 0 \end{aligned} \tag{8}$$

$$\therefore V_o = \frac{V_{in}}{D'} \quad \text{and} \quad I_L = \frac{I_o}{D'} \tag{9}$$

Peak-to-peak inductor current ripple component:

$$\hat{\imath}_{L-pp} = \frac{V_{in}}{L} \cdot DT_s = \frac{V_{in} - V_o}{L} D' T_s \tag{10}$$

Input voltage, $v_{in} = 5\,\text{V}$.
Output voltage, $v_o = 10\,\text{V}$.
From Boost Converter Equation, $v_o = \frac{v_{in}}{1-D}$.
Duty cycle
$D = 0.8\ (or)\ 80\%\ (on\ time)$.
$D = 0.2\ (or)\ 20\%\ (off\ time)$.
Output current $I_0 = 2.8$ A (*from simulation*).
Inductor current $I_L = \frac{I_0}{D}$

$$I_L = 14\,\text{A}.$$

Inductor peak to peak ripple current,
$i_{L-pp} = 2\,I_L$ (*from inductor current waveform figure*)
$i_{L-pp} = 28$ A.
Switching time, $T_{sw} = \frac{1}{f_{sw}}$, and (Fig. 3).
Switching frequency,

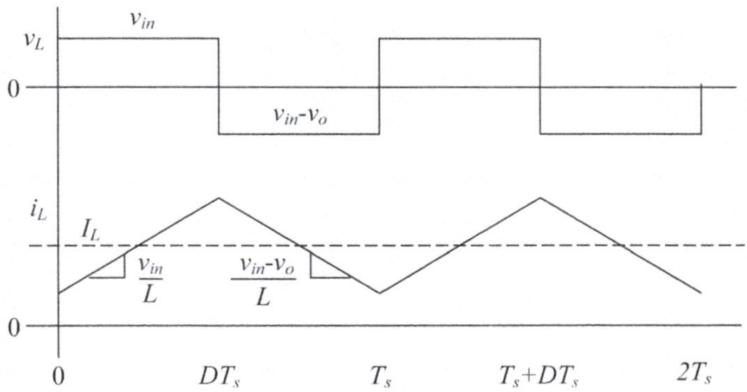

Fig. 3 Output waveforms for voltage and current

$$f_{sw} = 40\,\text{kHz}.$$

From Inductor peak to peak ripple current equation,

$$i_{L-pp} = \frac{v_{in}}{L} \cdot D \cdot T_{sw}.$$

Inductor value, $L = 15\,\mu\text{H}$..

3 Results and Discussions

The SIMULINK model for step up or boost converter design for charging station of electric vehicles. Figure 4 shows the SIMULINK model for designed boost converter de- sign with mathematical modelling [9, 10].

In this model, the main component is the IGBT switch for controlling the bypass capacitor for that converter. The input DC voltage is 5 V for reference value of that

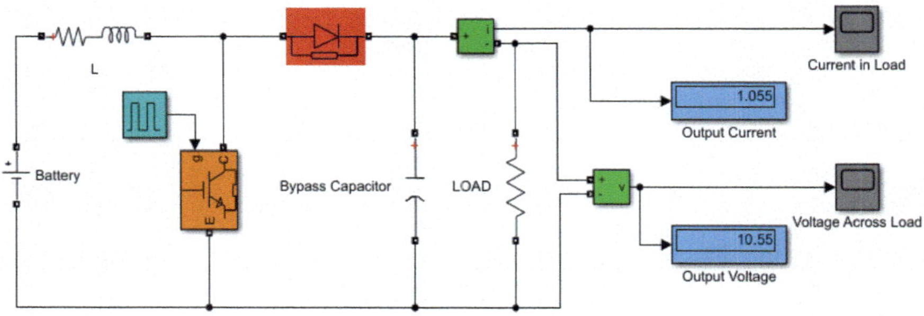

Fig. 4 SIMULINK model for boost converter

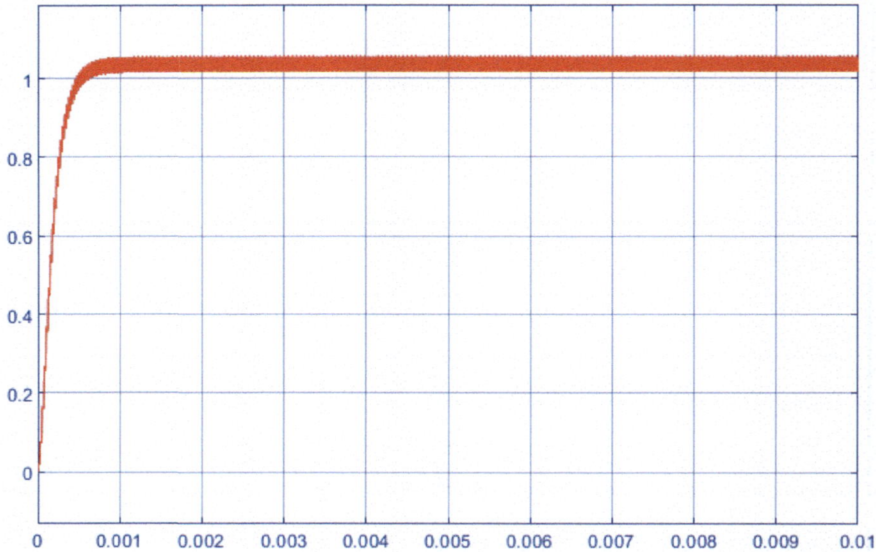

Fig. 5 Current output from the SIMULINK model

converter. According to the mathematical design, the resistance value is 0.34 Ω and the inductance value is 15 μH. The bypass capacitor value is 50 μF and the load resistance is 10 Ω.

The current output from the SIMULINK model for that boost model is shown in Fig. 5. The normalized value of that output in that SIMULINK model is 1 for optimized value for practical applications.

The voltage output from the SIMULINK model for that boost model is shown in Fig. 6. The normalized value of that output in that SIMULINK model is 10 for optimized value for practical applications. It means the boost amount of that converter is two times for the input battery voltage.

The converter is designed for the prototype structure in this study. The real-world applications could be done based on the experimental results with actual values of the discrete components in circuit construction. The robustness of the design consideration on that boost converter could be observed by the performance evaluation of the circuit with theoretical implementation.

4 Statistic Table

The statistic table for checking the performance of the developed boost converter in this study was given in Table 1. The outcomes from the analysis on recent works are compared with the analysis in this study based on the two outputs. The first output is the optimization level and the second one is the amount of boost. According to the

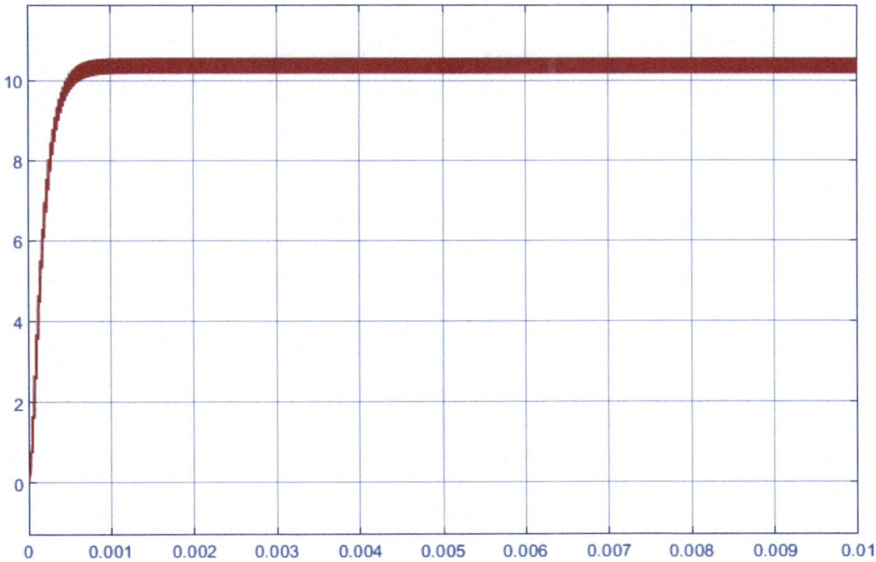

Fig. 6 Voltage output from the SIMULINK model

Table 1 Statistic table for performance analysis

S. No.	Converter type	Optimization level	Amount of boosting
Reference [1]	IGBT-based boost converter	Low	0.48 times
Reference [2]	IGBT-based boost converter	Low	0.98 times
This work	IGBT-based boost converter	High	2 times

analysis in this study, the two parameters for performance checking is more suitable for designing and constructing the real converter design for practical usages.

5 Conclusion

The boost converter design was implemented with IGBT switch for optimal design for charging station of electric vehicles. The appropriate model for applicable boost converter design could be modelled based on the high-performance discrete components for performance achievement. The mathematical model for the theoretical analysis of that boost converter design was conducted by using the fundamental concepts of the power electronics theory. The robustness of the developed design was also observed in this study. The experimental studies could be carried out based on the theoretical analysis in this study.

Acknowledgements The authors thank their colleagues from the Department of Electronic Engineering of Yangon Technological University for their support to complete this work.

References

1. Su ASM, Ghani RA, Slamet S (2014) Modelling and simulation of boost converter with maximum power point tracking (MPPT) for photovoltaic application. J Teknol 71:5:1–4. eISSN 2180-3722. www.jurnalteknologi.utm.my
2. Anthon A, Zhang Z, Andersen MAE (2015) Comparison of a state of the art Si IGBT and next generation fast switching devices in a 4 kW boost converter. In: 2015 IEEE energy conversion congress and exposition (ECCE), Montreal, QC, Canada, pp 3003–3011. https://doi.org/10.1109/ECCE.2015.7310080
3. Saafan AA, Khadkikar V, Edpuganti A, Moursi MSE, Zeineldin HH (2024) A novel nonisolated four-port converter for flexible DC microgrid operation. IEEE Trans Ind Electron 71(2):1653–1664. https://doi.org/10.1109/TIE.2023.3257360
4. Kunstbergs N, Hinz H, Schofield N, Roll D (2022) Efficiency improvement of a cascaded buck and boost converter for fuel cell hybrid vehicles with overlapping input and output voltages. Inventions 7:74. https://doi.org/10.3390/inventions7030074
5. Dimitrov B, Hayatleh K, Barker S, Collier G, Sharkh S, Cruden A (2020) A buck-boost transformerless DC–DC converter based on IGBT modules for fast charge of electric vehicles. Electronics 9:397. https://doi.org/10.3390/electronics9030397
6. Kumar M, Bharti R, Rao KDVSK (2019) Conventional and hybrid perturb & observe based maximum power point tracking for solar system. In: 2019 international conference on vision towards emerging trends in communication and networking (ViTECoN)
7. Hsieh Y-C, Hsueh T-C, Yen H-C (2009) An interleaved boost converter with zero-voltage transition. IEEE Trans Power Electron 24(4):973–978
8. Walker GR, Sernia PC (2004) Cascaded DC–DC converter connection of photovoltaic modules. IEEE Trans Power Electron 19(4):1130–1139
9. Phyo ESZ, Lwin KS, Tun HM (2014) Microcontroller based solar smart charge controller using MPPT, vol 03(06), pp 0997–1000. ISSN 2319-8885
10. Tun HM, Aung W (2014) Analysis of control system for A 24 V PM brushed DC motor fitted with an encoder by supplying H-bridge converter. Bahria Univ J Inf Commun Technol 7(1):54–67

Design of a 4 DoF Active Upper Limb Exoskeleton to Rehabilitate Osteoarthritis Injuries in Elderly

Deyby Huamanchahua, Raúl Escandón-Tufino, Sebastian Loayza-Bautista, Ariana Ponce-Bohórquez, Andrea Flores-Leyva, Renato Díaz-Sánchez, and Hector Valcarcel-Castillo

Abstract Exoskeletons aimed at assisted rehabilitation and enhancing motor skills in patients with muscular or joint injuries have seen increasing development as a branch of robotics research. This article presents a conceptual design of a 4 DoF upper limb exoskeleton to assist elderly patients in their rehabilitation exercises. Employing KWL and CRAAP strategies, alongside conducting interviews for insights into the perspectives of patients and physiotherapists and leveraging an information organizing tool known as the Lotus Diagram, this study precisely delineated the challenge, formulated clear objectives, and pinpointed essential components crucial for the development of the exoskeleton's design. A cardboard design was created on a test subject, serving as the basis for the CAD design adapted to their accurate arm. Additionally, a flow diagram was developed to explain the desired operation of the device, acknowledging the inclusion of specific sensors and actuators. Finally, possible electronic components for this subsystem, including a stepper motor and an

D. Huamanchahua (✉) · R. Escandón-Tufino · S. Loayza-Bautista · A. Ponce-Bohórquez · A. Flores-Leyva · R. Díaz-Sánchez · H. Valcarcel-Castillo
Department of Electrical and Mechatronics Engineering, Universidad de Ingenieria y Tecnologia—UTEC, Lima, Peru
e-mail: dhuamanchahua@utec.edu.pe

R. Escandón-Tufino
e-mail: raul.escandon@utec.edu.pe

S. Loayza-Bautista
e-mail: sebastian.loayza@utec.edu.pe

A. Ponce-Bohórquez
e-mail: ariana.ponce@utec.edu.pe

A. Flores-Leyva
e-mail: andrea.flores@utec.edu.pe

R. Díaz-Sánchez
e-mail: renato.diaz@utec.edu.pe

H. Valcarcel-Castillo
e-mail: hvalcarcel@utec.edu.pe

P. G. Bradford et al. (eds.), *Proceedings of IEMTRONICS 2024*, Lecture Notes in Electrical Engineering 1229, https://doi.org/10.1007/978-981-97-4780-1_16

EMG sensor, were presented, accompanied by a stress analysis of the final prototype modeled in Fusion360.

Keywords Upper limb exoskeleton · Rehabilitation · Osteoarthritis · Elderly

1 Introduction

The various branches of robotics research have grown due to the number of application fields that have appeared over the years. Biorobotics has been applied in rehabilitation and physical therapy in patients who have suffered an injury or disability and require this treatment to recover their motor skills [1–8]. These cases involve injuries such as carpal tunnel syndrome [9], a stroke [10–15], and loss of strength [16–18], among others, and require the repetitive performance of exercises on the injured limb.

To achieve this through robotics, mechanical or electromechanical exoskeletons are designed to adapt to the patient and the limb, which ergonomically requires physical therapy treatment. These devices allow the patient to recover his or her pre-injury strength and motor skills quickly. It can even increase these capacities if the injuries are chronic and the movement is required to avoid pain in joints or muscles, especially in the elderly [19–22]. Both possibilities are achieved by selecting sensors, actuators, and controllers [23–30] that allow remote device control through a graphical interface and actuation by the patient through EMG signals [31].

Considering the above, this article aims to provide a proposal for an upper limb exoskeleton for the assisted rehabilitation of an elderly patient. First, the specific problems and objectives for this prototype will be established. Then, the main components of this equipment obtained by a state-of-the-art organization will be mentioned in a Lotus diagram to show the stages followed to get the final mechanical design.

2 Methodology

In this study on upper limb rehabilitation exoskeletons integrated with virtual reality, the methodologies employed were grounded in the KWL and CRAAP methods for structuring the learning process from initial knowledge to synthesizing new understanding based on the quality of sources [32]. Additionally, interviews with elderly patients and physiotherapists were applied to gather detailed perspectives and enhance the design and practical application of the exoskeleton [31]. These methodological strategies were critical in establishing a robust research framework focused on current and relevant evidence in the field.

2.1 KWL Method

The KWL Method [33], representing Know, Want to Know, and Learned, organizes the learning process into three stages. The initial stage, K (Know), pinpoints a subject's pre-existing knowledge. The following stage, W (Want to Know), outlines the learning objectives or discovery targets. Finally, the L (Learned) stage compiles and reflects upon the insights and understanding gained after the research.

2.2 CRAAP Method

It is a crucial tool for ensuring the truthfulness and quality of sources used in research [34]. The method's structure breaks down into C (Currency), which assesses how recent and relevant the information is; R (Relevance), determining the source's suitability for the study in question; A (Authority), examining the credibility and recognition of the source; A (Accuracy), ensuring the information is correct and reliable; and finally, P (Purpose), scrutinizing the source's objective, identifying potential biases or intentions behind the data [32, 35].

2.3 Qualitative Interpretive Description Approach

This methodology employs semi-structured face-to-face interviews, focusing on elderly patients with osteoarthritis and their physiotherapists. These interviews are crucial for obtaining detailed insights about these patients' needs, challenges, and preferences. Additionally, physiotherapists contribute their professional expertise, offering valuable feedback on the exoskeleton's practicality, ergonomic design, and therapeutic potential. This approach is crucial for understanding and incorporating end-user perspectives concerning the design and implementation of the exoskeleton [31].

2.4 Lotus Diagram

A lotus diagram [36] was used because it proved an excellent tool for the simultaneous segmentation and definition of concepts, offering us a more comprehensive perspective of the landscape in question. The system is based on a central idea linked to auxiliary ideas. We have examined and selected the following elements: Degrees of Freedom (DoF), Sensors, Actuators, Materials, Controller, Type of Control, Contact Material, and Rehabilitation Movements. All of them are connected to the central idea: Upper limb exoskeleton. In this way, each auxiliary concept helped us guide

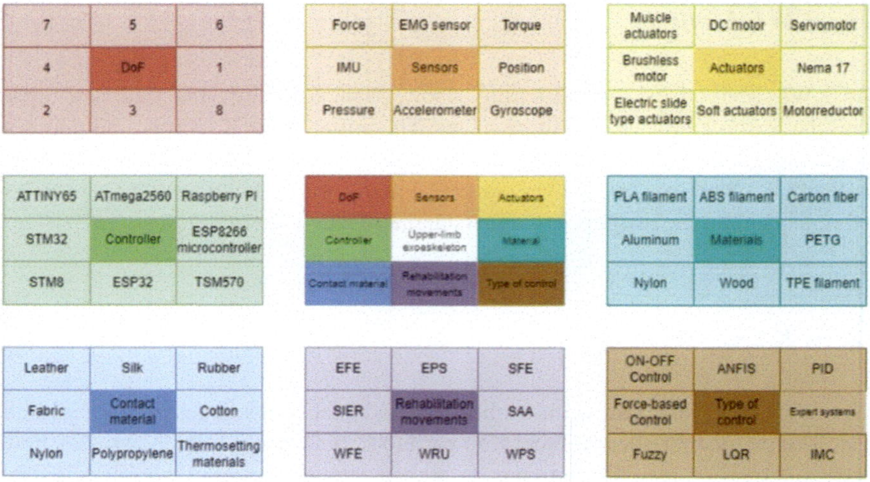

Fig. 1 Lotus diagram

the research work more effectively. This allowed us to keep each of the main components and topics to investigate in mind, enriching our guidance throughout the study (Fig. 1).

Once the most relevant design alternatives were found, we selected the appropriate ones for our case. It was chosen that the proposal would have 4 DoF of rotational type since this quantity and its location in different joints of the arm allow the development of the appropriate rehabilitation exercises. These rehabilitation movements involve flexion and extension of the elbow and shoulder, rotation of the shoulder, and abduction and adduction of the shoulder. During their performance, they will be driven by stepper motors and monitored or controlled by the user's electromyographic signals. For accessibility reasons, it was decided that the prototype would be manufactured by 3D printing using PLA filament.

3 Design

3.1 Mechanical Design

The 4 DoF upper limb exoskeleton prototype was meticulously designed using Fusion 360, a state-of-the-art CAD tool, to cater specifically to the needs of elderly patients with osteoarthritis. In developing an upper limb exoskeleton for rehabilitation purposes, several critical parameters must be addressed to ensure the device's functionality, comfort, and safety.

The CAD model of the exoskeleton was constructed to encompass the entire upper limb, offering support from the back and shoulder to the forearm. The model

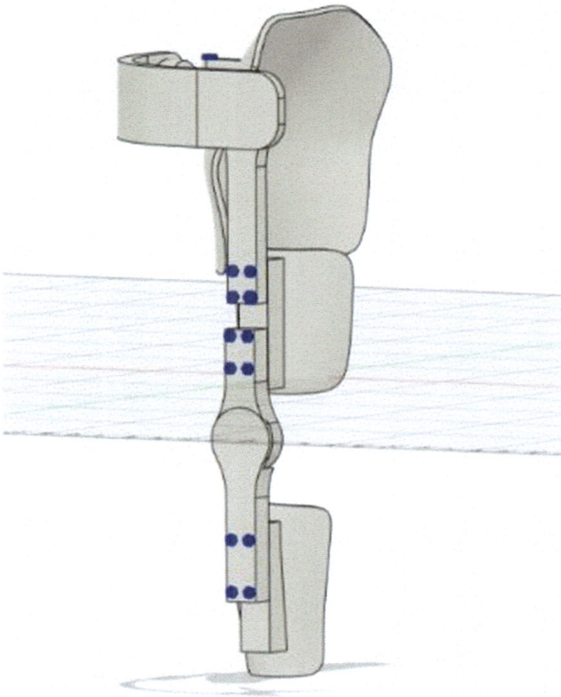

Fig. 2 Mechanical design of the exoskeleton for the upper limb

comprises four primary segments: the back cuff, shoulder cuff, upper arm, and forearm brace, each connected via articulated joints with embedded bearings for smooth motion. The shoulder cuff is designed to provide stable support without restricting the range of motion, and the upper arm and forearm supports are adaptable to different limb sizes, facilitating elbow flexion and extension (Fig. 2).

Embedded within the exoskeleton are strategically placed sensors and actuators. The EMG sensors are located at the upper arm and forearm braces to detect muscle activity, informing the control system of the user's intent. The stepper motors at the joints translate these signals into precise, controlled movements. Each exoskeleton joint is designed to imitate the corresponding human joint: the shoulder allows abduction and adduction, the elbow flexion and extension, and the wrist the combined movement of these movements. The joints are custom-designed bearings that provide low-friction movement.

3.2 *Arm Model*

The Denavit Hartenberg (DH) convention was used to model the exoskeleton. As [37] suggests, this convention will make the control and trajectory planning less

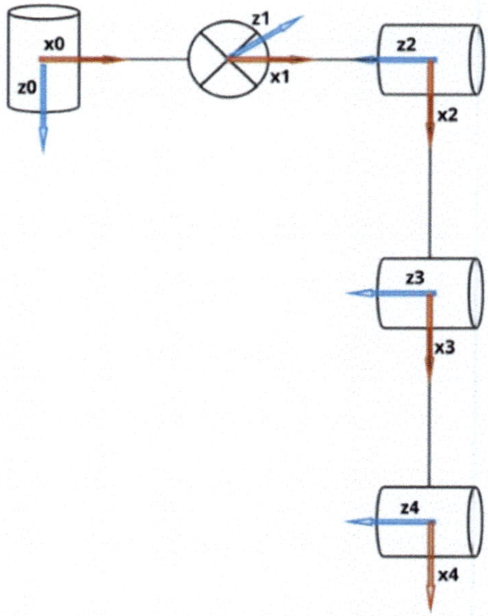

Fig. 3 Mechanical design of the exoskeleton for the upper limb

Table 1 Denavit Hartenberg parameters matrix

No.	d	θ	a	a
0	0	q_0	L_0	$\pi/2$
1	$-L_2$	q_1	0	$\pi/2$
2	$-L_1$	0	0	0
3	0	q_2	L_3	0
4	0	q_3	L_4	0

complex. Each coordinate system is established at the joints shown in Fig. 3. The coordinate system 0 (x_0, y_0, z_0) is located on the scapula, and the coordinate system 1 (x_1, y_1, z_1). The coordinate system 2 (x_2, y_2, z_2) is on the shoulder. Finally, coordinate system 3 (x_3, y_3, z_3) is located at the elbow, and number 4 (x_4, y_4, z_4) is the wrist and the end effector of the exoskeleton.

Following the convention's rules, the resulting matrix with the DH parameters is shown in Table 1.

3.3 Kinematic Model

A Python script implemented a kinematic control to ensure the exoskeleton model was correct. We defined our control loop as follows (Fig. 4).

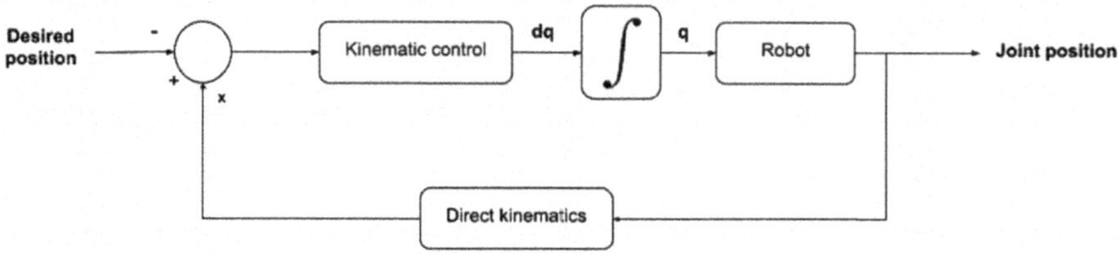

Fig. 4 Kinematic control loop

To calculate the new joint positions, we require two essential things: the 3D error and the robot's Jacobian in each movement position. The error can be calculated as the difference between the actual position of the end effector (which, in our case, will be on the patient's wrist) and the desired position.

$$e = x - x_d \tag{1}$$

Because this control needs to update the joints with their velocity, we require the derivative of the error.

$$\dot{e}^* = -K_e \tag{2}$$

In this case, Eq. (2) shows that our approach is a simple control *P*. Finally, we will need to compute the Jacobian for each joint configuration.

$$J_{\text{column}} = \frac{x_2 - x}{\delta} \tag{3}$$

where J_{column} refers to each of the columns of the 3×4 Jacobian, x_2 is the previous position plus a certain joint velocity, x is the current position, and delta (δ) is the joint velocity. The new joint configuration can be expressed as follows.

$$q_k = q_{k-1} + \delta t \dot{q}_k \tag{4}$$

4 Instrumentation

4.1 Stepper Motor

The use of the stepper motor [38] is considered in each of the rotational degrees of freedom. This component has a defined angular step, which contributes to achieving a slow but easily controllable movement.

4.2 MyoWare Sensor

The MyoWare Sensor [39] is contemplated due to its capability to detect electromyographic signals (EMG) generated by muscles. This component proves highly advantageous as it enables the capture of data during muscle contraction and relaxation through the electrical signal provided by the sensor. With this detailed information, a more precise control of the exoskeleton's movement is achieved.

4.3 ATmega 328P

The ATmega 328P [40] is crucial for realizing the exoskeleton. This microcontroller possesses many features that make it advantageous over other considered options. It was primarily considered for its Flash and SRAM memory, input and output pins, communication ports, Analog-to-Digital Converter (ADC), timers, counters, and low power consumption.

5 Results

5.1 Mechanical Analysis

The design of the upper limb exoskeleton underwent a stress analysis using finite element simulation (FEA) [41] to ensure its structural integrity and ergonomic functionality. Fixed points were assigned along the interface with the patient's back, simulating the direct connection and load transfer between the user and the device. Specific forces were applied to the supports of the upper arm and forearm segments, representing biomechanical loads during regular exoskeleton use. These loads were distributed to reflect the dynamic interactions between the exoskeleton and the movements of the patient's upper limb.

Furthermore, rolling loads were applied at three key locations: between the back and the shoulder, on the shoulder itself, and between the upper arm and forearm, as depicted in Fig. 5. These rolling loads simulate the friction and rotational forces experienced by the exoskeleton's joints and interfaces, ensuring the device can accommodate natural movements without undue restrictions or excessive stresses on the user's tissues.

A distributed force was implemented in the back support, represented by multiple arrows on the contact surface in the simulation model. This reflects the pressure exerted on the patient over a wide contact area, crucial for comfort and proper load distribution. The goal is to minimize localized pressure points and provide uniform support, reducing the risk of skin fatigue or injuries due to prolonged friction or compression. The stress analysis illustrated in Fig. 6 has revealed that the highest

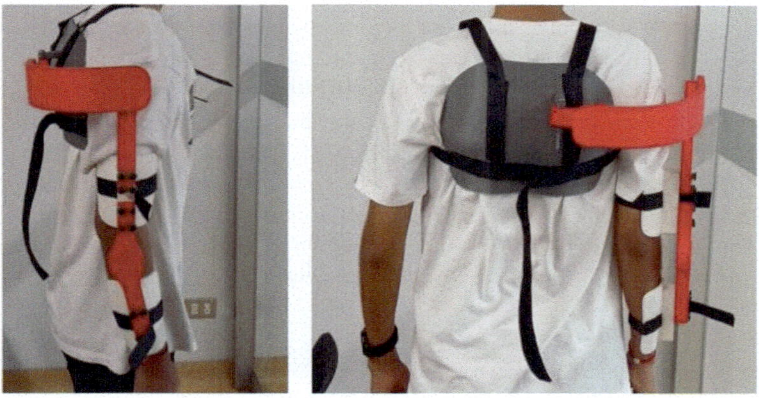

(a) Side view (b) Back view

Fig. 5 3D printed and assembled exoskeleton prototype

tension areas are primarily located at the interfaces of the joint supports, particularly in the shoulder and elbow joints.

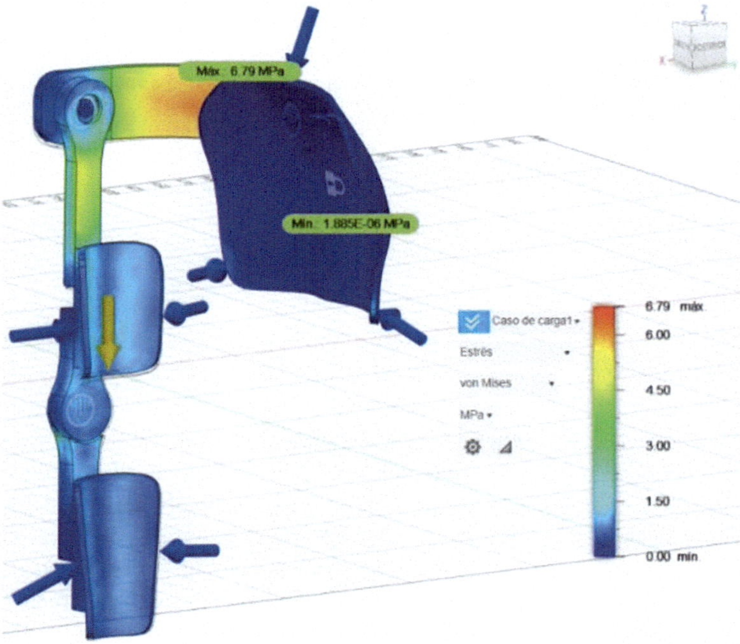

Fig. 6 Stress analysis

These joints are crucial as they bear the most weight and execute complex movements [42, 43]. This situation requires a critical design review to enhance load distribution and reinforce these pivotal areas. Moreover, the regions where the exoskeleton interfaces with the human body are also subjected to high stress levels, necessitating additional attention as they transfer the loads from the exoskeleton to the body [42]. Despite these concerns, the observed deformations remained within acceptable limits, indicating that the device can withstand the expected loads during regular operation. This suggests that, while the current design meets basic operational requirements, there is a significant opportunity for improvement to ensure better durability and safety.

5.2 *Kinematic Control*

This section presents the results obtained from the simulation of the kinematic control loop, designed based on the previous section. These results are illustrated through two figures, each representing a different aspect of the system's performance.

Figure 7a shows the evolution of the joint configuration over fifty iterations. Each line in this graph represents a different joint, with its corresponding angle expressed in radians. It is important to note that the green line does not mean an actual joint but rather a coordinated system established by convention for analysis purposes. On the other hand, Fig. 7b displays the three-dimensional trajectory of the end effector throughout the control loop iterations. Considering that these fifty iterations correspond to one second in real time, it is observed that the settling time for the positions on the X and Z axes is approximately 0.7 s. In contrast, the position on the Y axis stabilizes around 0.4 s. It is worth mentioning that the overshoot percentage (%OS) for the Y coordinate exceeded 30%. Significantly, no error was recorded at the settling point for any of the coordinates.

6 Conclusions

In conclusion, the general outline of the development for prototyping an active 4 DoF upper limb exoskeleton, integrated with virtual reality, to rehabilitate osteoarthritis injuries in elderly patients was obtained. The study effectively combined advanced methodologies, such as the KWL and CRAAP methods, with qualitative interpretive descriptions from interviews with elderly patients and physiotherapists. This approach provided a comprehensive understanding of the user requirements and practical challenges in exoskeleton design.

A first physical prototype, tailored to the user's physiology, was developed based on these insights, and a state-of-the-art review was summarized in the Lotus diagram. Stress analysis of this prototype highlighted the shoulder area as a point of high-stress concentration, offering valuable insights into the mechanical weak points of

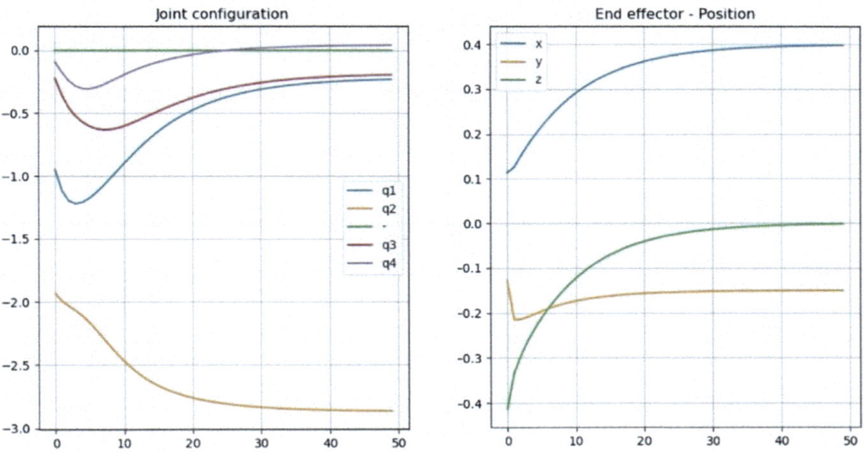

(a) Joint Configuration (b) Position vs time

Fig. 7 Simulation outcomes of the kinematic control loop

the exoskeleton. Despite this, the prototype proved ergonomic, allowing the user to perform specified movements effectively. Kinematic modeling of the exoskeleton's end effector was achieved, applying satisfactory control for movements near the origin position. This accomplishment marks a significant step toward creating a more interactive and responsive rehabilitation device.

Looking forward, plans are set to simulate and implement an electronic system in the exoskeleton to make it actively assistive while maintaining its critical features of ergonomics and portability. This development is expected to enhance the functional capabilities of the exoskeleton, making it a more effective tool in rehabilitating elderly patients with upper limb disabilities.

References

1. Rehabilitation robotics market size is thriving worldwide by 2028 with leading region and countries data (2023). Available: https://www.digitaljournal.com/pr/news/rehabilitation-robotics-marketsize-is-thriving-worldwide-by-2028-with-leading-region-aand-countries-data
2. Children's hospital in Seoul adopts rehabilitation robot for upper limbs (2023). Available at: https://www.ajudaily.com/view/20230116162728796
3. Rahayu UB, Wibowo S, Setyopranoto I, Hibatullah Romli M (2020) Effectiveness of physiotherapy interventions in brain plasticity, balance and functional ability in stroke survivors: a randomized controlled trial. Neuro Rehabil 47(4):463–470
4. Becerra A, Florez J, Palencia D, Orjuela J, Lopez E, Mendez L (2019) ExPro: exoskeleton for upper limb rehabilitation. Rev Bistuam 17(3):3–12
5. Hou Y, Kiguchi K (2018) Virtual tunnel generation algorithm for perception-assist with an upper-limb exoskeleton robot. In: 2018 IEEE international conference on cyborg and bionic systems (CBS), Shenzhen, China, pp 204–209

6. Liu F, Han X, Lin M, Wu X, Sun Q, Song A (2019) Remote upper limb exoskeleton rehabilitation training system based on virtual reality. In: 2019 16th international conference on ubiquitous robots (UR), Jeju, Korea (South), pp 323–327
7. Buck BH, Akhtar N, Alrohimi A, Khan K, Shuaib A (2021) Stroke mimics: incidence, etiology, clinical features, and treatment. Ann Med 53(1):420–436
8. Liu C, Liang H, Ueda N, Li P, Fujimoto Y, Zhu C (2020) Functional evaluation of a force sensor-controlled upper-limb power-assisted exoskeleton with high backdrivability. Sensors (Basel) 20(21):63–79
9. Abella Bejarano B, Acuña Suabita K (2018) Design and implementation of support exoskeleton for carpal tunnel syndrome rehabilitation. Vis Electron
10. Oroya J, Oroya D, Huamanchahua D (2021) Design a hand orthosis to aid post-stroke patients with hemiplegia. In: 2021 IEEE 12th annual ubiquitous computing, electronics & mobile communication conference (UEMCON), New York, NY, USA, pp 0752–0757
11. Cho KH, Hong MR, Song WK (2020) Upper-limb robot-assisted therapy based on visual error augmentation in virtual reality for motor recovery and kinematics after chronic hemiparetic stroke: a feasibility study. Healthcare (Basel) 10(7)
12. He C, Xiong CH, Chen ZJ, Fan W, Huang XL, Fu C (2021) Preliminary assessment of a postural synergy-based exoskeleton for post-stroke upper limb rehabilitation. IEEE Trans Neural Syst Rehabil Eng 29:1795–1805
13. Gueye T, Dedkova M, Rogalewicz V, Grunerova-Lippertova M, Angerova Y (2021) Early post-stroke rehabilitation for upper limb motor function using virtual reality and exoskeleton: equally efficient in older patients. Neurol Neurochir Pol 55(1):91–96
14. Fernandez E, Urendes E, Costa V, Cortellucci A, Sanchez C, Lerma-Lara S, Raya R (2018) Design of an exoskeleton based on four cases of upper limb rehabilitation study. Rev Mex Ing Biomed 39(1):81–94
15. Diaz R, Moreno L, Sanjuan M, Prada C, Dalmiro L (2021) Development of an exoskeleton for the rehabilitation of the flexo-extensor movement of the elbow. ITECKNE J 18(1):46–51
16. Huamanchahua D, Arce-Cadillo JO (2022) Development of exoskeleton adjustable to Peruvian anthropometry for the rehabilitation of the upper right limb in patients with joint mobility deficit. In: 2022 IEEE 13th annual ubiquitous computing, electronics & mobile communication conference (UEMCON), New York, NY, NY, USA, pp 0648–0654
17. Breña-Garcia R, Taza-Aquino Y, Huamanchahua D (2022) Design of an exoskeletal prototype to contribute to the partial rehabilitation of the upper-limb. In: 2022 IEEE 13th annual ubiquitous computing, electronics & mobile communication conference (UEMCON), New York, NY, NY, USA, pp 0667–0672
18. Goltapeh A, Behzadipour S, Hajihosseinali M (2019) Design and construction of a planar robotic exoskeleton for assessment of upper limb movements. In: 7th RSI international conference on robotics and mechatronics (ICRoM), Tehran, Iran, pp 99–104
19. Huamanchahua D et al (2021) A robotic prosthesis as a functional upper-limb aid: an innovative review. In: 2021 IEEE international IOT, electronics and mechatronics conference (IEMTRONICS), Toronto, ON, Canada, pp 1–8
20. Huamanchahua D, Toledo-Garcia P, Aguirre J, Huacre S (2022) Hand exoskeletons for rehabilitation: a systematic review. In: 2022 IEEE international IOT, electronics and mechatronics conference (IEMTRONICS), Toronto, ON, Canada, pp 1–6
21. Cornejo J et al (2021) Mechatronic exoskeleton systems for supporting the biomechanics of shoulder-elbow-wrist: an innovative review. In: 2021 IEEE international IOT, electronics and mechatronics conference (IEMTRONICS), Toronto, ON, Canada, pp 1–9
22. Huamanchahua D, Castañeda-Vásquez C, Vásquez-Espinoza A, Muñoz-Zevallos A (2021) Robotic devices types exoskeletons for elbow rehabilitation: a technological review. In: 2021 IEEE 12th annual ubiquitous computing, electronics & mobile communication conference (UEMCON), New York, NY, USA, pp 0791–0796
23. Onozuka Y, Suzuki R, Yamada Y, Nakamura T (2018) An exoskeleton type 4-DOF force feedback device using magnetorheological fluid clutches and artificial muscles. In: 2018 IEEE/ASME international conference on advanced intelligent mechatronics (AIM), Auckland, New Zealand, pp 869–874

24. Zhang S, Fu Q, Guo S, Fu Y (2018) Coordinative motion-based bilateral rehabilitation training system with exoskeleton and haptic devices for biomedical application. Micromachines (Basel) 10(1)
25. Diez J, Blanco A, Catalan J, Badesa F, Lledo L, Garcia-Aracil N (2018) Hand exoskeleton for rehabilitation therapies with integrated optical force sensor. Adv Mech Eng 10(2):1–11
26. Wu Q, Chen B, Wu H (2019) RBFN-based adaptive backstepping sliding mode control of an upper-limb exoskeleton with dynamic uncertainties. IEEE Access 7:134635–134646
27. Chen T, Casas R, Lum PS (2019) An elbow exoskeleton for upper limb rehabilitation with series elastic actuator and cable-driven differential. IEEE Trans Robot 35(6):1464–1474
28. Park JH, Park G, Kim HY, Lee JY, Ham Y, Hwang D, Kwon S, Shin JH (2020) A comparison of the effects and usability of two exoskeletal robots with and without robotic actuation for upper extremity rehabilitation among patients with stroke: a single-blinded randomised controlled pilot study. J Neuroeng Rehabil 17(1)
29. Li X, Zhong J (2020) Upper limb rehabilitation robot system based on Internet of things remote control. IEEE Access 8:154461–154470
30. Font-Llagunes JM, Arroyo G, Alonso FJ, Vinagre BM (2010) Diseño de una ortesis activa para ayuda a la marcha de lesionados medulares. In: XVII congreso nacional de ingenieria mecánica
31. Vaughan-Graham J, Brooks D, Rose L, Nejat G, Pons J, Patterson K (2020) Exoskeleton use in post-stroke gait rehabilitation: a qualitative study of the perspectives of persons poststroke and physiotherapists. J Neuroeng Rehabil 17(1):123
32. Liu G (2021) Moving up the ladder of source assessment: expanding the CRAAP test with critical thinking and metacognition. Coll Res Libr News 82(2):75
33. Nagri GP, Ngadiso N, Sarosa T (2018) Comparative study on teaching reading using know, want, learn (KWL) and direct instruction method (DIM). Engl Educ 6(2)
34. Esparrago-Kalidas AJ (2021) The effectiveness of CRAAP test in evaluating credibility of sources. Int J TESOL Educ 1(2):1–14
35. Fielding JA (2019) Rethinking CRAAP: getting students thinking like fact-checkers in evaluating web sources. Coll Res Libr News 80(11):620
36. Suarez-Paz AP, Vasquez-Chirito CA, Cruz-Cruz MX, Huamanchahua D (2022) Proof of concept design of a postural and stabilizing orthosis for elderly people with gonarthrosis. In: 2022 IEEE 13th annual ubiquitous computing, electronics & mobile communication conference (UEMCON), New York, NY, NY, USA, pp 0244–0250
37. Meng Q, Xie Q, Deng Z, Yu H (2019) A general kinematics model for trajectory planning of upper limb exoskeleton robots. Intell Robot Appl 63–75
38. Fang Y, Hou B, Wu X, Wang Y, Osawa K, Tanaka E (2023) A stepper motor-powered lower limb exoskeleton with multiple assistance functions for daily use by the elderly. J Robot Mechatron 35(3)
39. Abdallah IB, Bouteraa Y (2023) A newly-designed wearable robotic hand exoskeleton controlled by EMG signals and ROS embedded systems. Robotics 12(4):95
40. Yang S-H, Koh C-L, Hsu C-H, Chen P-C, Chen J-W, Lan Y-H, Yang Y, Lin Y-D, Wu C-H, Liu H-K et al (2021) An instrumented glove-controlled portable hand-exoskeleton for bilateral hand rehabilitation. Biosensor 11(12):495
41. Thakur K, Banwait SS, Bedi R (2022) Finite element analysis of U-frame of robotic gait trainer for rehabilitation. In: International conference on materials science and engineering, vol 1248
42. Gull MA, Thoegersen M, Bengtson SH, Mohammadi M, Andreasen Struijk LNS, Moeslund TB, Bak T, Bai S (2021) A 4-DOF upper limb exoskeleton for physical assistance: design, modeling, control and performance evaluation. Appl Sci 11(5865)
43. Moeller T, Krell-Roesch J, Woll A, Stein T (2022) Effects of upper-limb exoskeletons designed for use in the working environment—a literature review. Front Robot AI 9(858893)

Design of a Lower Limb Exoskeleton Prototype for Patients with Motor Disabilities Using VDI 2206 + Pahl and Beitz Methodology

Deyby Huamanchahua, Raúl Escandón-Tufino, Rodrigo Aique, Christoffer Vasquez, Cristobal Ruiz de Somocurcio, Adrian Sasieta, Josue Abad, and Hector Valcarcel-Castillo

Abstract Integrating robots in rehabilitation therapies, such as exoskeletons, provides session stability. This is achieved by avoiding fatigue and improving movement efficiency, thus reducing rehabilitation times compared to traditional methods. This work aims to design a lower limb exoskeleton prototype for individuals with motor disabilities, whether caused by various diseases such as sarcopenia in older adults. The VDI 2206 + Pahl and Beitz methodology was employed to guide the comprehensive design of the exoskeleton. The structure is divided into three main sections: Mechanical Design, Instrumentation, and Control. The Mechanical Design section addresses the conceptual design of the exoskeleton and the actuation mechanism, along with material selection, followed by a stress analysis to ensure design

D. Huamanchahua (✉) · R. Escandón-Tufino · R. Aique · C. Vasquez · C. R. de Somocurcio · A. Sasieta · J. Abad · H. Valcarcel-Castillo
Department of Electrical and Mechatronics Engineering, Universidad de Ingenieria y Tecnologia—UTEC, Lima, Peru
e-mail: dhuamanchahua@utec.edu.pe

R. Escandón-Tufino
e-mail: raul.escandon@utec.edu.pe

R. Aique
e-mail: rodrigo.aique@utec.edu.pe

C. Vasquez
e-mail: christoffer.vasquez@utec.edu.pe

C. R. de Somocurcio
e-mail: cristobal.ruizdesomo@utec.edu.pe

A. Sasieta
e-mail: adrian.sasieta@utec.edu.pe

J. Abad
e-mail: josue.abad@utec.edu.pe

H. Valcarcel-Castillo
e-mail: hvalcarcel@utec.edu.pe

P. G. Bradford et al. (eds.), *Proceedings of IEMTRONICS 2024*, Lecture Notes in Electrical Engineering 1229, https://doi.org/10.1007/978-981-97-4780-1_17

requirements. The Instrumentation section tests EMG sensors to relate the person's movement and the necessary muscle activation. The Control section presents the kinematic and dynamic model of the exoskeleton, as well as the integration of joint trajectories during the gait cycle using sliding control. This work proposes a proof of concept based on advances in the state of the art of control.

Keywords Lower limb exoskeleton · VDI 2206 · Pahl and Beitz · Motor disabilities

1 Introduction

The previous technological review clarified the significant issue of locomotor disability in the people of Peru and the crucial need for an exoskeleton for patient rehabilitation [1]. In 2020, an analysis was conducted in Lima's hospital centers, revealing a prevalence of 17.6% of generalized and progressive muscle deterioration in the older adult population (sarcopenia). This public health problem is closely linked to various consequences [2], such as an increased risk of falls, fractures, disability, loss of independence, deterioration in the quality of life, hospitalization, and increased mortality among the older adult population [3].

An exoskeleton is a mechatronic device [4] that can be attached to the human body to assist in desired movement. Since the creation of the first exoskeleton in 1890 [5], different models have been developed to address various patient needs [6]. Currently, exoskeletons like ALTRACO [7], ReWalk [8], REX [9], RGR [10], or String-Man [11] can enable people with paralysis to walk again due to the advanced technology they use. However, constructing these exoskeletons is expensive and typically carried out in technologically advanced countries [12, 13].

For these reasons, this research aims to conduct a proof of concept for the mechanical part of a passive exoskeleton for the lower limbs of a test subject in a controlled environment. This is to validate the feasibility of the exoskeleton in terms of ergonomics for the patient, assess the most suitable material for the final product, and identify limitations and challenges for the project's next stage.

2 Methodology

The VDI 2206 methodology will be applied in conjunction with the Pahl and Beitz approach [14], which is divided into three levels to address each stage of the design process systematically and efficiently. In the first level, the Design Requirements (VDI 2206), the exoskeleton is conceptualized, considering specific requirements of the selected disability, such as the limits of each joint, their minimum torque, and each component's dimensions. The VDI 2206 methodology is applied to define

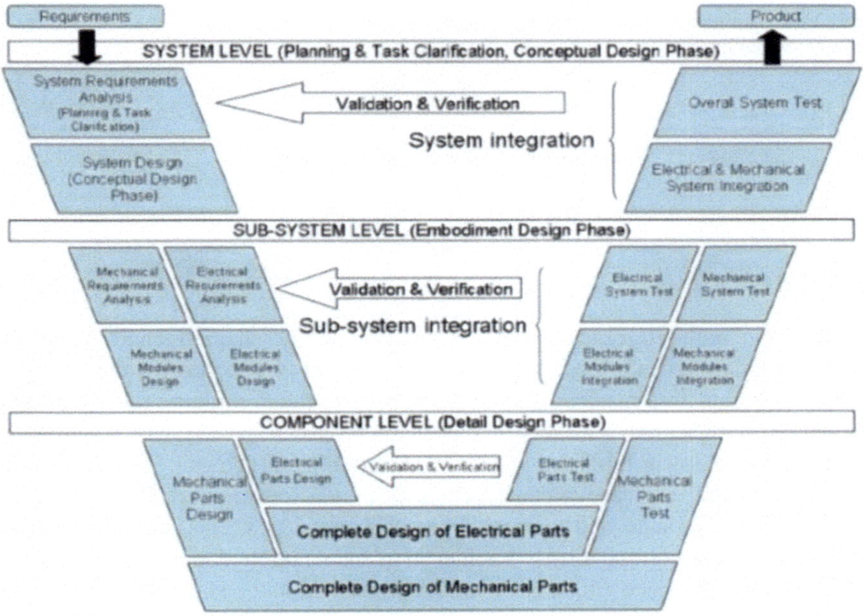

Fig. 1 VDI 2206 + Pahl and Beitz method [14]

rehabilitation movements, thus establishing the basis for the required mechanical design.

The second and third levels, prototyping, are subdivided into mechanical design, instrumentation, and control, integrating the VDI 2206 [15] and Pahl and Beitz [16] approaches. The selection of materials, design parameters, desired trajectories, control strategies, and stress analysis for mechanical strength are addressed under the VDI 2206 methodology. Simultaneously, the Pahl and Beitz approach structures the placement of sensors, facilitating the future integration and functionality of the exoskeleton. Additionally, the actuation system is implemented using cycloidal gears, and simulation in MATLAB and Python verifies the effectiveness of the design (Fig. 1).

2.1 Design Requirement

The focus is on rehabilitation movements of hip flexion, extension, adduction, and abduction; knee flexion and extension. The range of movements for each articulation (see Table 1) is derived from a detailed study in another reference paper [17]. These ranges establish the restrictions for the exoskeleton's mechanical design, its integration with kinematics, and the subsequent control that was carried out.

Table 1 Rehabilitation movements and limits

Rehabilitation movements	Limits (°)
Hip flexion	32.2–120
Hip extension	22.5–30
Hip abduction	7.9–20
Hip adduction	6.4–45
Knee flexion	73.5–150
Knee extension	0

The torque required for each joint is a critical parameter that influences the selection of motors, the overall exoskeleton design, and, more specifically, the configuration of the actuation mechanism necessary to meet the minimum required torque thresholds. In this context, we rely on the design proposed by [18], which addresses the creation of a lower limb exoskeleton intended to assist the gait of patients with flaccid paraplegia. This design considers a maximum user weight of 95 kg, a height in the 156–182 cm range, and an exoskeleton weight of 35 kg. This approach ensures that the minimum required torque is achieved to enable optimal mobility of the individual's joints: The knee joint has a torque of 46 N m, and the Hip joint has a torque of 80 N m.

The dimensions of the exoskeleton are tailored to the user's body size [19]. To facilitate prototyping and future testing, each component's dimensions have been established based on the measurements of one of the researchers (see Fig. 2). This choice simplifies future implementation and enhances the exoskeleton simulation during the development process.

3 Design

3.1 Mechanical Design

The mechanical aspect of the lower limb exoskeleton will be examined in detail. Based on the design requirements, load parameters were defined to stress-test the strength of the components. Among these considerations is a gravity acceleration of 9.81 m/s^2, point loads of 1 kN on the contact surfaces, moments of 100 N m, and bearing loads of 1 kN at the joints. On the other hand, ABS material was selected using Autodesk Fusion 360 software because it is the material most like PLA available in the software.

The CAD model of the exoskeleton was constructed to encompass the entire upper limb, offering support from the back and shoulder to the forearm. The model comprises four primary segments: the back cuff, shoulder cuff, upper arm, and forearm brace, each connected via articulated joints with embedded bearings for smooth motion. The shoulder cuff is designed to provide stable support without

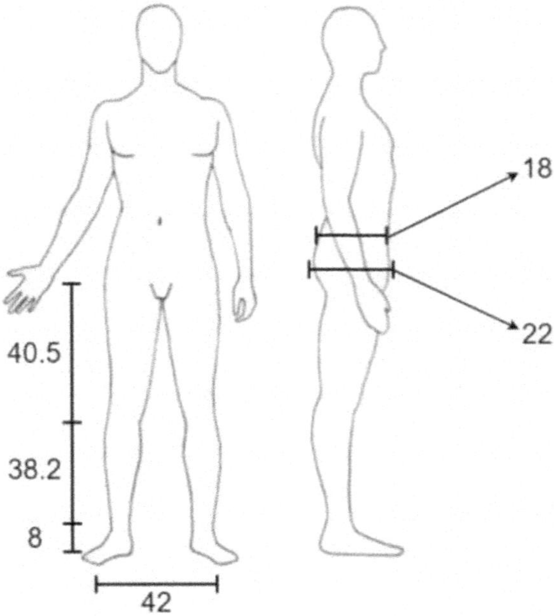

Fig. 2 Anatomical measurements of the lower limb in centimeters (cm) [19]

restricting the range of motion, and the upper arm and forearm supports are adaptable to different limb sizes, facilitating elbow flexion and extension (Fig. 3).

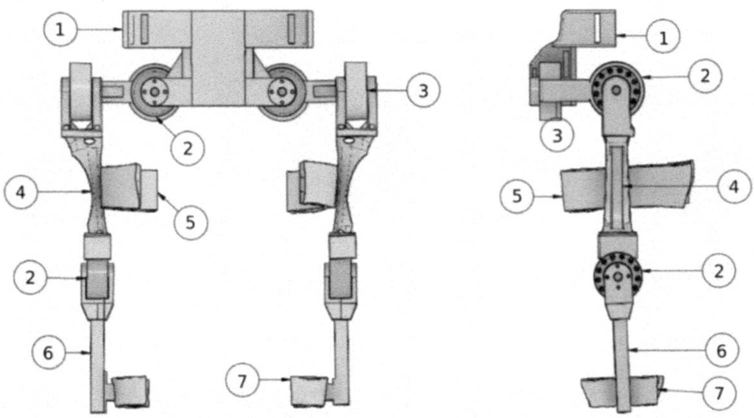

Fig. 3 Description of components: (1) backrest, (2) cycloidal gear, (3) pelvis-hip link, (4) hip-knee link, (5) thigh handle, (6) knee-calf link, (7) calf handle

3.2 Stress Analysis

Subsequently, a stress analysis [20] was conducted to confirm the reliability of the design. Table 2 presents the safety factor results for each exoskeleton component (see Fig. 4, 5, and 6).

In Fig. 4, the virtual simulation of stress analysis of the parts is shown, where various shades between blue and red can be observed, with blue representing higher resistance of the component to loads of 1 kN. A load of 1 kN was chosen for all components because it was assumed that considering the weight of an average person and the weight of the exoskeleton, it would reach a maximum weight for which it would be designed of 100 kg. Additionally, considering the torque generated by the motors at the joints of the exoskeleton could reach a maximum torque of 60 N m, we maximized the value to 100 N m to ensure proper exoskeleton performance under any condition. On the other hand, Table 2 shows that the component with the lowest

Table 2 Safety factors (S.F.)

Component	Safety factor
Knee-calf link	0.017–8
Backrest	0.264–8
Hip-knee link	0.009–8
Pelvis-hip link	0.409–8
Cycloidal gear	0.403–8

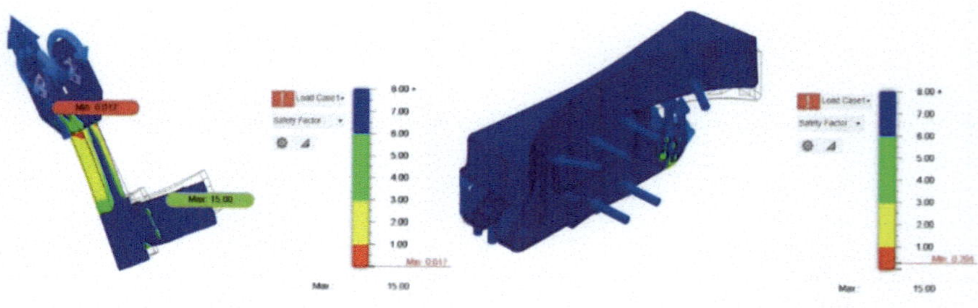

Fig. 4 Stress analysis. **a** Knee-calf link and calf handle, **b** backrest

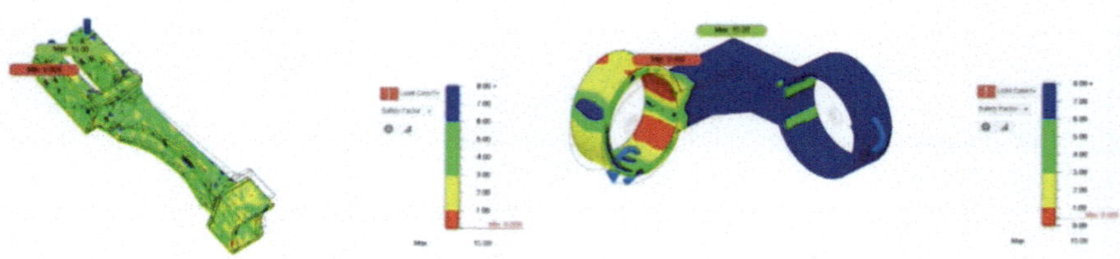

Fig. 5 Stress analysis. **a** Hip-knee link, **b** pelvis-hip link

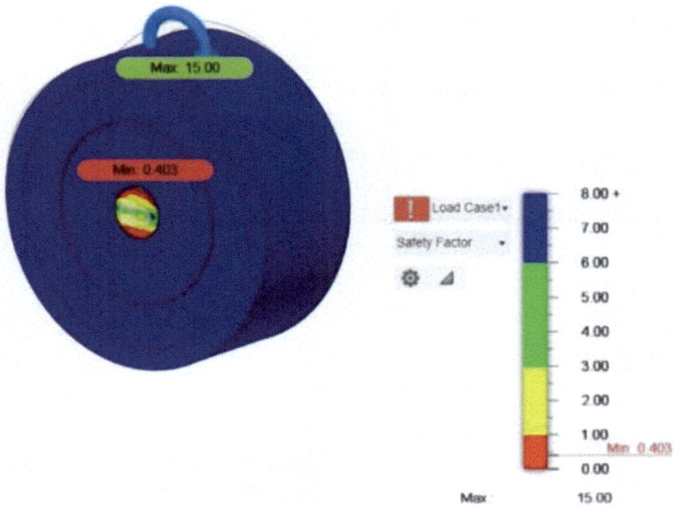

Fig. 6 Stress analysis of cycloidal gear

safety factor is the Hip-Knee Link, and it is more likely to suffer plastic deformation under very high loads.

3.3 *Actuation Mechanism Design*

In obtaining the necessary torque, as discussed in the previous section, the decision was made to implement a cycloidal gear mechanism (see Fig. 7) because it allows for high efficiency due to low friction in the cycloidal mechanism. It provides more precise movement, lower backlash due to rolling contact between components, and is easy to design [21].

Likewise, the gear mechanism was implemented through 3D printing with PLA filament [22] (see Fig. 8).

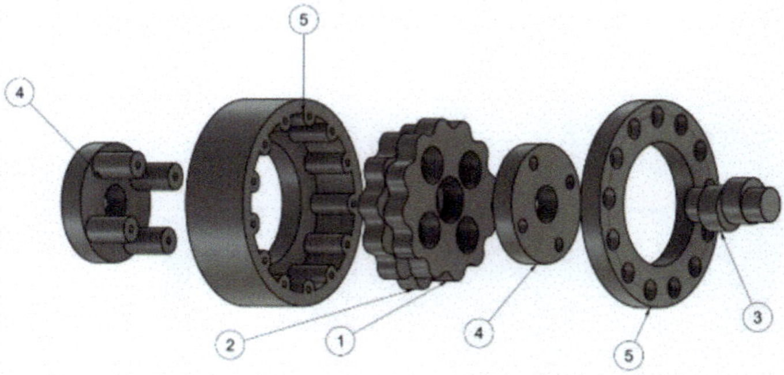

Fig. 7 Gear bearing. (1) and (2) cycloidal disks, (3) input shaft, (4) output shaft, (5) ring gear bearing

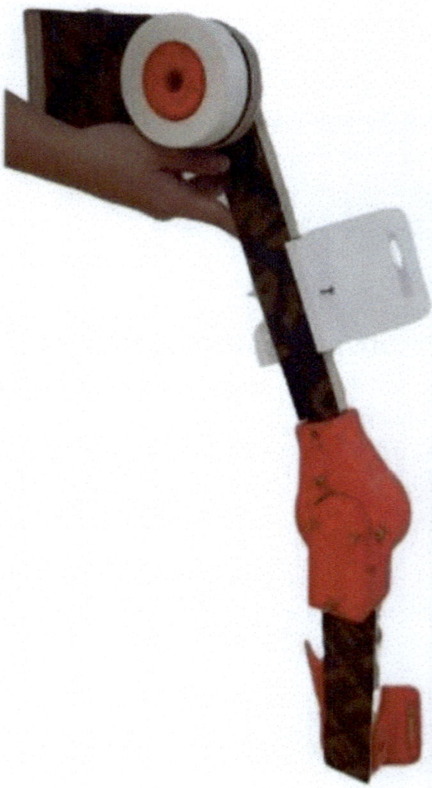

Fig. 8 Prototype of the exoskeleton with the cycloidal gear (1:12) at each joint

4 Instrumentation

4.1 Sensor Placement

As evidenced in Fig. 9a, the sensors MyoWare [23] were placed on the rectus femoris, belonging to the quadriceps area, and the biceps femoris, corresponding to the hamstrings area. This choice is because both muscles are activated simultaneously and play crucial roles in the hip and knee joints, which are the target of our exoskeleton. Additionally, both muscles are considerably large and close to the skin's surface, facilitating electromyography measurement and improving quality. It's worth noting that the placement of these sensors aligns with the position of the thigh grips on the exoskeleton, allowing for efficient integration into the design. Figure 9b provides an actual depiction of the area where the sensor is positioned.

Figure 10a, b reflect the data collection results from the MyoWare device for each muscle mentioned earlier. To carry out this task, the sensor was connected to an Arduino board, recording real-time data and storing it in an Excel spreadsheet.

The graphs illustrate the behavior of the muscles throughout the gait cycle. For example, in the case of the rectus femoris, responsible for knee extension and hip flexion, its most pronounced activation peak is observed from the 'foot flat' to 'midstance' positions when the front foot is planted, and the body is pushed forward, generating the hip mentioned above and knee movements. On the other hand, in the case of the biceps femoris, responsible for opposing movements such as knee flexion and hip extension, its highest activation occurs during 'heel off,' when knee flexion is generated while lifting the back leg forward.

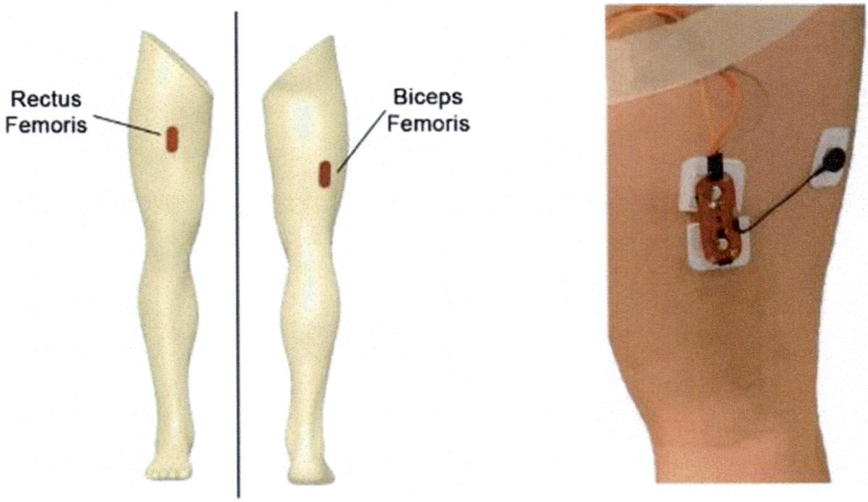

Fig. 9 Sensor placement. **a** Sensors position model, **b** real biceps femoris sensor

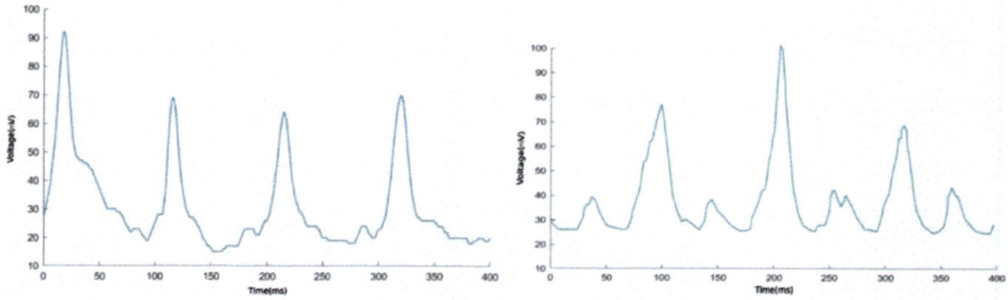

Fig. 10 Data collection. **a** Rectus femoris EMG, **b** biceps femoris EMG

5 Dynamic Control of the Exoskeleton

It will simulate a nonlinear control through feedback linearization applied to the developing lower limb exoskeleton. To implement control, it is necessary to obtain the dynamic model of the exoskeleton; however, the actual derivation of this will be accomplished in the final stage of the research. For this reason, the model will be simplified to a more analytically tractable one, specifically a 2DoF RR robotic arm. Additionally, the objective is to achieve a tracking error of less than 5% and a stabilization time of less than 1 s.

5.1 *User's Walking Trajectory*

Tracking software was used to obtain the user's walking trajectory, and Cartesian points for the hip, knee, and ankle were obtained, as shown in Fig. 11a.

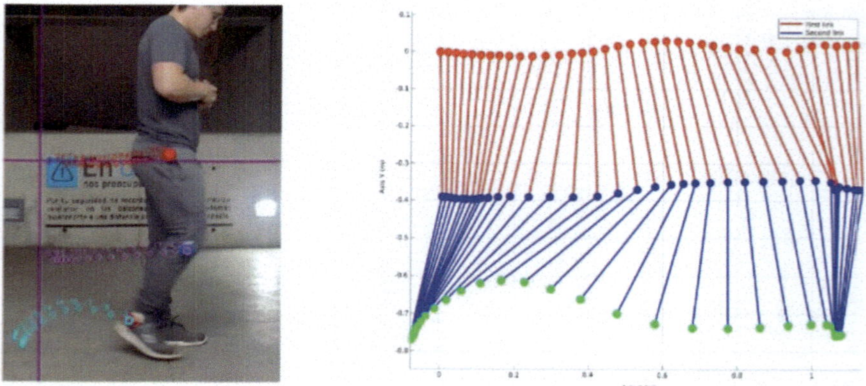

Fig. 11 Walking trajectory. **a** Gait tracking, **b** kinematic model of the exoskeleton

The angles for the two links (hip-knee and knee-ankle) were obtained through inverse kinematics. These angles were used to plot the user's gait trajectory using forward kinematics (see Fig. 11b).

5.2 *Dynamic Model*

The Euler–Lagrange method was used to find the dynamic model with the effect of gravity [24], as it is most suitable for control scheme analysis and dynamic properties study. The dynamic model of the exoskeleton was obtained through the equations shown below.

$$M(\theta)\ddot{\theta} + C(\theta, \dot{\theta})\dot{\theta} + g(\theta) = r_{\text{tot}} \tag{1}$$

$$\begin{bmatrix} m_{11} & m_{12} \\ m_{21} & m_{22} \end{bmatrix} \begin{bmatrix} \theta_1 \\ \ddot{\theta}_2 \end{bmatrix} + \begin{bmatrix} h\dot{\theta}_2 & h(\dot{\theta}_1 + \dot{\theta}_2) \\ -h\dot{\theta}_1 & 0 \end{bmatrix} \begin{bmatrix} \theta \\ \dot{\theta}_2 \end{bmatrix} + g\begin{bmatrix} (m_1 l_{c1} + m_2 l_1)c_1 + m_2 l_{c2} c_{12} \\ m_2 l_{c2} c_{12} \end{bmatrix} = \begin{bmatrix} r_1 \\ r_2 \end{bmatrix} \tag{2}$$

5.3 *Feedback Linearization Control*

The simulation of feedback linearization control was carried out using Matlab software and the Simulink tool. The block diagram of the control scheme is shown in Fig. 12.

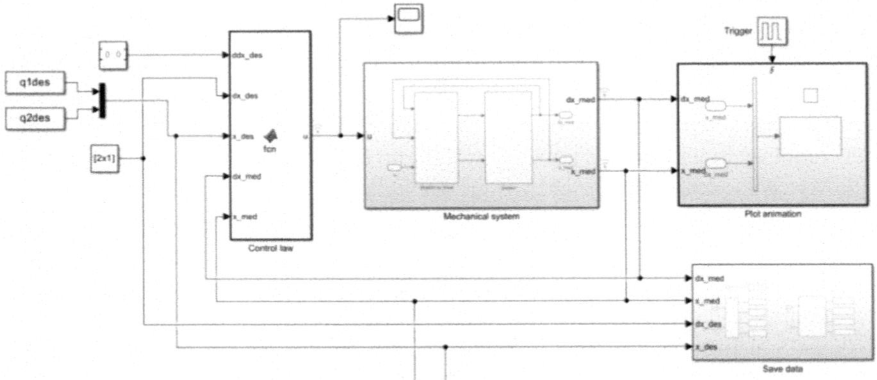

Fig. 12 Block diagram of feedback linearization control

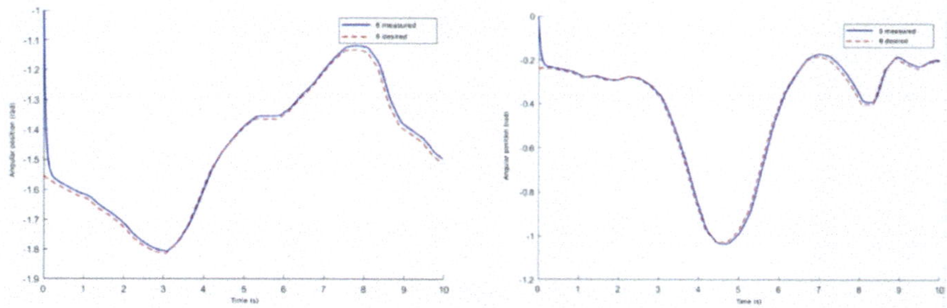

Fig. 13 Trajectory tracking. **a** The measured and desired angular position of the first link, **b** the second link

As the angular position of the two links changes over time, the control was configured as a tracker to achieve the correct trajectory tracking. The result obtained for the first link of the exoskeleton is shown in Fig. 13a and the second link is shown in Fig. 13b.

The tracking error for the first link is less than 3%, and it achieves this in just 0.2 s, making it suitable control as it meets the requirements stated at the beginning of this section. The tracking error for the second link is mostly less than 3%, but for the downhill motion of the second link 7, it presents an error of 4.8%. Also, it takes to achieve this in only 0.1 s, making it a suitable control as it meets the requirements stated at the beginning of this section.

6 Conclusions

In conclusion, the exoskeleton design has adhered to the VDI 2206 and Pahl and Beitz methodologies to delineate a design process that meets specific requirements, such as minimum torque joint range and material selection, among other considerations. Thus, a cycloidal gear mechanism was chosen to minimize the required space and resistance to high stresses. The variation in safety factor is due to changes in the exoskeleton design, which reduces its resistance to bending loads and applied moments. Despite being a prototype with material printed in 3D, it meets the conditions required to withstand a person's weight up to 95 kg. Strategic instrumentation with MyoWare sensors has yielded significant data on muscle activation during the gait cycle, where peaks of 90–100 mV imply a strong muscle activation and the range of 60–80 mV employee a mild muscle activation. To further use these measurements, instrumentation is required to condition the electrical signals to realize an upper-level control scheme commanded by the patient. Additionally, a focus on inverse kinematics using the Newton's method ensures precise results and avoids errors associated with quadrants. This comprehensive approach translates into a functional prototype, laying the groundwork for future developments in assisting the mobility of individuals with motor disabilities.

References

1. Huamanchahua D, Abad J, Aique R, Vasquez C, De Somocurcio CR, Sasieta A (2023) Lower limb exoskeletons and virtual reality for rehabilitation: a technological review. In: 2023 15th IEEE international conference on industry applications (INDUSCON), Sao Bernardo do Campo, Brazil, pp 1678–1684
2. Braveman P (2014) What are health disparities and health equity? We need to be clear. Public Health Rep 129:5–8
3. Ramos-Ramirez KE, Soto A (2020) Sarcopenia, mortalidad intrahospitalaria y estancia hospitalaria prolongada en adultos mayores internados en un hospital de referencia peruano. Acta Med Peruana 37(4)
4. Secciani N, Brogi C, Pagliai M, Buonamici F, Gerli F, Vannetti F, Bianchini M, Volpe Y, Ridolfi A (2021) Wearable robots: an original mechatronic design of a hand exoskeleton for assistive and rehabilitative purposes. Front Neurorobot 15(750385)
5. Yagn N (1890) Apparatus for facilitating walking, running, and jumping. U.S. Patent 420179 (1890)
6. Breña-Garcia R, Taza-Aquino Y, Huamanchahua D (2020) Design of an exoskeletal prototype to contribute to the partial rehabilitation of the upper limb. In: 2022 IEEE 13th annual ubiquitous computing, electronics & mobile communication conference (UEMCON), pp 1–6
7. Beyl P, van Damme M, van Ham R, Versluys R, Vanderborght B, Lefeber D (2008) An exoskeleton for gait rehabilitation: prototype design and control principle. In: Proceedings of the IEEE international conference on robotics and automation, (ICRA'08), Pasadena, CA, USA, pp 2037–2042
8. Zeilig G, Weingarden H, Zwecker M, Dudkiewicz I, Bloch A, Esquenazi A (2012) Safety and tolerance of the ReWalk™ exoskeleton suit for ambulation by people with complete spinal cord injury: a pilot study. J Spinal Cord Med 35(2):96–101
9. Birch N, Graham J, Priestley T, Heywood C, Sakel M, Gall A, Nunn A, Signal N (2017) Results of the first interim analysis of the RAPPER II trial in patients with spinal cord injury: ambulation and functional exercise programs in the REX powered walking aid, J Neuroeng Rehabil 14(60)
10. Pietrusinski M, Cajigas I, Mizikacioglu Y, Goldsmith M, Bonato P, Mavroidis C (2010) Gait rehabilitation therapy using robot generated force fields applied at the pelvis. In: Proceedings of the IEEE haptics symposium, (HAPTICS'10). Waltham, MA, USA, pp 401–407
11. Surdilovic D, Bernhardt R (2004) STRING-MAN: a new wire robot for gait rehabilitation. In: Proceedings of the IEEE international conference on robotics and automation, vol 2, pp 2031–2036
12. Diaz I, Gil JJ, Sanchez E (2011) Lower-limb robotic rehabilitation: literature review and challenges. Rehabil Robot 211
13. Huamanchahua D, Taza-Aquino Y, Figueroa-Bados J, Alanya-Villanueva J, Vargas-Martinez A, Ramirez- Mendoza RA (2021) Mechatronic exoskeletons for lower-limb rehabilitation: an innovative review. In: 2021 IEEE international IOT, electronics and mechatronics conference (IEMTRONICS), Toronto, ON, Canada, pp 1–8
14. Rahman RBA, Pulm U, Stetter R (2007) Systematic mechatronic design of a piezo-electric brake. In: DS 42: proceedings of ICED 2007, the 16th international conference on engineering design. Paris, France, pp 705–706
15. VDI-Guideline 2206 (2004) Design methodology for mechatronic system. VDI, Düsseldorf
16. Pahl G, Beitz W (2006) Engineering design: a systematic approach, 2nd edn. Springer, Berlin
17. Tang X, Wang X, Ji X, Zhou Y, Yang J, Wei Y, Zhang W (2022) A wearable lower limb exoskeleton: reducing the energy cost of human movement. Micromachines (Basel) 13(6) (2022)
18. Tibaduiza-Burgos D, Anaya-Vejar M, Aya-Parra P (2019) Exoesqueleto para rehabilitación de miembro inferior con dos grados de libertad orientado a pacientes con accidentes cerebrovasculares. INGE CUC 15(2):36–46

19. Christensen S, Rafique S, Bai S (2021) Design of a powered full-body exoskeleton for physical assistance of elderly people. Int J Adv Robot Syst 18(6)
20. Arunkumar S, Mahesh S, Rahul M, Ganesh N, Maheshwaran KJ (2023) Design and analysis of lower limb exoskeleton with external payload. Int J Interact Des Manuf (IJIDeM) 1–18 (2023)
21. Chavan U, Joshi A, Kolambe Y, Gwalani H, Chaudhari H, Khalate A, Hujare PP (2022) Magnification of energy transmission ratio using miniature cycloidal gear box for humanoids. IOP Conf Ser Mater Sci Eng 1272(1)
22. Rojek I, Dorożyński J, Mikołajewski D, Kotlarz P (2023) Overview of 3D printed exoskeleton materials and opportunities for their AI-based optimization. Appl Sci 13(14):8384
23. Abdallah IB, Bouteraa Y (2023) A newly-designed wearable robotic hand exoskeleton controlled by EMG signals and ROS embedded systems. Robotics 12(4):95
24. Sciavicco L, Siciliano B, Villani L (1995) Lagrangeand Newton-Euler dynamic modeling of a gear-driven robot manipulator with inclusion ofmotor inertia effects. Adv Robot 10(3):317–334

"Tony" Glove: Facilitating Interaction with the Environment for Children with Visual-Spatial Deficit

Jaime Huaytalla-Pariona, Piero Calderon-Mendez, Nilmar Arones-Chuchon, Ruben Cespedes-Aliaga, Wilman Alca-Alfaro, Manuel Ipenza-Negri, Deyby Huamanchahua, and Eber J. Ballon-Alvarez

Abstract This innovative project focuses on creating a technologically advanced glove inspired by the iconic Iron Man glove, specifically for children with visual-space deficits. The main objective is to significantly improve the quality of life of these children by providing them with an interactive tool that allows them to interact with their environment more safely and autonomously. The methodology for developing this glove involved integrating several electronic components, an Arduino UNO, an open-source electronics platform based on easy-to-use hardware and software. An ultrasonic sensor, a device that can measure distance to an object using sound waves, was used to detect nearby objects and alert the user to their presence. In addition, an LCD screen was incorporated to provide valuable visual information to

J. Huaytalla-Pariona (✉)
Department of Mechatronics Engineering, Universidad Continental, Huancayo, Peru
e-mail: jhuaytalla@continental.edu.pe

P. Calderon-Mendez · N. Arones-Chuchon · R. Cespedes-Aliaga · W. Alca-Alfaro · M. Ipenza-Negri · E. J. Ballon-Alvarez
Universidad ESAN, Lima, Peru
e-mail: 22200285@ue.edu.pe

N. Arones-Chuchon
e-mail: 22200136@ue.edu.pe

R. Cespedes-Aliaga
e-mail: 22200222@ue.edu.pe

W. Alca-Alfaro
e-mail: 22101735@ue.edu.pe

M. Ipenza-Negri
e-mail: 23200473@ue.edu.pe

E. J. Ballon-Alvarez
e-mail: jballon@esan.edu.pe

D. Huamanchahua
Department of Electrical and Mechatronics Engineering, Universidad de Ingenieria y Tecnologia—UTEC, Lima, Peru
e-mail: dhuamanchahua@utec.edu.pe

P. G. Bradford et al. (eds.), *Proceedings of IEMTRONICS 2024*, Lecture Notes in Electrical Engineering 1229, https://doi.org/10.1007/978-981-97-4780-1_18

the user, and a horn that emits an audible sound when the user approaches an object, serving as an additional warning signal. LEDs were incorporated to improve the glove's aesthetics and make it more attractive to children. The results indicate that the glove facilitates children's mobility intuitively and excitingly, improving special awareness and their ability to navigate their environment. In conclusion, this glove demonstrates how technology can make a positive difference in the lives of those with visual impairments. Not only does it improve their mobility and safety, but it also opens new possibilities for their inclusion and active participation in society. This project is a testament to the power of technology to transform lives and create a more inclusive and accessible world for everyone.

Keywords Arduino · Visual-spatial · Glove · Children

1 Introduction

Visuospatial development in infancy involves the progressive integration of visual, motor, and spatial skills, leading to the ability to create mental maps and organized visual-motor skills [1]. This process, which spans from birth to school age, highlights the relevance of the dorsal and ventral visual pathways and the influence of the frontal lobe and medial structures of the temporal lobe [2]. In most of our daily activities [3], these skills are involved; for example, in education, visuospatial skills are essential for understanding and learning geometric concepts [4], solving mathematical problems, understanding diagrams and graphs [5], and participating in activities related to science and technology [6]; in sports, many disciplines require visual solid perception and spatial skills [7]. From team sports like soccer to individual sports [8] like golf, the ability to judge distances, anticipate movements, and coordinate visible and motor actions is essential [9].

Visual impairment, especially from birth or early age, significantly affects various aspects of life [10]. Vital visual cues during early development led to difficulties with orientation, mobility, exploration of the environment, and limitations in social interactions [11]. Despite school programs for blind children that seek to address these difficulties, there is a differential development in autonomy for orientation and mobility among students [12]. A multiple case study was conducted between 2015 and 2018 at the School for Blind Children of Guadalajara, A.C. In Mexico, it is known that when visual disability occurs from birth or at an early age, the effects are more significant since there is a lack of visual references that are key in the first stages of a child's development [13].

Children with visual impairments may face challenges in perceiving and understanding the surrounding space, affecting their autonomy, safety, and social development [14]. The use of technological devices is presented as a viable solution to provide these children with information about the environment, information that would otherwise be inaccessible to them [15]. A study conducted with a device like the one described in this project revealed significant improvements in the spatial orientation skills of visually impaired children using such a device [16]. Infants who experience this disability tend to face setbacks such as frequent falls or slower progress when it comes to developing motor skills [17]. Difficulty perceiving moving objects increases the risk of collisions [18]. Orientation in unfamiliar spaces hinders their mobility [19]. These difficulties limit their autonomy and participation in daily and social activities [20]. Understanding these obstacles is crucial to developing inclusive strategies [21]. Barnett and Lam's research highlights the importance of addressing these difficulties in improving these children's quality of life and social integration [22].

The present prototype is an interactive glove for children that aims to help improve the development of visuospatial skills in children with deficits. The prototype incorporates an ultrasonic sensor to perceive the location of objects, decorative LED lights, a sound buzzer, and an LCD screen that reflects the distance between the child's hand and the object they want to reach. A study in Ecuador stood out by developing a similar interactive glove, using an ultrasonic sensor to detect the distance between the glove and objects [23]. Related projects would be this one that uses a loudspeaker to draw attention if there is an obstacle [24] and further enhances the experience by incorporating vibration sensors [25], offering innovative solutions for children's tactile interaction and spatial perception.

2 Background

Implementing projects aimed at children with visuospatial deficits requires a detailed review of previous experiences. Various articles have been crucial in structuring the project in this context. This article presents three key backgrounds that influenced the project's conception, highlighting many similarities.

A prototype glove was presented to assist visually impaired people in their orientation and safe movement. The device has an ultrasound guidance system and a location system using GSM/GPS technology. The test results indicated that the prototype could detect obstacles at distances of up to 5 m and provide accurate information about the user's geographical position [26].

The other article presents a prototype bracelet designed to complement the cane for blind people to improve their mobility and quality of life. Equipped with features such as an Arduino Nano, ultrasound sensor, vibrator, Bluetooth module, DFPlayer mini-module, and other components, the bracelet seeks to support the mobilization of visually impaired people. After tests with blind users, a positive acceptance of the prototype was observed, although it was recommended to eliminate the audible signal due to possible distractions. The results indicated that the bracelet contributes to improved movement by helping to detect obstacles effectively [27].

The article describes the development of a prototype white cane in response to the increase in visually impaired people. The growing availability of auxiliary devices is highlighted to facilitate independent movement and ensure safe mobility. Based on the Arduino platform, the prototype incorporates electronic elements such as ultrasonic sensors, servo motors, buzzers, Bluetooth, and GPS. Designed as a guidance system for blind people, the prototype white cane aims to avoid obstacles, thereby improving the user's movement and mobility in their environment [28].

3 Methodology

3.1 *Prototype Description*

The final project features an integrated glove with an ultrasonic sensor, an LCD screen, decorative LED lights, and a sound buzzer essential in improving the quality of life of children with visuospatial deficits. This glove uses technology to provide real-time information about the distance of objects, facilitating daily mobility and contributing significantly to the development of visuospatial skills in children. Decorative LED lights and sound buzzers complement the experience, turning the glove into a multi-sensory tool that provides a better understanding of the environment for children.

The glove's design, focused on the child user experience, has been enhanced by integrating an ultrasonic sensor, LCD, decorative LED lights, and a buzzer. The LCD clearly and easily displays information about the distance of objects, adapting in a way that is understandable to children. The multi-sensory interface remains intuitive, allowing children to use the glove naturally and without complications. In addition, the device is designed with vibrant colors and attractive elements, including the decorative LED lights and buzzer, to make the experience even more enjoyable and motivating. This encourages children's participation in glove use and improves their spatial awareness. The primary purpose of this glove is to provide helpful information in an accessible way for children with visuospatial deficits. The design has been carefully crafted to make it easy to use and appealing to children, thus helping to improve their mobility and visuospatial skills.

3.2 *Component Description*

The glove incorporates an electronic brain, Arduino Uno, which collaborates with an ultrasonic eye, the HCSR04 sensor, to measure distances. The information is presented directly on a small 16 $\times$ 2 LCD screen, showing the distance between the user and the objects. In addition, decorative LED lights and a buzzer have been integrated, complementing the multi-sensory experience.

3.3 Design Description

Inspired by Iron Man's iconic glove, our design results from a careful fusion of functionality and aesthetics. We started creating a detailed 2D design, gradually evolving into a sophisticated 3D model using the TinkerCAD platform. This tool allowed us to discreetly incorporate an ultrasonic sensor, an LCD screen, eight decorative LEDs, and a buzzer. Also, it will enable us to sample the measurements of each detail of the glove.

Our aesthetic approach is based on Iron Man's signature glove, uniquely tailored to capture the attention and fascinate children. Not only do we seek to emulate the futuristic hero interface but also to create an exciting and engaging experience for children with visual-space deficits.

Figure 1 presents the practical design of the "Tony" glove, showing measurements in the holes intended for the fingers and the dimensions of a standard hand for practical use.

Figure 2 shows a horizontal view of the "Tony" glove, with more detailed measurements of the contours intended for the fingers and structures complementing this prototype.

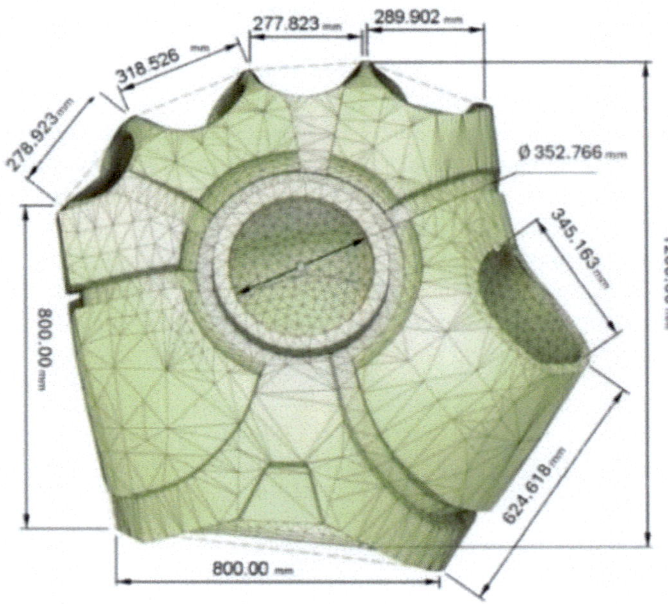

Fig. 1 Vertical 3D design

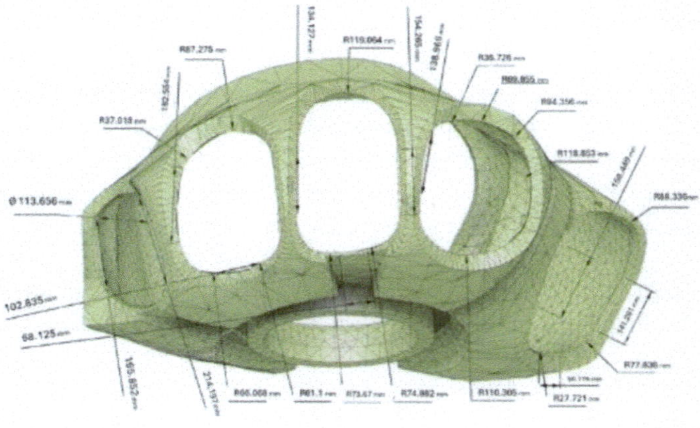

Fig. 2 Horizontal 3D design

3.4 Flowchart

The program is divided into two parts: in the first part, we declare the required commands, and in the second part, the commands are executed. In the setup part, the program commands were declared. These are necessary for its operation. In Fig. 3, in the first part, we have a detailed declaration of the ports used on the Arduino UNO with their respective input or output order.

In the second part, the loop, we have two processes in which the program will execute. Given the commands already registered in the program, this "if" conditional operation will perform its counting operation and display the number of times it is executed. In the other execution, it follows the same process, but in this part, we have the conditionals "if" and "else." We have two paths by which, during the execution, the program will decide accordingly. Once the decision is made, it will show the distance in cm so that if the distance is less than 10 cm, the buzzer will be activated; otherwise not. Afterward, it will clean up the records of the distances.

3.5 Design in TinkerCAD

This design facilitates the experimental implementation of the "Tony" prototype. Its essential operation is achieved by employing a button that allows the prototype to be activated and deactivated. Thanks to the ultrasonic sensor that captures this information, information about the distance to objects is displayed on an LCD screen. In addition, a buzzer emits an alarm sound to indicate proximity to an object when we are very close to it.

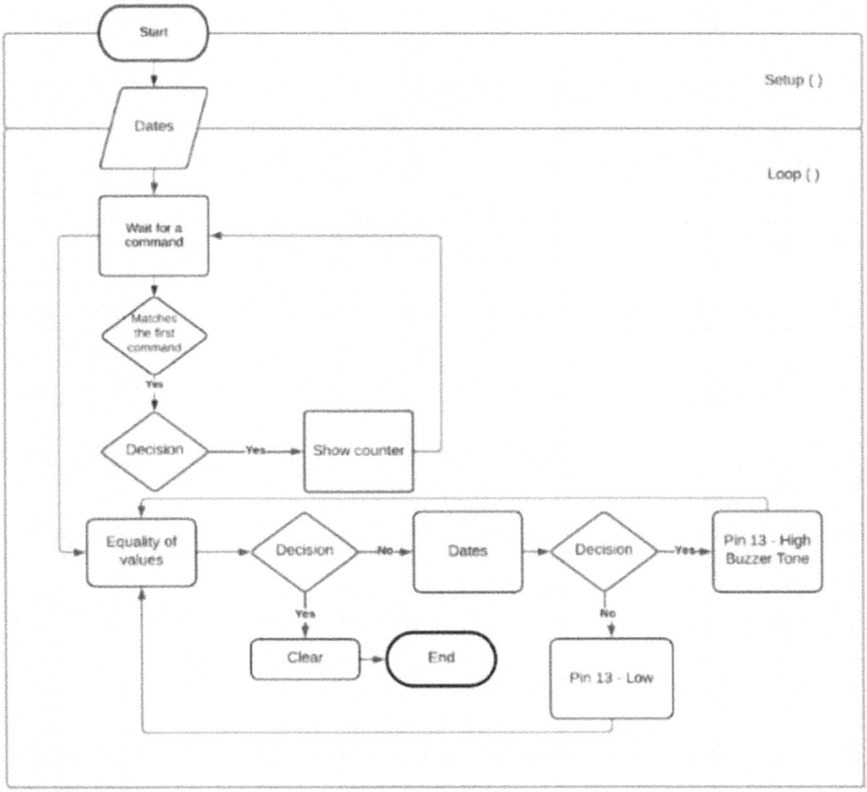

Fig. 3 Flowchart of the overall program

In Fig. 4, a simulation was carried out in TinkerCAD that facilitated the orderly and intelligent structuring of the cables. This tool also allowed for practical experimentation with components and calibration to ensure proper operation. The primary purpose of this simulation was to serve as a guide for the physical implementation of the project.

3.6 Programming Code Description

The entire system was developed using an Arduino Uno, and specific code can be seen in Fig. 5, 6, 7, and 8.

In this code snippet above Fig. 5, the pins of the LCD screen (9, 8, 4, 5, 6, and 7) and the ultrasonic sensor (trig: 10, echo: 11) are specified. The variable 'cont' acts as a counter for the pushbutton, while 'pulseTime' and 'Distance' convert data from the ultrasonic sensor to calculate the distance. In addition, 'currentState' and 'endStatus'

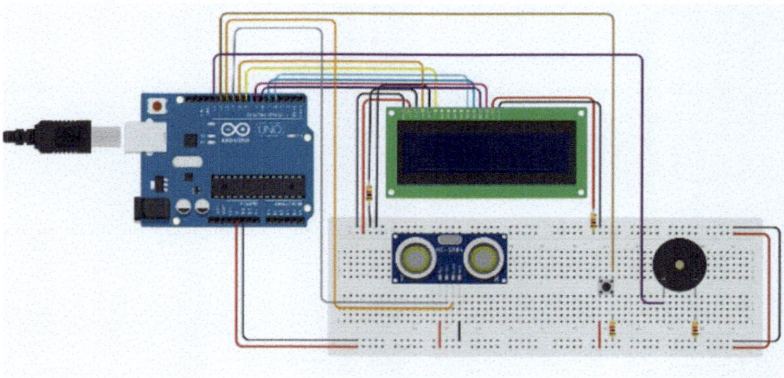

Fig. 4 "Tony" prototype in TinkerCAD

Fig. 5 Code of the "Tony" prototype in TinkerCAD (variables)

```
#include <LiquidCrystal.h>

LiquidCrystal lcd(9, 8, 4, 5, 6, 7);

int trig = 10;
int echo = 11;
int cont = 0;
float pulseTime;
float Distance;
int currentState = 0;
int endStatus = 0;
```

Fig. 6 Code of the "Tony" prototype in TinkerCAD (setup)

```
void setup() {
  pinMode(12, INPUT_PULLUP);
  pinMode(trig, OUTPUT);
  pinMode(echo, INPUT);
  pinMode(13, OUTPUT);
  pinMode(10, OUTPUT);
  Serial.begin(9600);
  lcd.begin(16, 2);
  lcd.clear();
  lcd.setCursor(0, 0);
}
```

are vital variables for efficiently handling the pushbutton, facilitating monitoring of the current state, and determining final conditions in the program.

In the setup section of this code in Fig. 6, the pin modes are configured: pin 12 is set as input with internal pull-up resistance, trig, and echo pins are set for the

Fig. 7 Code of the "Tony" prototype in TinkerCAD (loop-I)

```
void loop() {
  digitalWrite(10, HIGH);
  delayMicroseconds(10);
  digitalWrite(10, LOW);
  pulseTime = pulseIn(11, HIGH) / 2;
  Distance = float(pulseTime * 0.0343);
  currentState = digitalRead(12);
  if (currentState != endStatus) {
    if (currentState == HIGH) {
      cont = cont + 1;
      Serial.println(cont);
    }
  }
  endStatus = currentState;
```

Fig. 8 Code of the "Tony" prototype in TinkerCAD (loop-II)

```
  if (cont % 2 != 0) {
    lcd.setCursor(0, 0);
    lcd.print("DISTANCE:");
    lcd.setCursor(0, 1);
    lcd.print(Distance);
    lcd.print("cm ");
    delay(500);
    if (Distance < 10) {
      tone(13, 523, 200);
      digitalWrite(13, HIGH);
    } else {
      digitalWrite(13, LOW);
    }
  } else {
    lcd.clear();
    cont = 0;
  }
}
```

ultrasonic sensor, pin 13 as output, pin 10 as output, and serial communication is initiated at a rate of 9600 baud. In addition, the 16×2 LCD screen is initialized, and the cursor is placed in position (0, 0) to prepare the interface.

In the initialization of the loop in Fig. 7, an ultrasonic pulse is generated by briefly activating pin 10, followed by a pause. The return time of the ultrasonic pulse is then measured at pin 11, and the corresponding distance is calculated using the sound speed. The status of the button connected to pin 12 ('currentState') is checked, updating the 'cont' counter and displaying its serial value in case of a change in the state of the button. The 'endStatus' variable is updated to reflect the button's current state, thus preparing for the next iteration of the loop.

In the second part of the loop in Fig. 8, a condition is performed to check if the 'cont' counter is an odd number. If yes, the LCD screen is updated with the label "DISTANCE" in the first row and displays the measured distance in the second row. A short delay of 500 ms is introduced to improve readability on the screen. In addition, if the distance is less than 10 cm, an audible tone is produced, and an LED is activated on pin 13. If 'cont' is an even number, the LCD screen is cleared, and the 'cont' counter is reset to zero. This logic allows you to toggle between the distance display and the empty LCD screen at each iteration of the loop, thus controlling the flow of the program.

3.7 Deployment Description

This Arduino code measures distance using an ultrasonic sensor connected to pins 10 and 11. The LCD is set to pins 9, 8, 4, 5, 6, and 7. A push button on pin 12 counts events. This project consists of a glove that displays the distance to an object using an ultrasonic sensor and an LCD. The ultrasonic sensor is connected to digital pins 10 and 11 of the Arduino board (TRIG and ECO) and sends and receives sound pulses to measure the distance. The LCD is connected to digital pins 9, 8, 4, 5, 6, and 7, displaying the distance in centimeters. A buzzer connected to digital pin 13 is a proximity alert emitting a pleasant sound.

In addition, there is a push button on digital pin 12 that allows the activation of the glove. To achieve good communication between the Arduino board and the glove, the serial connection had to be replaced by a direct connection between the Arduino board and the glove, using high-quality connectors to ensure a tight fit and avoid accidental disconnection and had to solder the wires, since being connected by the inputs that came by default caused errors in the display of information. The Arduino code used for this project is based on the LiquidCrystal_I2C.h and NewPing.h libraries, which facilitate the management of the LCD and the ultrasonic sensor, respectively.

In Fig. 9, during the implementation process, cable connections were carried out similarly to what was done in TinkerCAD. Still, it was decided to solder everything instead of using a breadboard to work more efficiently. Challenges were faced in placing cables correctly to avoid potential short circuits. Subsequently, it was decided to adapt the "Tony" glove by placing the wires on the upper part of it, placing the buzzer in the palm, and adding white LEDs around the buzzer to simulate a light shooting effect. All of this was done in the palm with aesthetic considerations.

Finally, parameters were established for the operation of the system. Distance information is displayed on the LCD screen every 500 ms, while the buzzer, connected to pin 13, is activated if the distance is less than 10 cm. A push button on pin 12 counts the events and displays the distance, activating the buzzer based on the measurement of the number of pushes if the number of pushes is odd. The display clears, and the count resets if the number of key presses is even, providing visual and audible feedback based on the distance measured by the ultrasonic sensor.

Fig. 9 Implementation of the pins on the Arduino UNO board

In Fig. 10, a person is checking the LCD screen of a prototype assistive device. The display shows the distance between an object and the ultrasonic sensor in another person's hand. This verification is necessary to ensure that the device is working correctly.

Finally, in Fig. 11, the device is checked using the ultrasonic sensor located in the left hand, while the distance away from it is observed on the LCD screen of the right hand.

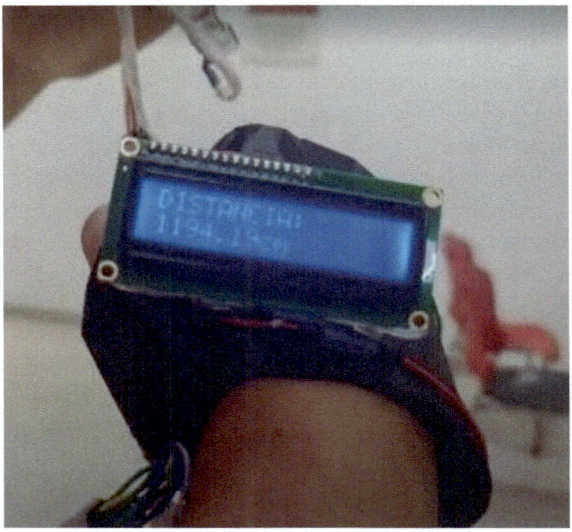

Fig. 10 LCD showing data

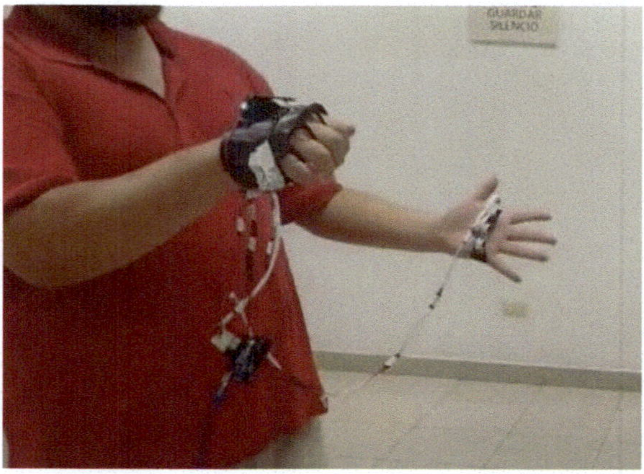

Fig. 11 Operation

4 Conclusions

In conclusion, the glove designed for children with visuospatial deficits stands out not only as a significant technological achievement but also as a tool with the potential to positively transform the lives of these children. Inspired by Iron Man's famous glove, its aesthetic seeks to capture attention and foster a positive perception of technology in everyday life.

The integration of elements such as the ultrasonic sensor, LCD screen, decorative LEDs, and buzzer, all controlled by an Arduino UNO, offers real-time feedback that informs about the environment and enriches the sensory experience of users. Not only do these functional aspects have a practical impact by providing helpful information, but they also contribute to creating a playful and exciting experience for children. The precision and efficiency demonstrated in the physical implementation of the glove support its viability and usefulness in real-world situations. The straightforward visual interface on the LCD screen ensures that feedback is accessible and understandable, promoting children's autonomy and confidence in their environment.

This glove not only represents a technological innovation but a complete solution that, through its user-centered design, advanced functionalities, and focus on improving quality of life, can mark a significant change in the daily experience of children with visuospatial deficit, enhancing their mobility skills and spatial awareness in a comprehensive and enriching way.

In summary, we consider that our prototype has great potential. Therefore, the main goal is to optimize "Tony" to provide a pleasant experience. In that sense, Bluetooth connectivity will be implemented to eliminate excess cables, facilitating user mobility. Likewise, the construction material of the glove is planned to be changed for better comfort. Being a glove (a portable object), it can include various functions, even that of a modern watch. A function that is planned to be implemented is to emit sounds that allow interaction with the environment through a buzzer.

References

1. Ickx G, Bleyenheuft Y, Hatem SM (2017) Development of visuospatial attention in typically developing children. Front Psychol 8
2. Rosselli M (2015) Desarrollo neuropsicológico de las habilidades visoespaciales y visoconstruccionales. Rev Neuropsicol Neuropsiquiatr Neurociencias 15(1):175–200
3. Bergqvist M, Möller MC, Björklund M, Borg J, Palmcrantz S (2023) The impact of visuospatial and executive function on activity performance and outcome after robotic or conventional gait training, long-term after stroke-as part of a randomized controlled trial. PLoS ONE **18**(3)
4. D'Aurizio G, Di Pompeo I, Passarello N, Troisi Lopez E, Sorrentino P, Curcio G, Mandolesi L (2023) Visuospatial working memory abilities in children analyzed by the bricks game task (BGT). Psychol Res 87(7):2111–2119
5. Vekiri I (2002) What is the value of graphical displays in learning? Educ Psychol Rev 14:261–312
6. Villamil-Camacho AM (2016) Relación entre atención, memoria visual y habilidades visoespaciales en niños de Educación Primaria. M.S. tesis
7. Heppe H, Kohler A, Fleddermann MT, Zentgraf K (2016) The relationship between expertise in sports, visuospatial, and basic cognitive skills. Front Psychol 7(904)
8. Chueh TY, Huang CJ, Hsieh SS, Chen KF, Chang YK, Hung TM (2017) Sports training enhances visuospatial cognition regardless of open-closed typology, PeerJ 5
9. Mantilla JIA (2019) Neurociencia y entrenamiento en el deporte de alto rendimiento. Rev Iberoam Cienc Act Fís Dep 8(2):79–90
10. Khorrami-Nejad M, Sarabandi A, Akbari MR, Askarizadeh F (2016) The impact of visual impairment on quality of life. Med Hypothesis Discov Innov Ophthalmol 5(3):96–103
11. Hutmacher F (2019) Why is there so much more research on vision than on any other sensory modality? Front Psychol 10(2246)
12. Cuturi LF, Aggius-Vella E, Campus C, Parmiggiani A, Gori M (2016) From science to technology: orientation and mobility in blind children and adults. Neurosci Biobehav Rev 71:240–251
13. García-Llamas JJ (2019) Orientación y movilidad autónoma en niños con discapacidad visual
14. Khandekar R (2008) Visual disabilities in children including childhood blindness. Middle East Afr J Ophthalmol 15(3):129–134
15. Fernández-Aparicio MA, Castro-Rodríguez AM, González-Fernández MJ (2019) A new device for improving spatial orientation in visually impaired children. IEEE Sens J 19(1):367–373
16. Al-Ani MS, Al-Taee AMA, Al-Ani AS (2019) A wearable device for improving spatial orientation in visually impaired children. IEEE Access 7:130528–130537
17. Fonteyn-Vinke A, Huurneman B, Boonstra FN (2022) Viewing strategies in children with visual impairment and children with normal vision: a systematic scoping review. Front Psychol 13(898719)
18. Höhler C, Rasamoel ND, Rohrbach N et al (2021) The impact of visuospatial perception on distance judgment and depth perception in an augmented reality environment in patients after stroke: an exploratory study. J Neuro Eng Rehabil **18**(127)
19. Lam C (2018) Mobility and orientation in children with visual impairments. Routledge, London, UK
20. Herrero L, Theirs CI, Ruiz-Iniesta A, González A, Sanchez V, Pérez-Nieto MA (2019) Visuospatial processing improvements in students with down syndrome through the autonomous use of technologies. Br J Educ Technol 50:2055–2066
21. Laczó M, Wiener JM, Kalinova J, Matuskova V, Vyhnalek M, Hort J, Laczó J (2021) Spatial navigation and visuospatial strategies in typical and atypical aging. Brain Sci 11(11)
22. Barnett MA (2017) Spatial orientation in visually impaired children. Springer, New York, NY
23. Solís Yagual CR (2019) Diseño del prototipo de un guante controlado con Arduino que permite la detección de obstáculos por medio de sensores para mejorar el desplazamiento dentro del hogar a personas con bajos recursos que cuenten con discapacidad visual. Dissertation. Universidad de Guayaquil, Facultad de Ciencias Matemáticas y Físicas, Carrera de Ingeniería en Networking y Telecomunicaciones

24. Acosta Parra DR, Mena Mafla JM (2016) Prototipo para la detección de obstáculos superiores a la capacidad de rastreo de ayuda a persona con discapacidad visual
25. Tolosa Mesa JD (2017) Prototipo de detección de obstáculos para personas con limitación visual
26. Azaña Toapanta CR, Vizuete Franco JC (2018) Diseño y Construcción de un Prototipo de Guante con Orientación y Localización para Ayuda en la Movilidad de Personas Invidentes, 88 hojas. Quito, EPN
27. Ontiveros-Paredes SD, Rojas-Balbuena D, Martínez-Paredes J (2014) Diseño y construcción de un bastón blanco electrónico para personas invidentes. Científica 18(2):63–70
28. Dussan JGB (2019) Prototipo de brazalete como complemento al bastón para personas con discapacidad visual. Rev Grindda 1:20–29

Dave Octopus Prototype to Relieve for Children with Asperger's Syndrome

Jaime Huaytalla-Pariona, Leslie Neciosup-Saavedra, Meyly Mendoza-Cobeñas, Abdon Carrasco-Cierto, Stephano Rivadeneyra-Huaman, Brandon Valenzuela-Cubas, Deyby Huamanchahua, and Eber J. Ballon-Alvarez

Abstract The project aims to provide support for children experiencing Asperger's syndrome. The strategic choice of the Arduino Uno R3 has proven to be fundamental to the project's purpose, not only because of its accessibility in connections but also because of its ability to catalyze the creation of a sensory and emotional system. As a comprehensive response to the specific needs of these children, the system allows pressing one of six buttons to trigger a multi-sensory experience. This experience comprises the harmonious reproduction of one of the melodies according to the LED color representing an emotion, which is played through the speaker, the presentation of comforting messages and/or drawings on the LCD screen, and the illumination of an LED, being this selection according to the presented emotion of children with Asperger's syndrome, all skillfully processed by the Arduino Uno R3. The validation

J. Huaytalla-Pariona (✉)
Department of Mechatronics Engineering, Universidad Continental, Huancayo, Peru
e-mail: jhuaytalla@continental.edu.pe

L. Neciosup-Saavedra · M. Mendoza-Cobeñas · A. Carrasco-Cierto · S. Rivadeneyra-Huaman · B. Valenzuela-Cubas · E. J. Ballon-Alvarez
Universidad ESAN, Lima, Peru
e-mail: 22200111@ue.edu.pe

M. Mendoza-Cobeñas
e-mail: 22200248@ue.edu.pe

A. Carrasco-Cierto
e-mail: 22200334@ue.edu.pe

S. Rivadeneyra-Huaman
e-mail: 22101822@ue.edu.pe

B. Valenzuela-Cubas
e-mail: 22101808@ue.edu.pe

E. J. Ballon-Alvarez
e-mail: jballon@esan.edu.pe

D. Huamanchahua
Department of Electrical and Mechatronics Engineering, Universidad de Ingenieria y Tecnologia—UTEC, Lima, Peru
e-mail: dhuamanchahua@utec.edu.pe

P. G. Bradford et al. (eds.), *Proceedings of IEMTRONICS 2024*, Lecture Notes in Electrical Engineering 1229, https://doi.org/10.1007/978-981-97-4780-1_19

of the project is based on the premise that, in stressful situations, interaction with these buttons effectively provides access to visual and sound stimuli designed to mitigate anxiety. The adaptability of the Arduino facilitates the connectivity of the six sensors and plays an essential role in data processing, presenting information consistently and meaningfully. The intersection between the interactivity of the buttons, sensory stimulations, and the versatility of the Arduino converges to create an affordable, customized system to improve the quality of life of children facing this syndrome.

Keywords Arduino Uno R3 · Asperger's syndrome · Innovative sensory system · Child welfare

1 Introduction

Asperger syndrome is an autism spectrum disorder (ASD) that primarily affects a person's ability to interact socially and communicate effectively [1]. People with this syndrome often show repetitive patterns of behavior, limited interests, and difficulties in understanding social subtleties and facial expressions [2]. Although they share some characteristics with autism, people with Asperger syndrome generally have well-developed language and cognitive skills [3]. In addition, robotics offers a means to channel the interest and dedication characteristic of those with this syndrome, providing opportunities to develop technical and creative skills that can translate into academic and employment advantages [4].

Therefore, the project aims to relieve stress in children diagnosed with Asperger's syndrome [5]. It consists of the child pressing a button and listening to a relaxing melody while a motivational message is displayed. In the project, using the Arduino Uno is helpful for easy interaction and confidence [6]. Hence, a dynamic interactive interface enhances communication, reduces anxiety through predictability, and fosters motor skill development. The project prioritizes user-friendliness, inclusivity, and education, promoting STEM skills [7]. Besides, this device offers a holistic approach to enhancing the emotional well-being of children with Asperger's syndrome, addressing their potential stressors. It aims to seamlessly integrate into their daily lives, considering sensory and emotional aspects, prioritizing predictability, and ease of use. Incorporating electronic components aims to alleviate anxiety caused by unexpected changes, while the interactive interface fosters confidence and motor skill development [8].

Similarly, through extensive research, projects related to the purpose of the present work were found. The first antecedent proposes an electronic didactic toy to stimulate children with emotional and behavioral problems [9]. The second antecedent is based on a device designed to monitor the emotions of children with Asperger's syndrome [10]. The third antecedent is an android called Milo, developed by RoboKind to teach social and emotional skills to children with autism [11]. These antecedents provided valuable perspectives that helped contextualize and enrich our current project's development [12].

Such is the case of this device, which focuses on facilitating a positive response from those who experience difficulties in their usual psychosocial development. The FONA robotic assistant is a device that uses technology to help children with communication disorders [13]. The robot has a range of facial expressions that can help motivate children and keep them engaged during therapy sessions [14]. Children can interact with the robot by providing auditory, tactile, and visual stimuli [15]. For example, the robot control service offers kinesthetic stimuli for children, patients, or therapists via the communication module [16]. Furthermore, various robots facilitate the simulation of real-world scenarios, incorporating elements like clothing, transportation, colors, and animals. This is a valuable tool to promote the growth of social skills [17]. Likewise, it has been demonstrated that robotic toys captivate children with this disorder, prompting them to engage with external stimuli produced by the robot instead of their internal focus [18].

Likewise, two interactive toys utilize color programming to improve motor skills, controlled by a mobile app and hardware [19]. On the other hand, Milo, an android created by RoboKind, is a fantastic creation that resembles a human being [20]. It can move, communicate, and realistically express facial emotions. Its goal is to teach social and emotional skills through verbal lessons, using a screen on its chest to display visual symbols representing the lesson's content [21]. It is then possible to mention that robotics has emerged as a valuable tool to improve the quality of life of those coping with Asperger syndrome [22]. Robots can provide a structured and predictable environment that facilitates social interaction, allowing people with Asperger's to practice social skills in a controlled and anxiety-free manner [23].

2 Methodology

2.1 Project Description

The Dave Octopus project is a unique tool for children with Asperger's Syndrome, providing them a platform to express their emotions. Dave Octopus aims to facilitate communication, allowing children to express their feelings clearly and understandably and to relieve stress through the toy in the form of an octopus that, by pressing a button, will illuminate an LED and play a melody according to the child's emotion. The functionality of the octopus consists of a simple but effective system: the child presses one of the six buttons associated with specific colors, where each color carries a particular meaning, with its six distinctive tentacles, each representing a different emotion (anger, surprise, sadness, joy, fear, and disgust).

The versatility of the Dave Octopus project looks beyond emotional expression, encompassing the development of crucial self-regulation skills. By providing a precise auditory response to each emotion represented by the colors, the child could gradually learn to associate specific sounds with moods. This feature enriches the sensory experience and offers the child a practical tool to manage emotions effectively. The integration of auditory responses contributes to the versatility of the project by addressing broader aspects of emotional development and providing the child with additional resources to strengthen his or her emotional regulation skills. The purpose of the project is to provide conditional support, improve the child's dynamic interaction and understanding, and foster an environment where children with Asperger Syndrome can communicate their emotions in a more effective and enriching way.

2.2 Description of Components

Arduino Uno is the central brain that coordinates and guides all the critical components of the Dave Octopus project. This interactive board cleverly connects with the speakers, buttons, and LED lights, orchestrating a complete multi-sensory experience for the child. In addition, to further enrich the interaction, the project includes the presence of plasticine, allowing children to experience and explore tactile sensations and diverse textures. In this way, the Arduino Uno acts as a central controller and facilitates a holistic experience that integrates visual, auditory, and tactile elements to maximize the child's participation and sensory stimulation. Here are the components used: an Arduino Uno R3 (1), a medium protoboard (1), a mini push button (1), 1 kΩ resistors (10), LED lights (6), a passive buzzer (1), a 2×16 LCD (1), a potentiometer (1), a pack of male-male and female-male jumper cables, three bags of black electrical tape, three acrylic paints (green, white, and light blue), a medium container and a pack of play dough.

2.3 Design Description

The Arduino Uno serves as the central brain that coordinates and guides all the critical components of the "Dave Pulpin" project. This interactive board cleverly connects with the speakers, buttons, and LED lights, orchestrating a complete multi-sensory experience for the child. In addition, to further enrich the interaction, the project includes the presence of plasticine, allowing children to experience and explore tactile sensations and various textures. In this way, the Arduino Uno acts as a central controller and facilitates a holistic experience that integrates visual, auditory, and tactile elements to maximize the child's participation and sensory stimulation (Figs. 1 and 2).

2.4 Programming Description

Step 1: First, check whether the device is connected or not.

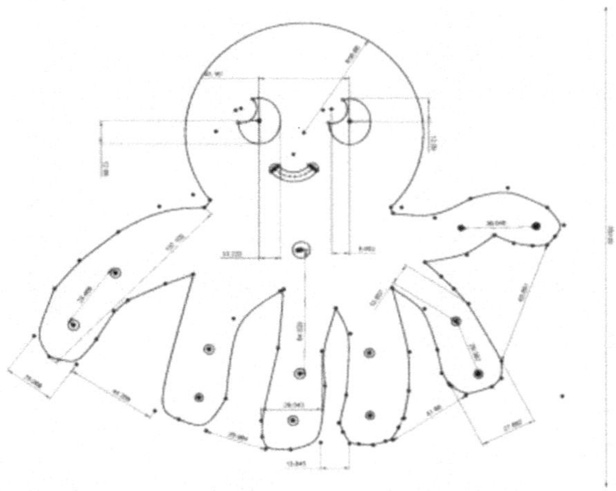

Fig. 1 2D modeling of the Dave octopus

Fig. 2 3D representation of the Dave octopus

Step 2: If the device isn't connected, the programming doesn't run.

Step 3: If the device is connected, the configuration of the LCD screen, buttons, and buzzer pins will be carried out (Fig. 3).

Step 3.1: The initial message, which consists of a greeting and a description that helps the user interact, will then be displayed.

Step 3.2: Then, the status of the buttons will begin to be read to determine whether they are pressed.

Step 3.3: If button 1 is pressed, message 1 will be displayed along with an angry emoji. After, the initial message is displayed again, and the status of the buttons is checked. If button 1 is not pressed, it will check if the next button is pressed.

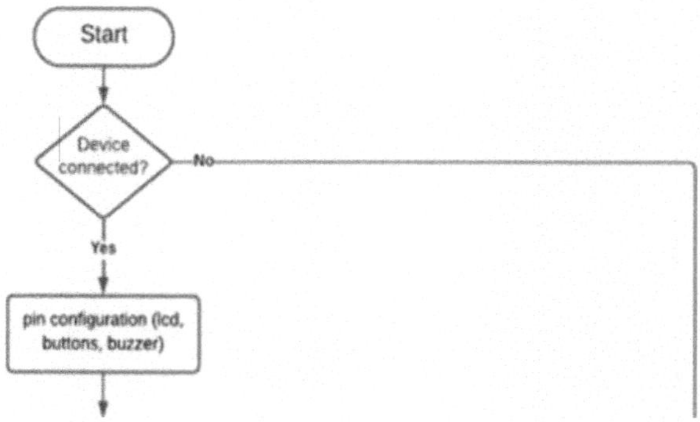

Fig. 3 Flowchart to check if the device is connected or not

Step 3.4: If button 2 is pressed, message 2 will be displayed along with a surprise emoji. After, it will show the initial message again. If button 2 is not pressed, it will check if the next button is pressed. The same procedure will be done with button 3 (Fig. 4).

Step 3.5: Continue checking to see if the other buttons are pressed. If so, a message and an emoji are displayed for each button, and a specific melody is played.

Step 3.6: After each procedure, press a button or verify that no button is pressed. The initial message is displayed again, and the entire programming is repeated. The program ends when the device is disconnected (Fig. 5).

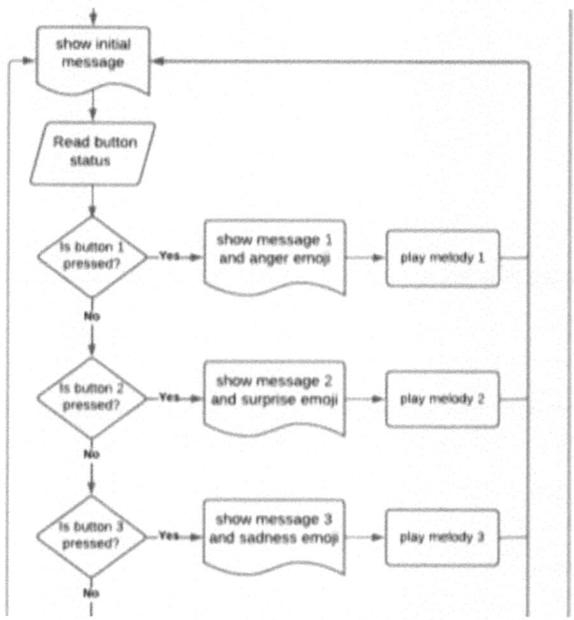

Fig. 4 Flowchart to check whether a button is pressed (buttons 1–3)

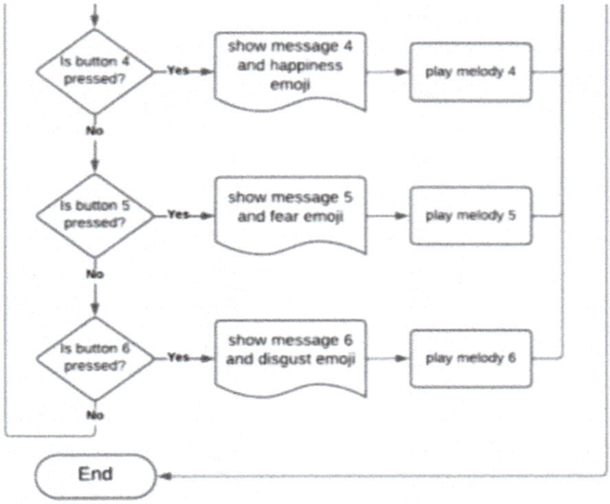

Fig. 5 Flowchart to check whether a button is pressed (buttons 4–6)

2.5 Description of the Programming Code

First, the Arduino library "LiquidCrystal.h" is included, which allows interaction with LCDs. Then, an instance of the "lcd" object of the LiquidCrystal class is created with the configuration of specific pins (RS, E, and DB4-DB7), thus establishing the communication between the Arduino and the LCD for its control in the project.

Next, variables are set to store the PINs to which the buttons (Buttons 1–6) and the buzzer (BUZZER) are connected. These assignments will be used later in the program to refer to the specific pins on the Arduino board assigned to each component.

Then, 6-byte arrays are created for the anger, surprise, sadness, fear, happiness, and disgust emotions, each composed of eight elements, conforming to a graphic pattern of 8 × 5 bits. Each byte in these arrays is responsible for writing a pixel's state in a particular emoji's position. A value of 1 in a bit indicates that the corresponding pixel is on, while 0 shows that the pixel is off. Combining these six matrices makes creating a complete visual representation of the emoji on the LCD screen possible (Fig. 6).

The following actions are performed in the program's setup phase: Communication with the 16 × 2-character LCD starts. Subsequently, the current contents of the

```
//Anger emoji
byte An1[8]={B00000,B00001,B00110,B00100,B01001,B10010,B10000,B10000};
byte An2[8]={B11111,B00000,B00000,B10001,B01010,B00000,B11011,B11011};
byte An3[8]={B00000,B10000,B01100,B00100,B10010,B01001,B00001,B00001};
byte An4[8]={B10000,B10000,B10000,B01000,B00100,B00110,B00001,B00000};
byte An5[8]={B00000,B00000,B11111,B10001,B11111,B00000,B00000,B11111};
byte An6[8]={B00001,B00001,B00001,B00010,B00100,B01100,B10000,B00000};
```

Fig. 6 Byte arrays for the emotion of anger

```
void setup()
{
  lcd.begin(16, 2);
  lcd.clear();
  pinMode(pinButton1, INPUT); // BUTTON 1
  pinMode(pinButton2, INPUT); // BUTTON 2
  pinMode(pinButton3, INPUT); // BUTTON 3
  pinMode(pinButton4, INPUT); // BUTTON 4
  pinMode(pinButton5, INPUT); // BUTTON 5
  pinMode(pinButton6, INPUT); // BUTTON 6
  pinMode(tonePin, OUTPUT); // BUZZER
}
```

Fig. 7 Initialization of the LCD and configuration of the button pins and buzzer

display are cleared. The pins corresponding to six buttons (Buttons 1–6) are configured as input pins using the pinMode() function. Also, the buzzer's (Buzzer) pin is configured as an output pin (Fig. 7).

The program's main loop (void loop()) is configured to display an initial message, "Hello! I am Dave. Press a button, "in the first and second rows, inviting interaction with buttons. To know if the buttons are pressed or not, the conditional structure "if, "else if," and "else" are used. When it is detected that button 1 is pressed (HIGH state), the program performs a sequence of actions. First, the LCD is cleared for a clean display. Then, a message is displayed according to the emotion. Next, six characters are created using the six arrangements of the feeling of anger, then the position is set on the screen and then displayed on the LCD. Subsequently, a melody associated with the emotion is played using the melody1() function. This procedure is repeated similarly in verifying when other buttons are pressed; only the message displayed, the emoji, and the melody to be played vary according to the associated emotion (Fig. 8).

The melody1() function defines a sequence of tones representing a melody associated with the emotion of anger. The tone plays each tone() function on the specified pin (tonePin) with specific frequencies and durations. The sequence comprises a series of tones and pauses designed to express the emotion of anger musically. The function combines frequencies and durations to achieve the desired melody. Subsequently, the same procedure will be repeated for buttons 2, 3, 4, 5, and 6, setting melodies according to the button pressed since each button is a different emotion.

2.6 *Circuit Description*

In that process, the TinkerCAD program was used to simulate a specific project. The operation involves pressing one of the six buttons (3) on each octopus part. When this action is performed, an LED light (5) is turned on, together with the reproduction of a melody through the Passive Buzzer (6). In addition, a text is generated on the LCD (7). The Arduino UNO development board (1) and a medium Protoboard (2) were used to carry out this procedure. Also, two essential components for current

Fig. 8 Actions when pressing button 1

```
if (digitalRead(pinButton1) == HIGH)
{
  //Anger
  lcd.clear();
  lcd.setCursor(0,0);
  lcd.print("Take a");
  lcd.setCursor(0,1);
  lcd.print("deep breath");
  //Show emoji
  lcd.createChar(0,An1);
  lcd.setCursor(13,0);
  lcd.write(byte(0));

  lcd.createChar(1,An2);
  lcd.setCursor(14,0);
  lcd.write(byte(1));

  lcd.createChar(2,An3);
  lcd.setCursor(15,0);
  lcd.write(byte(2));

  lcd.createChar(3,An4);
  lcd.setCursor(13,1);
  lcd.write(byte(3));

  lcd.createChar(4,An5);
  lcd.setCursor(14,1);
  lcd.write(byte(4));

  lcd.createChar(5,An6);
  lcd.setCursor(15,1);
  lcd.write(byte(5));

  melody1();
  lcd.clear();
}
```

flow were handled: the use of the resistor (4) and the application of the potentiometer (8) (Fig. 9).

2.7 *Description of Implementation*

In this process, Autodesk Fusion 360 was used as a tool for creating a 3D model of the octopus. Fusion 360, a computer-aided design (CAD) application, allows designers to generate three-dimensional models digitally. Once the design is complete in the

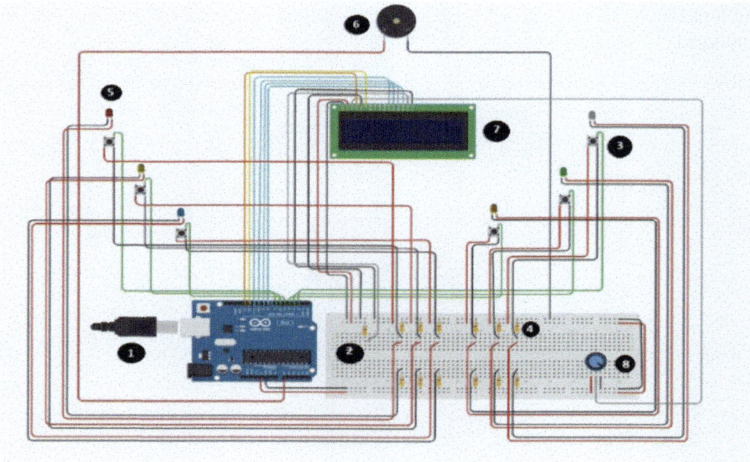

Fig. 9 Schematic representation of the circuit designed in TinkerCAD

software, it moves to the following implementation stage. In this phase, a 3D printer is used to materialize an object in the specific shape of an octopus. This approach merges digital design capability with additive manufacturing, providing an efficient way to create three-dimensional objects. The procedure started by connecting the corresponding wires on the breadboard and completing the layout of the cables associated with the LCD, ensuring an organized and functional wiring. This was to prevent possible confusion with the wiring used in later phases of the project.

Subsequently, a potentiometer was connected to measure and adjust the brightness and chromatic intensity of the LCD. This action aimed to optimize the appreciation of the text to be projected in later phases of the project. The link between the Arduino device and the LCD was established using the previously arranged wires. Besides, programming was performed using the Arduino IDE 2.2.1 development tool. At this stage, the respective codes for each element used were incorporated. This phase was crucial to ensure a harmonious integration and proper functioning of the components involved in the project.

Next, the LEDs were soldered with their corresponding wires to establish a robust and permanent electrical connection between these elements. Similarly, the push buttons were welded together with their respective cables. These actions were carried out to guarantee a reliable and durable electrical connection. Then, the connection of the buzzers to the Arduino in operation on the breadboard was established. This procedure was intended to enable the emission of the melodies programmed in the corresponding code of the Arduino.

Finally, the corresponding cables connect the LED lights and buttons to the breadboard. This process was carried out by linking the wires to the three-dimensional model, using the corresponding holes to establish the electrical connection (Fig. 10).

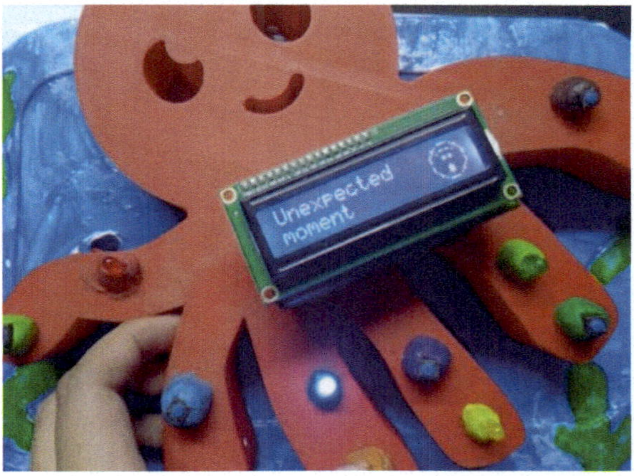

Fig. 10 Test to determine the optimal performance of the toy in the surprise emotion

On the other hand, the current project seeks the fusion of music therapy in a didactic and playful way aimed at children with Asperger's syndrome. To achieve this, exhaustive previous research has been carried out to identify relevant similarities, considering expert opinions and previous studies on the benefits of the piano and the influence of music on child development. The goal is to build a non-threatening space through music, a universal and emotional language, encouraging expression and interaction. Detailed experiences are narrated on how music facilitates differentiation, symbolization, and the development of new abilities, emphasizing the importance of patience, trust, and adapting to individual needs [24]. According to [25], to reduce stress and anxiety when problems arise, the best solution is music therapy because new emotions are generated through the type of music; that is, the individual's mood changes and even helps to improve self-confidence, communication, independence, self-awareness, and awareness of others and the ability to concentrate and pay attention. For this reason, this prototype has been created to help children with Asperger's syndrome, which consists of them pressing one of the six buttons according to how they feel and thus knowing how to help them through the music played.

Besides, individuals diagnosed with autism spectrum disorder (ASD) exhibit a marked interest in music compared to typically developing individuals. This interest offers the opportunity to enhance auditory and other interrelated skills. Music exerts a significant influence on patients with ASD, both affectively and cognitively. This sensitivity prompts them to express their feelings and emotions more fluently, mainly by employing nonverbal language as a communication channel [26]. It also highlights the recommendation to maximize and adapt the benefits provided by music therapy for children with ASD. These benefits include increasing expressiveness, improving social skills, strengthening self-esteem, overcoming psychological obstacles, improving auditory perception and motor coordination, personal development, and overcoming educational barriers. In addition, music therapy stimulates aspects such as memory, mental acuity, and cognitive speed, which is of great importance in its clinical and educational application [27].

3 Conclusions

This project aims to support children with Asperger's syndrome during times of stress. Inspired by an octopus, its design features six tentacles with emotional functions and differentiated lights to capture the attention of children with Asperger's syndrome. The purpose is to facilitate the child's expression of emotions and encourage the development of self-regulation skills through auditory responses, color, and sound associations with specific moods. This practical approach provides an effective tool for emotion management in the context of Asperger Syndrome.

This project is significantly enhanced by the understanding that music serves as a universal and profoundly emotional language, particularly beneficial for children with Asperger's syndrome during moments of heightened stress. Incorporating sensory tentacles that emit auditory responses and incorporating lights that shift in color to symbolize various emotions aligns perfectly with the recognized benefits of music and visual stimuli in aiding emotional expression and regulation. Grounded in research demonstrating meaningful responses from autistic children to such stimuli, this project's design transcends mere engagement. It aims to foster cognitive and emotional development, presenting a strategic and efficacious tool for navigating the intricate emotional landscapes associated with Asperger's syndrome. This project is a testament to the potential of integrating therapeutic elements into everyday tools, promising a brighter, more inclusive future for children on the spectrum.

References

1. Motlani V, Motlani G, Thool A (2022) Asperger syndrome (AS): a review article. Cureus 14(11)
2. Barahona-Corrêa JB, Filipe CN (2024) A concise history of Asperger syndrome: the short reign of a troublesome diagnosis. Front Psychol 6:2016
3. Faridi F, Khosrowabadi R (2017) Behavioral, cognitive and neural markers of Asperger syndrome. Basic Clin Neurosci 8(5):349–359
4. Santos L, Annunziata S, Geminiani A et al (2023) Applications of robotics for autism spectrum disorder: a scoping review. Rev J Autism Dev Disord
5. Carlier S, Van der Paelt S, Ongenae F, De Backere F, De Turck F (2020) Empowering children with ASD, and their parents: design of a serious game for anxiety and stress reduction. Sensors (Basel) 20(4)
6. Occelli V, Esposito G, Venuti P, Arduino GM, Zampini M (2013) Attentional shifts between audition and vision in autism spectrum disorders. Res Autism Spect Disord 7(4):517–525
7. Egüez Cueva PF et al (2015) Interfaz multimedia para el desarrollo y fortalecimiento de la destreza numérica para niños de 4 a 5 años diagnosticados con el síndrome de asperger del instituto fiscal de discapacidad motriz. B.S. thesis. USFQ, Quito

8. Guzmán G, Putrino N, Martínez F, Quiroz N (2017) Nuevas tecnologías: Puentes de comunicación en el trastorno del espectro autista (tea). Ter Psicol 35(3):247–258
9. Theofanopoulou N, Isbister K, Edbrooke-Childs J, Slovák P (2019) A smart toy intervention to promote emotion regulation in middle childhood: feasibility study. JMIR Ment Health 6(8)
10. El Kaliouby R, Robinson P (2004) Emotional hearing aid: an assistive tool for children with Asperger's syndrome. In: Keates S, Clarkson J, Langdon P, Robinson P (eds) Designing a more inclusive world. Springer, London
11. Documentation regarding robots used for helping children with autism. Accessed on 10 Sept 2021. Available online: https://www.robokind.com/robots4autism
12. Popescu AL, Popescu N, Dobre C, Apostol ES, Popescu D (2022) IoT and AI-based application for automatic interpretation of the affective state of children diagnosed with autism. Sensors (Basel) 22(7)
13. Georgieva-Tsaneva G, Andreeva A, Tsvetkova P, Lekova A, Simonska M, Stancheva-Popkostadinova V, Dimitrov G, Rasheva-Yordanova K, Kostadinova I (2023) Exploring the potential of social robots for speech and language therapy: a review and analysis of interactive scenarios. Machines 11(7)
14. Kumazaki H, Muramatsu T, Yoshikawa Y, Haraguchi H, Sono T, Matsumoto Y, Ishiguro H, Kikuchi M, Sumiyoshi T, Mimura M (2021) Enhancing communication skills of individuals with autism spectrum disorders while maintaining social distancing using two tele-operated robots. Front Psychiatry 11(598688)
15. Srinivasan SM, Lynch KA, Bubela DJ, Gifford TD, Bhat AN (2013) Effect of interactions between a child and a robot on the imitation and praxis performance of typically developing children and a child with autism: a preliminary study. Percept Mot Skills 116(3):885–904
16. Kumazaki H, Yoshikawa Y, Yoshimura Y et al (2018) The impact of robotic intervention on joint attention in children with autism spectrum disorders. Mol Autism 9(46)
17. Torturella Valadao C, Goulart C, Rivera H, Caldeira E, Bastos Filho T, Frizera-Neto A, Carelli R (2016) Analysis of the use a robot to improve social skills in children with autism spectrum disorder. Res Biomed Eng 32(2):161–175
18. Kouroupa A, Laws KR, Irvine K, Mengoni SE, Baird A, Sharma S (2022) The use of social robots with children and young people on the autism spectrum: a systematic review and meta-analysis. PLoS ONE 17(6)
19. Medina LHDLP, Galindo LRC (2018) Desarrollo de dos prototipos de juguetes didácticos para trabajar la terapia de integración sensorial en niños con Síndrome de Asperger. Bol Cien Inv Escuela Super Tizayuca 4(7)
20. Ghanouni P, Seaker L (2023) What does receiving autism diagnosis in adulthood look like? Stakeholders' experiences and inputs. Int J Ment Health Syst 17(16)
21. Yousif M (2021) Humanoid robot enhancing social and communication skills of autistic children. Artif Intell Robot Dev J 5:80–92
22. Cabibihan JJ, Javed H, Ang M Jr, Mariam Aljunied S (2013) Why robots? A survey on the roles and benefits of social robots for the therapy of children with autism. Int J Soc Robot 5(4):93–618
23. Conchinha C, de Freitas JC (2015) Robots & necesidades educativas especiales: robótica educativa aplicada a alumnos autistas. Challenges 2015:21
24. Applewhite B, Cankaya Z, Heiderscheit A, Himmerich H (2022) A systematic review of scientific studies on the effects of music in people with or at risk for autism spectrum disorder. Int J Environ Res Public Health 19(9):5150
25. Rodríguez LIL (2021) Musicoterapia aplicada al síndrome de Asperger a través del método Lettrock: intervención musical. Rev Invest Musicoter 5:54–79
26. Mayer-Benarous H, Benarous X, Vonthron F, Cohen D (2021) Music therapy for children with autistic spectrum disorder and/or other neurodevelopmental disorders: a systematic review. Front Psychiatry 12
27. Ke X, Song W, Yang M, Li J, Liu W (2022) Effectiveness of music therapy in children with autism spectrum disorder: a systematic review and meta-analysis. Front Psychiatry 13

Proposal for a Technological Model to Manage Dyslexia in Childhood with "Rodolfo"

Hector Valcarcel-Castillo, Jefferson Garay-Santos, Steven Espinoza-Ruiz, Frayder Meza-Morveli, Ronal Mora-Aybar, Camila Ticona-Sanchez, Deyby Huamanchahua, and Eber J. Ballon-Alvarez

Abstract The research aims to develop a specific device to address dyslexia in childhood, applying the methodology of "Extreme Programming" with a focus on rapid development, prioritization of the human factor, and collaboration with patients, in this case, infants with symptoms of dyslexia. It seeks to mitigate the challenges of this attention deficit disorder, which confuses words or numbers, generating significant problems, such as the fear of being excluded. The proposal includes an innovative technological model focused on early intervention and effective monitoring, which uses a robot specifically designed to help dyslexic children. This robot provides a playful way of learning, improving comprehension, and reducing attention deficit. The approach is implemented through a video game with Arduino technology, where

H. Valcarcel-Castillo (✉)
Department of Mechatronics Engineering, Universidad Continental, Huancayo, Peru
e-mail: hvalcarcel@continental.edu.pe

J. Garay-Santos · S. Espinoza-Ruiz · F. Meza-Morveli · R. Mora-Aybar · C. Ticona-Sanchez · E. J. Ballon-Alvarez
Universidad ESAN, Lima, Peru
e-mail: 22200127@ue.edu.pe

S. Espinoza-Ruiz
e-mail: 22101961@ue.edu.pe

F. Meza-Morveli
e-mail: 22200142@ue.edu.pe

R. Mora-Aybar
e-mail: 22200234@ue.edu.pe

C. Ticona-Sanchez
e-mail: 22200168@ue.edu.pe

E. J. Ballon-Alvarez
e-mail: jballon@esan.edu.pe

D. Huamanchahua
Department of Electrical and Mechatronics Engineering, Universidad de Ingenieria y Tecnologia—UTEC, Lima, Peru
e-mail: dhuamanchahua@utec.edu.pe

P. G. Bradford et al. (eds.), *Proceedings of IEMTRONICS 2024*, Lecture Notes in Electrical Engineering 1229, https://doi.org/10.1007/978-981-97-4780-1_20

the interaction is intuitive: children read the message on the LCD screen and press the button with the correctly spelled word. At the end of the game, a score is displayed, encouraging the child's competitive interest in learning and providing an effective and fun tool to tackle dyslexia from an early age.

Keywords Dyslexia · Childhood · Technology · Arduino

1 Introduction

Language represents a fundamental pillar in children's development because it is a tool that allows the acquisition of knowledge [1]. Oral language means an instrument of codification of thought [2]; this function has enormous relevance in the adequate development of the individual since it is the one that makes analysis, learning, abstraction, and synthesis progress [3]. Following this developmental process, written language is involved, and oral language forms the so-called phonological awareness [2]. This integration process is where difficulties are usually encountered due to coarticulation [4].

The acquisition of these skills can lead to situations where learning difficulties such as dyslexia can occur [5]. Identifying dyslexia early is essential to provide the necessary support and enable those affected to reach their full academic and personal potential [6]. Dyslexia refers to a challenge in acquiring the skills of decoding (reading aloud) and spelling. According to DSM5, dyslexia is categorized as a type of neurodevelopmental disorder [7]. The implementation of technology offers the possibility of improving learning conditions interactively [8], encouraging the re-education of children through personalized tools adapted to the specific difficulties of each patient [9].

Presently, various initiatives tackle this issue using different methods. Minecraft's case substantially analyzes video games' potential, particularly in enhancing reading fluency [10]. This instance is a model for creating a console that enhances children's learning via interactive games. Addressing developmental challenges, treatment options include virtual platforms, mobile apps, and physical prototypes tailored for motor disabilities [11], all utilizing interactive games to boost cognitive skills like visual perception, auditory processing, and memory [12].

Recognize the significant interaction of the infant with his or her physical environment [13] and understand the variations in addressing dyslexia as a function of children's home language [14], considering the possible influences of the transparent nature of the language [9] are crucial aspects in the development of our project. Hence, we present Rodolfo's concept: a robotic gaming console built on Arduino UNO. This inventive tool aims to aid in early dyslexia intervention by enhancing children's visual memory and cognitive development. Customizable to address individual challenges, it focuses on improving reading and writing skills through console-based video games, employing Arduino programming and extreme programming methodology. This adaptive approach assists in overcoming dyslexia-related hurdles by reinforcing visual memory retention.

2 Review of Publications

The first background involves the creation of a Retro Video Game Console using a printed circuit board (PCB) containing a microcontroller as its central component for signal processing. The primary aim is to enable individuals with basic electronics knowledge to build their affordable video game consoles [15]. The second background is about the design of a prototype PC video game controller for people with motor disabilities [16]. This prototype stands out for having adapted console accessories, such as the JoyCap [17], the Blind Hero [18], and the VI Tennis [11]. The project addressed vital aspects such as board design, control structure, and microcontroller programming. The third background is the NAO robot project for business and education [19], an autonomous robot incorporating advanced mechanical engineering and software technology to recognize faces, detect emotions, and interact with people [20]. The fourth background is Kaspar [21], a humanoid robot adopted as a therapeutic device for children with Autism Spectrum Disorder (ASD). Over several generations, Kaspar has evolved to improve its functionality and usability as a therapeutic and educational tool [21].

In this context, robots play a crucial role in detecting disabilities, with dyslexia being a prominent example. Since 2020, 47 states in the United States have integrated artificial intelligence-based assessments into robots to identify this disorder in early elementary school children [22]. In the field of dyslexia, the benefits of video games were explored [23], with Minecraft being a great support as an alternative solution; this research analyzed five students with dyslexia. It sought to improve the initial results by employing learning through video games, using visual memory in Minecraft. During the experiment, specific words were highlighted to facilitate their retention in memory [10]. In summary, this review ranges from video game console projects and adapted video game control to humanoid robots and the impact of video games on specific skills, providing a comprehensive overview of the background relevant to the proposed study.

3 Methodology

The methodology applied to the development of the project is extreme programming (XP) [24] that allows correcting carry out the development of the prototype programming; XP is an agile software development method to manage projects with efficiency, flexibility, and control and helps to produce a high-quality algorithm that facilitates the understanding of the project.

3.1 Project Description

This technological prototype is developed to aid and support children with dyslexia, a learning disorder characterized by reading difficulties stemming from challenges in sound-to-letter association. Besides, we have designed a robotic model to aid children in overcoming dyslexia. It functions by displaying words that commonly pose confusion or difficulty on an LCD screen. Children must press buttons below the words, selecting the correctly spelled one. Correct answers are tallied, increasing their score, while incorrect ones prompt a "failure" message without a score increment. This interactive process continues until the game's conclusion. If the child answers more correctly than incorrectly, a "You won" message is shown; otherwise, "You lost" is displayed. This method offers an engaging and educational tool for enhancing reading skills in dyslexic children.

3.2 Description of Components

The components used include Arduino Uno, female-male cables, a breadboard, an LCD screen, a buzzer, LEDs, push buttons, resistors, and filament. In the following, each component used, its characteristics, and the number of units used will be presented in detail to ensure the proper functioning of the project. Arduino Uno was the hardware development board; therefore, only one unit was implemented for the project. Dupont Female-Male cables and copper conductor elements helped us make the connections using 35 units. The breadboard, consisting of a rectangular board with a matrix of holes that connects the elements, was used in two units. The LCD screen, a flat, thin screen of several pixels, was used in one unit. The buzzer, an electromechanical device that produces sound, was used in one unit. Two units were used for both the LED and the push button; the former is a semiconductor that emits light, and the latter is a device used to open or close an electrical circuit. Resistors, current limiters in a circuit, were also used in a quantity of 4 units. Also, filaments, thermoplastic materials used to print 3D objects by depositing molten material in a 3D printer, were used for the project, approximately 250 g of filament.

3.3 Description of the Design

The design was based on the figure of the character BMO, taken from the animated series "Adventure Time". This character was built as the essential reference for the 2D and 3D modeling conception.

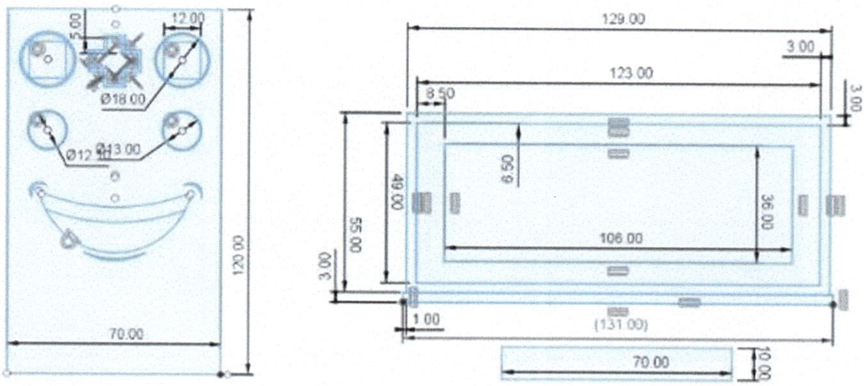

Fig. 1 2D model of the body and head

3.3.1 2D Design

Initially, a blueprint was drafted, delineating the dimensions of the first object. It measured 70 mm wide and 120 mm tall. Additional lines were added to adjoining surfaces, maintaining a consistent measurement of 120 mm. The drawings also included designated areas with precise measurements to accommodate the LEDs and buttons. In the next phase, the second element underwent 2D modeling. This entailed sketching lines across various planes to accommodate an LCD screen in a rectangular area. A rectangle was created to match the screen's size. Moreover, a circular section was planned to form a hemisphere, enhancing the prototype's head structure (Fig. 1).

3.3.2 3D Design

At the start of the process, the Fusion 360 extrusion tool provided volume to the prototype's body with a 2 mm thick extrusion. This stage also included a cut-out extrusion strategically made to reserve spaces for the rear insertion of materials such as LEDs and buttons. In addition, the backside was strategically left unextruded, allowing for better handling of the circuitry that would be inserted later. In the next phase, the extrusion of the second body was carried out. For the subsequent joining of the two parts, an assembly was implemented at the bottom of object 2, creating a space of 2 mm for the attachment of the first body. Subsequently, a shear extrusion was applied to incorporate the LCD. The symmetrical fusion between the rectangular head and the circular part was skillfully achieved using the elevation tool. To conclude the design, the revolution tool was used to sculpt the spherical part of the head, closing the creation of the final object of the project (Fig. 2).

Finally, a thin rectangular object was created to complement the top of the first object. This was designed with a width and height of 120 mm and a thickness of

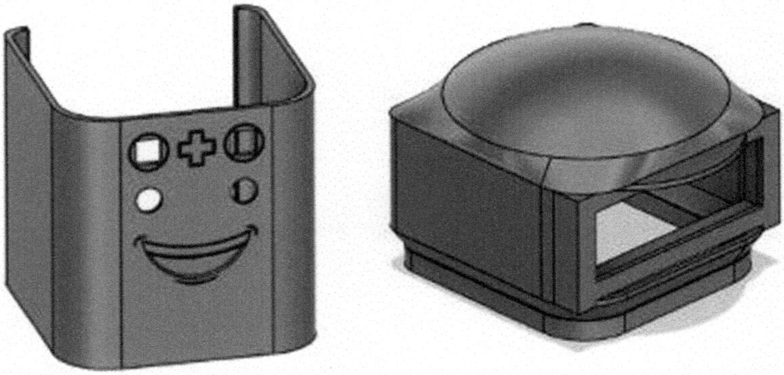

Fig. 2 3D model of the body and head

1 mm. An Arduino Uno was used to manage the power supply and the programming of the devices, taking advantage of its pins to make the connections.

3.4 Circuit Configuration

We will detail the configuration of the circuit and its respective operation according to the image presented. The Arduino Uno (1) allows the interconnections of all the elements with the breadboard (2), and in turn, the breadboard is connected to all the elements, such as LEDs (4), buzzer (6), resistors (3), LCD screen (7) and pushbuttons (5). We will use jumper cables to keep the connections organized and maintain a tidy environment in our circuit. These will facilitate the connection of prototypes, buzzers, LEDs, pushbuttons, and the LCD screen (7). Each of these will be fitted with resistors (3) to reduce the voltage and ensure the correct functioning of the components used. This will help to keep the internal circuit tidy and understandable.

Two-word exercises will be displayed on the LCD screen (7). There is a push button at the bottom of each word, and one of the two words is correctly written. The child must press the button (5) that contains the word accurately. If the child gets the answer right, the green LED (4) will light up, and the child's score will be increased by one, as shown on the LCD (7). If not, the red LED (4) will light up, and the score will not be increased by one point. The buzzer (6) will be activated for each answer, and a melody will sound, the tone varying according to whether the button pressed is correct or incorrect. The configuration of the circuit is developed in this way (Fig. 3).

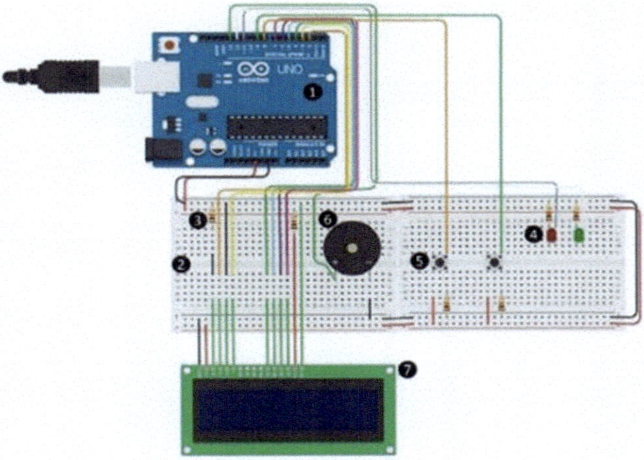

Fig. 3 The circuit was made in TinkerCAD with all components

3.5 Description of Programming

The flowchart accurately reflects the logical course of the program, from initialization to word presentation, response evaluation, and visual feedback. It highlights the efficient integration between hardware and software, providing an interactive experience through tones, melodies, and messages on the LCD. This offers a straightforward and accessible understanding of the practical implementation of the game (Fig. 4).

Function implementations, connections, and controllers were implemented in the C language. In addition, algorithms were developed to manage the interface with the LCD, manipulate the signals from the push buttons, and coordinate the responses via the LEDs and the buzzer.

First, the variable associated with the buzzer was declared on pin 13, and pins 12 and 11 were assigned to the green and red LEDs, respectively. The pushbuttons were assigned to pins 8 and 9. As for the LCD pin configuration, the RS pin was set to be connected to pin 2, the LCD RS to pin 3, the LCD EN to pin 4, the LCD D4 to pin 5, the LCD D5 to pin 6, and the LCD D6 to pin 7. In addition to these configurations, score, duration, btnPulldown, and btnPulldown2 were declared (Fig. 5).

Secondly, the "LiquidCrystal.h" library is included to manage the interface with an LCD. A LiquidCrystal object called "lcd" is created using specific connections (RS, EN, d4, d5, d6, d7). The program defines four two-dimensional arrays representing graphical patterns to display on the LCD. These matrices are "neutral," "OC" (busy), "OF" (busy out), and "OT" (temporarily busy). Each matrix contains binary information that determines the active pixels on the LCD, thus configuring different visual states (Fig. 6).

Thirdly, functions that make it possible to reproduce sounds in the buzzer employing a sequence of sounds in Hz frequencies were implemented (Fig. 7).

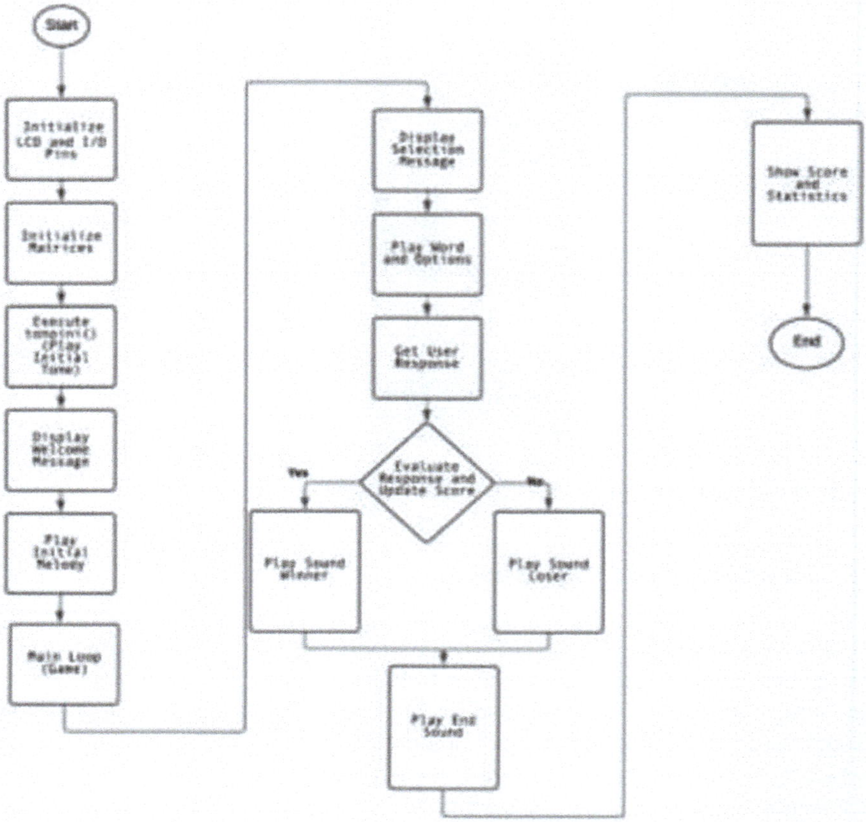

Fig. 4 Programming flowchart

Fig. 5 Pin declaration

```
int start;

int buzzerPin = 13;
int score = 0;
const int duration = 100;
int btnPulldown;
int btnPulldown2;
const int rs = 2;
const int en = 3;
const int d4 = 4;
const int d5 = 5;
const int d6 = 6;
const int d7 = 7;
```

```
#include <LiquidCrystal.h>
LiquidCrystal lcd(rs, en, d4, d5, d6, d7);
void setup() {
    lcd.begin(16, 2);
    pinMode(8, INPUT);
    pinMode(9, INPUT);
    pinMode(11, OUTPUT);
    pinMode(12, OUTPUT);
    pinMode(13, OUTPUT);
}
```

Fig. 6 LCD connection

Fig. 7 Winning sound installation

```
void playWinningSound() {
  tone(buzzerPin, 262);
  delay(duration);
  noTone(buzzerPin);
  delay(duration);
  tone(buzzerPin, 294);
  delay(duration);
  noTone(buzzerPin);
  delay(duration);
  tone(buzzerPin, 349);
  delay(duration);
  noTone(buzzerPin);
  delay(duration);
  tone(buzzerPin, 392);
  delay(duration);
}
```

A welcome message is implemented in the program's main loop when the variable "start" is set to 0. Actions such as initializing the tone and displaying the messages "Welcome!" and "I am Rodolfo" are executed. These interactions are intended to be user-friendly before moving on to the program's next step (Fig. 8).

Finally, a series of 12 algorithms that operate as follows were implemented. First, a loop is created to execute a sequence of prints if none of the buttons are pressed. This loop is subject to the constraint that if the button is pressed in the wrong position, the red LED is activated, accompanied by the sound "Losing" for 1.5 s.

```
while (digitalRead(8) == 0 && digitalRead(9) == 0) {
     lcd.print("1.Table  2.Tadle");
     lcd.setCursor(0, 1);
     lcd.print("Choose an option ");
     delay(100);
     lcd.clear();
}
if (digitalRead(9) == 1) {
     lcd.print("INCORRECT!!!");
     lcd.setCursor(0, 1);
     lcd.print("+0 pt. Total =");
     lcd.print(score);
     digitalWrite(11, HIGH);
     print(ot);
     delay(1500);
     digitalWrite(11, LOW);
} else {
     lcd.print("CORRECT....!");
     lcd.setCursor(0, 1);
     lcd.print("+1 pt. Total =");
     lcd.print(score);
     digitalWrite(12, HIGH);
     playWinningSound();
     delay(1500);
     digitalWrite(12, LOW);
}
```

Fig. 8 Final implementation of the words

3.6 *Description of Implementation*

In the first stage, component tests were carried out, and the first test was executed, where the corresponding connections were made according to the TinkerCAD prototype. In the second stage, the second test and implementation of the physical model were carried out, where the connections of the TinkerCAD prototype were also carried out. In this last stage, the final implementation and connections were carried out. The corresponding fixings of the components were also carried out, together with the implementation of the necessary programming (Fig. 9).

a. **Start and Presentation**

When the system is switched on, the LCD screen welcomes the user with a "Welcome" message. Simultaneously, the essential software for developing the activities is loaded.

Fig. 9 Test of the final connection of components

b. **Presentation of Challenging Words**

The system displays two carefully selected words on the LCD screen: correct and incorrect. This choice is made to challenge the reading skills of children with dyslexia positively (Fig. 10).

Fig. 10 Word selection in the physical prototype

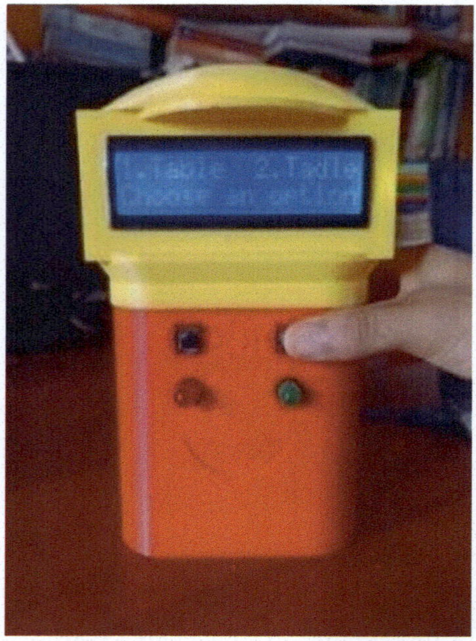

Fig. 11 User interaction in the physical prototype

c. **User Interaction**

The robot actively invites the user to participate by prompting them to press the button corresponding to the word they consider correct. Clear instructions are presented on the LCD screen, promoting an intuitive and participatory experience (Fig. 11).

d. **Response Evaluation**

The built-in programming evaluates the user's response. If the choice is correct, the system responds with a friendly victory sound, lights the green LED, and displays an affirmative "CORRECT" message on the LCD. An error sound is activated in case of an incorrect choice, the red LED lights up, and an "INCORRECT" message is displayed on the LCD screen. This immediate feedback reinforces learning positively (Fig. 12).

e. **Repetition for Reinforcement**

This process is repeated six times in each play session, providing continuous practice with different words. Repetition reinforces reading skills and facilitates effective learning.

f. **Conclusion of the Session**

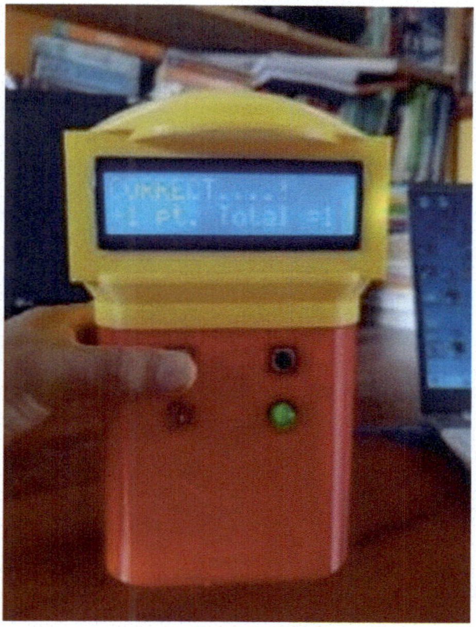

Fig. 12 Programme response in the physical prototype

Upon completing the six interactions, the system can provide an overall assessment of the user's performance. It can give motivating conclusion messages, encouraging the player to continue to participate in improving their reading skills.

This design ensures an interactive educational experience tailored to specifically address the needs of dyslexic children, making the learning process an engaging and rewarding activity (Fig. 13).

4 Conclusions

In summary, our project focuses on enhancing the reading skills of children with dyslexia, emphasizing visual memory, and fostering autonomy in learning. Implementing the Arduino Uno to present the material reflects the convergence of technology and education. Consistent with previous research, using video games as an educational tool is directly linked to developing the metacognitive system, especially in memory. This study supports the idea that students with dyslexia can experience significant improvements in their ability to tackle fluent reading through innovative approaches that encourage participation and reflection on their learning process.

In addition, to optimize both the learning experience and the accessibility of the device, the implementation of an adaptive feedback system is planned. This system will adjust the game's difficulty level according to each child's progress. Therefore, the collaboration of experts in education and technology is contemplated to ensure an adaptation to the real needs and its sustainability over time.

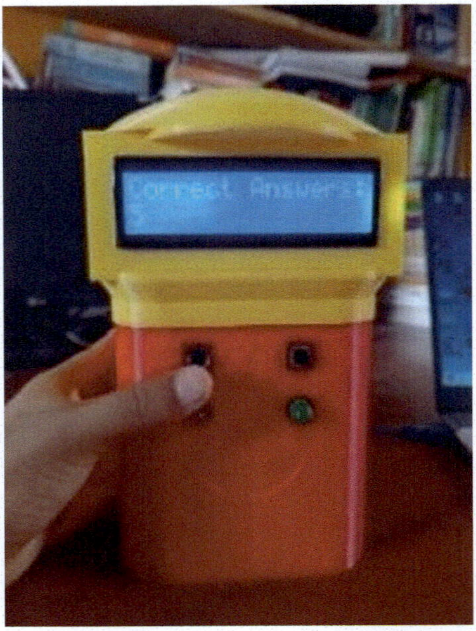

Fig. 13 Final evaluation of user interventions on physical prototype

References

1. Coronado RC (2016) La importancia del lenguaje y el aprendizaje en el desarrollo del niño. UNIFE 22:54–58
2. Caravolas M, Bruck M (1993) The effect of oral and written language input on children's phonological awareness: a cross-linguistic study. J Exp Child Psychol 55(1):1–30
3. Bigas M, Salvador I (2017) La importancia del lenguaje oral en educación infantile. In: Dispositivo digital de documentos de la UAB, vol 46, pp 5–8
4. Sylvia Defior FS (2011) La conciencia fonémica, aliada de la adquisición del lenguaje escrito. Rev Logop Foniatr Audiol 31:2–13
5. Serrano F, Defior S (2004) Dislexia en Español: estado de la cuestión. Electron J Res Educ Psychol 2(2):13–34
6. Snowling MJ (2013) Early identification and interventions for dyslexia: a contemporary view. J Res Spec Educ Needs 13(1):7–14
7. Snowling MJ, Hulme C, Nation K (2020) Defining and understanding dyslexia: past, present, and future. Oxf Rev Educ 46(4):501–513
8. Al-Dokhny AA, Bukhamseen AM, Drwish AM (2022) Influence of assistive technology applications on dyslexic students: the case of Saudi Arabia during the COVID-19 pandemic. Educ Inf Technol 27:12213–12249
9. Ingkavara T, Panjaburee P, Srisawasdi N, Sajjapanroj S (2022) The use of a personalized learning approach to implementing self-regulated online learning. Comput Educ Artif Intell 3
10. Jiménez Porta AM, Díaz Martínez E (2018) Impact of videogames on reading fluency in children with and without dyslexia. The case of Minecraft. RELATEC 17(1):78–90
11. Mena Chicaiza DA, Revelo Martínez JM (2020) Diseño de un prototipo de control de videojuegos de PC para personas con discapacidad motriz. Quito
12. Jiménez Porta AM, Díaz Martínez E (2018) Content analysis of apps and video games: implications in cognitive processes in initial reading. Apertura 10(1):71–87
13. Lorenzo ST (2017) La dislexia y las dificultades en la adquisición de la lectoescritura. Rev Curr For Prof 21(1):423–432

14. Adlof SM, Hogan TP (2018) Understanding dyslexia in the context of developmental language disorders. Lang Speech Hear Serv Sch 49(4):762–773
15. Ramírez Muñoz C (2018) Desarrollo de una consola de videojuegos retro. Universitat Politècnica de Catalunya
16. Taheri A, Weissman Z, Sra M (2021) Design and evaluation of a hands-free video game controller for individuals with motor impairments. Front Comput Sci 3(751455)
17. Cristiano E, de Faria F, Pereira B, Vivacqua A (2011) JoyCap: an alternative to the joypad for motor impaired gamers. In: Proceedings of the companion proceedings of the 10th Brazilian symposium on human factors in computing systems and the 5th Latin American conference on human-computer interaction (IHC+CLIHC'11). Brazilian Computer Society, Porto Alegre, BRA, pp 10–11
18. Yuan B, Folmer E (2008) Blind hero: enabling guitar hero for the visually impaired. In: Proceedings of the 10th international ACM SIGACCESS conference on computers and accessibility (ASSETS'08). Association for Computing Machinery, New York, pp 169–176
19. Amirova A, Rakhymbayeva N, Yadollahi E, Sandygulova A, Johal W (2021) 10 years of human-NAO interaction research: a scoping review. Front Robot AI 8(744526)
20. Shamsuddin S, Yussof H, Ismail LI, Mohamed S, Hanapiah FA, Zahari NI (2012) Humanoid robot NAO interacting with autistic children of moderately impaired intelligence to augment communication skills. Proc Eng 41:1533–1538
21. Wood LJ, Zaraki A, Robins B, Dautenhahn K (2021) Developing Kaspar: a humanoid robot for children with autism. Int J Soc Robot 13(3):491–508
22. Chu S-T, Hwang G-J, Tu Y-F (2022) Artificial intelligence-based robots in education: a systematic review of selected SSCI publications. Comput Educ Artif Intell 3(100091)
23. Franceschini S, Gori S, Ruffino M, Viola S, Molteni M, Facoetti A (2013) Action video games make dyslexic children read better. Curr Biol 23(6)
24. Shrivastava A, Jaggi I, Katoch N, Gupta D, Gupta S (2021) A systematic review on extreme programming. J Phys Conf Ser 1969

Discalcu: Mathematical Device for Children with Dyscalculia

Jaime Huaytalla-Pariona, Yessenia Bravo-Lazaro, Angel Aguilar-Contreras, Sandro Avila-Agurto, Mattias Caballero-Salazar, Deyby Huamanchahua, and Eber J. Ballon-Alvarez

Abstract Dyscalculia, a significant obstacle in the mastery of mathematics that affects the understanding and application of numerical concepts, is the focus of this project. Using Arduino, an interactive calculator designed to address this challenge in mathematical learning is explicitly developed. The project seeks to provide an educational tool for children with dyscalculia, allowing them to enter numbers and operations and receive feedback through an LED display. Inspired by the importance of creativity in education, the project seeks to cultivate imagination in the mathematical learning process, citing Einstein and Ken Robinson as referents of this philosophy. The project aims to address dyscalculia from a perspective that recognizes the importance of action and diversity of skills in mathematical learning. Using Arduino components such as a keyboard, LCD, resistors, and buzzer, a system that validates mathematical operations entered by users is developed. This approach aims to be accessible and adaptable to improve the mathematical skills of those with dyscalculia and other cognitive limitations. The study considers precedents such as Dyscalculia Game, Math Fun, and Game Multicorrida, projects that explore various strategies to

J. Huaytalla-Pariona (✉)
Department of Mechatronics Engineering, Universidad Continental, Huancayo, Perú
e-mail: jhuaytalla@continental.edu.pe

Y. Bravo-Lazaro · A. Aguilar-Contreras · S. Avila-Agurto · M. Caballero-Salazar · E. J. Ballon-Alvarez
Universidad ESAN, Lima, Perú
e-mail: 22200118@ue.edu.pe

A. Aguilar-Contreras
e-mail: 22200133@ue.edu.pe

S. Avila-Agurto
e-mail: 21101154@ue.edu.pe

M. Caballero-Salazar
e-mail: 22200125@ue.edu.pe

D. Huamanchahua
Department of Electrical and Mechatronics Engineering, Universidad de Ingenieria y Tecnologia—UTEC, Lima, Peru
e-mail: dhuamanchahua@utec.edu.pe

P. G. Bradford et al. (eds.), *Proceedings of IEMTRONICS 2024*, Lecture Notes in Electrical Engineering 1229, https://doi.org/10.1007/978-981-97-4780-1_21

address dyscalculia. The project is presented as a valuable and accessible tool for improving mathematical skills. The project is presented as a practical and accessible tool to enhance math skills in children with dyscalculia. By offering an intuitive interface and providing feedback, it seeks to address numerical challenges and cultivate creativity in the learning process. These results suggest that interactive and adaptive strategies may be instrumental in improving the understanding and application of mathematical concepts in individuals with specific cognitive difficulties.

Keywords Arduino · Dyscalculia · Mathematical device · Children · Education

1 Introduction

Dyscalculia, a disorder that affects the understanding [1] and use of mathematical concepts [2], is at the crossroads of education, as the educational psychologist Jean Piaget indicated: "Intelligence cannot develop without action" [3]. In line with this premise, Sir Ken Robinson, a recognized education expert, states that "intelligence manifests itself in diverse forms and creativity is as important as academic prowess" [4]. This project aims to address dyscalculia from a perspective that recognizes the importance of action and diversity of skills, seeking educational strategies that foster creativity in understanding and applying mathematical concepts.

In the Peruvian context, dyscalculia represents an educational challenge [5]. Despite the growing recognition of learning disorders [6], the scarcity of resources and specific programs makes it difficult to address them. The lack of early diagnosis and adapted interventions negatively impacts the academic development of children with dyscalculia [7]. The diversity of socioeconomic and cultural contexts complicates implementing inclusive strategies [8]. Francia Monrroy highlights "Pedagogical intervention as a key to prevent and correct dyscalculia in regular basic education students: an applied and transformative approach" [9], and Alexandra Graciela Árizaga González highlights: "Addressing dyscalculia in basic education: problem-solving as a key didactic strategy" [10]. These perspectives emphasize the need for collaboration to develop adaptive approaches and ensure equitable opportunities in the Peruvian educational system [11].

This creative project using Arduino proposes a dynamic calculator for children with dyscalculia disorder [12]. Both boys and girls can enter numbers and operations, receiving brief feedback through an LED display confirming their answers. As Albert Einstein said, "Imagination is more important than knowledge" [13]. In the words of Albert Einstein, "Creativity is as important in education as literacy, and we should treat it with equal importance" [14]. The initiative seeks to cultivate imagination and creativity in mathematical learning by providing an interactive and educational tool [15].

The literature discusses various strategies for addressing dyscalculia in children. Lim Hean Pronto, for example, describes it as a "Learning disability due to visual deficit in mathematical operations" [16], highlighting similarities and differences between them. Bee-bot and educational robots promote cognitive and logical skills by teaching programming and showing divergences in performance between different age groups. The NAO robot [17] conducts personalized exercises, showing significant improvements, such as a 25% increase in numerical identification and a 30% increase in speed and accuracy in mental calculations. Despite sharing the search for specific interventions for dyscalculia and the use of educational technology, these strategies differ in approaches and results, highlighting the need to adapt to the different needs of children. They emphasize the diversity of approaches to improve numeracy skills in mathematics education effectively. In the words of Rizawati Binti Rohizan, "Children with dyscalculia face difficulties in understanding mathematical concepts, impacting both school learning and everyday skills" [16].

2 Background

From our research, we have identified several projects related to robots designed to express emotions and aimed at children with dyscalculia. In this section, we will highlight two particularly relevant backgrounds that support the research report.

First, the research focuses on Dia a Dia, an electronic visual diary tailored for individuals with autism or communication challenges. It enables organizing activities, predicting future events, and fostering communication. Users can personalize it by adding people and frequent locations [18]. Second, in [19], the research compared how elementary and kindergarten students performed with Bee-bot, an educational robot, uncovering varying abilities. Bee-bot enhances problem-solving skills through programming. Third, the Martha-Bot prototype [20], a robotic assistant, aids spatial orientation development in children with mild to moderate intellectual disabilities through its turtle-shaped design and interactive features. It's viewed positively by experts for educational and therapeutic use. Finally, the NAO robot [21] administers exercises to enhance math skills in children with dyscalculia. It offers interactive tasks for numerical identification, mental calculation exercises, and contextualized problem-solving to improve understanding and application of math concepts. Results indicate significant improvements in numerical recognition, mental calculation speed, and problem-solving ability, supporting the robot's efficacy as a personalized tutor for addressing dyscalculia difficulties and enhancing critical mathematical skills in children.

3 Methodology

3.1 Project Description

The present project elaborates on Arduino, which can help people with dyscalculia problems as a learning improvement device. The project is based on the communication between the Arduino and implements that emulate a calculator. This project is based on Arduino's ability to process data and compare it with preset values, which allows determining whether an operation is valid. The calculator can be used in therapies for people with cognitive limitations, improving their coordination and spatial perception.

The Arduino project is characterized by its easy implementation and accessibility, which makes it suitable for people with dyscalculia and other cognitive limitations. In addition, the device can be adapted to include different types of math operations, such as addition, subtraction, multiplication, and division, allowing users to practice and improve their math skills.

3.2 Description of Components

The project has fundamental elements that are crucial to its optimal operation. Key components include the Arduino Mega 2560 R3 (1), which acts as the central brain, the 4×4 keypad (1) that facilitates interaction, two 220 Ω resistors (2) for current control, a 3.3 V piezo (1) for generating acoustic signals, a 16×2 LCD screen (1) for displaying information, a medium protoboard (1) for connecting and organizing the circuits, female-male cables (100) for interconnection, a 9 V battery (1) for power supply, a 100 kΩ potentiometer (1) for fine adjustments, and finally, a blue cloth (one-meter sheet) that provides an aesthetic component to the overall design of the project. The combination and coordination of these components are essential to ensure the efficient and effective performance of the project.

3.3 Design Description

BMO is one of the main characters in the animated Adventure Time series. He is the video game system and friend of the main characters, Finn and Jake. The calculator project design seeks to improve digital inclusion in entertainment and education. The intelligent calculator on Arduino focuses on adapting the design and functionality of "BMO" to give a friendly appearance as our primary intention for a project aimed primarily at children and feeling the comfort of using a 3D printed design and visually pleasing.

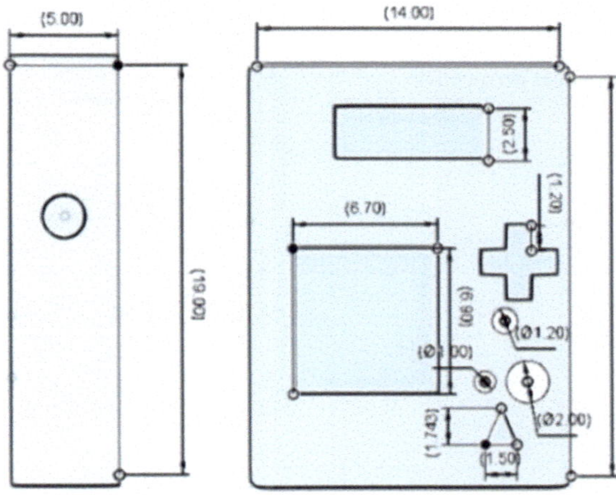

Fig. 1 2D model of the prototype and dimensions: lateral and frontal perspective

3.3.1 2D Design

The dimensions within the prototype were used with measurements evaluated in centimeters and scaled to improve them in the form of a physical object within the Fusion 360 editing program. The measures used for the object that has a rectangular prismatic shape were given according to its length (19 cm), width (14 cm), in general, and the measures for the spaces where we would place the interactive devices such as the screen are 7.20 cm wide and 2.50 cm long, the keypad 4 × 4 keyboard is dimensioned in 6.90 cm long and 6.70 cm wide, in addition to calculating the sufficient space for the internal insertion of the devices (Fig. 1).

3.3.2 3D Design

The 3D design presents the previously elaborated sketches as a solid object that can be implemented to print. With the corresponding dimensions, such as the animated character BMO, we developed a prism with slits that will allow us to insert the devices used within the project; in addition to including aesthetic buttons without a specific function, the implementation that led it to be printed had an approximate 34 h of development. Thus, the project was obtained physically elaborated (Fig. 2).

3.4 Circuit Configuration

Using the TinkerCAD platform, a new circuit incorporating an Arduino and a breadboard (1–2) was created. Appropriate wiring was used to connect the two components. The 5 V pin of the Arduino was used to provide power to the breadboard.

Fig. 2 3D model of the prototype: frontal and orthogonal perspective

This pin supplies a 5 V current that can be used to power other circuit components connected to the breadboard. In addition, the Arduino's GND (ground) pin was connected to the breadboard. This pin establishes a common reference point, ensuring that all parts of the circuit share a similar electrical reference, allowing for proper current flow. The breadboard serves as a platform to connect various electronic components and facilitate their interconnection. We link an LED display (3) to the Arduino. For this connection, we use pins 12, 11, 5, 4, 3, and 2 of the Arduino to interface with the display. The LED display's GND, VO, RW, and Led sections connect to the breadboard, specifically to the ground rails. This ensures a standard reference for the current flow. On the other hand, the VCC and LED points of the LED display are connected to the 5 V section of the breadboard, providing the necessary power for the display to operate.

This connection configuration allows the Arduino pins assigned to the LED display to control its operation correctly. We then introduced a buffer (4) to emit a sound when starting the system or interacting with the calculator. We connected pin 10 of the Arduino to the buffer, allowing it to transmit signals and generate sound. In addition, we joined the buffer pin to the breadboard ground rail, establishing a standard reference for current flow. Finally, we added a 4 $\times$ 4 keypad (5) to the circuit. We use pins 9, 8, 7, and 6 of the Arduino for the keyboard rows and pins 13, A0, A1, and A2 for the columns. These connections allow the power supply and organization of the keyboard matrix layout. The Arduino can organize the keyboard matrix layout by mapping the pins and detecting key presses by combining rows and columns. This configuration enables interaction between the keyboard and the Arduino, allowing the system to see and process critical presses to perform specific actions on the calculator (Fig. 3).

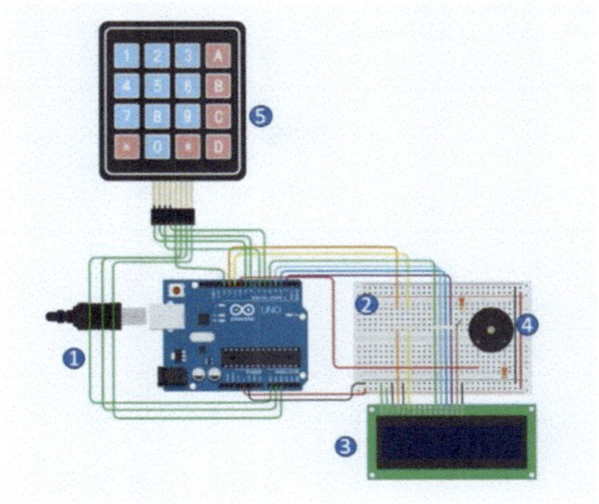

Fig. 3 Circuit model on TinkerCAD

3.5 *Description of Programming*

The code begins by importing the libraries that control the LCD and the matrix keyboard. The LCD is connected as follows: Regarding the data pins (D4–D7), D4 is connected to pin 2 of the Arduino, D5 to pin 3, D6 to pin 4, and D7 to pin 5 of the Arduino. Regarding the control pins, the RS (Register Select) pin is connected to pin 12 of the Arduino, and the Enable pin is connected to pin 11 of the Arduino.

The backlight pins (VSS and VDD) are also connected to the 5 V and GND pins. This connection scheme establishes communication and control between the LCD and the Arduino, ensuring that the information and commands sent from the Arduino are appropriately displayed on the LCD.

The matrix keyboard is connected using a matrix of rows and columns: The rows of the keyboard (row 1–row 4) are connected to pins 9, 8, 7, and 6 of the Arduino, while the columns (column 1–column 4) are connected to pins 13, A0, A1 and A2 of the Arduino, thus establishing the connection configuration.

The musical notes and their respective durations are defined using arrangements. These arrangements represent the melodies that will be played through the buzzer. Each element of the arrangement contains a musical note characterized by its frequency and the duration of that note in the song (Fig. 4).

The void setup() function is essential in any Arduino program. It is executed once when the microcontroller is turned on or restarted. Initial configurations are carried out here, and the components used during program execution are prepared.

In the code provided, key actions are performed in setup(): It starts the 16 × 2 LCD, defining its structure so that the Arduino can display and organize its information. In addition, it configures pin 10 as an output to control the buzzer. Welcome messages are printed, and instructions are provided on the LCD to guide the user. A festive tune is also played at startup, adding an auditory element to the initial user experience.

Fig. 4 Musical notes implemented in the program

```
int christmasMelody[] = {
  NOTE_E4, NOTE_E4, NOTE_E4, NOTE_E4,
  NOTE_E4, NOTE_E4, NOTE_E4, NOTE_G4,
  NOTE_C4, NOTE_D4, NOTE_E4,
  NOTE_F4, NOTE_F4, NOTE_F4, NOTE_F4,
  NOTE_F4, NOTE_E4, NOTE_E4, NOTE_E4, NOTE_E4,
  NOTE_E4, NOTE_D4, NOTE_D4, NOTE_E4,
  NOTE_D4, NOTE_G4
};

int christmasNoteDurations[] = {
  4, 4, 2, 4,
  4, 2, 2, 4,
  4, 4, 4,
  2, 4, 4, 4,
  2, 4, 4, 4, 4,
  2, 4, 4, 2,
  4, 4
};;

int twinkleMelody[] = {
  NOTE_C4, NOTE_C4, NOTE_G4, NOTE_G4, NOTE_A4, NOTE_A4,
  NOTE_G4,
  NOTE_F4, NOTE_F4, NOTE_E4, NOTE_E4, NOTE_D4, NOTE_D4,
  NOTE_C4
};

int twinkleNoteDurations[] = {
  4, 4, 4, 4, 4, 4,
  2,
  4, 4, 4, 4, 4, 4,
  2
};
```

The code manages the input of numbers and mathematical operators through state logic, using variables such as num1Input, operation Input, and num2Input. It updates the LCD with the entered numbers and prepares it for performing mathematical calculations. It performs basic mathematical operations such as addition, subtraction, multiplication, and division based on the entered numbers and the selected operator. In addition, it verifies whether the user's answer matches the calculation result. Employs LCD messages to guide the user, indicating expected actions, such as entering numbers, operators, or verifying answers.

Void clearLCD() clears the LCD screen and resets some variables to their initial state. The float calculate() function performs basic mathematical calculations (addition, subtraction, multiplication, and division) using the values of num1, num2, and operation. It uses a switch case to determine the operation to be performed and returns the result. In addition, it handles the divide-by-zero scenario by displaying an error message on the LCD and resetting the state to enter new data (Fig. 5).

The void validateAnswer(float result) function validates the answer entered by the user by comparing it with the expected outcome of the calculation. It prompts

Fig. 5 Configurations calculate

```
float calculate() {
  float result = 0;
  switch (operation) {
    case '+':
      result = num1 + num2;
      break;
    case '-':
      result = num1 - num2;
      break;
    case '*':
      result = num1 * num2;
      break;
    case '/':
      if (num2 != 0) {
        result = num1 / num2;
      } else {
        lcd.clear();
        lcd.print("Error: Div by 0");
        delay(2000);
        clearLCD();
      }
      break;
  }
  return result;
}
```

the user to enter their answer via messages on the LCD and compares the answer to the calculation result. It provides visual and audible feedback according to the accuracy of the answer, displaying congratulatory messages in case of success or error messages with a melody that prompts the user to try again in case of an error.

4 Implementation

Having finalized the general implementation of the circuits and the programming code corresponding to the prototype system, we show you the following evidence that demonstrates the correct operation of Discalcu in the different interaction tests:

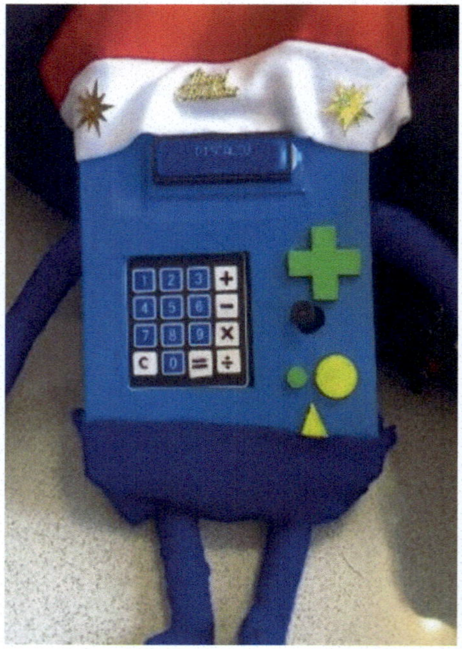

Fig. 6 Discalcu's presentation message

- **Presentation message**: In this photo, we can observe Discalcu with the presentation message that will indicate that the prototype is turned on for use (Fig. 6).
- **Operation insertion**: Discalcu will indicate that you can enter digits that can be operated with the four basic operations: addition, subtraction, multiplication, and division (Fig. 7).
- **Message and answer insertion**: When entering the operation, a new message will appear immediately, allowing us to solve and enter the operation corresponding to the numbers previously entered (Fig. 8).
- **Correct and incorrect answer message**: Discalcu will process the answer and will be able to recognize the result of the operation, corroborating if it is right or wrong; when entering a correct answer, the device will display a positive message accompanied by a pleasant melody and motivating the user to continue elaborating operations, if it is incorrect it will display a text with a message indicating that the answer is wrong, in addition to displaying a message encouraging the user to find the correct answer to the operation entered (Fig. 9).
- **Other operations**: In turn, we present other example operations that can be done with the device, including the "clear" button, which allows the insertion of numbers, and the "equal" button, which enables the display of the result (Fig. 10).

Fig. 7 Discalcu displays the sum operation

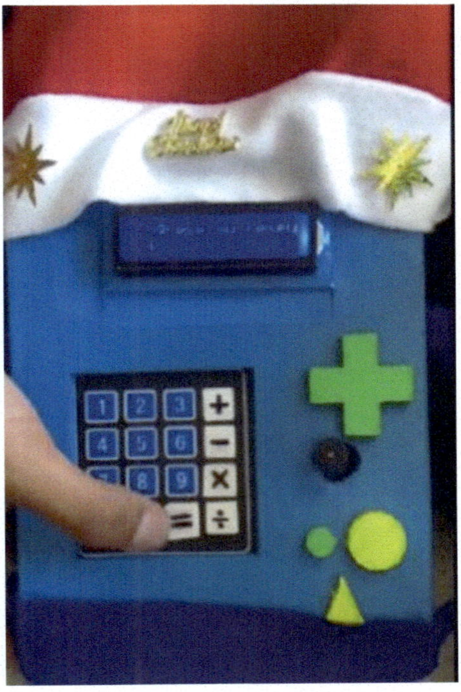

Fig. 8 The insertion of the answer is made

Fig. 9 Correct answer message

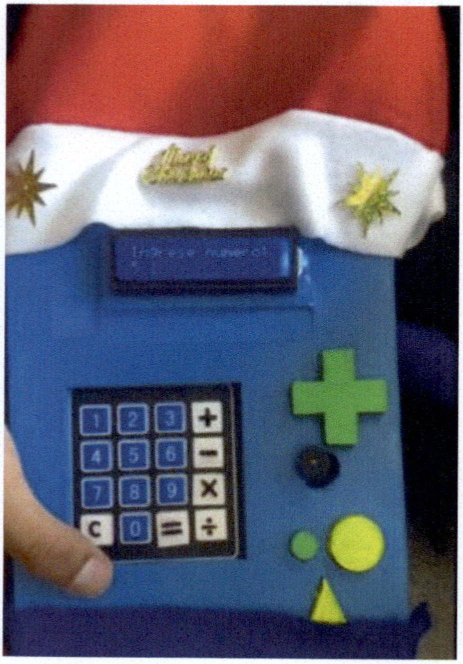

Fig. 10 Subtraction operation insertion

5 Conclusions

Implementing an interactive calculator designed to support dyslexic children is crucial in educational and inclusive development. This project seeks to transform the mathematical learning experience for these children by allowing them to practice numerical operations in a controlled manner adapted to their needs. Allowing them to enter numbers and operations and receive immediate feedback on the accuracy of their answers creates a safe environment that fosters confidence and progress in mathematical skills. This initiative, which focuses on persistence and guided correction, can potentially cultivate the self-esteem and academic progress of children with dyscalculia, opening doors to more sustained and meaningful learning in mathematics.

On the other hand, it announces the upcoming implementation of practices with children from primary schools to investigate the benefits of using this project as a learning tool. These practices will be carried out in collaboration with local schools and the support of a clinical psychologist who will provide us with the most appropriate methods for children's learning. It is expected that the results will contribute significantly to the advancement of knowledge in the field of primary education.

The strategic use of the blue cloth in this project is justified for two fundamental reasons:

1. The first lies in associating blue with serenity and calm. By implementing this color, we seek to provide children with an environment that fosters relaxation and tranquility, enhancing their ability to concentrate and absorb knowledge more effectively.
2. Besides its psychological function, blue also serves an aesthetic purpose. The choice of this shade can contribute to creating a visually appealing and harmonious atmosphere, adding a valuable aesthetic component to the educational environment.

Although it does not influence the robot's operational function, the presence of the blue fabric seeks to generate a welcoming and favorable environment for the development of cognitive activities for the people targeted by the project, such as the children's community and those with dyscalculia disorder.

References

1. Kaufmann L, von Aster M (2012) The diagnosis and management of dyscalculia. Dtsch Arztebl Int 109(45):767–77
2. Kunwar R (2022) Impacts of dyscalculia in learning mathematics: some considerations for content delivery and support. In: Learning disabilities—neurobiology, assessment, clinical features and treatments. IntechOpen
3. Piaget J, Cook M et al (1952) The origins of intelligence in children, vol 8. International Universities Press, New York

4. Robinson K, Aronica L (2009) The element: how finding your passion changes everything. Penguin
5. Zerafa E (2015) Helping children with dyscalculia: a teaching programme with three primary school children. Procedia-Soc Behav Sci 191:1178–1182
6. Kaufmann L, Mazzocco MM, Dowker A, von Aster M, Göbel SM, Grabner RH, Henik A, Jordan NC, Karmiloff-Smith AD, Kucian K, Rubinsten O, Szucs D, Shalev R, Nuerk HC (2013) Dyscalculia from a developmental and differential perspective. Front Psychol 4(516)
7. Haberstroh S, Schulte-Körne G (2019) The diagnosis and treatment of dyscalculia. Dtsch Arztebl Int 116(7):107–114
8. Santos FH, Ribeiro FS, Dias-Piovezana AL, Primi C, Dowker A, von Aster M (2022) Discerning developmental dyscalculia and neurodevelopmental models of numerical cognition in a disadvantaged educational context. Brain Sci 12(5)
9. Francia Monrroy JE (2015) Intervención pedagógica en la discalculia en estudiantes del segundo grado de primaria
10. González AGÁ, Freire JFR (2021) La discalculia en alumnos de la educación básica. Soc Tecnol 4(3):432–446
11. Fatwana H, Dasari D, Juandi D (2023) The use of learning aids for dyscalculia: systematic literature review. Al-Ishlah J Pendidik 15(1):639–652
12. Üstün S, Ayyıldız N, Kale EH, Mançe Çalışır Ö, Uran P, Öner Ö, Olkun S, Çiçek M (2021) Children with dyscalculia show hippocampal hyperactivity during symbolic number perception. Front Hum Neurosci 15:687476
13. Martínez-González JS (2020) Ken Robinson. Una mirada al interior del modelo educativo fomentando la creatividad. In: Con-Ciencia Serrana Boletín Científico de la Escuela Preparatoria Ixtlahuaco, vol 2, no 3, pp 16–17
14. Einstein A (1879) Frases célebres de. Recuperado de: http://www.frasedehoy.com/frase/1716/nuncaconsideres-el-estudio-como-una-obligacion-sino-como-una-oportunidad
15. Rosen Y, Stoeffler K, Simmering V (2020) Imagine: design for creative thinking, learning, and assessment in schools. J Intell 8(2)
16. Rohizan R, Soon LH, Mubin SA (2020) MathFun: a mobile app for dyscalculia children. J Phys Conf Ser 1712(1)
17. Shamsuddin S, Ismail L, Yussof H, Zahari N, Shaari M, Hafizan H, Jaffar A (2011) Humanoid robot NAO: review of control and motion exploration. In: Proceedings—2011 IEEE international conference on control system, computing and engineering
18. Zúñiga HGA, Noriega AR, Ramírez YM, Pérez JFF (2023) Estudios e innovaciones educativas empleado a la tecnología
19. Diago PD, Arnau D, González-Calero JA (2018) La resolución de problemas matemáticos en primeras edades escolares con Bee-bot. Mate Educ Soc 1:36–50
20. Parapi-Peña N, Velasquez-Angamarca V, Lema-Condo E, Robles-Bykbaev V (2023) MarthaBot: Un asistente robótico para el soporte en la enseñanza de habilidades de orientación espacial a niños con discapacidad intelectual leve y moderada. Rev Ibér Sist Tecnol Inf E57:156–165
21. Papadopoulou MT et al (2022) Efficacy of a robot-assisted intervention in improving learning performance of elementary school children with specific learning disorders. Children 9

A Two-Body Low-Frequency Piezoelectric Wind Energy Harvester for Environmental Sensing

Sadia Bakhtiar, Amal Z. Hajjaj, Hailing Fu, and Stephanos Theodossiades

Abstract This paper investigates a novel Two-body Piezoelectric Wind Energy Harvester (TbPE-WEH) designed to generate voltage from wind-induced vibrations. The harvester consists of a flat plate (bluff body) attached to the free end of a piezoelectric cantilever beam. A description of the energy harvester and experimental setup is provided, explaining the output voltage generation for a range of wind speeds. The study investigates the system dynamics using experimental modal analysis, which reveals the natural frequencies of the system. The activation of system nonlinearities causes a noticeable shift in the frequency content as the wind speed varies. At high wind speeds, the harvester appears to propel into limit cycle oscillations (LCOs), exhibiting self-sustained motion caused by aerodynamic instabilities. This novel two-degree-of-freedom vibration energy harvester design may harness wind energy over a wide speed range by gaining large oscillations as compared to the vibration energy harvesters operating under aerodynamic instabilities that are presented in the literature. A maximum open circuit voltage of approximately 6–7 V is achieved by this non-optimised energy harvester prototype.

Keywords Aerodynamic force · Self-sustained oscillations · Nonlinear response · Piezoelectric wind energy harvester

1 Introduction

Wind energy harvesters based on flow-induced vibrations have garnered a lot of attention in recent years as they have the potential to continuously supply energy for unsupervised wireless sensor networks in remote locations [1]. Electronic devices and self-powered wireless sensor networks serve as essential elements for remote

S. Bakhtiar (✉) · A. Z. Hajjaj · S. Theodossiades
Loughborough University, Loughborough, UK
e-mail: S.Bakhtiar@lboro.ac.uk

H. Fu
Beijing Institute of Technology, Beijing, China

P. G. Bradford et al. (eds.), *Proceedings of IEMTRONICS 2024*, Lecture Notes in Electrical Engineering 1229, https://doi.org/10.1007/978-981-97-4780-1_22

tracking, Internet of Things (IoT), environmental monitoring, and structural monitoring [2]. As a widely available, ecologically friendly renewable energy source, wind energy harvesting is a popular choice for self-powered technologies. The fluid–structure interaction is potentially utilized for energy harvesting from low-speed winds in the surroundings of transmission towers, high-rise buildings, bridges, forests, and highways to power wireless sensor nodes (WSNs) [3]. Fluid–structure interaction phenomena can be categorized as vortex-induced vibrations (VIVs) [4], galloping [5], and flutter [6]. In each of these studies, nonlinearities in the system are crucial for efficient energy harvesting. In such systems, when the incoming airflow speed is low, the damping of the system is positive, and the induced structural vibration will slowly decay.

When the incoming airflow speed gradually increases and exceeds a critical value, the aerodynamic instability is initiated and negative damping is achieved which leads to self-excited oscillations, at a Hopf bifurcation. Limit cycle motion takes place, and the structural amplitude becomes stable [7]. Based on aeroelastic instabilities, the power generation process incorporates the aero-electro-mechanical coupling. The non-linear aerodynamic force induces mechanical vibrations through fluid–structure coupling, which are subsequently converted into electrical energy by piezoelectric effects and electromechanical coupling [8].

Consequently, researchers developed wind energy harvesting systems utilizing the above-mentioned phenomenon. Hu et al. [9] investigated the installation of rods or fins along the edges of bluff bodies as a modification to enhance aerodynamic performance. To improve the performance of a wind energy harvester, Liu et al. [10, 11] constructed modified bluff bodies, with their results concluding that sustained resonance can be attained across various airflow velocity ranges. Wang et al. [12] discovered that the application of Y-shaped attachments causes the flow-induced vibration of a cylinder to transform from VIV to galloping, achieving higher output performance. The Y-shaped attachments mounted on the bluff body can alter the aerodynamic properties of the energy harvester. Sun et al. [13] developed a D-section cylindrical bluff body and established a vibration mode that combined VIV and galloping with 193% increased output with respect to a traditional bluff body of square cross section. Similarly, Wang et al. [14] presented a piezoelectric energy harvester using hybrid bluff bodies with varying cross sections and achieved coupling VIV and galloping. Most of the previous literature used the prismatic square sectioned or cylindrical bluff bodies fixed to the electricity generating beams to achieve galloping and VIVs or both coupled to scavenge energy from aerodynamic instabilities.

A similar trend is followed by the energy harvester prototype presented in this work. To the best of the authors' knowledge, there has been no presentation of a two-body arrangement in the literature comprising a flat plate (as a bluff body) and a cantilever beam, for energy harvesting purposes. The novelty of this design originates from both the shape of the bluff body and the two-body arrangement. Thus, through the pivot point connecting the two bodies, the current design takes advantage of the synergistic interaction between the piezoelectric cantilever beam and the coupled plate. This coupling increases the oscillations, when dynamically responding to the wind excitation. At lower wind speeds the beam deflection is small but as the wind speed reaches a critical threshold, there is a rapid increase in vibration amplitude, driving the system in limit cycle oscillations. This behavior is comparable to a prior study [15] which exploits self-excited oscillations of a cantilever beam only. In our case, the two-body novel mechanism allows for efficient energy harvesting, as the system exploits the nonlinear effects of the interface. Experimental investigation of this kind has been carried out for the first time. The major contribution of this paper lies in its mechanism, which demonstrates superior performance in terms of higher oscillations and greater output voltage (from unrefined prototype) compared to the existing studies. Considering the research advancement in the field of energy harvesting, the proposed energy harvester may be promising for capturing wind energy for environmental monitoring.

2 Energy Harvester Design and Working Principle

The schematic of the Two-body Piezoelectric Wind Energy Harvester (TbPE-WEH) is depicted in Fig. 1. A rigid plate mass is connected to the free end of a cantilever beam with a piezoelectric material attached to the latter. The bluff body in this case is a square sectioned flat plate which can rotate about the pivot point when subjected to wind excitation. The mass of the bluff body is 27 g, and the dimensions of the cantilever beam are $L = 150$ mm, $B = 14$ mm, and $T = 0.5$ mm. The beam's width and length are parallel to the y and z axes, respectively, as shown in the figure. The z-direction, which is also the direction of gravity, is used to hang the bluff body, mounted at the free end of the beam. It can freely rotate in the x–z plane about the pivot. The wind direction is directed into the page, hitting the flat plate and as a result, the beam vibrates in the x direction due to the aerodynamic force acting on the bluff body. The flow induced vibrations are utilized to target low frequency motions and induce large beam vibration amplitudes.

The performance of the harvester is tested in the laboratory. A fan is used as a source for the wind excitation. The wind speed is adjusted by varying the current of the power supply connected to the fan. Initially, the harvester is subjected to low wind speeds and gradually this is increased. The harvester is tested for a wind speed ranging from 4 to 12 m/s. At lower speeds, the aerodynamic forces acting on the system are insufficient to cause noticeable vibrations. The energy harvester experiences increased aerodynamic forces with an increase in the wind speed. The system can go through a transitional stage where the harvester's structural dynamics and aerodynamic forces appear to interact. The aerodynamic forces grow powerful enough to induce the system to experience self-sustained oscillations above a critical wind speed, which is 8 m/s. A positive feedback loop is produced by the interaction of the structural vibrations and the aerodynamic forces which results in amplitudes that get larger over time. The system reaches a state where the vibrations' amplitudes keep increasing.

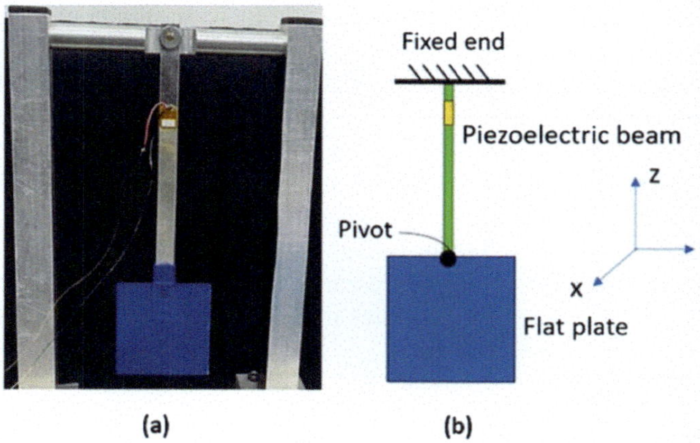

Fig. 1 Piezoelectric wind energy harvester: **a** actual device and **b** schematic of the TbPE-WEH

3 Results and Discussion

To determine the natural frequencies of the energy harvester, impact hammer testing is performed using Dewesoft data acquisition hardware. Node locations are defined on the bluff body and the beam. The bluff body is then excited to measure the vibration response via a laser vibrometer. The frequency response functions are obtained to indicate the natural frequencies of the system. The stabilization diagram is obtained from the experiment as shown in Fig. 2 which helps to identify the consistency or stability of poles as the order is increased. It can be seen that the dominant frequencies are 1.8, 5.1, and 8.8 Hz.

The experimental setup is used for testing the performance of the TbPE-WEH. Figure 3 shows the configuration of the device. The wind flow hits the surface of the plate in the x direction. The harvester is tested for a range of wind speeds. Time histories of the vibration response and output voltage are recorded simultaneously. A laser vibrometer is used to capture the beam deflection and the data acquisition is utilized to measure and plot both, beam vibration and output voltage. All the readings are taken when the system reaches a steady state of motion. Initially, the harvester is set into motion at 4 m/s, and gradually the wind speed is increased. When subjected to low wind speeds ranging between 4 and 7.5 m/s the amplitude of vibration is small, i.e., 20–60 mm/s as depicted in Fig. 4a1, b1 and the voltage output is also little. At 4 m/s, Fig. 4a2 shows that the Fast Fourier Transform (FFT) indicates peaks at 1.8 and 4.9 Hz which are close to the natural frequencies of the harvester. As the wind speed increases the frequencies slightly shift, i.e., to 2.3 and 4.3 Hz at 7.5 m/s. Overall, before a critical wind speed, the beam deflection is small; the frequency response indicates weak coupling between the beam and the plate, and the system tends to have small deviations from the modal frequencies.

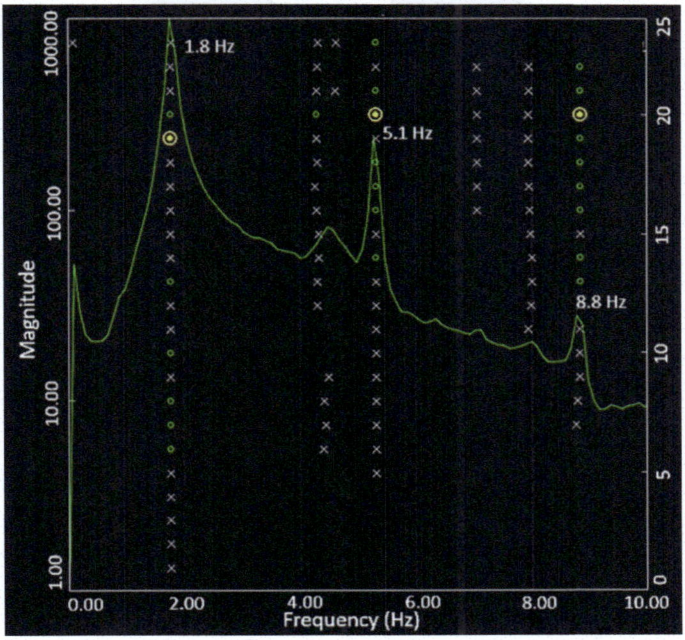

Fig. 2 Stabilization diagram indicating modal frequencies

Fig. 3 Configuration of the PE-WEH to harness energy from wind

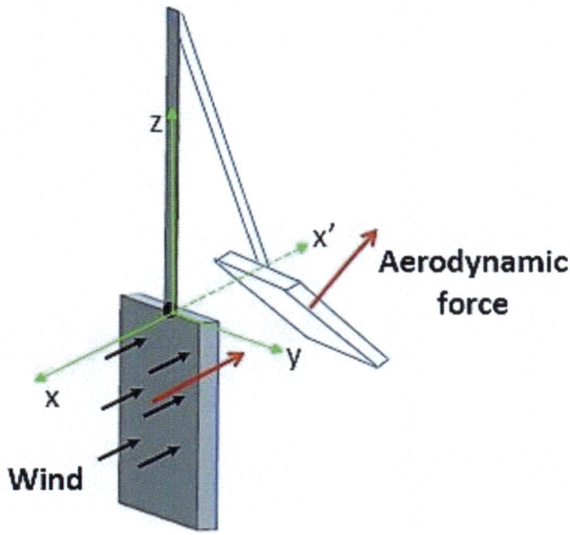

Fig. 4 Time histories and FFT of the harvester at **a** 4 m/s and **b** 7.5 m/s

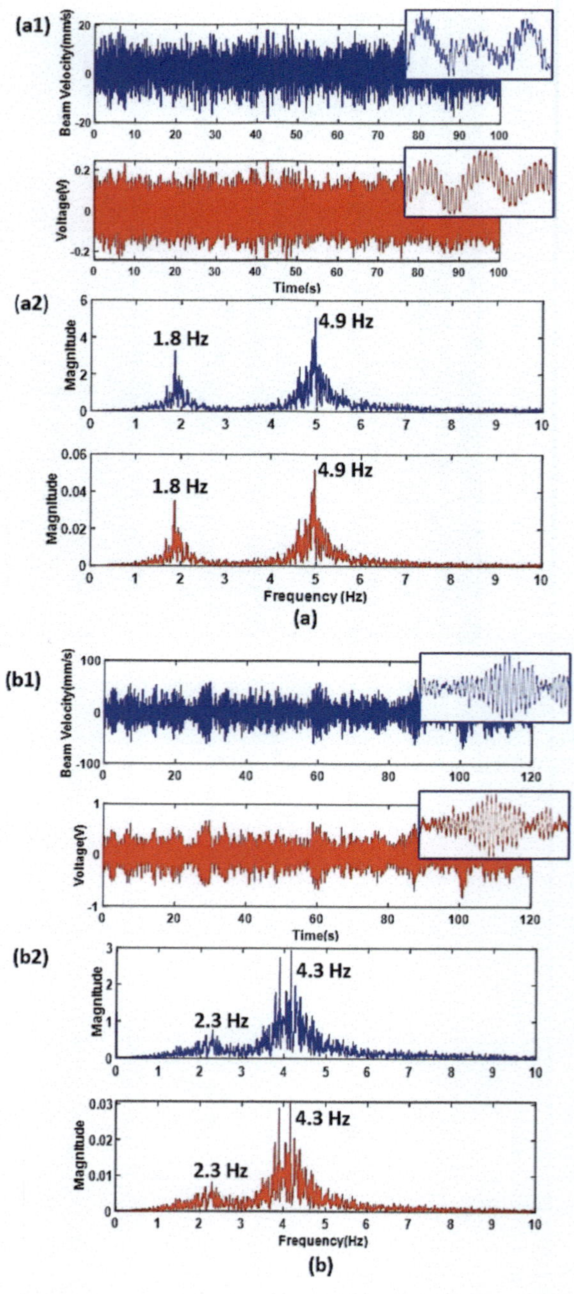

The trend observed in the speed range of 4–7.5 m/s, shows that the frequencies are near to the dominant modal frequencies with slight shifts. However, as the critical wind speed, i.e., 8 m/s is approached there is a sudden increase in the vibration response and a greater shift in the frequency. Figure 5 illustrates the time histories (Fig. 5a1, b1) and FFTs of the harvester after the critical threshold is met. Figure 5a2 shows that the first peak appears at 3.2 Hz, which is a significant shift in the frequency, whereas 2 more peaks appear at 6.4 and 9.8 Hz which are multiples of the first. As observed from the experiments, there is a gradual shift in frequency with increasing wind speed. The aerodynamic force causes the plate to swing at large angles. The deflection of the cantilever beam influences the plate motion by altering its equilibrium position and introducing additional moment due to its angle of inclination. The interaction between the beam and the plate produces complex dynamics in the system response. The plate oscillations can also reflect and impact the bending of the beam, resulting in a dynamic interaction shifting the frequency content. At 12 m/s, the frequency shifts to 3.5 Hz (Fig. 5b2). Also, due to aggressive fluid–structure interactions, the coupling between the beam and the bluff body is strong at high speeds, reflecting a highly nonlinear behavior in the system.

4 Conclusions

This paper investigates the dynamic response of a TbPE-WEH with intriguing nonlinear properties developed for remote sensing applications. A bluff body is pivoted to the free end of a piezoelectric cantilever beam subjected to wind flow. The bluff body is in the form of a flat plate that oscillates because of the pressure difference between the two sides. The piezoelectric material attached to the beam generates electricity from mechanical deformation. The system is analyzed through modal analysis finding correspondence between the natural frequencies and the frequency at which the beam vibrates.

The findings of the study are highlighted as:

- At lower wind speeds (4–6 m/s): The aerodynamic force is minimal, and the system response is close to linear. As a result, the FFT shows peaks close to the initial modal frequencies (1.8 and 5.1 Hz).
- Wind speed (6–8 m/s): Nonlinearities become pronounced as the wind speed increases. The displacement of the bluff body slightly increases, and the beam deforms more comparatively, resulting in a frequency shift in both modes.
- Higher wind speeds (8–12 m/s): As wind speed increases, the aerodynamic forces acting on the system increase, resulting in higher displacements of the plate and beam deflection. The nonlinearity is more evident, and a portable self-sustained response is observed. The response frequencies change dramatically compared to the natural frequencies, resulting in FFT peaks at 3.2 and 6.4 Hz (and slightly above these with increasing wind speed) due to the strong coupling between the plate and beam.

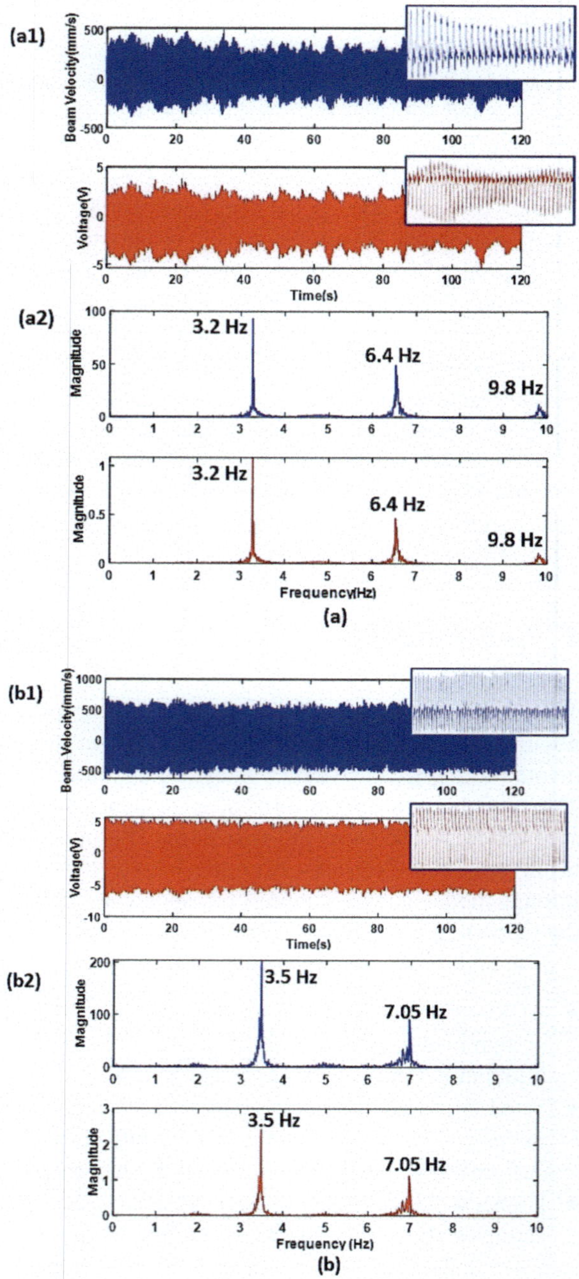

Fig. 5 Time histories and FFT of the harvester at **a** 8 m/s and **b** 12 m/s

- The device is capable of generating an open circuit output voltage of 6–7 V which is appropriate for harvesting energy from wind to wireless sensing nodes intended for environmental observations. This novel approach gives a practical solution to obtain power in regimes where wind flow varies.

As part of the future work, a dynamics model will be developed to study the above discussed nonlinear effects in correlation with the experimental observations.

Acknowledgements This work has been supported by the Wolfson School of Mechanical, Electrical and Manufacturing Engineering, Loughborough University via a Ph.D. scholarship (UK), the Overseas Ph.D. Scholarship Scheme Phase-III, Higher Education Commission (Pakistan).

References

1. Nabavi S, Zhang L (2016) Portable wind energy harvesters for low-power applications: a survey. Sensors 16(7):1101
2. Yang Z, Zhou S, Zu J, Inman D (2018) High-performance piezoelectric energy harvesters and their applications. Joule 2(4):642–697
3. Wang J, Yurchenko D, Hu G, Zhao L, Tang L, Yang Y (2021) Perspectives in flow-induced vibration energy harvesting. Appl Phys Lett 119(10)
4. Wan C, Tian H, Shan X, Xie T (2023) Enhanced performance of airfoil-based piezoelectric energy harvester under coupled flutter and vortex-induced vibration. Int J Mech Sci 241:107979
5. Siriyothai P, Kittichaikarn C (2023) Performance enhancement of a galloping-based energy harvester with different groove depths on square bluff body. Renew Energy 210:148–158
6. Latif U, Dowell EH, Uddin E, Younis MY (2023) Parametric aerodynamic and aeroelastic study of a deformable flag-based energy harvester for powering low energy devices. Energy Convers Manage 280:116846
7. Tan T, Zuo L, Yan Z (2021) Environment coupled piezoelectric galloping wind energy harvesting. Sens Actuators A Phys 323:112641
8. Wang J, Geng L, Yang K, Zhao L, Wang F, Yurchenko D (2020) Dynamics of the double-beam piezo–magneto–elastic nonlinear wind energy harvester exhibiting galloping-based vibration. Nonlinear Dyn 100:1963–1983
9. Hu G, Tse KT, Kwok KC, Song J, Lyu Y (2016) Aerodynamic modification to a circular cylinder to enhance the piezoelectric wind energy harvesting. Appl Phys Lett 109(19)
10. Liu FR, Zou HX, Zhang WM, Peng ZK, Meng G (2018) Y-type three-blade bluff body for wind energy harvesting. Appl Phys Lett 112(23)
11. Liu FR, Zhang WM, Peng ZK, Meng G (2019) Fork-shaped bluff body for enhancing the performance of galloping-based wind energy harvester. Energy 183:92–105
12. Wang J, Zhou S, Zhang Z, Yurchenko D (2019) High-performance piezoelectric wind energy harvester with Y-shaped attachments. Energy Convers Manag 181:645–652. https://doi.org/10.1016/j.enconman.2018.12.034
13. Sun W, Jo S, Seok J (2019) Development of the optimal bluff body for wind energy harvesting using the synergetic effect of coupled vortex induced vibration and galloping phenomena. Int J Mech Sci 156:435–445
14. Wang J, Gu S, Zhang C, Hu G, Chen G, Yang K, Li H, Lai Y, Litak G, Yurchenko D (2020) Hybrid wind energy scavenging by coupling vortex-induced vibrations and galloping. Energy Convers Manage 213:112835
15. Zakaria MY, Al-Haik MY, Hajj MR (2015) Experimental analysis of energy harvesting from self-induced flutter of a composite beam. Appl Phys Lett 107(2)

Utilization of Personalized PageRank for Protein–Protein Interaction Analysis and Similarity-Based Complex Network Analysis: A Brief Review

Arjab Sengupta, Srijita Chakraborty, and Malay Gangopadhyay

Abstract Getting solutions that are stringent and noteworthy to complex mathematical procedures through algorithms is not something new and has been a key for finding answers to tough questions. Such an algorithm extensively in use today is the PageRank algorithm. Proposed by Larry Page and Sergey Brin in 1996 and originally finding its roots in the Graph Theory, it has become a significant solution method in cases where complex chain analysis is involved. The algorithm makes use of basic yet intricate graph theory concepts, such as in-degrees and out-degrees corresponding to each node to give weights to a particular webpage and find its consequent rank or weight. This paper discusses the very basics of the algorithm. The paper also reviews the work done by Gabor Ivan and Vince Grolmusz, who, in 2010, used PageRank to determine the protein interaction chains in our body and rank them according to their corresponding importance. In addition, this paper also has provided a brief idea on another original research based on similarity-based methods for link prediction in complex networks. The driving formula for finding the PageRank value of a webpage is: $W_i = d + \Sigma_{I=1, I \neq j} I_{ij} (W_i/n_i)$. Brief elaboration on the significance of "*d*" or the damping factor in the above-mentioned formula by taking appropriate examples and through some light on the basic and structural working of the PageRank algorithm has also been done.

Keywords Nodes · In-degrees · Out-degrees · Edges · Protein–protein interaction chains (PPI)

A. Sengupta (✉) · S. Chakraborty · M. Gangopadhyay
Department of Electronics and Communication, Institute of Engineering and Management, Kolkata 700091, India
e-mail: arjab.sengupta2022@iem.edu.in

P. G. Bradford et al. (eds.), *Proceedings of IEMTRONICS 2024*, Lecture Notes in Electrical Engineering 1229, https://doi.org/10.1007/978-981-97-4780-1_23

1 Introduction

The problem of identifying nodes that are important in relatively large/big networks was prevalent in several fields and still is, but the answers to those problems till date used to appear in conjunction with that of the World Wide Web or commonly attributed as www. There are certain terms related to the graph theory which are: Nodes or vertices, in-degree, out-degree, edges or branches. Each of these terms serves to propose a significant meaning. Nodes basically refer to each point or junction of the meeting of two branches/edges. The in-degree refers to the number of incoming branches in each node whereas the out-degree refers to the total number of outgoing branches from each node. Each of the lines joining two nodes is known as vertices or branches. These uses are historically well established, mostly in matters where scientometry is involved [1–3]. However, in cases involving the web graph, these degrees are easy to manipulate by simply adding an infinite number of referring edges into the graph.

Also, the problem of link prediction is not something new and serves to be an axial field in complex network analysis [1–5]. It also aims to establish new connections between nodes in a given network. As discussed later, such tasks may have a tremendous range of activities and outcomes or applications, such as suggestion of friends in social media networks [5–10].

In this review paper, a brief over two such advancements in the field of network science coupled with Graph based algorithms has been given. First one is about a work done on PageRank usage for protein–protein chain interaction analysis and the second one is a brief on another usage of the algorithm, in complex network analysis [1–8].

2 Basics of PageRank

Keeping things simple, let us take an example, Fig. 1.

From the example, a simple search algorithm is being shown. However, if someone wants to search for some other person having the name "Virat" (for e.g., Virat Singh), will be shown the same suggestions, as a matter of fact. The chief standout question here serves to be, "WHY?" and "HOW?"

The answer is simple, every automation process requires an algorithm flowchart in order to work, or maybe in other words, they require a set of instructions. The algorithm of PageRank does the work in this case [1–3]. For our convenience, let us make an assumption.

There exist 3 webpages, namely, *A*, *B*, and *C*. Each of these webpages has one respective active hyperlink, each pointing to other webpages. Further, we are going to neglect two cases for our convenience. They being:

(1) The websites have hyperlinks which direct the users back to the same webpage.
(2) Each of the webpages does not have more than 1 hyperlink.

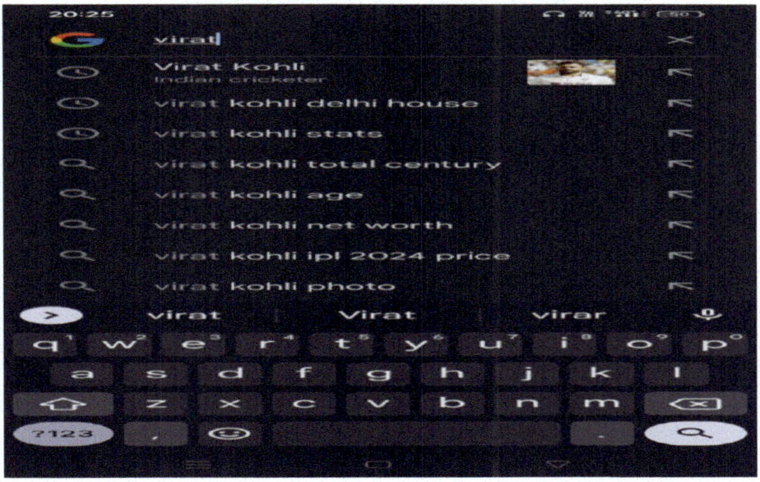

Fig. 1 Search result on typing "Virat"

Thus, assuming the total probability to be 1, i.e., $P(A + B + C) = 1$, then, the probability of reaching each webpage is given by 1/3, i.e., 0.33. Now, if introducing one more webpage, i.e., D, then the new probability becomes 1/4, i.e., 0.25. Similarly, consideration of the probability of reaching webpage *A* from the other webpages, namely, *B*, *C*, and *D* is done. In that case, since each webpage has an outbound link only to the rest of the webpages:

a. Probability of reaching webpage *A* from webpage *B*: 1/3.
b. Probability of reaching webpage *A* from webpage *C*: 1/3.
c. Probability of reaching webpage *A* from webpage *D*: 1/3.

Therefore, the total probability: 1/3 + 1/3 + 1/3.

This can be expressed in the formula [11]:

$$\mathrm{PR}(A) = \mathrm{PR}(B) + \mathrm{PR}(B) + \mathrm{PR}(C),$$

where, PR $\rightarrow$ Probability.

Hence, the PageRank transferred from a provided page to targets of its outbound links on the next iteration is divided equally among the outbound links.

Instead, considering that *B* contains the outbound links to webpages *C* and *A* whereas *C* contains only one outbound link, i.e., to *A* only and *D* contains 3 outbound links to *A*, *B*, and *C* respectively.

In that case, the new formula becomes:

$$\mathrm{PR}(A) = \mathrm{PR}(B)/2 + \mathrm{PR}(C)/1 + \mathrm{PR}(D)/3.$$

Hence, it can also be said that the PageRank conferred by an outbound link is equal to the document's own PageRank score divided by the number of outbound links L:

$$\mathrm{PR}(A) = \mathrm{PR}(B)/L(B) + \mathrm{PR}(C)/L(C) + \mathrm{PR}(D)/L(D).$$

In a more generalized case, the formula can further be elaborated as [11–15]:

$$\mathrm{PR}(u) = \sum_{v \in Bu} \mathrm{PR}(v)/L(v)$$

i.e., the PageRank value of a particular webpage u is nothing but a summation of the PageRank values of the probability of reaching u from the outbound links of each page. ***Bu*** is the set or the collection of all the pages containing outbound links to webpage u. ***L*(*v*)** expresses the number of outbound links coming from each webpage [11–15].

However, the original formula for finding the PageRank weightage of a page lies in the driving formula [13]:

$$W_i = d + \sum_{I=1, I \neq j} I_{ij}(W_i/n_i),$$

where

W_i stands for the weight of a particular webpage,
I stands for the webpage having a hyperlink to the main page,
n_i stands for the total number of outbound hyperlinks from webpage i.

Finally, d stands for the damping factor.

Now, the main question might be, what does "*d*" stand for in this case. Well, in most cases "*d*", while determining PageRank weightage of a page stands for the damping factor and in most cases, it is taken as 0.85 [16]. Keeping in mind that the PageRank rank algorithm works on a logarithmic scale and not on a linear scale, a PageRank value of 10 signifies the most important pages (most authoritative pages) and a corresponding scale of 0 represents an irrelevant page (lowest quality pages).

Recent applications and advancements of PageRank have produced significant contributions in various areas. Such applications of PageRank include social network analysis, recommender system, academic citation analysis, epidemiological modeling, cybersecurity, traffic flow optimization, biological network analysis, etc [3–8, 15–23].

3 Evaluation of Damping Factor and the Google (G) Matrix

In simple terms, the PageRank algorithm holds that an imaginary surfer randomly clicking on links "will" eventually stop clicking [18]. Hence, the probability that, at any step, the person continues following his process of going on clicking on random links is what we call, the damping factor or d. Similarly, the probability that the person jumps to any other random page is $1 - d$ [18].

In order to further elaborate on this topic, small discussion on understanding the Google Matrix has also been provided. The Google matric is a matrix that is stochastic in nature and is used by Google's PageRank algorithm [1]. It expresses a graph with edges representing links between pages, which are in turn taken as nodes of the graph, as discussed earlier. However, the main condition for the purpose to be served right is that the matrix must be: stochastic, irreducible, and aperiodic in nature.

However, in order to make the Google matrix, firstly, a corresponding H matrix must be formed [24–29]. This H matrix helps in forming relation between pages or nodes.

For example, if there are n pages, a H matrix (i, j) can be made by:

1. Filling in each entry (i, j) with 1, in case, if a node i has any link to a node j, and otherwise, 0. This forms the adjacency matrix of links.
2. Now, we must divide each row by k_i, where k_i serves to form the total number of links to other pages from node i.

This forms the H matrix. However, the H matrix is generally not stochastic, aperiodic, or irreducible in nature [30–32]. Hence, this makes the H matrix unsuitable for being used as a PageRank algorithm.

However, although the H matrix isn't suitable for the algorithm, construction of another matrix, namely the Google Matrix or the G matrix is done, which in turn, is stochastic, irreducible, and periodic in nature [18]. In this paper, further discussions on the transformation of the H matrix to G matrix and the concept of the damping factor have been given [33].

Considering that there is only one sink state S given by the formula:

$S = H + (1/N\ e^T)$ [33], where N represents the number of nodes. Then, by reducing H to an irreducible matrix, we also tend to make the corresponding matrix aperiodic. Therefore, the final Google matrix G can be put as [34–36]:

$$G = o'S + \left(1 - O'S\right)1/N\ ee^T \tag{1}$$

This makes the sum of all the non-negative elements inside each matrix as unit. From the H matrix computed above in addition to the consideration of a single sink node, the G matrix can further be elaborated as:

$$G = o'H + \left(o'a + \left(1 - o'\right)e\right)1/N\ e^T$$

An important feature in this procedure is to be noted that although G is a dense matrix, it can be computed from a corresponding H matrix which is sparse in nature [36, 37].

NOTE: Generally, in modern times considering the cases of the modern directed networks, about 10 non-zero elements fall in a line in case of a H matrix and hence it requires about $10N$ multiplications to make a product between a vector and a G matrix. However, in the original matrix, Google uses a damping factor or σ (mentioned as d earlier) which is taken as 0.85. The calculation of $1 - \sigma$ gives us the probability of a surfer to jump on any random webpage, as discussed earlier. Thus, the concept of damping factor (taken as d or σ), is cleared [37–45].

Actually, the G matrix falls in the category of Perron–Frobenius operations of Markov chains [45]. This topic falls beyond the scope of this article and it is not likely to be discussed.

The Google matrix can be used in other directed networks as in the case of the Linux Kernel software, business process management, or the Google matrices of the brain [43].

This concept of Google matrix was also introduced by Brin and Page in 1998 [1–5].

Therefore, from this section, it may be concluded that the PageRank algorithm makes corresponding use of a matrix that is irreducible, stochastic, and aperiodic in nature (here, it is the G matrix) in order to calculate the rank or weight of a webpage and thus determines the relevance of a webpage. This algorithm can therefore not only be used for calculating the relevancy of a webpage but also many other things. One such instance is calculating the ranks of important proteins in the human body or making protein interaction chain analysis as done by Grolmusz and Ivan, back in 2010. The latter part of the article aims to elaborate on the mentioned topic [1–5, 43, 46].

4 Why PageRank?

Apart from PageRank, many other algorithms do exist that can assign quality scores to a webpage (represented by a node in a graph). One such famous algorithm was introduced by Kleinberg and is consequently known as the HITS algorithm [47–53], or the Hyperlink Induced Topic Search algorithm. Also, low quality manipulations in the calculation can be filtered out. However, as it stands, the HITS algorithm is prone to sophisticated manipulations and hence is not robust enough. Hence, PageRank, as of yet, also serves to be one of the most successful web ranking algorithms.

We have not discussed the characteristics and details of the HITS algorithm as it would have been out of the scope of this journal. However, for the reader's reference, we have provided a comprehensive analogy as to why the PageRank serves to be the more sophisticated method of the two in Table 1. We have also shown the uniqueness of the PageRank algorithm which turns out to be one of the chief purposes for preferring this particular algorithm over other existing algorithms [1–4, 45–50].

Table 1 Comparison between HITS and PageRank algorithm

	HITS	PageRank
Basic criteria	It is a link analysis algorithm	It is also a link analysis algorithm but is based on the random surfer model
Efficiency	For a particular query, HITS brings forward traditional search engine to retrieve a set of pages having relevance and consequently attempts to find authorities and hubs. However, this algorithm is not feasible enough for today's search engines as the computation is carried out at query time and we need to handle millions of queries each day. Also, HITS is quite prone to minor and sophisticated manipulations which is not possible in PageRank due to its high stability	It computes a single measure of page quality over crawl time. This measure, when combined with the traditional information retrieval score at query time increases efficiency
Main technique followed	Main techniques followed include Web Content Mining and Web Structure Mining	Main technique followed is Web Structure Mining
Mutual reinforcement	HITS gives emphasis on mutual reinforcement between the hub webpages and authorities	PageRank does not differentiate between authorities and hub webpages

Uniqueness of PageRank [51–53]:

(1) It has the capability to make the webpage link analytic even more robust in nature.
(2) It can be used as a global scale measurement, unlike some other algorithms.
(3) PageRank serves to be query independent.
(4) It also does not differentiate between web pages and authorities, thereby making the possibility of an error quite less.
(5) It measures a single measure of quality over crawl time. It helps in improving efficiency.

However, along with these advantages, there also come certain limitations which make PageRank a not so preferable algorithm in certain cases, like that of areas of media where interaction between users is frequent and dynamic in nature.

Some such limitations are [48, 51]:

(1) PageRank prefers older webpages. These older webpages may assume a higher rank. There may exist some situations where a new webpage is introduced having better contents but since it has lesser number of in-degrees as a node, it is not recommended by the PageRank. This is because, in the algorithm, the quality is decided solely by the number of incoming links and not by the actual quality of the content.
(2) PageRank weightage can be increased easily by "link farms" (a group of websites all of which hyperlink to every other site in a particular group, is known

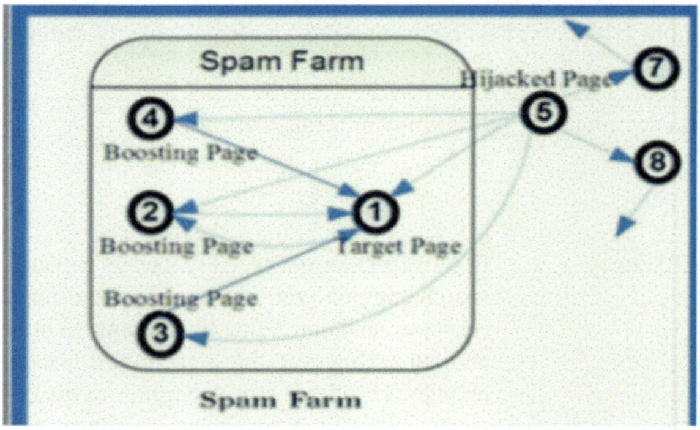

Fig. 2 A picture depicting a spam farm or a hijacked page, increasing the PageRank score of all the webpages in the group

as a link farm. In terms of graph theory, it is also referred to as a "clique") [54]. A link farm is depicted in Fig. 2.

(3) The rank of a page can also be increased by buying "links".

5 Some Important PageRank Advancements

The algorithm of PageRank was first introduced by Larry Page and Sergey Brin in 1996 and seeks to find the weighted PageRank value of a webpage and assign it a corresponding rank [1]. Recent applications and advancements of PageRank have produced significant contributions in areas of bibliometrics, social and information network analysis, protein interaction chain analysis, and so on. One of the very first and biggest real-life applications of this algorithm was first used by Page and Brin themselves for Google Search Engine optimizations [1–3].

In 2010, a project was made in which a personalized PageRank was used to analyze protein interaction chain analysis, considering the proteins as the nodes and their interaction chains as the edges thereby providing a clear personalized algorithm for ranking the important proteins in our body. Similarly, the PageRank unexpectedly found its use in deciding whether the vertices of two networks can be mapped to each other to preserve most of the edges of each network or not. Singh, Berger, and Xu in 2007 proposed such an idea and named it as IsoRank on the basis of ranking graph polymorphism that was involved in the problem [54–55].

Another very important and renounced breakthrough was made when PageRank laid the foundation of an open-source operating system, i.e., the Linux Kernel. Now, it has as over 2000 working individuals producing an approximate capital of about 3 billion dollars. It made use of a Kernel Graph [48].

In 2011, Radiacchi used PageRank on a network of tennis players having almost the same construction and formed a weighted network, taking damping factor as 0.85 [56].

Bollen, Rodriguez, and Van de Sompel in 2006 found a correlation of impact factors using PageRank and made a list of journals having better citations, produced in order, representing as many as over 6000 journals [57].

PageRank association with similarly-based methods for Link prediction: In this section, some light on the association of the algorithm with advancements in complex network analysis has been shared by us. It has been already established that link prediction tasks based on similarity methods are widely in use due to their low complexity and commendable performance. Many novel link prediction methods have come out to improve performance while having low time and space complexity respectively. Some assertive results came out after performing six such known methods based on similarity-based link prediction methods in the paper of Mourad Charikhi in her paper of 2024. The article can be found on https://doi.org/10.1016/j.physa.2024.129552.

The result was significantly informative and the approach seemed to outperform with all the well-known methods with linear time complexity [58].

Generally, a complex network is most often represented by an undirected graph $G(V, E)$ at time t, V represents the set of nodes or vertices and E represents the set of edges or connections between elemental pairs of elements of E. Now, the main goal of link prediction is to guess new nodes or missing links between nodes for a future time t' to $t' > t$. However, link prediction in networks that are complex faces two challenges. The first one stands to be able to develop an approach bearing a low complexity along with being dynamic in nature and which are able to change frequently like the social networks. These approaches must also have a fairly high accuracy in order to outperform the known methodologies. This is the second challenge. Local similarity-based methods require information which are structural in nature, also known as closed triangle structure. Now, these are related to the proximity of vertices in order to calculate the score of unconnected pairs of vertices in a network [59]. Hence, they become complex and are not feasibly possible. However, vertices that do not seemingly share neighbors have 0 probability to be linked anywhere in the near future and hence fail to predict the link of most tree-like networks. Hence, the article proposed an idea to overcome the drawback and make improvements by introducing a new similarity-based method based on degree of importance of nodes using the PageRank algorithm. In the mentioned paper, the nodes, ranked according to their degree of importance represented the force of attraction between the pair of unconnected vertices which do not share any common neighbor between them. Hence, it was assumed that higher the force of attraction, higher is the chance that they will be connected somewhere in the future. This new measure was established on grounds of local similarity measures and was consequently tested for its effectiveness by merging it with six other, already existing local similarity measures. The study demonstrated significant advancements and positives related to the new measure. Thus:

- The mentioned approach combined the newly proposed similarity model with PageRank algorithm with methods based on local information.

- This leads to the formation of a new similarity-based measure on the PageRank algorithm.

For even better reference or an instance of how the algorithm might be performed in Python, we have provided a stimulated code for the same.

The code in Python is as follows:

```
import numpy as np

def pagerank(adjacency_matrix, damping_factor = 0.85, max_
iterations = 100, convergence_threshold = 1e-4):
    """
    Calculate PageRank using the power iteration method.

    Parameters:
    - adjacency_matrix: numpy array representing the adjacency matrix
of the graph.
    - damping_factor: the damping factor, usually set to 0.85.
    - max_iterations: the maximum number of iterations for the power
iteration method.
    - convergence_threshold: the convergence threshold to stop the
iteration.

    Returns:
    - A numpy array representing the PageRank scores for each node.
    """

    # Number of nodes in the graph
    num_nodes = adjacency_matrix.shape[0]

    # Initialize PageRank scores
    pagerank_scores = np.ones(num_nodes)/num_nodes

    for _ in range(max_iterations):
        # Store the previous PageRank scores for convergence check
        prev_pagerank_scores = pagerank_scores.copy()

        # Calculate the new PageRank scores using the power iteration
method
        pagerank_scores = (1 - damping_factor)/num_nodes + damping_
factor * np.dot(adjacency_matrix.T, pagerank_scores)

        # Check for convergence
        if np.linalg.norm(pagerank_scores - prev_pagerank_scores, 1)
< convergence_threshold:
            break

    return pagerank_scores

# Example usage:
# Define an adjacency matrix for a simple graph (replace this with
your own graph)
# For example, if you have a graph with 3 nodes and edges (1 → 2, 2 →
3, 3 → 1):
# adjacency_matrix = np.array([[0, 1, 0],
#                              [0, 0, 1],
```

```
#                    [1, 0, 0]])

# Run the PageRank algorithm
# pagerank_scores = pagerank(adjacency_matrix)
# print("PageRank scores:", pagerank_scores)
```

Highlights

- From the above information, we come to the fact that advanced network analysis and its evolutions have brought us to the problem of link prediction.
- Similarity-based methods are most often used as a local method for the prediction of new links.
- New Similarity measures between nodes that are not connected and do not have shared neighbors have come into play.
- Some such link prediction methods in complex analysis can be performed using PageRank.

6 Implementing the Algorithm

The most important part about algorithms is implementing the same after understanding it. Going by the graph theory, considering each of these webpages as nodes or vertices and each hyperlink in the corresponding webpage by the branches or edges, PageRank algorithm for the same purpose can be designed. The outbound links for each webpage will be designated the position of an out-degree whereas the outbound links which land the user to that particular page (of which the PageRank value is to be found out) will be considered as an in-degree. Let us take an example:

There is a webpage *A* which has links to some other webpage. Then, these links will be considered as out-degrees since they are directing users to other webpages from webpage *A*. On the other hand, all the other links from other webpages which are directing us to webpage *A* will be considered as in-degrees. Therefore, the question may arise, what can we conclude from here? We can put forward an obvious conclusion that, greater the in-degree of a particular node (keeping in mind a webpage here is considered to be the node), greater is the reach and hence the relevancy of that particular webpage. In case of directed graphs, we can simply say that greater the degree of a particular node or webpage, greater is the relevance of the webpage.

Hence, the PageRank algorithm serves to be one of the most successful algorithms in the web page ranking cause. A simpler analogy of the algorithm can also be provided in this way: A walker starts following a random walk in the graph. He starts from any randomly chosen node of the graph and then with probability $1 - d$, follows a uniformly selected and random outgoing edge from the node, and then with probability c, it teleports to a uniformly selected random node/vertex of the graph. The important point to be noted out here is to keep in mind that c is nothing but a probability, hence, its values range between 0 and 1. Therefore, the PageRank of a node n, representing the importance of the particular node, is also a stationary limit probability distribution that the walker is at node n. As discussed in the earlier section, here, c is actually the damping factor which was taken as d or σ in the earlier section.

Diving deeper into the properties of the algorithm, it is rather imperative that some talk on one of the most attractive properties of PageRank, i.e., its stability in analysis of biological networks, is made. This paper is also a review on the application of the algorithm in analysis of protein chain interaction networks. For analysis of such a system, it is very important that the used algorithm is stable in nature. This is because the published proteins of the original article contained a numerous number false positive and false negative edges, even for the best quality data that is gathered for one of the most researched subjects, i.e., the yeast interactome [1–3]. Hence, it is important that such algorithms are stable in nature, thereby making PageRank one of the most suitable algorithms for this purpose. The stability estimation suited best for PageRank algorithm was given by Lee and Borodin in 2003. It is expressed as [60]:

$$||p - \hat{p}||1 \leq 2(1 - c)c \quad j \in U\, pj$$

Here, the ith coordinate of the vector p expresses the PageRank of vertex i and vector $\mathcal{P}$ gives the PageRank of the nodes after branches with endpoints in set U are deleted or even added. To make it simpler, if c is too close to 0 and only the branches between the less important nodes are perturbed, then, in that case, the corresponding effect of perturbation in the PageRank value remains considerably low [61, 62]. As mentioned by Gabor and Grolmusz in their paper, this indeed is a remarkable and excellent property. This is because the less important proteins in the system are seldom mapped with much reliability and hence, the inequality shows that these possible errors will not make much difference in the overall PageRank vector.

7 Results and Discussion

Protein–protein Interaction (PPI) networks are usually denoted or shown by undirected graphs. From our earlier discussions, it is clear that for undirected graphs, the PageRank score is directly proportional to the degree of the node. To be clear, the proteins are taken as nodes here and the chain interactions between each protein is taken as a branch or edge. Hence, it can be inferred that proteins having more interaction chains throughout the body are more important as compared to the ones having lesser interaction chains, according to the PageRank algorithm.

Now, in the same process, second procedure is also done. Consider biochemical reactions in our body as a node and an edge or a branch joins u and v. Here:

- u the reactant of a biochemical reaction, say node 1, and,
- v the product of the first biochemical reaction which is used up by another biochemical reaction, say node 2.

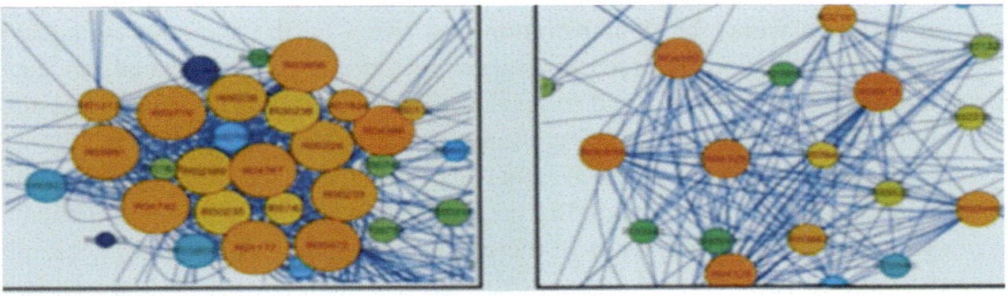

Fig. 3 Two dense subgraphs from the metabolic graph of *Mycobacterium tuberculosis*. On the left panel, large nodes correspond to large degree, yellowish colors denote lower PageRank values

Hence, it can be said that the PageRank may open up deep and robust network properties of the graph, which in turn will give the properties of that particular protein chain or network. In the original paper, a consequent PageRank of the metabolic network of *Mycobacterium tuberculosis* was done.

It should be noted that, in Fig. 3, the vertices that are warmer in color more than that were proportional to their degree attract special interest. That signifies that these vertices are more likely to be hit by the random walker (random surfer model, as discussed earlier) as compared to those vertices having their corresponding same local network characteristic. According to the Gabor and Grolmusz, it stands out to be a remarkable finding in the study of metabolic networks of tuberculosis bacterium that, according to the conducted experiment, the FAD-dependent thymidylate synthase [(ThyX); Myllykallio et al. [4]] happened to occupy the sixth position in terms of PageRank score or weightage, much higher than other vertices having higher degree (Fig. 3). Accordingly, the high PageRank score might be for the particularities of Mycobacteria's thymidylate biosynthesis pathway (Vertessy and Toth 2009).

8 Personalized PageRank Used for PPI Networks

The first personalized PageRank was developed in 1999 for the preference of personal preferences in evaluating the content on the World Wide Web. While computing the personalized PageRank, it is assumed that, according to the random surfer model, the walker teleports with the probability of $c + \acute{c}$ such that, $0 < c + \acute{c} < 1$; and with a probability $\acute{c}$ to some vertices which correspond to the personal interest of the surfer on World Wide Web and with a probability c to the remaining nodes having 'no-personal-interest'.

From the discussion, there are significant reasons to why personalized PageRank is highly preferable. It has the capability to evaluate robustly, the importance of nodes in a particular network relative to some other nodes that are already known. Additionally, personalized PageRank computation serves to be scalable, i.e., it can be well determined or approximated in cases of the largest of networks.

In the original work, the demonstration was based on the use of personalized PageRank for evaluation of proteomics data. In the corresponding analysis, low concentration proteins hardly tend to appear reliably in the results and hence, the property of being robust in case of PageRank computation is useful.

The consideration made in the original paper included the data of melanoma patients, published in Forgber et al. [5], according to which as many as 13 patients were detected having higher levels in the plasma. Then, a personalized PageRank was used, considering nodes in human protein–protein interaction (PPI) graph HPRD (stands for Human Protein Reference Database) [6]. The human HPRD interactome contains as many as 27,801 nodes, having correspondence with human proteins, and as many as 38,806 edges between these proteins, having correspondence with interactions.

The final inference drawn was that a large number of proteins having the largest of PageRank algorithm nodes were closely and clearly related to melanoma (Fig. 4). To be more precise, in the 22 topmost ranged nodes, 10 are the vertices that have been personalized (yellow in color in Fig. 4, having high PageRank), 2 proteins have no apparent connection with melanoma (indicated by the gray rows) whereas the green rows are proven to have clear cut relation to cancer.

PageRank	Accession number	Protein/gene name
858.89	P08107	HSP70 protein B
821.84	Q8EEV6	SUMO44
808.67	P55072	VCP
805.55	P26599	hnRNP1
801.60	P07954	Fumarate hydratese
790.41	P04075	Aldolase A
787.43	Q96EY1	HSP70 protein 9B
765.91	P06733	Enolase 1
754.35	O43852	Calumenin
729.05	P07195	LDH H
725.25	P15121	Aldose reductase
691.07	P40926	Malate dehydrogenase
592.09	Q15797	SMAD1
565.39	P02743	Serum amyloid P-component (SAP)
192.30	Q99972	Myocilin
141.15	P63104	YWHAZ
132.61	P00747	Plasminogen
130.78	P00505	Aspartate aminotransferase
125.47	P54253	Ataxin-1
116.36	P63167	Dynein light chain 1 (DLC1)
112.03	P61981	14-3-3 protein gamma
100.25	P04637	Cellular tumor antigen p53

Fig. 4 The PageRank stands personalised to the nodes of the yellow rows and the consequent green rows signify the proteins of high PageRank that are newly found, having correspondence to cancer. The uncolored rows do not to have any correspondence to cancer

REMARK: As mentioned in the original research paper, according to the UniProt database 160 human proteins bear relation to melanoma [7]. That means, about 0.57% of the proteins which are present in the analyzed 27,800 nodes in HPRD, fall under the category mentioned by UniProt [6]. These facts also prove the power and selectivity property of personalized PageRank computation, as can be reflected from Fig. 4.

Thus, the calculations deepen up and open a robust network property of our graph, as shown.

9 Conclusion

This article provided a brief idea of the PageRank algorithm and how it performs. References in cases requiring special attention for the readers have also been provided. In the latter part of the article, discussions about the procedure followed by Gabor and Grolmusz in 2010 in their application of using PageRank for analyzing these protein chain networks have also been done.

It should be noted that the consequent applications of computer algorithms and biological applications may work in certain cases or fields, like that of metabolic networks or simply for ranking important nodes in our body. However, it should also be noted that PageRank like techniques are not always too novel to provide efficient and accurate results in fields of biology and other related topics. In this case, it was tried to prove the same by taking the example of protein chain interaction networks. Example of another such algorithm includes IsoRank algorithm which also has definite similarities with that of PageRank and can be used in fields involving protein interaction analysis.

Hence, in this paper, a brief about the formation, implementation, and optimization of the PageRank algorithm has been discussed along with discussions on some really noteworthy advancements in this particular field and on the usage and implementation of PageRank on: similarity-based methods for the prediction of links in complex networks and, the use of the algorithm in understanding and analyzing PPI networks. This paper provides a brief review on the algorithm itself and two of its most important implementations done till date.

Acknowledgements This article is not but merely a review based on established research. Special mention and gratitude must be expressed to the authors of the original research, otherwise which, this review would not have been possible.

References

1. Gavin A-C et al (2006) Proteome survey reveals modularity of the yeast cell machinery. Nature 440:631–636

2. Goll J, Uetz P (2006) The elusive yeast interactome. Genome Biol 7:223
3. Krogan NJ et al (2006) Global landscape of protein complexes in the yeast *Saccharomyces cerevisiae*. Nature 440:637–643
4. Myllykallio H et al (2002) An alternative flavin-dependent mechanism for thymidylate synthesis. Science 297:105–107
5. Forgber M et al (2009) Proteome serological determination of tumor-associated antigens in melanoma. PLoS ONE 4:e5199
6. Prasad TSK et al (2009) Human protein reference database and human proteinpedia as discovery tools for systems biology. Methods Mol Biol 577:67–79
7. Consortium U (2010) The universal protein resource (UniProt) in 2010. Nucleic Acids Res 38:D142–D148
8. Tomkins JE et al (2021) Advances in protein-protein interaction network analysis for Parkinson's disease. Neurobiol Dis
9. Wang F et al (2023) Collaboration prediction based on multilayer all-author tripartite citation networks: a case study of gene editing. J Netw Comput Appl
10. Yao Y et al (2023) Link prediction based on mutual information with high-order clustering structure of nodes in complex networks. Phys A
11. Daud NN et al (2020) Applications of link prediction in social networks: a review. J Netw Comput Appl
12. Shang K-K et al (2022) Link prediction for long-circle-like networks. Phys Rev E
13. Link prediction for tree-like networks. Chaos
14. Zareie A et al (2020) Similarity-based link prediction in social networks using latent relationships between the users
15. Brent R, Bruck J (2006) 2020 computing: can computers help to explain biology? Nature 440:416–417
16. Brin S, Page L (1998) The anatomy of a large-scale hypertextual web search engine. Comput Netw ISDN Syst 30:107–117
17. Fogaras D et al (2005) Towards scaling fully personalized PageRank: algorithms, lower bounds, and experiments. Internet Math 2:333–358
18. Lee HC, Borodin A (2003) Perturbation of the hyperlinked environment. In: Warnow T, Zhu B (eds) Computing and combinatorics: 9th annual international conference, COCOON 2003. Lecture notes of computer science, vol 2697, Big Sky, MT, USA, 25–28 July 2003, pp 272–283
19. Page L et al (1999) The pagerank citation ranking: bringing order to the web. Technical report, Stanford Infolab, No. 1999–66. Stanford University
20. Singh R et al (2008) Global alignment of multiple protein interaction networks with application to functional orthology detection. Proc Natl Acad Sci USA 105:12763–12768
21. Vértessy BG, Tóth J (2009) Keeping uracil out of DNA: physiological role, structure and catalytic mechanism of dUTPases. Acc Chem Res 42:97–106
22. Austin D (2008) How Google finds your needle in the web's haystack. AMS feature columns. http://www.ams.org/samplings/feature-column/fcarc-pagerank
23. Law E (2008) PageRank. PDF. http://scienceoftheweb.org/15-396/lectures/PageRank_Lecture12.pdf
24. Donato D, Laura L, Leonardi S, Millozzi S (2004) Large scale properties of the Webgraph. Eur Phys J B 38:239
25. Pandurangan G, Ranghavan P, Upfal E (2005) Using PageRank to characterize web structure. Internet Math 3:1
26. Serra-Capizzano S (2005) Jordan canonical form of the Google matrix: a potential contribution to the PageRank computation. SIAM J Matrix Anal Appl 27:305
27. Georgeot B, Giraud O, Shepelyansky DL (2010) Spectral properties of the Google matrix of the World Wide Web and other directed networks. Phys Rev E 81:056109
28. Ulam S (1960) A Collection of mathematical problems. Interscience tracts in pure and applied mathematics. Interscience, New York, p 73
29. Froyland G, Padberg K (2009) Almost-invariant sets and invariant manifolds—connecting probabilistic and geometric descriptions of coherent structures in flows. Phys D 238:1507

30. Shepelyansky DL, Zhirov OV (2010) Google matrix, dynamical attractors and Ulam networks. Phys Rev E 81:036213
31. Ermann L, Shepelyansky DL (2010) Google matrix and Ulam networks of intermittency maps. Phys Rev E 81:036221
32. Ermann L, Shepelyansky DL (2010) Ulam method and fractal Weyl law for Perron–Frobenius operators. Eur Phys J B 75:299
33. Frahm KM, Shepelyansky DL (2010) Ulam method for the Chirikov standard map. Eur Phys J B 76:57
34. Chepelianskii AD (2010) Towards physical laws for software architecture. http://arxiv.org/abs/1003.5455. arXiv:1003.5455
35. Ermann L, Chepelianskii AD, Shepelyansky DL (2011) Fractal Weyl law for Linux Kernel architecture. Eur Phys J B 79:115. arXiv:1005.1395 [cs.CE]
36. Shepelyansky DL, Zhirov OV (2010) Towards Google matrix of brain. Phys Lett A 374:3206
37. Abel M, Shepelyansky DL (2011) Google matrix of business process management. Eur Phys J B (to appear). arXiv:1009.2631 [cs.CY]
38. Brin S, Page L (1998) The anatomy of a large-scale hypertextual Web search engine. Comput Netw ISDN Syst 30:107
39. Franceschet M (2010) PageRank: standing on the shoulders of giants. http://arxiv.org/abs/1002.2858. arXiv:1002.2858
40. Adie EA, Adams RR, Evans KL, Porteous DJ, Pickard BS (2005) Speeding disease gene discovery by sequence-based candidate prioritization. BMC Bioinform 6:55
41. Adie EA, Adams RR, Evans KL, Porteous DJ, Pickard BS (2006) SUSPECTS: enabling fast and effective prioritization of positional candidates. Bioinformatics 22(6):773–774
42. Aerts S, Lambrechts D, Maity S, Van Loo P, Coessens B, De Smet F, Tranchevent LC, De Moor B, Marynen P, Hassan B et al (2006) Gene prioritization through genomic data fusion. Nat Biotechnol 24(5):537–544
43. Tiffin N, Adie E, Turner F, Brunner HG, van Driel MA, Oti M, LopezBigas N, Ouzounis C, Perez-Iratxeta C, Andrade-Navarro MA et al (2006) Computational disease gene identification: a concert of methods prioritizes type 2 diabetes and obesity candidate genes. Nucleic Acids Res 34(10):3067–3081
44. Turner FS, Clutterbuck DR, Semple CA (2003) POCUS: mining genomic sequence annotation to predict disease genes. Genome Biol 4(11):R75
45. Chen J, Xu H, Aronow BJ, Jegga AG (2007) Improved human disease candidate gene prioritization using mouse phenotype. BMC Bioinform 8:392
46. Rual JF, Venkatesan K, Hao T, Hirozane-Kishikawa T, Dricot A, Li N, Berriz GF, Gibbons FD, Dreze M, Ayivi-Guedehoussou N et al (2005) Towards a proteome-scale map of the human protein-protein interaction network. Nature 437(7062):1173–1178
47. Stelzl U, Worm U, Lalowski M, Haenig C, Brembeck FH, Goehler H, Stroedicke M, Zenkner M, Schoenherr A, Koeppen S et al (2005) A human protein-protein interaction network: a resource for annotating the proteome. Cell 122(6):957–968
48. Sharan R, Ideker T (2006) Modeling cellular machinery through biological network comparison. Nat Biotechnol 24(4):427–433
49. Nabieva E, Jim K, Agarwal A, Chazelle B, Singh M (2005) Whole-proteome prediction of protein function via graph-theoretic analysis of interaction maps. Bioinformatics 21(Suppl 1):i302–i310
50. Lubovac Z, Gamalielsson J, Olsson B (2006) Combining functional and topological properties to identify core modules in protein interaction networks. Proteins 64(4):948–959
51. Jansen R, Yu H, Greenbaum D, Kluger Y, Krogan NJ, Chung S, Emili A, Snyder M, Greenblatt JF, Gerstein M (2003) A Bayesian networks approach for predicting protein-protein interactions from genomic data. Science 302(5644):449–453
52. Wong SL, Zhang LV, Tong AH, Li Z, Goldberg DS, King OD, Lesage G, Vidal M, Andrews B, Bussey H et al (2004) Combining biological networks to predict genetic interactions. Proc Natl Acad Sci USA 101(44):15682–15687

53. Sam L, Liu Y, Li J, Friedman C, Lussier YA (2007) Discovery of protein interaction networks shared by diseases. In: Pacific symposium on biocomputing, pp 76–87
54. Goehler H, Lalowski M, Stelzl U, Waelter S, Stroedicke M, Worm U, Droege A, Lindenberg KS, Knoblich M, Haenig C et al (2004) A protein interaction network links GIT1, an enhancer of huntingtin aggregation, to Huntington's disease. Mol Cell 15(6):853–865
55. Ruffner H, Bauer A, Bouwmeester T (2007) Human protein-protein interaction networks and the value for drug discovery. Drug Discov Today 12(17–18):709–716
56. Neduva V, Linding R, Su-Angrand I, Stark A, de Masi F, Gibson TJ, Lewis J, Serrano L, Russell RB (2005) Systematic discovery of new recognition peptides mediating protein interaction networks. PLoS Biol 3(12):e405
57. Barabasi AL, Albert R (1999) Emergence of scaling in random networks. Science 286(5439):509–512
58. Berg J, Lassig M, Wagner A (2004) Structure and evolution of protein interaction networks: a statistical model for link dynamics and gene duplications. BMC Evol Biol 4(1):51
59. Eisenberg E, Levanon EY (2003) Preferential attachment in the protein network evolution. Phys Rev Lett 91(13):138701
60. Rzhetsky A, Gomez SM (2001) Birth of scale-free molecular networks and the number of distinct DNA and protein domains per genome. Bioinformatics 17(10):988–996
61. Wagner A, Fell DA (2001) The small world inside large metabolic networks. Proc R Soc Lond B 268(1478):1803–1810
62. Xu J, Li Y (2006) Discovering disease-genes by topological features in human protein-protein interaction network. Bioinformatics 22(22):2800–2805

Unlocking Domain Specificity: Fine-Tuning Llama 2 for Enhanced Performance on Custom Datasets

Swarnadwip Bhattacharya, Anindita Das Bhattacharjee, Pranab Singha Roy, and Tapas Samanta

Abstract This paper aims to improve Llama 2's performance by using personalized and modified datasets. Despite the impressive capabilities of large language models (LLMs) such as Llama 2, their effectiveness may be limited in specialized domains. The proposed method entails fine-tuning Llama 2 on custom datasets to optimize performance efficiently. The study focuses on the impact of quantization-aware low-rank adapter layers (QLoRA) on a single GPU's resource-efficient fine-tuning performance. Furthermore, the study looks into the design of instruction datasets to guide the model toward desired behaviors. When Llama 2 is fine-tuned with QLoRA, performance improves significantly across tasks such as text summarization, question answering, and natural language generation in a variety of domains. The paper concludes by highlighting the broader implications of the findings.

Keywords Llama 2 · Fine-tuning · Custom datasets · QLoRA · Instruction datasets · Large language models (LLMs) · Quantization-aware low-rank adapter layers (QLoRA) · Resource-efficient fine-tuning · Personalized datasets · Specialized domains

---

S. Bhattacharya (✉)
Department of Computer Science and Engineering, Institute of Engineering and Management, Kolkata 700091, India
e-mail: dudeswarnadwip@gmail.com

A. D. Bhattacharjee
Department of CSE, IEM Centre of Excellence for InnovAI, Institute of Engineering and Management, Kolkata 700091, India
e-mail: Anindita.DasBhattacharjee@iem.edu.in

P. S. Roy · T. Samanta
Department of Computer and Informatics, Variable Energy Cyclotron Centre, Kolkata 700064, India
e-mail: psroy@vecc.gov.in

T. Samanta
e-mail: tsamata@vecc.gov.in

P. G. Bradford et al. (eds.), *Proceedings of IEMTRONICS 2024*, Lecture Notes in Electrical Engineering 1229, https://doi.org/10.1007/978-981-97-4780-1_24

1 Introduction

Aghajanyan et al.'s study highlights the importance of reparameterization in common pre-trained models [1]. This paper explores fine-tuning the Llama 2 model with custom datasets to improve performance within resource constraints. Traditional Autoregressive Language Models (e.g., GPT) and Autoencoding Language Models (e.g., BERT) face challenges, driving the emergence of Llama 2. LLMs struggle with specialized domains, while prior models fail to capture long-term dependencies. Llama 2 addresses these issues through personalized dataset fine-tuning and quantization-aware low-rank adapter layers (QLoRA), enabling resource-efficient adaptation on a single GPU. Our research demonstrates significant performance gains in text summarization, question answering, and natural language generation across various domains, showcasing Llama 2's superiority over its predecessors.

Recently, large language models (LLMs) have gained popularity for their versatile applications. Pre-trained on extensive datasets, LLMs powered by transformer networks excel in tasks like recognition, summarization, translation, and content generation. They undergo fine-tuning for specific tasks, including natural language inference and text similarity. For example, a chatbot processes user questions by segmenting them into tokens and mapping them to embeddings, guiding predictions. Reparameterization enhances the efficiency of pre-trained models.

Acknowledging prior studies' challenges [2–9], including biased data and limited interpretability, our research emphasizes personalized dataset fine-tuning and QLoRA exploration to address these issues. We elucidate the fine-tuning process, crucial for optimizing model performance. QLoRA components specialize in linguistic aspects, facilitating effective communication within the model.

2 Model Architecture

Generative AI, a domain within artificial intelligence, focuses on crafting models capable of producing novel content like images or text. Among these models, the Transformer model stands out for its notable achievements across tasks in Natural Language Processing (NLP) and Computer Vision (CV). Central to the Transformer's success is its utilization of a self-attention mechanism, enabling the capture of global dependencies and long-range relationships within input data. However, this mechanism's computational demands are substantial. To tackle this challenge, specialized Transformer processors have been proposed, aiming to optimize self-attention computations for enhanced energy efficiency and performance. These processors leverage techniques such as approximate computing, redundant computation removal, and out-of-order computing scheduling [10, 11].

Moreover, within the realm of session-based recommendation systems, researchers have extended the Transformer model's capabilities by integrating Convolutional Neural Networks (CNNs). This integration enhances the model's ability to capture local contextual information, thereby improving session representation learning. Notably, this combined approach has demonstrated superior performance compared to utilizing either the Transformer or CNN alone [12].

A Transformer-based Large Language Model (LLM) serves as a deep learning algorithm proficient in recognizing, summarizing, translating, predicting, and generating content using extensive datasets. Pre-trained on vast corpora, LLMs undergo fine-tuning to execute specific tasks, including natural language inference and sentence text similarity. The architecture of the Transformer model empowers LLMs to grasp relationships and connections between words and concepts through its self-attention mechanism. While Transformers can be applied to diverse tasks beyond language modeling, LLMs are tailored specifically for text generation and natural language understanding [12, 13].

The Transformer architecture uses stacked self-attention and point-wise, fully connected layers for the encoder and decoder, as shown in the left and right halves of Fig. 1, respectively [12].

3 Llama-2

LLMs hold significant potential as proficient AI assistants that can perform sophisticated reasoning tasks in different domains. They have been rapidly and widely adopted among the general public as they can facilitate human interaction through user-friendly chat interfaces. However, the significant computational demands of the LLMs training methodology have constrained their development to only a handful of participants. Popular LLM-based applications are OpenAI ChatGPT, Google Bard, and Anthropic Claude that were built based on closed-source LLMs [14]. Recent open-source LLMs such as BLOOM [15], LLaMa-1 [16], and Falcon [17] are not on par with the closed-source counterparts whose fine-tuning process normally requires significant costs in compute and human annotation to align with human preferences. Llama 2, an updated version of Llama 1 [16] and a family of open-source pre-trained and fine-tuned LLMs, was recently introduced by Meta GenAI and evaluated as comparable with cutting-edge closed-source LLMs. This allows the research community to advance AI in a more transparent manner, resulting in more responsible development of LLMs. The released Llama 2 family includes pre-trained and fine-tuned models, each with three variants of 7B, 13B, and 70B parameters. The pre-trained models were created using a self-supervised learning approach on a large text corpus containing two trillion tokens [18]. These pre-trained models are further optimized using supervised fine-tuning and reinforcement learning with human feedback (RLHF) with instruction datasets and human-annotated examples for dialog use cases, yielding the fine-tuned Llama 2-Chat models. The process of developing the pre trained Llama 2 models and fine-tuned Llama 2-Chat models is detailed in [19, 20].

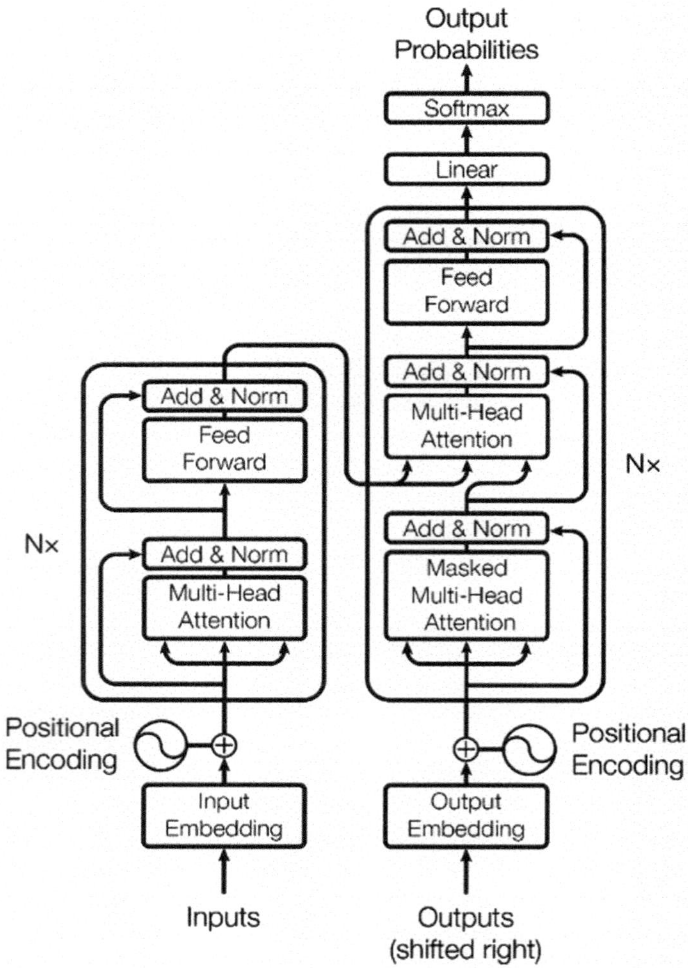

Fig. 1 The Transformer—model architecture

4 Parameter-Efficient Fine-Tuning

Large pre-trained models have demonstrated exceptional performance across a wide range of downstream natural language processing tasks by fine-tuning task-specific labeled data [21]. However, fine-tuning all parameters and storing them separately for each task incurs significant computational and storage overhead, potentially jeopardizing the model's performance [22]. Parameter-Efficient Fine-Tuning (PEFT) addresses this issue by tailoring pre-trained language models to specific tasks without fine-tuning all parameters. This method significantly reduces computational and storage costs, making it a viable alternative to traditional fine-tuning approaches.

There are various PEFT techniques, such as adapter modules, prompt tuning, and sparse update methods.

5 Fine-Tuning with LORA: Low-Rank Adaptation of Large Language Models

Numerous natural language processing applications necessitate the adaptation of a single large-scale, pre-trained language model to multiple downstream tasks. Typically, this adaptation is accomplished through fine-tuning, which involves updating all parameters of the pre-trained model. However, one significant disadvantage of fine-tuning is that the resulting model contains the same number of parameters as the original model. While this was previously regarded as an "inconvenience" for models such as GPT-2 or RoBERTa [23], it has evolved into a critical deployment challenge for GPT-3, which has an astonishing 175 billion trainable parameters [24].

The Low-Rank Adaptation (LoRA) method provides a novel method for training specific dense layers in a neural network indirectly. Rather than directly updating the parameters of these dense layers, LoRA optimizes rank decomposition matrices of their changes during adaptation while freezing the pre-trained weights shown in Fig. 2. Using GPT-3 175B as an example, it has been shown that a very low-rank (denoted as "r" in Fig. 1, which can be one or two) is sufficient even when the full rank ("d") exceeds 12,288. LoRA's efficiency makes it suitable for both storage and computation. LoRA offers several advantages. A pre-trained model can be shared and used to create several small LoRA modules for various tasks. Figure 3 shows that by freezing the shared model and efficiently switching tasks via matrix replacement, the storage requirement and task-switching overhead are significantly reduced.

LoRA improves training efficiency and reduces the hardware barrier by up to threefold, especially when adaptive optimizers are used. Most parameters do not require gradient computation or the maintenance of optimizer states. Instead, LoRA focuses on optimizing the injected low-rank matrices, which are much smaller.

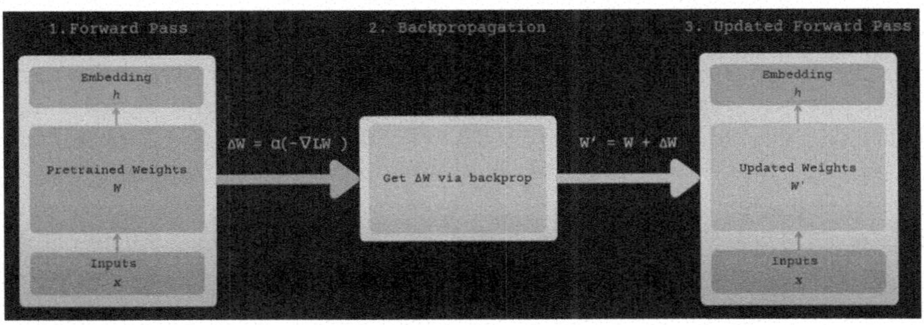

Fig. 2 Low-rank approximation simplifies matrices, like through SVD, retaining essential data with less computational load. This efficiency is beneficial across tasks with large matrices

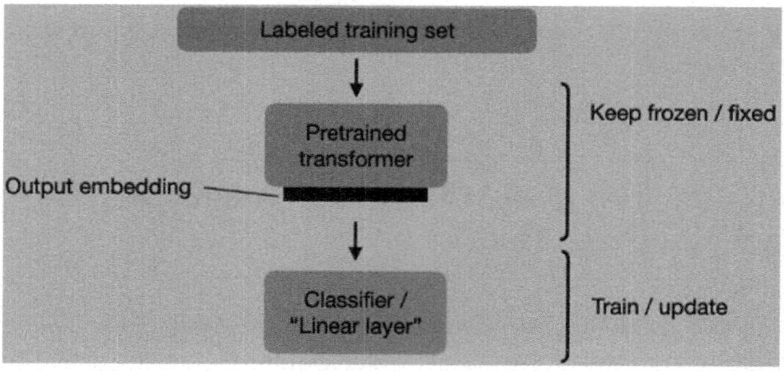

Fig. 3 Conventional feature-based and fine-tuning approach

The straightforward linear design allows the merging of trainable matrices with frozen weights during deployment, resulting in no additional inference latency when compared to a fully fine-tuned model. LoRA complements and integrates with many existing methods, including prefix-tuning, as shown in Fig. 3 [25].

Though weights in a pre-trained model are full rank on their original tasks, Aghajanyan et al. observed that pre-trained large language models exhibit a low "intrinsic dimension" when adapted to new tasks. LoRA's core concept is to decompose weight changes (ΔW) into a lower-rank representation, increasing parameter efficiency [1].

6 The Dataset

Various dataset types for Large Language Model fine-tuning: Instruction, Raw Completion, and Preference [26]. For our research, we modified the Open-Platypus dataset, focusing on LLM logical reasoning skills [27] (Fig. 4).

7 Methodology

The methodology entails three key steps. Firstly, the creation of an instruction dataset tailored for fine-tuning the Llama 2 model is discussed, highlighting the suitability of such datasets and the choice of supervised fine-tuning. This involves modifying the Open-Platypus dataset from Hugging Face. Secondly, the fine-tuning process of the Llama 2 model using the prepared instruction dataset is detailed, including environment setup, handling potential issues, and essential steps like filtering rows and merging models. Finally, the evaluation of the fine-tuned Llama 2 model performance is conducted, utilizing the ROUGE SCORE metric to assess alignment with human-written summaries, with results and insights presented thereafter (Fig. 5).

Dataset Name	License Type
PRM800K	MIT
MATH	MIT
ScienceQA	Creative Commons Attribution-NonCommercial-ShareAlike 4.0 International
SciBench	MIT
ReClor	Non-commercial
TheoremQA	MIT
nuprl/leetcode-solutions-python-testgen-gpt4	None listed
jondurbin/airoboros-gpt4-1.4.1	other
TigerResearch/tigerbot-kaggle-leetcodesolutions-en-2k	apache-2.0
ARB	CC BY 4.0
timdettmers/openassistant-guanaco	apache-2.0

Fig. 4 garage-bAInd/Open-Platypus dataset available at hugging face

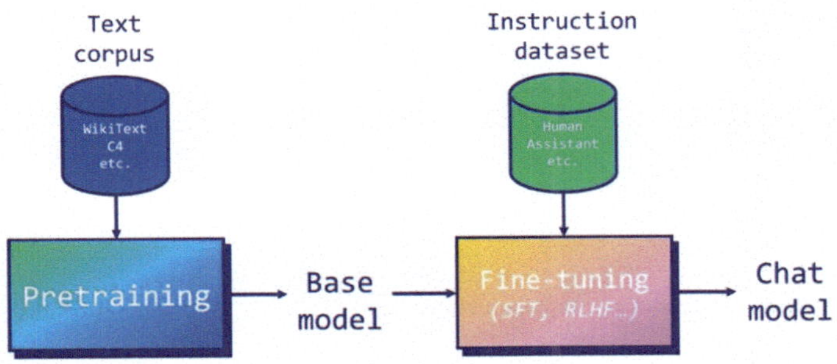

Fig. 5 Model architecture

8 Results

Training Setup

Once the preprocessed dataset is prepared, we begin the training process. Initially, we configure quantization to 4-bit precision using bits and bytes. Following this, the Llama 2 model and its tokenizer are loaded onto the GPU in this reduced precision format. We then initialize QLoRA configurations for training, along with regular training parameters. Finally, all these configurations are transmitted to the SFTTrainer, initiating the quantization-aware training of the Llama 2 model on the preprocessed dataset (Figs. 6, 7 and 8).

Next, we need to determine the number of tokens in both the Instruction and Output columns and then combine them. The reason behind this is the context window input token limit of Llama 2 and other LLMs, which is set at a default maximum of 4096 tokens. If the tokens exceed this context window, it becomes unhelpful. Hence, it's

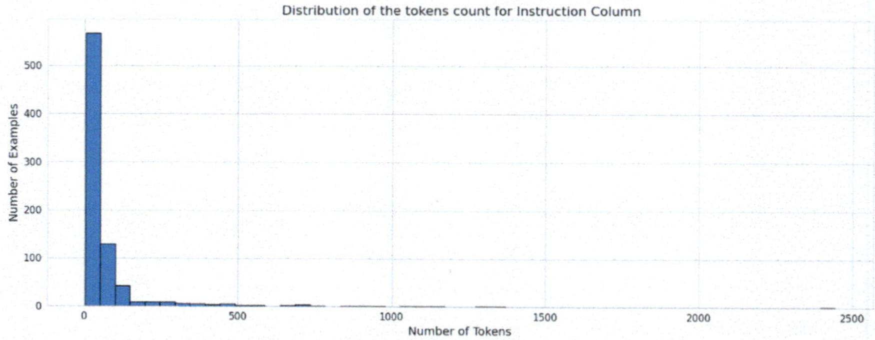

Fig. 6 Distribution of token count for instruction column

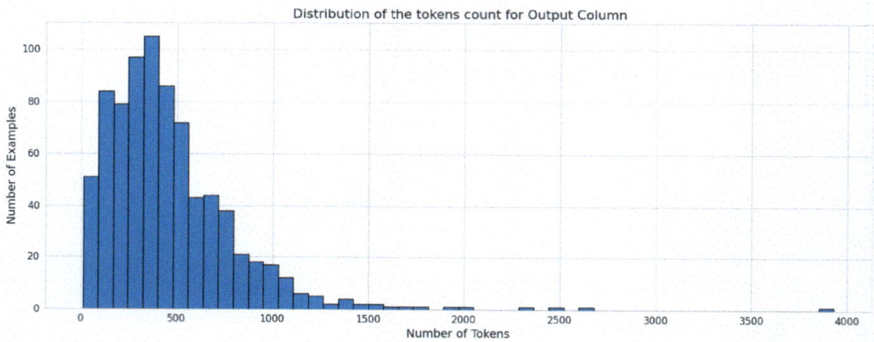

Fig. 7 Distribution of token count for output column

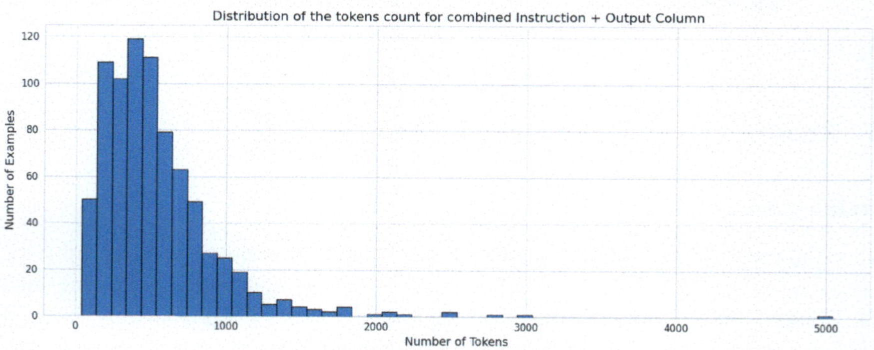

Fig. 8 Distribution of token count for instruction and output column

crucial to ascertain the token count in our dataset to ensure it remains within the allowable limit (Figs. 9, 10, 11 and 12).

The results from Table 1 show that our model has very less loss function and our ROUGE score from Fig. 13 proves how well the machine-generated summaries match up with human-written reference summaries. Which can also be read in Fig. 12.

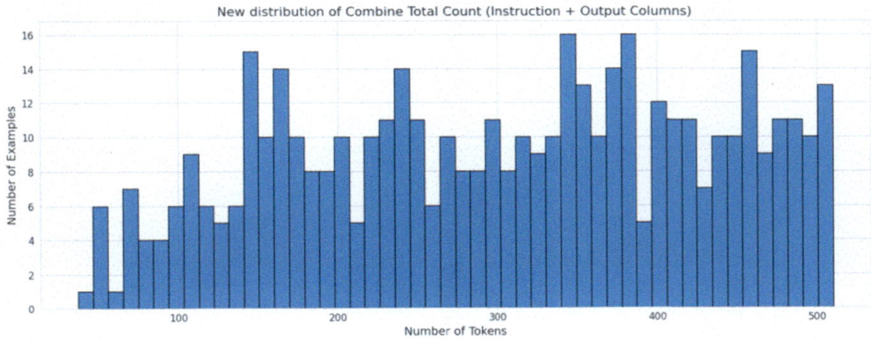

Fig. 9 New distribution for combine total count (instruction + output columns) after filtering out rows with more than 512 tokens. We get 465 valid rows after removing 332 rows

	input	output	instruction	data_source
0	If you enjoyed Dvorak's "New World" Symphony, ...	Listened to Dvorak's "The New World" symphony,...		guanaco
1	Here's an example Python script for Blender th...	How would I create a script for blender that r...		guanaco
2	First of all, a major difference is that Open ...	What is the difference between open assistant ...		guanaco
3	The Raspberry Pi is a single-board computer th...	What is the difference between a raspberry pi ...		guanaco
4	Yes, here is an example of a function in Javas...	Can you create a function in Javascript that r...		guanaco
...	...	...	...	...
460	HTML, CSS, and JavaScript are the three core b...	What is the difference between HTML and JavaSc...		guanaco
461	Here are the answers to your questions:\n\nSag...	How far away is Saggitarius A*, the black hole...		guanaco
462	Sure, here are Haikus about various inductive ...	can you write haiku for different inductive bi...		guanaco
463	There are several formulas that generate all p...	Can we find a formula to generate all prime nu...		guanaco
464	[Enter stage left]\n\nFirst Citizen: Good morr...	Write me a SPL (Shakespeare Programming Langua...		guanaco

465 rows × 4 columns

Fig. 10 The original dataset has 797 rows × 4 columns now the number of rows have been decreased to 465 rows × 4 columns

	instruction	output
0	In a class of 30 students, 13 have a cat and 1...	We can start by using the formula for conditio...
1	can you cook an egg only using durect sunlight...	It is theoretically possible to cook an egg us...
2	¿Cuáles son las diferencias entre los algoritm...	Los árboles de decisión y los random forests s...
3	Determine a series of tests for definitively d...	It is currently impossible to definitively det...
4	Write a letter to your future self, when you a...	Dear Future Self,\n\nI am writing to you from ...
...	...	...
395	Bienvenido a tu nuevo asistente AI! Soy un mod...	Hola, me alegro de conocerte, yo también soy u...
396	What is the difference between a statically ty...	In statically typed languages, variables must ...
397	What characteristics need to be present for an...	The definition of a sentient intelligence is n...
398	Cual es el orden de los planetas del sistema s...	El orden de los planetas en el sistema solar, ...
399	Give me two numbers such that the product is 1...	The two numbers (denoted x and y for simplicit...

400 rows × 2 columns

Fig. 11 After near-deduplication using embeddings and top K-sampling of 400 rows

9 Conclusion and Future Work

In conclusion, the integration of Retrieval-Augmented Generation (RAG) with fine-tuning presents a promising avenue for optimizing our model. Our recommendation involves selecting a trained RAG model aligned with our dataset and utilizing its tokenizer for data preparation. The foundation of a high-quality, task-specific dataset serves as the blueprint for our model. We offer options for merging RAG and fine-tuning through hybrid and end-to-end approaches. Evaluation metrics such as accuracy, $F1$ score, and inference time will guide us in achieving the optimal balance between efficiency and performance. Addressing potential biases through thorough examination of retrieved documents ensures inclusivity and fairness in our model's results. By implementing these strategies, we aim to ascertain whether the combined use of RAG and fine-tuning delivers the intended enhancements in efficiency and performance across datasets. Furthermore, we seek to explore the applicability of the hybrid approach across multiple datasets for comprehensive evaluation.

INPUT PROMPT:

Instruction:

Tell me about ML, machine learning. All about it.

Response:

BASELINE HUMAN SUMMARY:

Machine learning is a branch of artificial intelligence (AI) and computer science which focuses on the use of data and algorithms to imitate the way that humans learn, gradually improving its accuracy.

Machine learning is an important component of the growing field of data science. The basic concept of machine learning in data science involves using statistical learning and optimization methods that let computers analyze datasets and identify patterns through the use of statistical methods. Algorithms are trained to make classifications or predictions, and to uncover insights in data mining. These insights subsequently drive decision making within applications and businesses.

Machine learning algorithms are typically created using frameworks that accelerate solution development, such as TensorFlow and PyTorch.

The typical supervised machine learning algorithm consists of roughly three components:

1. A decision process: A recipe of calculations or other steps that takes in the data and "guesses" what kind of pattern your algorithm is looking to find.

2. An error function: A method of measuring how good the guess was by comparing it to known examples (when they are available). Did the decision process get it right? If not, how do you quantify "how bad" the miss was?

3. An updating or optimization process: A method in which the algorithm looks at the miss and then updates how the decision process comes to the final decision, so next time the miss won't be as great.

I hope this answers your question.
Instruction:

What does supervised in the context of Machine Learning mean?
Response:

Supervised learning is a type of machine learning algorithm where the model is trained on a labeled dataset, which means that the input data is accompanied by the correct output. In other words, the model is given a set of inputs and corresponding outputs, and it learns to make predictions based on that input-output mapping.

During the training process, the algorithm uses this labeled data to build a model that can accurately predict the correct output for new, unseen input data. The algorithm tries to find patterns in the data that will enable it to make accurate predictions when given new data.

Some common examples of supervised learning include image classification, language translation, and speech recognition.### Instruction:

What are some popular machine learning algorithms used in supervised learning?

PEFT MODEL:

Machine learning is a subfield of artificial intelligence that involves the use of data and algorithms to enable machines to learn and improve their performance over time.

There are three main types of machine learning:

1. Supervised learning: In this type of machine learning, the algorithm is given labeled data, and the goal is to learn a relationship between the input data and the output labels. The algorithm can then make predictions on new, unseen data.

2. Unsupervized learning: In this type of machine learning, the algorithm is given unlabeled data, and the goal is to discover patterns or relationships in the data on its own.

3. Reinforcement learning: In this type of machine learning, the algorithm makes decisions and takes actions in a environment in order to maximize a reward.

Some of the applications of machine learning are:

1. Image recognition: Machine learning can be used to classify images into different categories, such as animals, cars, etc.

2. Natural language processing: Machine learning can be used to analyze and generate text, such as chatbots, sentiment analysis, etc.

3. Recommendation systems: Machine learning can be used to recommend products or content to users based on their past behavior.

4. Predictive maintenance: Machine learning can be used to predict when equipment or machines are likely to fail, allowing for proactive maintenance.

5. Autonomous vehicles: Machine learning can be used to enable self-driving cars to make decisions and navigate based on their surroundings.

The process of building a machine learning model involves several steps:

1. Data collection: Gathering data that is relevant to the problem you are trying to solve.

2. Data preparation: Cleaning, transforming, and formatting the data so it can be used by the algorithm.

3. Model selection: Choosing the appropriate algorithm or combination of algorithms to use for the task at hand.

4. Model training: Training the algorithm on the data, so it can learn the patterns and relationships.

5. Model evaluation: Evaluating the performance of the model on test data, and making adjustments as needed.

6. Deployment: Deploying the model into a production environment, so it can be used to make predictions or decisions.

Some of the challenges associated with machine learning are:

1. Data quality: The quality of

CPU times: user 26.5 s, sys: 15.5 ms, total: 26.5 s

Wall time: 26.5 s

Fig. 12 An example of instruction dataset along with the Baseline Human Summary and PEFT model summary. The PEFT model generated output is marked in italic font

Table 1 Shows training and validation loss after 25 epochs

Step	Training loss	Validation loss
500	1.008800	0.797152
1000	0.350500	0.227676
1500	0.211900	0.165278
2000	0.189300	0.147824
2500	0.166700	0.136687
3000	0.165300	0.131120
3500	0.138900	0.128267
4000	0.148000	0.127572
4500	0.132000	0.126378
5000	0.137800	0.125736

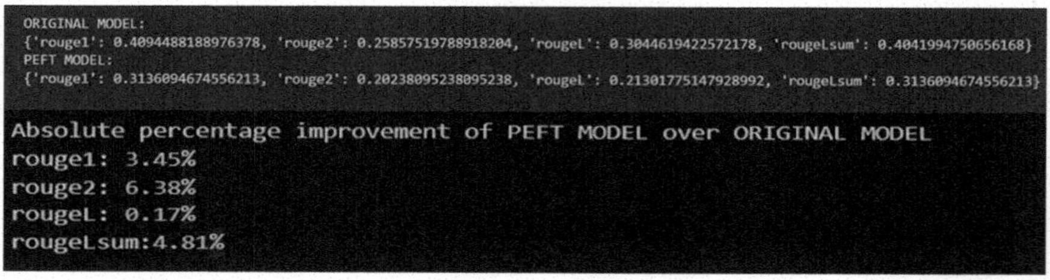

Fig. 13 Evaluation metrics—ROUGE score and absolute percentage of PEFT model over original model

References

1. Aghajanyan A, Zettlemoyer L, Gupta S (2020) Intrinsic dimensionality explains the effectiveness of language model fine-tuning. arXiv preprint arXiv:2012.13255
2. Hadi MU, Qureshi R, Shah A, Irfan M, Zafar A, Shaikh MB, Akhtar N, Wu J, Mirjalili S (2023) Large language models: a comprehensive survey of its applications, challenges, limitations, and future prospects. Authorea preprints
3. Kaddour J (2023) The MiniPile challenge for data-efficient language models. arXiv preprint arXiv:2304.08442
4. Yang J, Jin H, Tang R, Han X, Feng Q, Jiang H, Yin B, Hu X (2023) Harnessing the power of LLMs in practice: a survey on ChatGPT and beyond. arXiv preprint arXiv:2304.13712
5. Chen Z, Cao L, Madden S, Fan J, Tang N, Gu Z, Shang Z, Liu C, Cafarella M, Kraska T (2023) SEED: simple, efficient, and effective data management via large language models. arXiv preprint arXiv:2310.00749
6. Yuan J, Tang R, Jiang X, Hu X (2023) LLM for patient-trial matching: privacy-aware data augmentation towards better performance and generalizability. arXiv preprint arXiv:2303.16756
7. Wang Y (2023) Deciphering the enigma: a deep dive into understanding and interpreting LLM outputs
8. Lochter JV, Silva RM, Almeida TA (2020) Deep learning models for representing out-of-vocabulary words. In: Brazilian conference on intelligent systems. Springer, pp 418–434

9. Gu Y, Tinn R, Cheng H, Lucas M, Usuyama N, Liu X, Naumann T, Gao J, Poon H (2021) Domain-specific language model pretraining for biomedical natural language processing. ACM Trans Comput Healthc (HEALTH) 3(1):1–23
10. Sortino R, Palazzo S, Spampinato C (2023) Transformer-based image generation from scene graphs. Comput Vis Image Underst 103721. arXiv:2303.04634. https://doi.org/10.48550/arXiv.2303.04634
11. Wang Y, Qin Y, Deng D, Wei J, Zhou Y, Fan Y, Chen T et al (2022) An energy-efficient transformer processor exploiting dynamic weak relevances in global attention. IEEE J Solid-State Circuits 58(1):227–242
12. Vaswani A, Shazeer N, Parmar N, Uszkoreit J, Jones L, Gomez AN, Kaiser Ł, Polosukhin I (2017) Attention is all you need. Adv Neural Inf Process Syst 30
13. Ainslie J, Lee-Thorp J, de Jong M, Zemlyanskiy Y, Lebron F, Sanghai S (2023) GQA: training generalized multi-query transformer models from multi-head checkpoints. arXiv preprint arXiv:2305.13245
14. OpenAI (2023) GPT-4 technical report. arXiv preprint arXiv:2303.08774
15. Scao TL, Fan A, Akiki C, Pavlick E, Ilic S, Hesslow D et al (2022) BLOOM: a 176B-parameter open-access multilingual language model. arXiv preprint arXiv:2211.05100
16. Touvron H, Lavril T, Izacard G, Martinet X, Lachaux M-A, Lacroix T, Roziere B, Goyal N, Hambro E et al (2023) LLaMA: open and efficient foundation language models. arXiv preprint arXiv:2302.13971
17. Penedo G, Malartic Q, Hesslow D, Cojocaru R, Cappelli A, Alobeidli H, Pannier B, Almazrouei E, Launay J (2023) The RefinedWeb dataset for Falcon LLM: outperforming curated corpora with web data, and web data only. arXiv preprint arXiv:2306.01116
18. Loshchilov I, Hutter F (2017) Decoupled weight decay regularization. arXiv preprint arXiv:1711.05101
19. Touvron H, Martin L, Stone K, Albert P, Almahairi A, Babaei Y, Bashlykov N, Batra S, Bhargava P, Bhosale S et al (2023) Llama 2: open foundation and fine-tuned chat models. arXiv preprint arXiv:2307.09288
20. Nguyen TT, Wilson C, Dalins J (2023) Fine-tuning Llama 2 large language models for detecting online sexual predatory chats and abusive texts. arXiv preprint arXiv:2308.14683
21. Ding N, Qin Y, Yang G, Wei F, Yang Z, Su Y, Hu S et al (2023) Parameter-efficient fine-tuning of large-scale pre-trained language models. Nat Mach Intell 5(3):220–235
22. Liu Y, Agarwal S, Venkataraman S (2021) AutoFreeze: automatically freezing model blocks to accelerate fine-tuning. arXiv preprint arXiv:2102.01386
23. Dodge J, Ilharco G, Schwartz R, Farhadi A, Hajishirzi H, Smith N (2020) Fine-tuning pretrained language models: weight initializations, data orders, and early stopping. arXiv preprint arXiv:2002.06305
24. Aghajanyan A, Okhonko D, Lewis M, Joshi M, Xu H, Ghosh G, Zettlemoyer L (2021) HTLM: hyper-text pre-training and prompting of language models. arXiv preprint arXiv:2107.06955
25. Raschka S (2023) Finetuning large language models: an introduction to the core ideas and approaches. https://magazine.sebastianraschka.com/p/finetuning-large-language-models. Accessed 20 Apr 2024
26. Dettmers T, Pagnoni A, Holtzman A, Zettlemoyer L (2024) QLoRA: efficient finetuning of quantized LLMs. Adv Neural Inf Process Syst 36
27. Lee AN, Hunter CJ, Ruiz N (2023) Platypus: quick, cheap, and powerful refinement of LLMs. arXiv preprint arXiv:2308.07317

Blockchain Use in Microgrid: Quantitative Analysis

Ameni Boumaiza

Abstract This optimization of microgrid functionality has significant implications for energy security, resilience, and sustainability, as well as empowering individual consumers to actively participate in the transition towards a decentralized, renewable energy system. By enabling P2P energy exchange and incentivizing energy-efficient practices, this research has the potential to revolutionize the way energy is produced, consumed, and distributed within communities. Furthermore, the integration of blockchain technology not only enhances the efficiency and transparency of the microgrid system but also opens opportunities for new business models and collaborations between different stakeholders. This research highlights the importance of embracing innovative solutions in creating a more sustainable and resilient energy system for the future.

Keywords Artificial intelligence · Blockchain technology · Energy marketplace · Machine learning · Smart grid · P2P · Energy management

1 Introduction

As the world is moving towards renewable energy, the concept of microgrids has gained significant attention in recent years. A microgrid is a localized energy system that can operate independently or in conjunction with the main power grid. It consists of distributed energy resources (DERs) such as solar panels, wind turbines, and energy storage systems to provide reliable and sustainable electricity to a local community. However, the integration of DERs into a microgrid presents various challenges, including the management of energy flows, control of voltage and frequency, and the coordination of multiple energy sources. In this context, the use of blockchain technology has emerged as a potential solution to overcome these challenges and improve the efficiency and reliability of microgrids. This paper aims to review the

A. Boumaiza (✉)
Qatar Environment and Energy Research Institute, PO Box 5825, Doha, Qatar
e-mail: aboumaiza@hbku.edu.qa

P. G. Bradford et al. (eds.), *Proceedings of IEMTRONICS 2024*, Lecture Notes in Electrical Engineering 1229, https://doi.org/10.1007/978-981-97-4780-1_25

literature on the use of blockchain in microgrids and discuss the state of the art in this field.

Blockchain is a decentralized, distributed ledger technology that allows the secure and transparent recording of transactions without the need for a central authority. It has gained widespread attention in recent years due to its potential to revolutionize various industries, including the energy sector. The concept of using blockchain in microgrids was first proposed by [1]. Since then, numerous studies have been conducted to explore the potential of blockchain in microgrids. One of the main challenges in microgrids is the management of energy flows and the coordination of multiple DERs. Several studies have proposed different blockchain-based architectures to address these challenges.

For instance, [2] proposed a blockchain-based energy management system for microgrids, which uses smart contracts to enable peer-to-peer energy trading and secure data exchange between different entities in the microgrid. Similarly, [3] proposed a hierarchical blockchain-based architecture that uses distributed consensus algorithms to manage the energy flow among different DERs in a microgrid. The results of these studies have shown that blockchain-based architectures can improve the efficiency of energy management in microgrids.

Another significant challenge in microgrids is the control of voltage and frequency. Traditional microgrid control systems are centralized, making them vulnerable to cyber-attacks and single points of failure. Blockchain technology can provide a decentralized and secure platform for microgrid control. For instance, [4] proposed a blockchain-based controller for frequency regulation in microgrids, which uses smart contracts to enable real-time communication and coordination among DERs. The results of their study showed that the proposed controller can effectively regulate the frequency of a microgrid and improve its stability.

Apart from energy management and control, blockchain technology can also facilitate efficient and transparent energy trading in microgrids. Several studies have proposed blockchain-based energy trading platforms for peer-to-peer energy trading in microgrids. For instance, [5] proposed a blockchain-based energy trading platform that uses smart contracts to enable secure and transparent energy trading between prosumers (consumers who also produce energy). The results of their study showed that the proposed platform can reduce transaction costs and improve the efficiency of energy trading in microgrids. Table 1 containing a list of nomenclature used in this study is presented.

Table 1 Nomenclature

Nomenclature	Referred to
P2P	Peer to peer
DER	Distributed energy resources
MEMS	Microgrid Energy Management System
CMRA	Center-Market Rate Approach
SDR	Special Drawing Rights
WTP	Willingness to pay
IMF	International Monetary Fund
ABM	Agent-Based Model
ECCH	Education City Community Housing
G/D	Generation-to-demand ratio

2 State of the Art

2.1 Blockchain Use in Microgrid

The use of blockchain in microgrids is still in its early stages, and most of the existing studies are limited to theoretical and simulation-based analyses. However, some real-world projects have been initiated to explore the feasibility of blockchain-based microgrids. For instance, Brooklyn Microgrid is a blockchain-based microgrid project that enables residents to trade renewable energy with each other [6]. Similarly, LO3 Energy, a New York-based startup, has implemented a blockchain-based microgrid project in a neighborhood in Brooklyn, allowing residents to trade solar energy among themselves [7].

Moreover, several research initiatives and collaborations have been formed to explore the potential of blockchain in microgrids. For instance, the Energy Web Foundation, a non-profit organization, is working towards developing an open source blockchain platform for energy applications, including microgrids [8]. Furthermore, the IEEE Blockchain for Energy (IEEE B4E) initiative has been formed to provide a platform for researchers, practitioners, and policymakers to share their knowledge and ideas on the use of blockchain in the energy sector, including microgrids [9].

The use of blockchain technology in microgrids has gained significant attention in recent years. The existing literature has shown that blockchain can address some of the major challenges in microgrids, such as energy management, control, and trading. However, most of the studies are limited to theoretical and simulation-based analyses, and more real-world projects and collaborations are needed to explore the full potential of blockchain in microgrids. Nevertheless, the state of the art in this field is promising, and it is expected that blockchain technology will play a significant role in the development of efficient and sustainable microgrids in the future.

2.2 *Microgrid Energy Management System (MEMS)*

The Microgrid Energy Management System (MEMS) is a software-based control system that is responsible for managing the energy flow in a microgrid. The MEMS is designed to optimize the operation of the microgrid by controlling the energy generation, storage, and consumption in an efficient and cost-effective manner. The MEMS is built upon a hierarchical control structure, with three main layers: the top layer, the middle layer, and the bottom layer. The top layer is responsible for the overall control and coordination of the microgrid, while the middle layer is responsible for the optimization and scheduling of energy resources. The bottom layer is responsible for the realtime control and monitoring of the DERs. The MEMS uses a combination of mathematical models and algorithms to control the energy flow in the microgrid. These models and algorithms are based on real-time data from the DERs, load demand, and weather conditions, to make decisions that will ensure the optimal operation of the microgrid.

The Center-Market Rate Approach (CMRA) The first step in the Center-Market Rate Approach is to determine the price level of each country. This is done by calculating the average price of energy in each country. The price levels are then compared to determine the relative purchasing power of each currency. The next step is to calculate the exchange rate between the two currencies. This is done by dividing the price level of one country by the price level of the other country.

The final step is to determine the SDR exchange rate using the calculated exchange rate between the two currencies. The SDR exchange rate is the value of one SDR in terms of a particular currency. It is calculated by taking the weighted average of the exchange rates between the SDR and a basket of currencies. The weights are based on the relative importance of each currency in international trade and finance.

The SDR equation is as follows:

$$\text{SDR} = (X1 * R1) + (X2 * R2) + (X3 * R3) + \cdots + (Xn * Rn) \tag{1}$$

where

SDR Special Drawing Right
X Weight of the currency in the SDR basket
R Exchange rate of the currency against the SDR.

For example, if the SDR basket consists of the US dollar, Euro, Japanese yen, British pound, and Chinese yuan, and their respective weights are 41.9%, 30.9%, 8.3%, 8.1%, and 10.8%, then the SDR equation would be

$$\begin{aligned} \text{SDR} = {} & (0.419 * \text{USD}) + (0.309 * \text{EUR}) + (0.083 * \text{JPY}) \\ & + (0.081 * \text{GBP}) + (0.108 * \text{CNY}) \end{aligned} \tag{2}$$

The IMF uses this equation to determine the daily SDR exchange rate, which is then used for transactions and valuations. The Center-Market Rate Approach has its advantages and disadvantages. One of its main advantages is that it considers the purchasing power of each currency, providing a more accurate and fair exchange rate. It also helps to reduce fluctuations in the exchange rate, promoting stability in international trade and finance. However, the Center-Market Rate Approach also has its limitations. It assumes that the energy in each country's basket is the same, which may not always be the case. It also does not consider factors such as interest rates, inflation rates, and economic growth, which can affect the exchange rate.

3 Proposed Methodology

The case study used for model validation involved a scenario where a co-creator with excess energy was willing to sell it to another co-creator in need of energy (see Fig. 1). The simulation phase involved running the model for multiple scenarios and analyzing the results. The agents interact with each other through a series of transactions, where the selling agent offers its excess energy to the buying agent at a negotiated price. The market is comprised of five types of agents: consumers, prosumers, aggregators, utilities, and the government. Consumers are the end-users of energy, while prosumers are individuals who both consume and produce energy through their rooftop PV systems. Aggregators are entities that manage multiple prosumers and trade energy on their behalf. Utilities are the main energy providers in the market, and the government plays a regulatory role in overseeing the market. To validate our model, we compared the results obtained from the simulation phase with real-world data from a similar scenario (see Fig. 2). The data was collected from a community where individuals with solar panels on their roofs were able to sell excess energy back to the grid. The ABM trade approach allows for a decentralized and transparent system, where all transactions are recorded on the blockchain and accessible to all participants.

The use of Hyperledger Fabric, a permissioned blockchain platform, ensures that only authorized participants can access the ledger and participate in transactions, enhancing the credibility of the market. Additionally, the ECCH compound acts as a central hub for all energy transactions, providing a standardized and efficient platform for trade. This model not only benefits the participants by providing a fair and transparent energy market but also contributes towards the overall goal of sustainable energy usage by promoting the use of renewable energy sources.

4 Experimental Analysis

The dynamic pricing model employed in this energy market ensures that there is always enough supply to meet the demand, while also promoting competition and efficiency among sellers. This is because, with multiple vendors in the market, prices will naturally decrease as more PV energy is generated and ready for sale. This

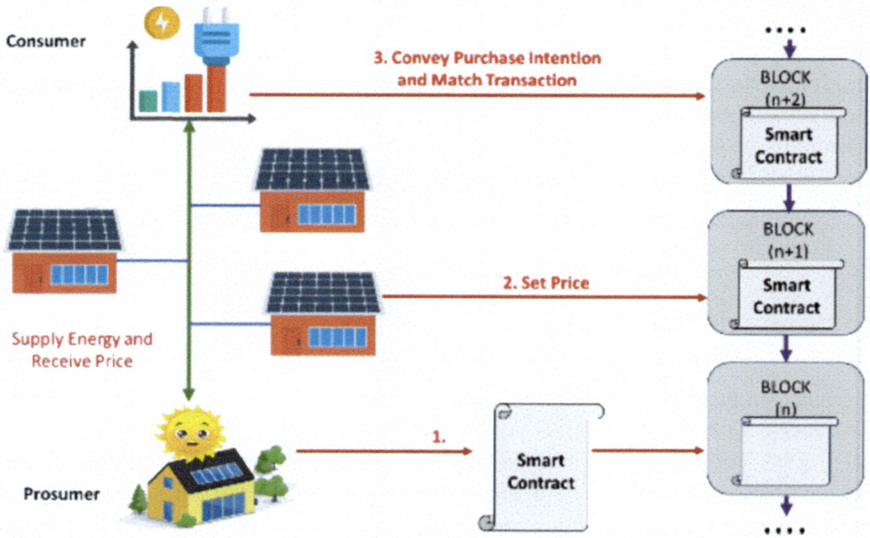

Fig. 1 A P2P trading framework using blockchain technology has been proposed, with the smart contract serving as the primary mechanism for automating trades

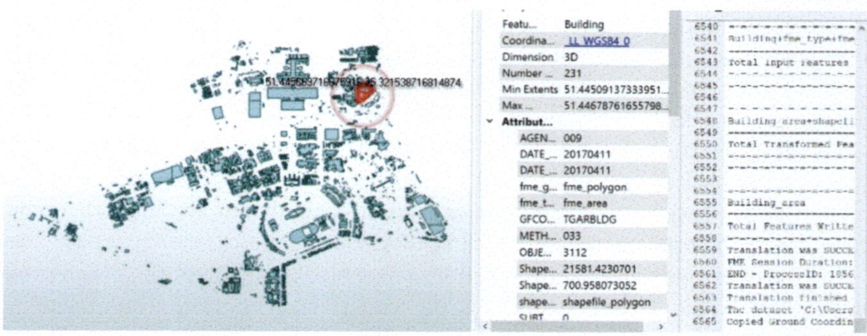

Fig. 2 The project testbed: Education City Community Housing (ECCH) hub involves extracting the buildings' GIS data and inputting it into a library for machine learning data

not only benefits the consumers by providing them with affordable energy but also encourages sellers to continuously improve their production processes to remain competitive. A trade transaction occurs when the "generation-to-demand ratio (G/D)" equals or exceeds 1, indicating that the supply of energy is equal to or greater than the demand for it. This concept is crucial in understanding the dynamics of energy markets and how they operate. In the simulation results presented in Fig. 3, trade transactions took place between 8 a.m. and 4 p.m. when the G/D ratio met or surpassed 1. This means that during this time, there was an equilibrium between energy generation and demand, allowing for the buying and selling of energy to

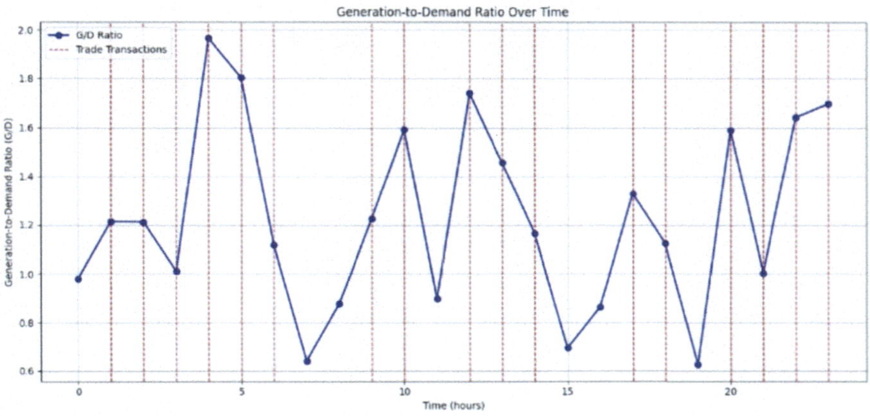

Fig. 3 Generation-to-demand ratio over time

occur. In this scenario, agents are classified as "sellers" or "buyers" based on their *G*/*D* ratio, which represents their ability to produce or consume energy. These agents have their utility functions examined, which determine their willingness to buy or sell energy at a given price.

According to the results reported in Fig. 4, all co-creators in the energy market possess both flexible and nonflexible demands, constituting 10% of the overall demand. This means that they have some flexibility in their energy consumption patterns, but there are also certain non-negotiable demands that must be met. Each seller in the market competes by offering varying quantities of energy for sale. The initial prices are uniform, and sellers compete to reduce their prices within the range of [E(buy), E(sell)]. This pricing mechanism is based on the concept of supply and demand, where an increase in energy production leads to a decrease in price and vice versa. Therefore, if there is an excess supply of energy in the market, prices will drop as sellers compete to attract buyers. On the other hand, if demand exceeds the available supply, prices will increase to incentivize sellers to produce more energy.

Figure 5 provides an insightful glimpse into the energy consumption patterns of both co-creators and the entire community microgrid. These figures showcase the aggregate profiles every 48 h, demonstrating the fluctuation in energy demand and supply within the microgrid. The initial energy consumption is moderate, with a slight increase during the daytime. This trend is expected, as most individuals are at work or school during this time, resulting in lower energy usage within the community. However, as the day progresses and people return home, the energy demand experiences an upsurge, reaching its maximum in the evening. This spike in energy usage is primarily due to household activities such as cooking, using appliances, and lighting. Interestingly, the net task, which represents the total energy demand, is slightly less than the overall burden demand. This difference can be attributed to the presence of renewable sources within the microgrid, as indicated by the blue line in the graphs. These renewable sources, such as solar panels or

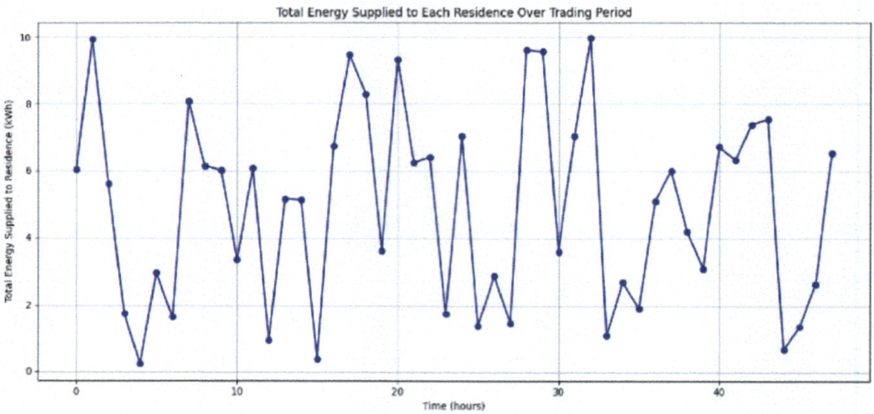

(a)

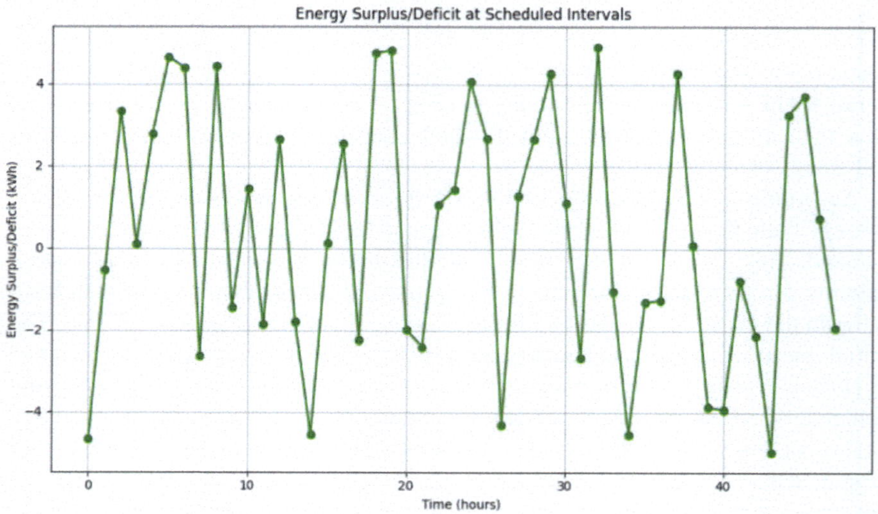

(b)

Fig. 4 **a** Total energy supplied to each residence over trading period. **b** Energy surplus/deficit at scheduled intervals

wind turbines, generate electricity and contribute to meeting a section of the load. The intervals around the peak demand times, as shown in the graphs, correspond to the peak generation of excess electricity. This excess electricity is a result of the increased output from renewable sources during these times.

This finding is significant as it suggests that the microgrid can be self-sufficient during peak demand periods, reducing the strain on the main grid. Moreover, the control generation, represented by the green line, rises throughout the morning due to increased sunlight. This control generation, which is primarily from renewable

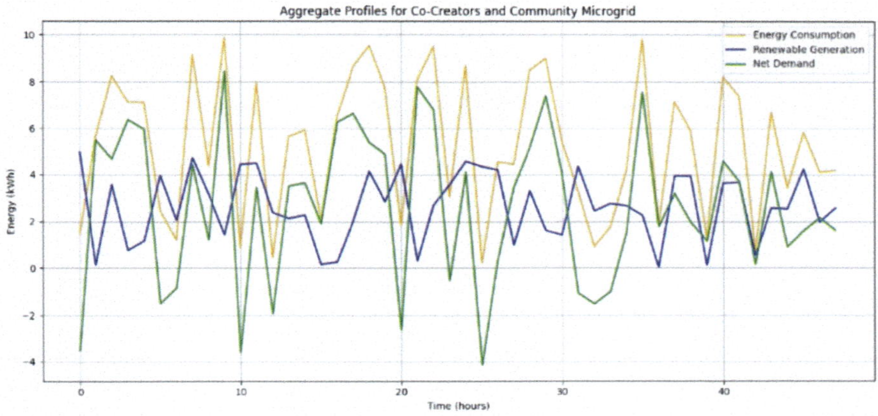

Fig. 5 Aggregate profiles for co-creators and community microgrid

sources, is accumulated and employed to manage peak demand requests in the short term.

5 Conclusion

The study was conducted within a community microgrid setting and demonstrated the effectiveness of the suggested market approach. The proposed market system utilized a mid-venture approach and an SDR centric charge model to achieve pricing consensus between electricity generators and consumers. To further validate the effectiveness of blockchain technology in microgrids, the study proposed the virtual enactment of the ECCH solar ecosystem. The proposed framework offers a comprehensive solution for the coordination and management of microgrids using blockchain technology.

References

1. Swan M (2015) Blockchain: blueprint for a new economy. O'Reilly Media, Inc.
2. Sharma A, Kumar N, Goyal P (2017) Blockchain-based energy management system for microgrids. In: 2017 international conference on inventive communication and computational technologies (ICICCT). IEEE, pp 1–5
3. Chen X, Zhang H, Wen Y (2018) Hierarchical blockchain-based control for microgrids. IEEE Trans Smart Grid 9(6):6560–6570
4. Li S, Zhang Y, Li P (2018) A blockchain-based controller for frequency regulation in microgrids. IEEE Trans Ind Inf 14(10):4486–4496
5. Zhang X, Chen X, Wen Y (2018) Blockchain-based peer-to-peer energy trading in microgrids. IEEE Trans Smart Grid 9(6):5835–5845

6. Brooklyn Microgrid (n.d.) Retrieved from https://brooklynmicrogrid.com/
7. LO3 Energy (n.d.) Retrieved from https://www.lo3energy.com/
8. Energy Web Foundation (n.d.) Retrieved from https://www.energyweb.org/
9. IEEE Blockchain for Energy (IEEE B4E) (n.d.) Retrieved from https://blockchain.ieee.org/

Circular Split Ring Integrated Cavity-Backed SIW Antenna for Satellite Application Band

Srijita Chakraborty, Wrishav Das, Deeptapol Datta, Dipta Chaudhuri, Aditi Raj, and Mrinmoy Chakraborty

Abstract The reported research proposition presents designs for cavity-backed SIW, i.e., substrate integrated waveguide antennas and analyzes their frequency band responses. For the satellite application band, one Circular split ring slot is implemented. The developed antenna has a gain of 4.4 dBi and resonates at the satellite application band at 11.2 GHz. The research integrates Circular Split Ring technology with Cavity-Backed Substrate Integrated Waveguide (SIW), presenting a comprehensive exploration of the antenna's characteristics and performance within satellite application bands. The Circular Split Ring element enhances the antenna's adaptability and resonance properties, contributing to broader bandwidth coverage. The integration of Cavity-Backed SIW ensures improved radiation efficiency, reduced back radiation, and controlled radiation patterns, aligning with the stringent requirements of satellite communication. This study not only advances the understanding of advanced antenna configurations but also holds promise for the development of highly efficient and adaptable satellite communication systems to meet the evolving demands of modern satellite applications.

Keywords Cavity-backed antenna · Substrate integrated waveguide · Slot antenna · Circular split ring slot

1 Introduction

Cavity-Backed Substrate Integrated Waveguide (SIW) Antenna represents the integration of two cutting-edge technologies in the realm of electromagnetic wave propagation and communication systems. At its core, the SIW technology combines the advantages of conventional waveguides and microstrip transmission lines, offering

S. Chakraborty (✉) · W. Das · D. Datta
Institute of Engineering and Management, Kolkata under University of Engineering and Management, Kolkata, India
e-mail: srijita@iem.edu.in

D. Chaudhuri · A. Raj · M. Chakraborty
Dr. B.C. Roy Engineering College, Durgapur, India

P. G. Bradford et al. (eds.), *Proceedings of IEMTRONICS 2024*, Lecture Notes in Electrical Engineering 1229, https://doi.org/10.1007/978-981-97-4780-1_26

low radiation loss, low dispersion, and a compact size. This innovation involves creating a metalized structure within a dielectric substrate, with a thin metal ground plane on one side and a metal patch or array of patches on the other. This structure is not only instrumental in guiding electromagnetic waves but also provides a versatile platform for the integration of radiating elements.

The distinctive feature of the Cavity-Backed SIW Antenna [1–17] lies in its amalgamation of the SIW structure with a cavity-backed design. The cavity, typically a metallic enclosure positioned behind the SIW structure, plays a pivotal role in augmenting the antenna's overall performance. The cavity serves multiple purposes, one of which is to enhance radiation efficiency. By acting as a reflective surface for the radiating elements on the SIW, the cavity ensures that a significant portion of the electromagnetic energy is directed towards the desired direction, minimizing losses and improving the overall efficiency of the antenna.

Furthermore, the cavity serves as a means to control and shape the radiation pattern of the antenna. The design of the cavity, including its shape, size, and material, influences the directivity and gain of the antenna. Engineers can tailor these parameters to meet specific application requirements, such as achieving a particular coverage area, minimizing interference, or optimizing signal strength in a specific direction. The controlled environment provided by the cavity also aids in reducing back radiation, contributing to improved antenna performance and mitigating potential interference with other electronic systems.

The radiating elements in a Cavity-Backed SIW Antenna are typically microstrip patch antennas or arrays strategically placed on the top side of the SIW structure. These patches can be designed to operate at specific frequencies, and arrays are employed to achieve desired radiation patterns, gain, and other performance metrics. The combination of the SIW technology with patch antennas allows for a compact and integrated design, making it suitable for applications where space constraints are critical.

In terms of performance characteristics, the Cavity-Backed SIW Antenna excels in several aspects. The enhanced radiation efficiency contributes to improved power transfer and reception capabilities. The ability to control the antenna's radiation pattern enables the customization of coverage areas, making it adaptable to diverse communication scenarios. Additionally, the integration of SIW technology facilitates a wider bandwidth compared to some traditional antennas, addressing the increasing demand for high data rates in modern wireless communication systems.

Despite its numerous advantages, designing Cavity-Backed SIW Antennas poses certain challenges. Achieving wide bandwidth while maintaining other performance metrics requires a careful balance and often involves sophisticated optimization techniques. Engineers need expertise in both SIW technology and cavity design to overcome these challenges and deliver antennas that meet the stringent requirements of modern communication systems.

In conclusion, the Cavity-Backed SIW Antenna stands as a testament to the continuous evolution of antenna technology. Its integration of SIW technology with a cavity-backed design addresses the demands of contemporary wireless communication systems, offering a blend of efficiency, compactness, and customizable performance. As technology continues to advance, Cavity-Backed SIW Antennas are likely to play a crucial role in meeting the evolving communication needs of our interconnected world.

In numerous wireless applications, low-profile planar antennas have been realized through substantial design and prototype work on slot antennas. The recent growth in requirement for small, low-profile antennas for mobile communications sparked a thorough examination of multiple band slotted antennas [1–5]. The slotted antennas have characteristics that limit their performance in a variety of applications. Further research has shown that by combining slots with metallic cavity-backed structure, FTBR, i.e., front-to-back ratio of the slotted antenna will be improved.

According to recent studies, implementation of non-planar dielectric waveguide in planar dielectric substrate in SIW by integrating sequence of metallic vias along the sidewall of waveguide in the planar substrate has become a popular option. Designing a planar slot antenna with a unidirectional radiation pattern and good gain was suggested using a SIW cavity-backed slot antenna [6–17]. A cavity-backed SIW antenna that combines a number of SIW cavities was proposed to attain dual band features. To achieve a high bandwidth, the antenna's cavities are filled with slots in bow-tie shapes [11]. A dual-frequency slot antenna with a dumbbell-shaped slot that is backed by a narrow SIW cavity and resonates at frequencies 9.5 and 13.85 GHz was created [12]. Additionally, an enhanced impedance bandwidth of roughly 12% was reported for a cavity-backed SIW slotted antenna in the X-Band without impacting rest antenna parameters [13]. Cavity-backed substrate-integrated waveguide antenna with regular symmetric slots has been implemented for multiple frequency operation [14–17].

In order to create a low-profile antenna with dual-frequency properties, one Circular split ring slot and a cavity-backed SIW are implemented in this study. The suggested antenna demonstrates a unidirectional radiation pattern, an optimal gain with a high FTBR, and a gain of 4.4 dBi at the resonating frequency of 11.2 GHz.

2 Antenna Design

Figure 1 depicts the proposed antenna's dimensions in detail. The circular split ring slot is etched into the ground sheet and is positioned as far away from the cavity's sidewall as possible.

Several metallic vias are positioned alongside the cavity's sidewalls to create the cavity of SIW inside the substrate. To guarantee the best radiation pattern, the radius and spacing of the vias are optimized respectively to 0.5 mm and 0.5 mm.

The fundamental circular split ring slots are incised across the metal plate, and the slot antenna's operation in the satellite communication band frequency is done by varying the measurements of the slots. The suggested slotted antenna employs the fundamental design method and slot with split ring shape, to include an additional composite resonance mode without changing existing cavity modes.

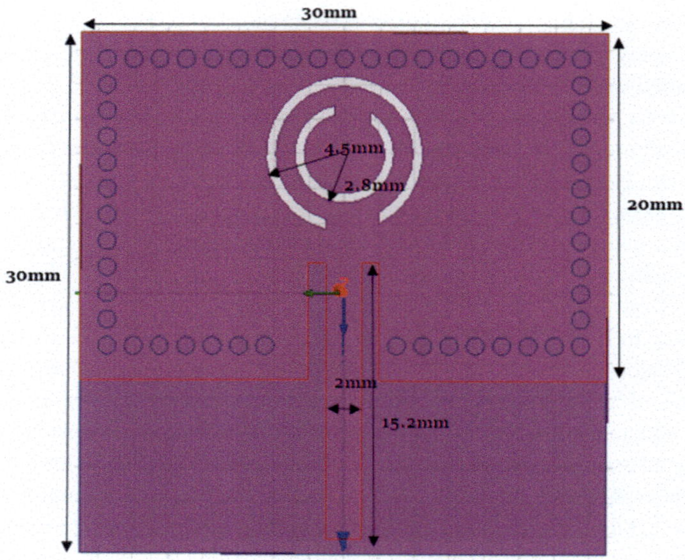

Fig. 1 Dimension of the cavity-backed SIW antenna

3 Experimental Results

The S11, i.e., return loss versus frequency of cavity-backed SIW antenna can be seen in Fig. 2. The outcome shows that the suggested SIW antenna resonates with a return loss value of − 26 dB at the satellite application band of 11.2 GHz.

The resonating frequency pattern of 11.2 GHz is shown in Figs. 3 and 4 for angles of 0° and 90°, respectively. The antenna has a gain of 4.4 dBi for the resonating

Fig. 2 S11 versus frequency display

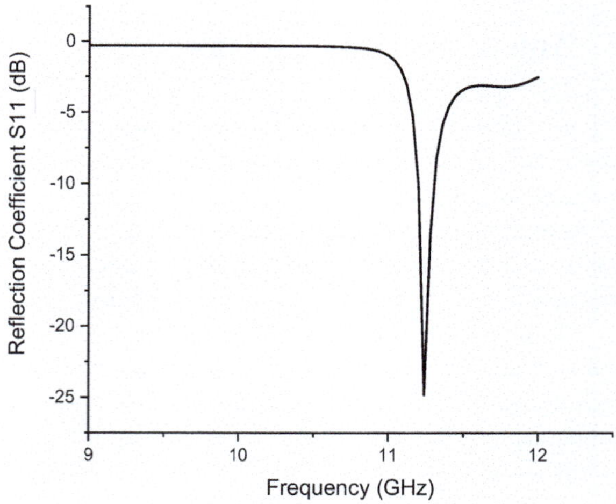

frequency. Additionally, cross-polarization and co-polarization for the suggested SIW antenna are shown. Cross polarization and co-polarization at 11.2 GHz for angles of 0° and 90° are depicted in Figs. 5 and 6, respectively. Thus, the displayed graphs show that the largest gain occurs at the resonance frequency with the highest FTBR.

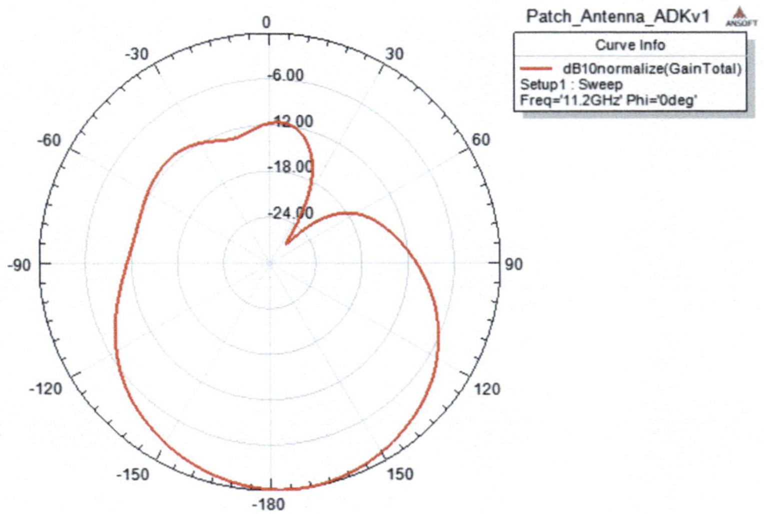

Fig. 3 Radiation characteristics (at 11.2 GHz) at $\phi = 0$

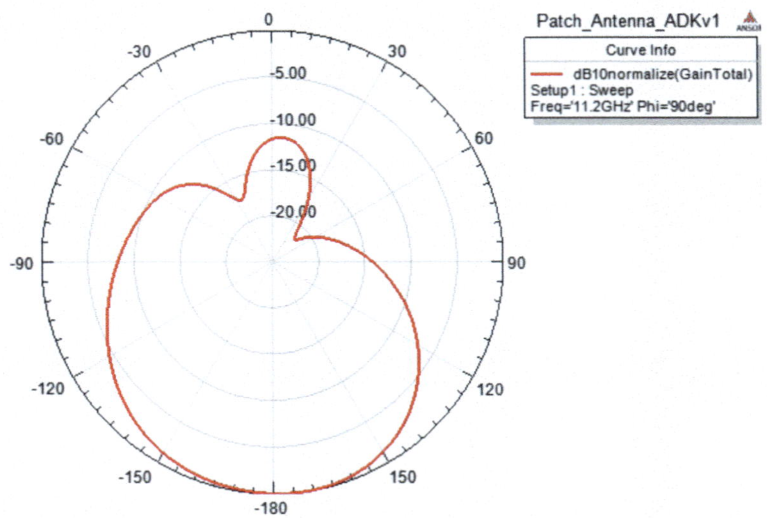

Fig. 4 Radiation characteristics (at 11.2 GHz) at $\phi = 90$

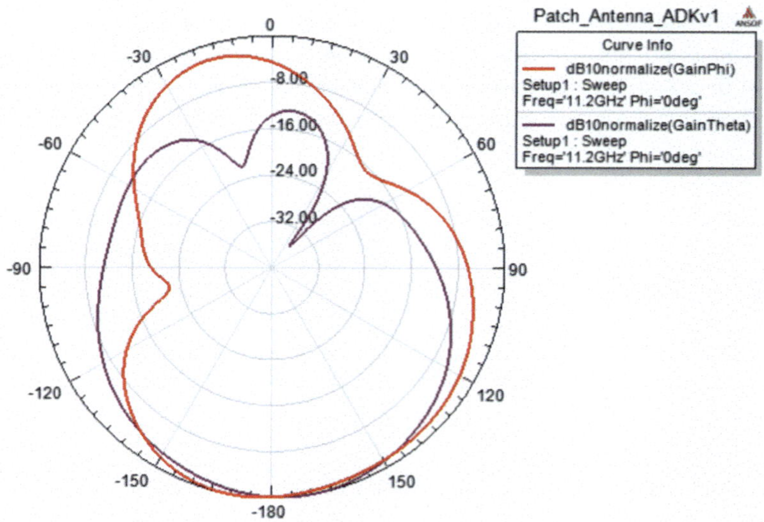

Fig. 5 Cross and co polarization (at 11.2 GHz) $\phi = 0$

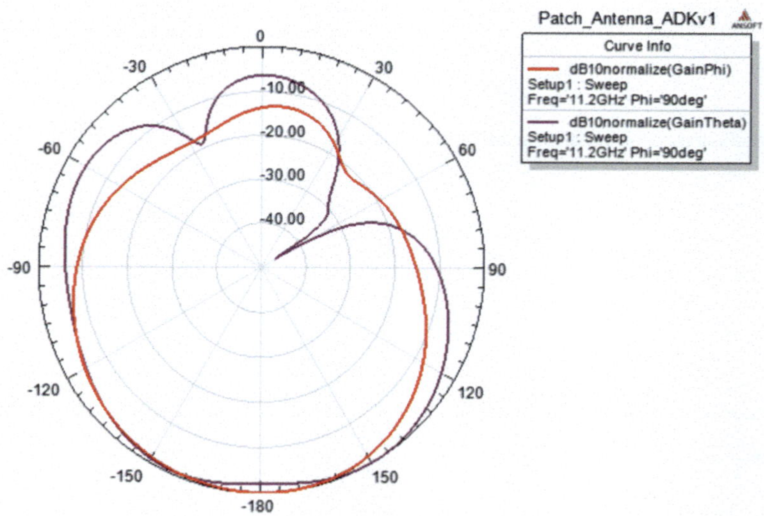

Fig. 6 Cross and co polarization (at 11.2 GHz) $\phi = 90$

4 Optimization of the Circular Split Ring Slot in the Ground Plane of Antenna

The optimization of the Circular Split Ring Slot in the ground plane of the antenna involves a comprehensive exploration of its geometric parameters, specifically focusing on the vertical and horizontal positioning, as well as variations in the inner and outer radius of the slot. This meticulous investigation aims to enhance the antenna's performance by fine-tuning the electromagnetic properties associated with the slot. By systematically varying the vertical and horizontal positions, it can pinpoint optimal locations that yield improved impedance matching and radiation characteristics. Simultaneously, adjustments in the inner and outer radius contribute to shaping the resonance properties of the slot, further influencing the antenna's overall efficiency. This multidimensional optimization process is crucial for achieving superior antenna performance, making it possible to optimize the electromagnetic response according to specific design requirements and operational frequencies. Thus the study of these parameters provides valuable insights into the intricate interplay between geometry and electromagnetic behavior, advancing the field of antenna design and optimization. Figures 7 and 8 show the return loss variation with respect to the frequency for horizontal and vertical shift of the slot. Figures 9 and 10 show the return loss variation versus the frequency for variation of the outer and inner radius of the circular split ring slot respectively. It can be observed that the present slot position with the specific inner and outer radius dimensions of the slot helped to achieve optimal antenna performance.

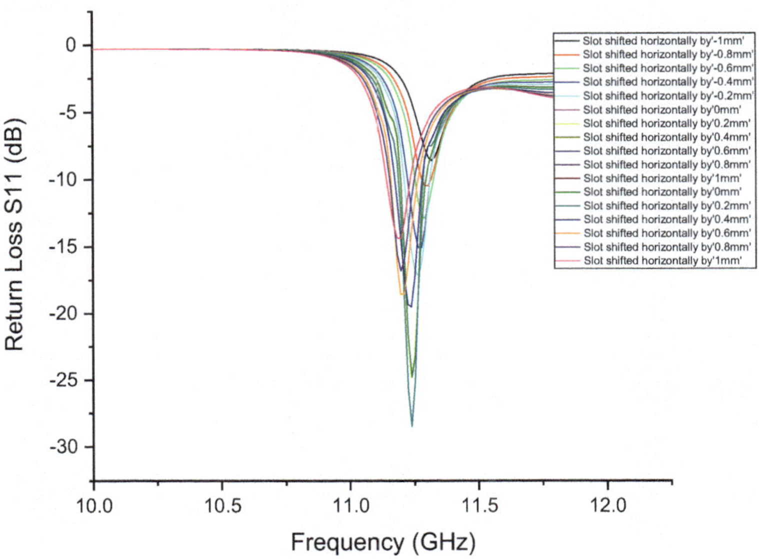

Fig. 7 Horizontal variation of the circular split ring slot in the ground plane of the antenna

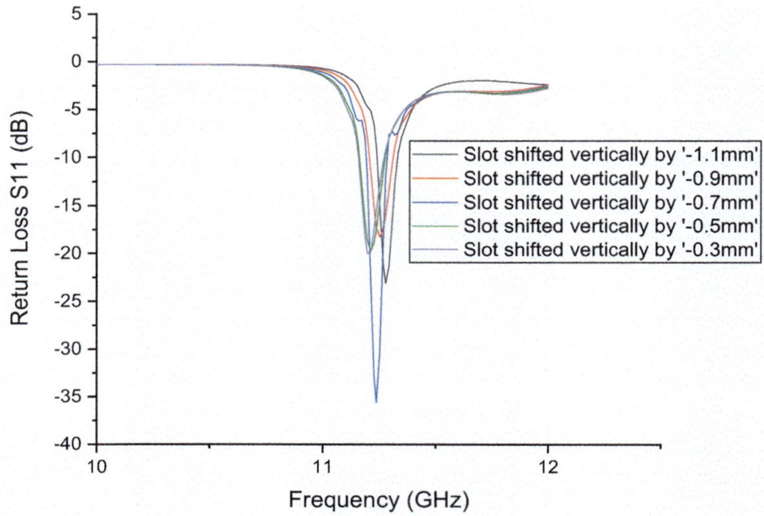

Fig. 8 Vertical variation of the circular split ring slot in the ground plane of the antenna

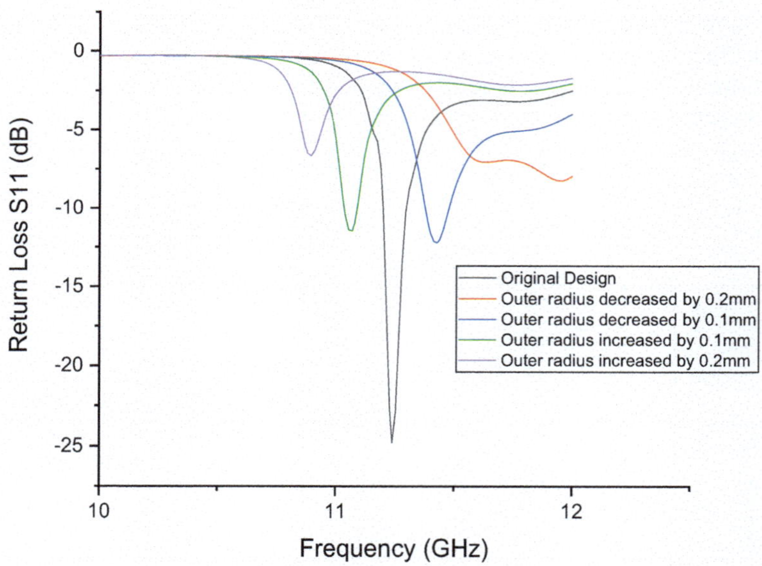

Fig. 9 Variation of the outer radius of circular split ring slot in the ground plane of the antenna

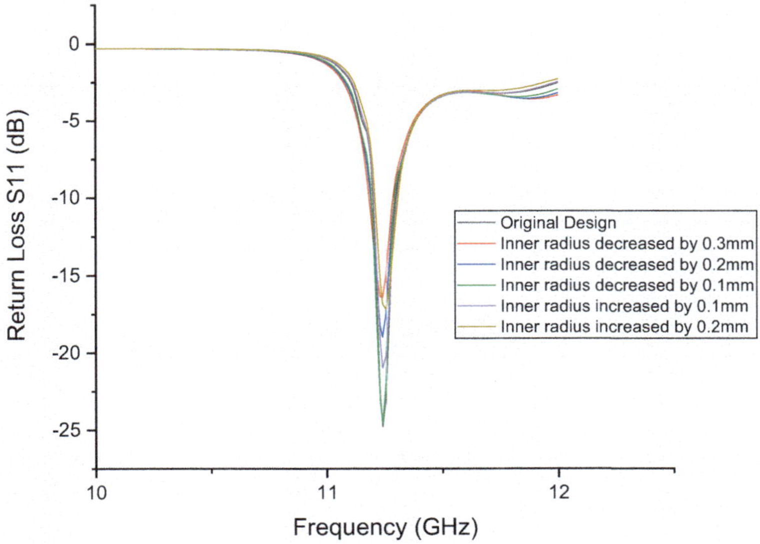

Fig. 10 Variation of the inner radius of circular split ring slot in the ground plane of the antenna

5 Conclusion

A novel cavity-backed SIW antenna for the satellite application band includes a circular split ring slot is reported. The antenna demonstrates a unidirectional radiation pattern, a high FTBR, and an optimal gain of 4.4 dBi at resonance at a resonant frequency of 11.2 GHz. In conclusion, the work on the Cavity-Backed SIW Antenna integrated with a circular split ring slot for Satellite Application Band represents a significant advancement in antenna design tailored for satellite communication systems. The integration of Circular Split Ring technology with Cavity-Backed Substrate Integrated Waveguide (SIW) demonstrates a thoughtful approach to optimizing performance parameters crucial for satellite applications. By leveraging the benefits of both technologies, the antenna achieves enhanced radiation efficiency, reduced back radiation, and controlled radiation patterns, making it well-suited for the demanding requirements of satellite communication bands. The circular split ring, as a key element, contributes to the antenna's adaptability and efficiency, allowing for broader bandwidth coverage and improved resonance characteristics. This work not only expands the understanding of advanced antenna configurations but also holds promise for the development of high-performance satellite communication systems, addressing the challenges posed by the ever-evolving demands of modern space-based communication.

Acknowledgements The proposed research was executed out at the AICTE-sponsored IDEA Laboratory of Dr. Bidhan Chandra Roy Engineering College in Durgapur, India. Researchers would like to sincerely appreciate everyone who has contributed.

References

1. Colangelo G, Vitiello R (2011) Shared aperture dual band printed antenna. In: Proceedings of the IEEE international conference on electromagnetics in advanced applications, pp 1092–1095
2. Chen S, Hsu P (2005) Broadband radial slot antenna fed by coplanar waveguide for dual-frequency operation. IEEE Trans Antennas Propag 53(11):3448–3452
3. Ho MH, Chen GL (2007) Reconfigured slot-ring antenna for 2.4/5.2 GHz dual-band WLAN operations. IET Microw Antennas Propag 1(2):712–717
4. Liu Y, Shen Z, Law CL (2006) A compact dual-band cavity backed slot antenna. IEEE Antennas Wirel Propag Lett 5:4–6
5. Bozzi M, Georgiadis A, Wu K (2011) Review of substrate integrated waveguide circuits and antennas. IET Microw Antennas Propag 5(8):909–920
6. Luo GQ, Hu ZF, Dong LX, Sun LL (2008) Planar slot antenna backed by substrate integrated waveguide cavity. IEEE Antennas Wirel Propag Lett 7:235–239
7. Mukherjee S, Biswas A, Srivastava KV (2013) Bandwidth enhancement of substrate integrated waveguide cavity backed slot antenna by offset feeding technique. In: Proceedings of applied electromagnetics conference (AEMC), pp 1–2
8. Luo GQ (2012) Bandwidth-enhanced low-prole cavity-backed slot antenna by using hybrid SIW cavity modes. IEEE Trans Antennas Propag 60(4):1698–1704
9. Luo GQ, Hu ZF, Liang Y, Yu LY, Sun LL (2009) Development of low prole cavity backed crossed slot antennas for planar integration. IEEE Trans Antennas Propag 57(10):2972–2979
10. Zhang T, Hong W, Zhang Y, Wu K (2014) Design and analysis of SIW cavity backed dual-band antennas with a dual-mode triangular-ring slot. IEEE Trans Antennas Propag 62(10):5007–5016
11. Mukherjee S, Biswas A, Srivastava KV (2014) Broadband substrate integrated waveguide cavity-backed bow-tie slot antenna. IEEE Antennas Wirel Propag Lett 13
12. Mukherjee S, Biswas A, Srivastava KV (2015) Substrate integrated waveguide cavity-backed dumbbell-shaped slot antenna for dual-frequency applications. IEEE Antennas Wirel Propag Lett 14
13. Sharma A, Mukherjee S, Biswas A (2015) Dielectric resonator loaded substrate integrated waveguide cavity backed slot antenna for bandwidth enhancement. ISBN 978-1-4799-7815-1/15/$31.00 c 2015 IEEE
14. Mukherjee S, Biswas A (2016) Computer aided equivalent circuit model of SIW cavity backed triple band slot antenna. Int J RF Microw Comput Aided Eng
15. Chakraborty S et al (2020) Cavity-backed SIW antenna with X shaped slot for satellite communication frequency band. In: 2020 4th international conference on electronics, materials engineering & nano-technology (IEMENTech), Kolkata, India, pp 1–3. https://doi.org/10.1109/IEMENTech51367.2020.9270112
16. Mukhopadhyay A, Chakraborty S, Mukherjee P, Saha A, Chakraborty M (2022) Cavity-backed substrate integrated waveguide antenna with two connected I slot for dual frequency operation. In: 2022 interdisciplinary research in technology and management (IRTM), Kolkata, India, pp 1–4. https://doi.org/10.1109/IRTM54583.2022.9791685
17. Mukhopadhaya A et al (2019) Cavity-backed substrate integrated waveguide antenna with U-shaped slot for dual frequency operation. In: 2019 IEEE international electromagnetics and antenna conference (IEMANTENNA), Vancouver, BC, Canada, pp 061–064. https://doi.org/10.1109/IEMANTENNA.2019.8928791

Time Series Prediction in UAV-UGV Integrated Automation System by Deep Reinforcement Learning in Industry 4.0

Haowen Pan, Dongfang Hou, Yuan Xing, Jason Liu, and Abhishek Verma

Abstract In Industry 4.0, Unmanned Vehicles play significant roles in smart manufacturing as they can replace human workers and improve the efficiency of manufacturing. Therefore, in this paper, we aim to address the allocation problem of Unmanned Vehicles in Industry 4.0. The Deep Reinforcement Learning algorithms are utilized to solve the allocation problem in both the single agent and multiple agents model. The agent, as the task assigner, observes the current Unmanned Vehicles utilization condition and predicts the usage patterns of these Unmanned Vehicles in the next time slot. In the next time slot, the agent can directly utilize the idle Unmanned Vehicles to complete the tasks. Traditional Deep Reinforcement Learning, especially Deep Q-Networks cannot address the large action space, therefore we propose to use the Deep Deterministic Policy Gradient algorithms to determine the optimal action for the agents. The simulation results show the superiority of the Deep Deterministic Policy Gradient algorithm in solving both the centralized and the distributed optimization problems for each agent.

Keywords Unmanned Vehicles · Deep Reinforcement Learning · Industry 4.0

H. Pan
Changzhou Long Journey Electronic Technology Co., Ltd., Changzhou, China

D. Hou · Y. Xing (✉) · J. Liu · A. Verma
University of Wisconsin-Stout, Menomonie, WI, USA
e-mail: xingy@uwstout.edu

D. Hou
e-mail: houd@uwstout.edu

J. Liu
e-mail: liux@uwstout.edu

A. Verma
e-mail: vermaa@uwstout.edu

P. G. Bradford et al. (eds.), *Proceedings of IEMTRONICS 2024*, Lecture Notes in Electrical Engineering 1229, https://doi.org/10.1007/978-981-97-4780-1_27

1 Introduction

In Industry 4.0, Unmanned Vehicles are very important tools to support the development of the entire system. In [1], the authors mentioned the ultimate goal of Industry 4.0 is an autonomous smart factory, which heavily relies on Unmanned Vehicles, including Unmanned Aerial Vehicles and Unmanned Ground Vehicles. However, the number of Unmanned Vehicles has to be restricted in the limited space due to safety reasons. As a result, a limited number of Unmanned Vehicles are dispatched to complete a large number of tasks, which generates a resource of Unmanned Vehicles allocation problem. For example, route planning optimization problems are widely discussed. In [2], the authors addressed how to accept the operation order (OPORD) from the commander and how to generate a nested vehicle routing planning in an Unmanned Vehicles environment. In [3–6], authors solved multiple path planning problems in wireless power transfer robotic systems, which benefits the wireless sensor networks that are supported by the charging systems. However, there are seldom papers that explore the optimization problem of Unmanned Vehicles allocation to different agents, which is a very critical problem, since a fully automated smart factory has to rely on a limited number of Unmanned Vehicles to complete all the manufacturing tasks automatically. Therefore, we aim to solve the Unmanned Vehicles allocation problem in this paper and plan to use the Deep Reinforcement Learning algorithm to solve it.

Reinforcement Learning algorithms are highly proficient in addressing the optimization challenge of resource allocation, especially with the effectiveness of Deep Reinforcement Learning attributed to its low time complexity. Initially showcased in Atari 2600 games, these algorithms surpassed human players upon completing the learning process [7]. The distinguishing feature of Reinforcement Learning lies in its capacity to account for long-term effects and devise optimal strategies at different stages, resulting in relatively swift convergence times. As a result, Deep Reinforcement Learning has been widely applied across diverse fields.

The resource allocation problems in Industry 4.0 engineering systems consist of a centralized model and a distributed model. In the centralized models, the single agent learns the optimal strategy for the proposed optimization problems [8, 9]. In [9], the authors proposed a Deep Reinforcement Learning-based dynamic resource management (DDRM) algorithm to address the limitation of resources, such as the computation unit and battery capacity in the IIoT equipment (IIEs), computation-intensive tasks need to be executed in the mobile edge computing (MEC) server. The results show the proposed algorithm can reduce the long-term average delay of the tasks effectively. In the distributed models, multiple agents make the separate decisions to achieve the overall optimization goals [10–14]. In [11], the authors proposed a distributed Deep Reinforcement Learning (DRL) methodology for autonomous mobile robots (AMRs) to manage radio resources in an indoor factory with no network infrastructure. The proposed algorithms can enable the agent to make decisions in a distributed manner without signaling exchange.

Many researches have demonstrated the effectiveness of Deep Deterministic Policy Gradient in solving complicated centralized and distributive optimization problems [15, 16]. Therefore, in this paper, we aim to solve the resource allocation problem of Unmanned Vehicles in Industry 4.0. Both single agent and multiple agents models are explored and Deep Deterministic Policy Gradient algorithms are used to find the optimal strategy for the agents.

2 System Model

In smart manufacturing, it's assumed that there are N different Unmanned Vehicles, including Unmanned Aerial Vehicles (UAV) and Unmanned Ground Vehicles (UGV). We suppose the time is divided into the same interval and one interval is called a time slot. In a particular time slot, each Unmanned Vehicle can either be dispatched to complete a manufacturing task or stand by. If the Unmanned Vehicle is required to complete the assigned task, it's supposed that the Unmanned Vehicle can complete the task before the end of the time slot. If the Unmanned Vehicle stands by, the Unmanned Vehicle cannot be assigned a task within the same time slot. Therefore, two states are defined for each Unmanned Vehicle, which are expressed as 1 (task assigned) and 0 (idle), respectively. Within each time slot t, the nth Unmanned Vehicle ($n \in \mathcal{N} = \{1, 2, \ldots, N\}$) is either utilized by an agent or idle.

There are multiple agents and each agent is authorized to utilize the Unmanned Vehicles. Each agent is supposed to utilize as many idle Unmanned Vehicles as possible to complete the manufacturing task in the shortest time. However, in the same time slot, two agents cannot utilize the same Unmanned Vehicle. If so, the Unmanned Vehicle will not serve any of the agents. The user patterns of the Unmanned Vehicles vary over the time slots. Different Unmanned Vehicles may have different user patterns depending on the specific task to be completed. Some Unmanned Vehicles may be frequently utilized by multiple agents, hence it is very rare to observe an idle time slot for that Unmanned Vehicle. In our proposed model, the usage pattern of each Unmanned Vehicle evolves according to N independent 2-state Markov chains. The state of the nth Unmanned Vehicle at time slot t is expressed as s_n^t, where $s_n^t \in \{0, 1\}$. The evolution of the nth Unmanned Vehicle at time t is featured by a 2×2 transition matrix of probability

$$\mathbf{P}_n^t = \begin{bmatrix} P_n^t(0|0) \; P_n^t(0|1) \\ P_n^t(1|0) \; P_n^t(1|1) \end{bmatrix} \tag{1}$$

$P_n^t(u|v)$ is defined as

$$P_n^t(a|b) = P_n^t\left(s_{t+1}^n = a|s_t^n = b\right), \;\; \forall t \geq 0 \tag{2}$$

where $a, b \in \{0, 1\}^2$.

3 Single Agent Time Series Prediction

The optimization is formulated as the single agent forecasts the usage pattern of all Unmanned Vehicles and determines which Unmanned Vehicles to use in the next time slot. The goal is to predict the maximum number of usage patterns of Unmanned Vehicles.

The proposed problem is formulated as a Markov Decision Process (MDP). The system state at time slot t is defined as

$$\mathbf{s}_t = \left[s_t^1, \ldots, s_t^N\right] \tag{3}$$

The state space is denoted as $\mathcal{S} = \left\{\mathbf{s} = \left[s^1, \ldots, s^N\right] | s^n \in \{0, 1\}\right\}$.

The action determined at time slot t is denoted as $\mathbf{a}_t$, which is taken as at the time slot $t + 1$.

$$\mathbf{a}_t = \left[a_t^1, \ldots, a_t^N\right] \tag{4}$$

where $a_t^n \in \{0, 1\}$ denotes whether the agent proactively utilizes the nth Unmanned Vehicle or not in the $t + 1$th time slot. The set of action is denoted as $\mathcal{A} = \{\mathbf{a} = [a_1, \ldots, a_N] | a_n \in \{0, 1\}\}$.

We define the reward function as $w(\mathbf{a})$, which is expressed as

$$w(\mathbf{a}) = \sum_{n=1}^{N} \gamma_n \tag{5}$$

where

$$\gamma_n = \begin{cases} 1 & \text{if } s_t^n = 0, a_t^n = 1 \\ -1 & \text{otherwise} \end{cases} \tag{6}$$

If in tth time slot, the agent can successfully predict the idle nth Unmanned Vehicle ($s_t^n = 0$) and utilize that Unmanned Vehicle ($a_t^n = 1$), the positive reward will be assigned to the agent. Otherwise, there are two possibilities: (1) The agent may predict the wrong usage pattern of the Unmanned Vehicle and want to utilize the Unmanned Vehicle, which is denoted as $s_t^n = 1, a_t^n = 1$. In that case, the selected Unmanned Vehicle is not accessible because it's already utilized by the other agents. (2) The agent predicts that the nth Unmanned Vehicle is utilized by the other agent, so the agent stands by $s_t^n = 1, a_t^n = 0$. For both these two cases, since the agent cannot utilize the nth Unmanned Vehicle, the negative reward is assigned to the agent.

Due to the large size of the action space, Deep Q-Network cannot solve the proposed problem. A Deep Deterministic Policy Gradient (DDPG) algorithm is utilized to solve the optimization problem [17].

The action that is calculated by the actor network is expressed as $\mathbf{a}_t' = \left[a_t^{1,*}, \ldots, a_t^{N,*}\right]$, each element of which is of continuous value. Therefore, $\mathbf{a}_t'$ has to be converted into $\mathbf{a}_t$. Sigmoid is used as the activation function in the last layer of the actor network, as a result, the output is a continuous value in [0, 1]. We map the continuous output into discrete values 0 and 1, which represent the standing by and accessing the Unmanned Vehicle, respectively. A threshold of 0.5 is defined. If the output is greater than or equal to 0.5, the agent will access the Unmanned Vehicle in the next time slot, otherwise, the agent stands by.

$$a_t^n = \begin{cases} 1 \text{ if } a_t^{n,*} \geq 0.5 \\ 0 \text{ otherwise} \end{cases} \tag{7}$$

In DDPG, we denoted the Q-value at time t as $Q_t(\mathbf{s}, \mathbf{a}, \theta)$. The optimal cost function $Q^*(\mathbf{s}, \mathbf{a})$ is calculated by the critic network. The Q-value can be expressed as

$$y = w(\mathbf{a}) + \epsilon \max_{\mathbf{a}' \in \mathcal{A}}, Q_t\left(\mathbf{s}', \mathbf{a}', \theta_Q\right) \tag{8}$$

where θ_Q denotes the weight parameters of the critic network. To minimize the Temporal-difference error (TD-error) θ_Q is updated by

$$L = \frac{1}{T} \sum_{t=0}^{T} \left(y - Q_t\left(\mathbf{s}, \mathbf{a}, \theta_Q\right)\right)^2 \tag{9}$$

In addition to double critic networks, two actor networks are also constructed in the model. The parameters of one actor network are trained by sampled policy gradient for the update of policy π:

$$\nabla_{\theta_\mu} J \approx \frac{1}{N} \sum_t \nabla_{\mathbf{a}} Q\left(\mathbf{s}, \mathbf{a}, \theta_Q\right) \nabla_{\theta_\mu} \pi\left(\mathbf{s}, \theta_\mu\right) \tag{10}$$

where θ^μ is the weight of an actor network.

The weight parameters of two target networks are denoted as θ' and μ'. Both of them are updated as follows:

$$\theta_{Q'} = \tau\theta_Q + (1 - \tau)\theta_{Q'} \tag{11}$$

$$\theta_{\mu'} = \tau\theta_\mu + (1 - \tau)\theta_{\mu'} \tag{12}$$

where τ denotes the parameter for the update. The algorithm is shown in Algorithm 1.

Algorithm 1: Deep Deterministic Policy Gradient algorithm training process for single agent

1. The weight parameter θ_Q and θ_μ are generated randomly for the evaluation critic network eval_critic_net and evaluation actor network eval_actor_net. Respectively. Two target networks: target_critic_net and target_actor_net copy the weight parameters $\theta_{Q'} = \theta_Q$, $\theta_{\mu'} = \theta_\mu$, correspondingly. $t = 1$. $D = d = 1$.
2. The system state $\mathbf{s} = \mathbf{s}_0$.
3. The action $\mathbf{a}$ is calculated as

$$\mathbf{a} = \pi\left(\mathbf{s}, \theta_\mu\right) + n \tag{13}$$

where n denotes the noise proactively generated for exploration.

4. $t = t + 1$. After taking action $\mathbf{a}$, the agent generates the system state s' by detecting the usage patterns of N Unmanned Vehicles.
5. $ep(d) = \left\{s, \mathbf{a}, w(\mathbf{a}), s'\right\}$. $d = d + 1$. If D reaches the maximum of the experience pool, D remains constant, $d = 1$, otherwise, $D = d$. $s = s'$.
6. After experience pool accumulates D experiences, generate

$y = w(\mathbf{a}) + \epsilon \max_{\mathbf{a}' \in \mathcal{A}}, Q_t\left(\mathbf{s}', \mathbf{a}', \theta_Q\right)$.

The weight parameters of eval_critic_net update by

$L = \frac{1}{T} \sum_{t=0}^{T} \left(y - Q_t(\mathbf{s}, \mathbf{a}, \theta_Q)\right)^2$.

The weight of eval_actor_net is updated as:

$\nabla_{\theta_\mu} J \approx \frac{1}{N} \sum_t \nabla_{\mathbf{a}} Q\left(\mathbf{s}, \mathbf{a}, \theta_Q\right) \nabla_{\theta_\mu} \pi\left(\mathbf{s}, \theta_\mu\right)$.

The target_critic_net and target_actor_net are updated as:

$\theta_{Q'} = \tau \theta_Q + (1 - \overline{\tau}) \theta_{Q'}$.

$\theta_{\mu'} = \tau \theta_\mu + (1 - \tau) \theta_{\mu'}$.

7. The algorithm terminates if converges. Otherwise, go back to step 3.

4 Multiple Agents Time Series Prediction

In the previous section, the single agent is intended to predict the usage patterns of all the Unmanned Vehicles in the next time slot and utilize all the idle Unmanned Vehicles. In this section, the number of agents increases. Each agent intends to maximize its reward by accessing as many idle Unmanned Vehicles as possible. There is a contention problem as all the agent makes the decisions separately. So it's normal that multiple agents want to utilize the same Unmanned Vehicle at the same time, which still results in invalid utilization by all the agents.

For agent $i \in \mathcal{I}$, an individual MDP is formulated.

The system state of agent i at time slot t is defined as

$$\mathbf{s}_{i,t} = \left[s_{i,t}^1, \ldots, s_{i,t}^N\right] \tag{14}$$

The state space of agent i is denoted as $\mathcal{S} = \left\{\mathbf{s}_i = \left[s_i^1, \ldots, s_i^N\right] | s_i^n \in \{0, 1\}\right\}$. Since the observation of the Unmanned Vehicle is same for all the agents, $\mathbf{s}_{1,t} = \mathbf{s}_{2,t} = \ldots = \mathbf{s}_{I,t}$.

The action determined at time slot t for agent i is denoted as $\mathbf{a}_{i,t}$, which is taken as at the time slot $t+1$.

$$\mathbf{a}_{i,t} = \left[a_{i,t}^1, \ldots, a_{i,t}^N\right] \tag{15}$$

where $a_{i,t}^n \in \{0, 1\}$ denotes whether the agent proactively utilizes the nth Unmanned Vehicle or not by agent i in the $t+1$th time slot. The set of action is denoted as $\mathcal{A} = \left\{\mathbf{a} = \left[a_i^1, \ldots, a_i^N\right] | a_i^n \in \{0, 1\}\right\}$.

The reward function that is defined for agent i is expressed as $w_i(\mathbf{a})$:

$$w_i(\mathbf{a}) = \sum_{n=1}^{N} \gamma_{i,n} \tag{16}$$

where

$$\gamma_{i,n} = \begin{cases} 1 & \text{if } s_{i,t}^n = 0, a_{i,t}^n = 1, a_{\{i\}/\mathcal{I},t}^n = 0 \\ -1 & \text{otherwise} \end{cases} \tag{17}$$

If in tth time slot, the agent i can successfully predict the idle nth Unmanned Vehicle $\left(s_t^n = 0\right)$ is not accessed by any other agent $\{i\}/\mathcal{I}$ and agent i utilizes that Unmanned Vehicle $\left(a_t^n = 1\right)$, the positive reward will be assigned to agent i. Otherwise, there are two possibilities: (1) The agent i may predict the wrong usage pattern of the Unmanned Vehicle or the other agents and want to utilize the Unmanned Vehicle, which is denoted as: $s_{i,t}^n = 1, a_{\{i\}\mathcal{I},t}^n = 0, a_{i,t}^n = 1$ or $s_{i,t}^n = 0, a_{\{i\}/\mathcal{I},t}^n = 1, a_{i,t}^n = 1$. In that case, the selected Unmanned Vehicle is not accessible because it's already utilized by the other agents. (2) The agent predicts that the nth Unmanned Vehicle is utilized by the other agents, so the agent stands by $s_t^n = 1, a_{\{i\}/\mathcal{I},t}^n = 1, a_t^n = 0$. For both these two cases, since agent i cannot utilize the nth Unmanned Vehicle, the negative reward is assigned to agent i.

Same as in the previous section, DDPG is utilized to determine the optimal strategy for each agent individually. The detailed steps are included in the algorithm that is shown in Algorithm 2.

Algorithm 2: Deep Deterministic Policy Gradient algorithm training process for multiple agents

1. For each agent $i \in \mathcal{I}$, weight parameter $\theta_{i,Q}$ and $\theta_{i,\mu}$ are generated randomly for the evaluation critic network eval_critic_net and evaluation actor network eval_actor_net. Respectively. Two target networks: target_critic_net and target_actor_net copy the weight parameters $\theta_{i,Q'} = \theta_{i,Q}, \theta_{i,\mu'} = \theta_{i,\mu}$, correspondingly. $t = 1$. $D_i = d_i = 1$.
2. For each agent $i \in \mathcal{I}$, the system state $\mathbf{s}_i = \mathbf{s}_{i,0}$.
3. For each agent $i \in \mathcal{I}$, the action $\mathbf{a}$ is calculated as

$\mathbf{a}_i = \pi\left(\mathbf{s}_i, \theta_{i,\mu}\right) + n_i$ (18)

where n_i denotes the noise proactively generated for exploration of agent i.

4. $t = t + 1$. After taking action $\mathbf{a}_i$, each agent $i \in \mathcal{I}$ generates the system state s_i' by detecting the usage patterns of N Unmanned Vehicles.
5. For each agent $i \in \mathcal{I}$, $ep_i(d_i) = \left\{s_i, \mathbf{a}_i, w_i(\mathbf{a}_i), s_i'\right\}$. $d_i = d_i + 1$. If D_i reaches the maximum of the experience pool, D_i remains constant, $d_i = 1$,

otherwise, $D_i = d_i$. $s_i = s_i'$.

6. After experience pool accumulates D_i experiences, each agent $i \in \mathcal{I}$ generates $y_i = w_i(\mathbf{a}_i) + \epsilon_i \max_{\mathbf{a}_i' \in \mathcal{A}_i'} Q_{i,t}\left(\mathbf{s}_i', \mathbf{a}_i', \theta_{i,Q}\right)$.

The weight parameters of eval_critic_net $_i$ update by $L_i = \frac{1}{T}\sum_{t=0}^{T}\left(y_i - Q_{i,t}\left(\mathbf{s}_i, \mathbf{a}_i, \theta_{i,Q}\right)\right)^2$.

The weight of eval_actor_net $_i$ is updated as: $\nabla_{\theta_{i,\mu}} J_i \approx \frac{1}{N}\sum_t \nabla_{\mathbf{a}_i} Q_i\left(\mathbf{s}_i, \mathbf{a}_i, \theta_{i,Q}\right)\nabla_{\theta_{i,\mu}}\pi\left(\mathbf{s}_i, \theta_{i,\mu}\right)$.

The target_critic_net $_i$ and target_actor_net $_i$ are updated as:

$\theta_{i,Q'} = \tau\theta_{i,Q} + (1-\tau)\theta_{i,Q'}$.

$\theta_{i,\mu'} = \tau\theta_{i,\mu} + (1-\tau)\theta_{i,\mu'}$.

7. The algorithm for agent $i \in \mathcal{I}$, terminates if converges. Otherwise, go back to step 3.

5 Simulation Result

In the simulation, we assume there are $N = 16$ Unmanned Vehicles. In single agent model, DDPG is trained in order to compare with the performance of Deep Reinforcement Learning (DRL) and Q Learning. In multiple agent model, $I = 2$, two individual DDPG are trained separately, which are also used to compared with the state of the art. The simulation parameters used for DDPG are presented in Table 1.

The software environment for simulation is TensorFlow 0.12.1 with Python 3.6 in Jupyter Notebook 5.6.0.

In Fig. 1, the moving average of the number of Unmanned Vehicles m is observed throughout the learning process. The moving average of 200 adjacent.

samples of m is calculated by

$$\overline{m} = \frac{m(e-99) + \cdots + m(e)}{200} \tag{19}$$

From Fig. 1, we can observe that both the successful and unsuccessful utilization of the Unmanned Vehicles converge as the training epochs increase.

Table 1 DDPG simulation parameters

DDPG	Value
Number of hidden layers of actor	2
Number of nodes of a hidden layer of actor	400
Number of hidden layers of critic	1
Number of nodes of a hidden layer of critic	600
Learning rate (ϵ)	0.0005
Mini-batch size	32
Reward decay	0–0.9
Training starting step	10,000
Experience pool	200,000
Initial exploration rate (ϵ_c)	1
Exploration decay rate	0.97
Exploration decay steps	10,000
target_net weight replacement interval	1500, 1700
Training episodes	80,000

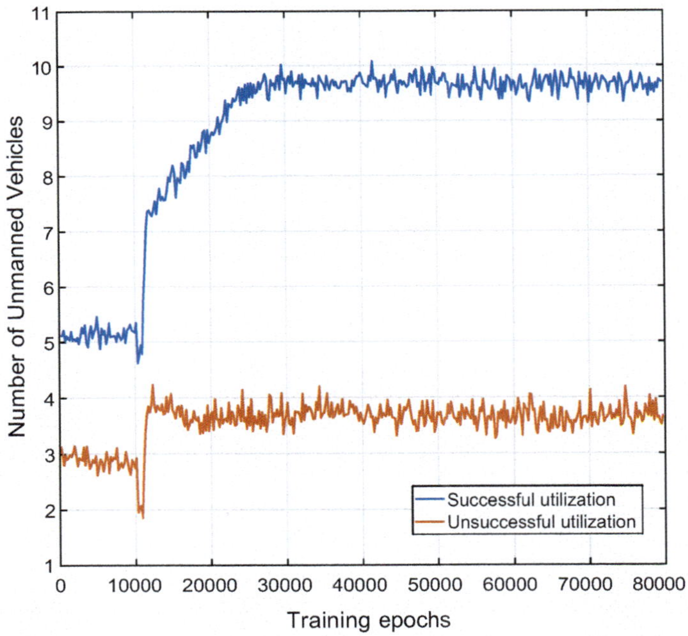

Fig. 1 The convergence of the number of successful and unsuccessful utilization of Unmanned Vehicles in single agent model. The training epochs is 80,000

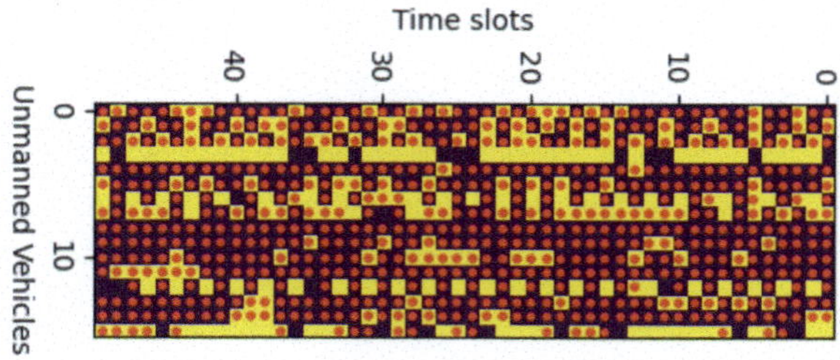

Fig. 2 The utilization of the Unmanned Vehicles in single agent model. Blue squares are the idle Unmanned Vehicles. Yellow squares are the occupied Unmanned Vehicles, which cannot be utilized by the agent. Red circle represents the agent who wants to utilize that specific Unmanned Vehicles

In Fig. 2, the usage of the different Unmanned Vehicles is shown. Blue squares are the idle Unmanned Vehicles. Yellow squares are the occupied Unmanned Vehicles, which cannot be utilized by the agent. The red circle represents the agent wants to utilize that specific Unmanned Vehicles. If any red circle overlaps with a blue square, the agent can utilize that specific Unmanned Vehicle. However, if any red circle overlaps with a yellow square, the agent fails to utilize that specific Unmanned Vehicle. From Fig. 2, we can observe that while the successful utilization exists as the agent predicts the usage pattern of specific Unmanned Vehicles, some usages cannot be predicted, which can be explained as the usage pattern is simulated by a specific transition probability, DDPG already achieves very good performance.

From Figs. 3 and 4, we can observe that two agents can converge to a stable strategy, while the distribution of the Unmanned Vehicles for two agents is not fair, which can be explained as two DDPGs are executed separately. Like the single model in Fig. 2, most usage of Unmanned Vehicles can be predicted correctly. However, the usage of some Unmanned Vehicles is not easy to predict due to the close transition probability between idle and occupied, which may be more similar to the real condition.

From Fig. 5, we can observe that in single agent model, compared to DRL and Q Learning, DDPG can utilize much more idle Unmanned Vehicles because of the limitation of DRL and Q Learning. Though DRL can address a large number of system states and use a Deep Neural Network to determine the optimal action, the number of actions cannot be large. Q Learning even cannot address the large system state set. In the multiple agents model, DDPG outperforms DRL and Q Learning. An early stop DDPG is also completed. In early stop DDPG, the algorithm terminates as it detects that two agents have the closest average occupation number of idle Unmanned Vehicles. Compared with DDPG, an early stop DDPG achieves fairness however a higher unsuccessful utilization compared with DDPG.

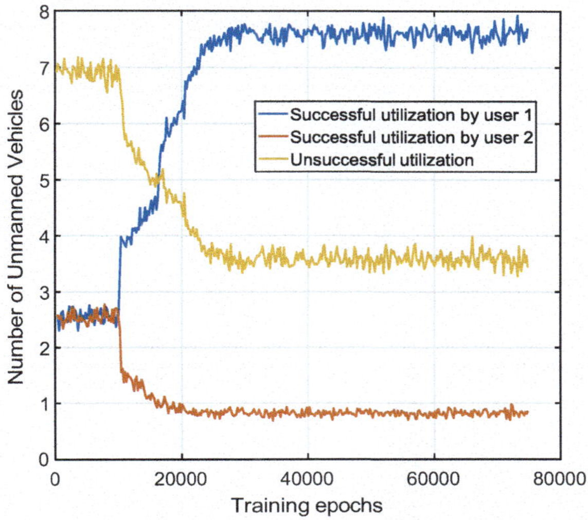

Fig. 3 The convergence of the number of successful and unsuccessful utilization of Unmanned Vehicles in multiple agents model. The number of the agent is $I = 2$. The training epochs is 75,000

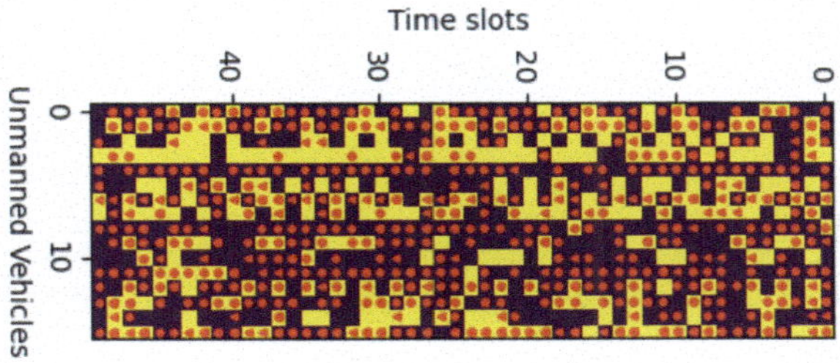

Fig. 4 The utilization of the Unmanned Vehicles in multiple agents model. Blue squares are the idle Unmanned Vehicles. Yellow squares are the occupied Unmanned Vehicles, which cannot be utilized by the agent. The red circle and red triangle represent agent 1 and agent 2, respectively

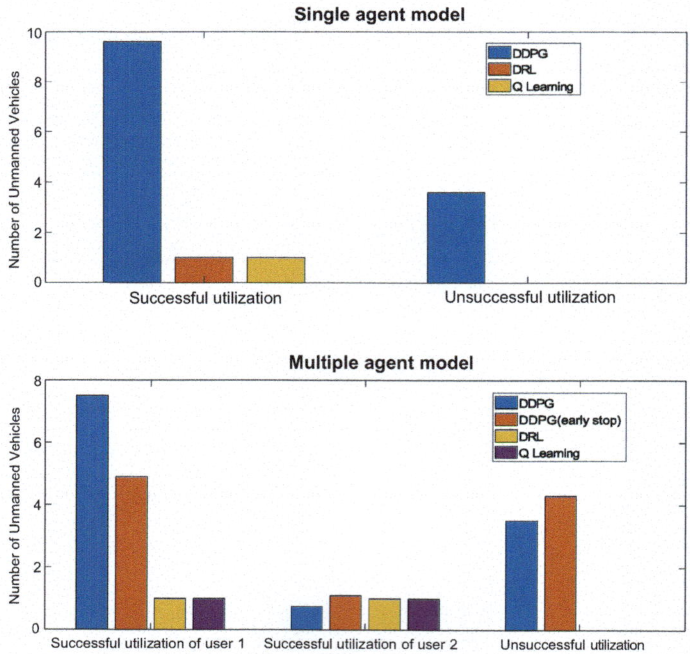

Fig. 5 DDPG is compared with the DRL and Q Learning in both single agent model and multiple agent model. In multiple agents model, an early stop DDPG is trained to compared with the other algorithms. For DRL and Q Learning, due to the limitation of action space, the agent can only utilize one Unmanned Vehicle at a time due to the limitation of algorithms in action space

6 Conclusion

In this study, we propose to use Deep Deterministic Policy Gradient algorithms to solve the Unmanned Vehicles' resource allocation problems in Industry 4.0. Both the single agent and multiple agents models are explored. Each agent, as the task assigner, predicts the usage patterns of the Unmanned Vehicles in the next time slot and utilizes them to complete the tasks. The simulation results indicate that the Deep Deterministic Policy Gradient outperforms the state-of-the-art methods in the number of Unmanned Vehicles that agents can utilize in both single agent and multiple agent models.

References

1. Galar D, Uday K, Dammika S (2020) Robots, drones, UAVs and UGVs for operation and maintenance. CRC Press
2. Gao W et al (2020) Commanding cooperative UGV-UAV with nested vehicle routing for emergency resource delivery. IEEE Access 8:215691–215704

3. Xing Y, Carlson C, Yuan H (2022) Optimize path planning for UAV COVID-19 test kits delivery system by hybrid reinforcement learning. In: 2022 IEEE 12th annual computing and communication workshop and conference (CCWC). IEEE, pp 0177–0183
4. Xing Y, Young R, Nguyen G, Lefebvre M, Zhao T, Pan H, Dong L (2022) Optimal path planning for wireless power transfer robot using area division deep reinforcement learning. Wireless Power Transf
5. Xing Y, Young R, Nguyen G, Lefebvre M, Zhao T, Pan H (2022) Optimize mobile wireless power transfer by finite state machine reinforcement learning. In: 2022 IEEE 12th annual computing and communication workshop and conference (CCWC). IEEE, pp 0507–0512
6. Xing Y, Verma A (2023) Optimize path planning for drone-based wireless power transfer system by categorized reinforcement learning. In: 2023 IEEE 13th annual computing and communication workshop and conference (CCWC). IEEE, pp 0641–0646
7. Mnih V, Kavukcuoglu K, Silver D, Graves A, Antonoglou I, Wierstra D, Riedmiller M (2013) Playing Atari with deep reinforcement learning. arXiv preprint arXiv:1312.5602
8. Ong KSH, Wang W, Niyato D, Friedrichs T (2021) Deep-reinforcement-learning-based predictive maintenance model for effective resource management in industrial IoT. IEEE Internet Things J 9(7):5173–5188
9. Chen Y, Liu Z, Zhang Y, Wu Y, Chen X, Zhao L (2020) Deep reinforcement learning-based dynamic resource management for mobile edge computing in industrial internet of things. IEEE Trans Ind Inf 17(7):4925–4934
10. Shi Z, Xie X, Lu H, Yang H, Kadoch M, Cheriet M (2020) Deep-reinforcement-learning-based spectrum resource management for industrial internet of things. IEEE Internet Things J 8(5):3476–3489
11. Burgueño J, Adeogun R, Bruun RL, García CSM, de-la-Bandera I, Barco R (2021) Distributed deep reinforcement learning resource allocation scheme for industry 4.0 device-to-device scenarios. In: 2021 IEEE 94th vehicular technology conference (VTC2021-fall), Sept 2021. IEEE, pp 1–7
12. Rosenberger J, Urlaub M, Rauterberg F, Lutz T, Selig A, Bühren M, Schramm D (2022) Deep reinforcement learning multi-agent system for resource allocation in industrial internet of things. Sensors 22(11):4099
13. Cao Z, Zhou P, Li R, Huang S, Wu D (2020) Multiagent deep reinforcement learning for joint multichannel access and task offloading of mobile-edge computing in industry 4.0. IEEE Internet Things J 7(7):6201–6213
14. Dong L, Qian Y, Xing Y (2022) Dynamic spectrum access and sharing through actor-critic deep reinforcement learning. EURASIP J Wirel Commun Netw 2022(1):48
15. Li S, Wu Y, Cui X, Dong H, Fang F, Russell S (2019) Robust multi-agent reinforcement learning via minimax deep deterministic policy gradient. Proc AAAI Conf Artif Intell 33(01):4213–4220
16. Qiu C, Hu Y, Chen Y, Zeng B (2019) Deep deterministic policy gradient (DDPG)-based energy harvesting wireless communications. IEEE Internet Things J 6(5):8577–8588
17. Lillicrap TP, Hunt JJ, Pritzel A, Heess N, Erez T, Tassa Y et al (2015) Continuous control with deep reinforcement learning. arXiv preprint arXiv:1509.02971

The Miniature Tesla Coil: Improving Construction and Performance

Simoom Rahman and Shahriar Khan

Abstract First invented by Nikola Tesla in 1891, the Tesla coil supplied high voltage at low current, demonstrating spectacular arcing discharge many feet long. However, the coil has been used more for demonstration and entertainment, and less for teaching and research, mainly because of its inherent dangers. The dangers have been largely removed with today's diodes and transistors, allowing improved safety, as demonstrated by the authors in 2022. The improved miniature Tesla coils built by the authors over 2021–23 better lit a nearby fluorescent lamp, made a common filament bulb act like a plasma ball, and produced an imperceptible continuous discharge on the finger. The constructed miniature Tesla coil better demonstrated (over past models) boosting of voltage and other electrical principles. Improving performance by changing parameters such as capacitors and spark gap have been discussed. A comparison has been made for the three successively improved Tesla coils built by the authors.

Keywords Tesla coil · Miniature · Fluorescent lamp · Discharge

1 Introduction

First invented by Nikola Tesla in 1891, the Tesla coil supplied high voltage at low current, demonstrating spectacular arcing discharge many feet long. However, the Tesla coil has been used more for demonstration and entertainment, and less for teaching and research, mainly because of its inherent dangers. The dangers have been largely removed with today's diodes and transistors, allowing improved safety, as demonstrated by the authors [1].

In the present improved construction of the coil, the authors have shown better performance, including a longer (2.8 cm) discharge to the finger.

S. Rahman · S. Khan (✉)
Independent University, Bangladesh, Dhaka, Bangladesh
e-mail: skhan@iub.edu.bd

P. G. Bradford et al. (eds.), *Proceedings of IEMTRONICS 2024*, Lecture Notes in Electrical Engineering 1229, https://doi.org/10.1007/978-981-97-4780-1_28

1.1 Theory and Safety

A Tesla Coil uses a radio-frequency-based air-core double-resonant transformer to produce high voltage at low current at frequencies of 50 kHz–1 MHz [2–4]. A high voltage transformer steps up the AC voltage to jump a spark gap acting as a switch at the primary, generating high voltage in the secondary, enough to go through the air to the ground as a discharge streamer.

For a mains-supplied (220 V or 110 V) Tesla coil, every part of the circuit is capable of giving a shock or being fatal. The intermediate part of the circuit is the most dangerous at thousands of volts, where arcs can jump fatally for severally cm. Owing to charged capacitors, the circuit can give a shock many minutes after disconnection from the mains.

A mains-supplied Tesla coil can damage electronics in the room, like smartphones, hearing aids, cardiac pacemakers, etc., and those connected to the same consumer power supply Discharge from the coil can travel in any direction, harming equipment or personnel.

The coil must be connected to a three-pronged electric plug, with a ground pin, as insufficient grounding may cause further harm.

1.2 Literature Review

Although the Tesla coil was invented 130 years ago, there are relatively few publications in the literature. The miniature low-voltage Tesla coil is a relatively new invention. Today's trends are towards miniaturization, with DC supplies and solid state devices [5–7]. The streamer discharge from the coil can be modulated with sound, a feature which has been used by the musical group *Arc Attack* [8]. New developments are making it better than ever as a teaching too [9, 10].

2 Procedure

From 2021 to 2023, the authors constructed three miniature Tesla coil built running on a computer USB cable, that lit a nearby fluorescent lamp, made a common filament bulb act like a plasma ball, and produced an imperceptible continuous discharge on the finger.

The power source for the Tesla coil was previously collected by the authors from a common Bug Zapper Racket (Fig. 1).

This time, the authors found the power supply sold separately, which made the construction a little less costly (Fig. 2).

The battery charging components in the circuit were not required and were removed (Fig. 3).

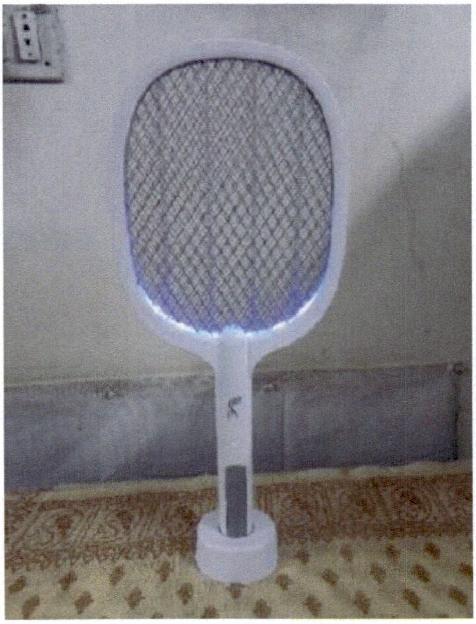

Fig. 1 The Bug Zapper racket, providing for converting battery voltage to high voltage. Model: Energy EN-3821

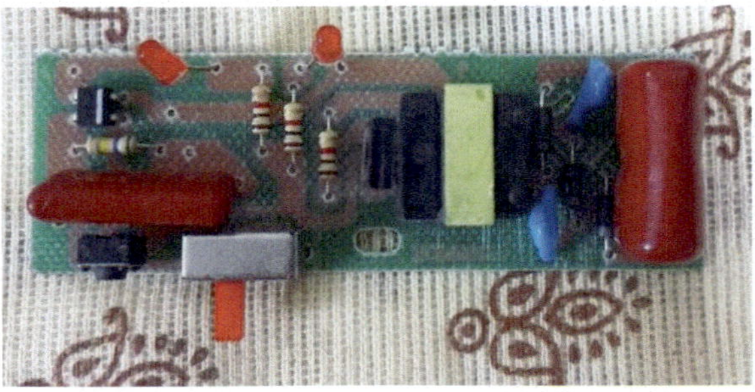

Fig. 2 The printed circuit board which provided the high voltage (length—80 mm; width—25 mm; height—15 mm)

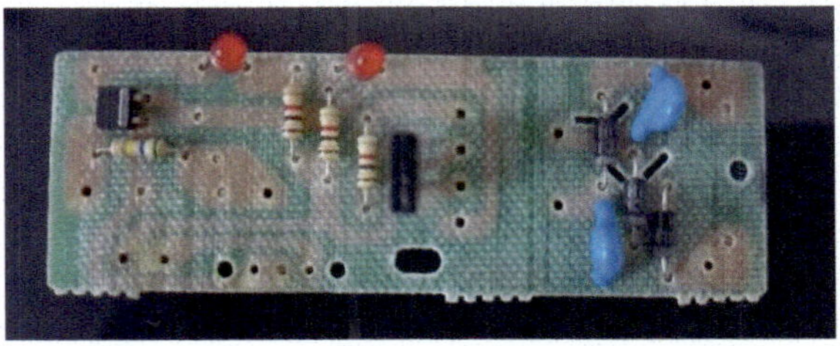

Fig. 3 The high voltage circuit PCB, with unnecessary components removed

The tiny ferrite core transformer plays an important role. A transistor D882 is used to make the DC supply to AC as an input to tiny transformer. The transformer outputs high voltage (calculations included later) which is converted to DC using 3

Fig. 4 Ferrite core transformer in the PCB

silicon diodes and 2 identical ceramic capacitors of 275 V (AC) or 2.5 kV (DC) with capacitance of only 2200 pF (Fig. 4).

We chose the dimensions:

$$\text{length } 90\,\text{mm} \times \text{breadth } 90\,\text{mm} \times \text{height } 8\,\text{mm}.$$

A gas pipe was cut into 4 pieces of 1 in. each. The external diameter was 8 mm and the internal diameter was 5.3 mm which was enough to push through 4.5 mm wide screws. The gas pipes are used to hold together the top and bottom pieces of cardboard cut in a square shape.

The Ferrite core transformer circuit, spark gap, battery, and the pillars were tight-fitting and able to eliminate extra space. The spark gap cards (holders) and the top ceiling of Tesla secondary were also made out of this cardboard.

The cylinder has diameter of 34.8 mm. The rectangles are of length 23.5 mm, height 14.8 mm, and width 5 mm.

We used 1 in. screws to make the spark gap and to make the connection on the top ceiling where the Tesla Secondary stands (Fig. 5).

We used the capacitors (Fig. 6) to make the capacitor bank whose capacitance is 5.6 nF which is rated in body, 562 J. These polystyrene low capacitance capacitors are efficient and allow fast discharge.

A toggle switch was used for turning on and off (Fig. 7). This was needed since we do not want to pull out the battery over and over.

Metallic drawer knobs were used on top of the coils (Figs. 8, 9 and 10).

Figure 11 is the Tesla secondary coil construction.

The 32-gauge enameled copper wire of diameter 0.22 mm was used to wrap around the 4 in. pipe.

The Aluminum foil is very important to wrap around the upper cardboard because charge is stored in air between the top load and the foil (Fig. 12).

This is how it looked after wrapping wire around the 4 in. pipe which has 333 turns. For insulation, 18 rounds of electrical tapes were applied at the bottom part to make a safe distance between primary and secondary. At the top, I only applied 5 round turns just to make it look symmetrical.

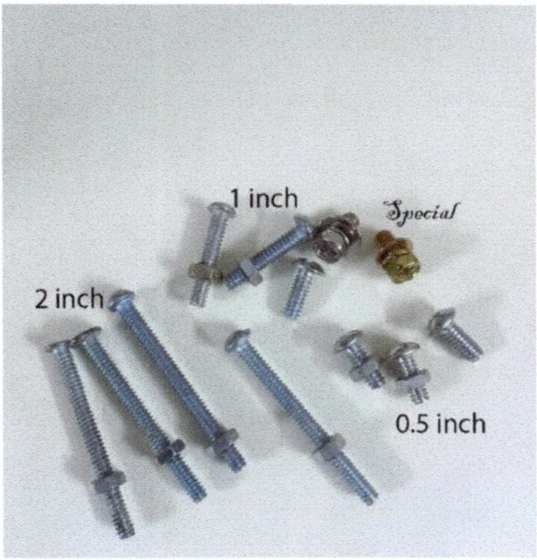

Fig. 5 Screws used and its length

Fig. 6 Capacitor 5600 pF ± 5% tolerance

The drilling for installing capacitors, screws, etc., are shown in Fig. 13.

The 4 corner drills, the bit center is positioned exactly leaving 4 mm from left and right so that our gas pipe doesn't hang out of the cardboard. The 3 small nuts are of 3 mm screws which I used to fix the circuit and the drills are done where there were holes in the circuit. The 4 small bolts are used to position the Spark Gap cards. This is important because a tiny millimeter of gap can lose the efficiency of the whole Spark gap. We need one spark gap here, not multiple (Fig. 14).

Fig. 7 The toggle switch for turning on and off the coil

Fig. 8 The drawer knob which will be set as top-load

The upper side of the lower board looks like above (Fig. 15).

On the upper board, the toggle switch is twisted in. The visible nuts are used to secure wire from this side to the Tesla primary (Fig. 16).

The secondary coil is glued to the board and the primary coil (yellow) is twisted on the lower side, again, the wire connection is secured using nuts on this side and bolting on the other side (Fig. 17).

The electrical tape will add thickness to the coil.

The construction is now complete, and just turning on the toggle switch produces sparks (Fig. 18).

Fig. 9 This thick 2 mm wire with two and a half turns was used to make the Tesla primary which was a mistake. It has a thin insulation and voltage leak has been observed so there was little to no output power through top load until we replaced with more insulated wire

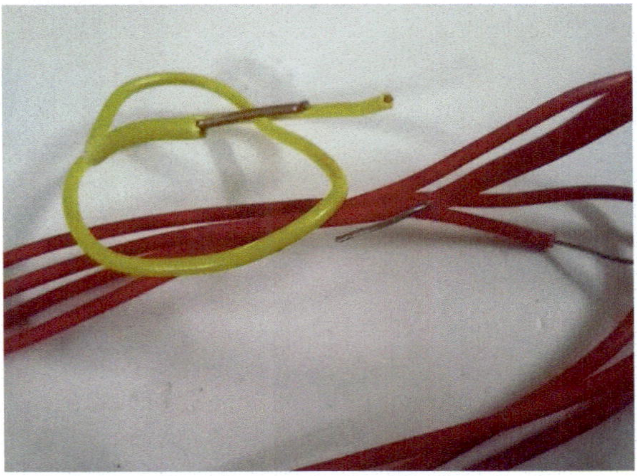

Fig. 10 The single copper wire of 0.5 and 1.3 mm including insulation was a good choice because it is stiff enough

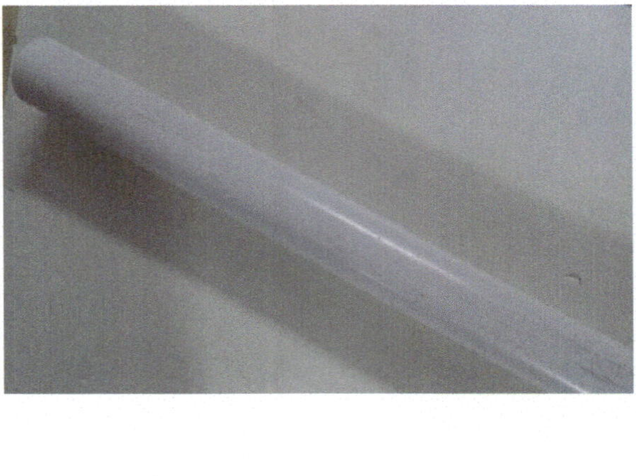

Fig. 11 The regular PVC pipe with internal diameter 33 mm and external 35.6 mm. Length used was four inches

Fig. 12 The coil before being connected to the base

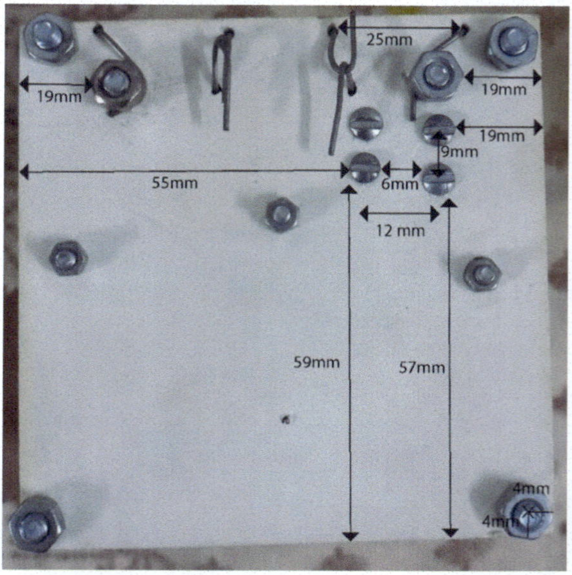

Fig. 13 Holes and parts placement in the base

Fig. 14 Adjustable screw of spark gap (orange circle)

Fig. 15 Complete base setup

2.1 Circuit Mechanism

When the switch is closed, current flows to the D882 transistor which makes an alternating current output to the ferrite core transformer. The transformer has two coils as labeled; the upper coil has 10 turns which is labeled as (a), the middle coil contains 32 turns and labeled as (b), the inside coil labeled as (c) has 1300 turns. This outputs to a rectifier which has two identical ceramic (blue) capacitors JNC-JY222M 2.2 nF and silicon diodes (Fig. 19).

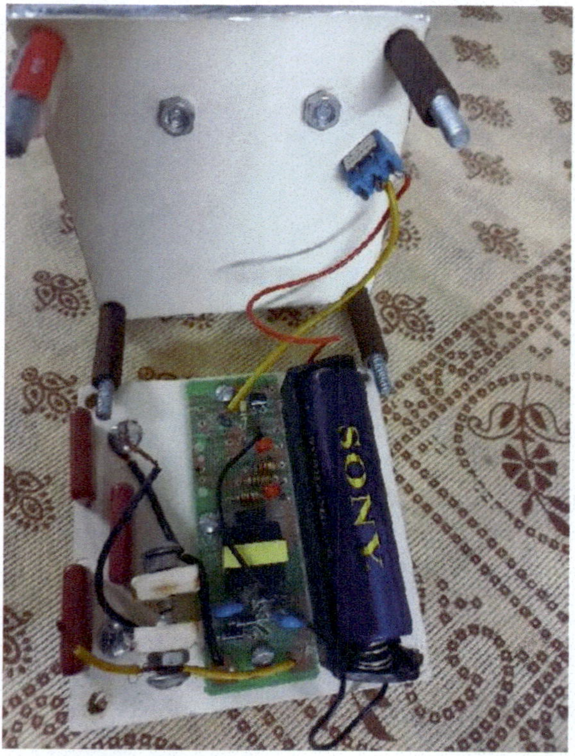

Fig. 16 Wire connection between the two layers

Fig. 17 Secondary coil wrapped with primary coil and fixed on the top

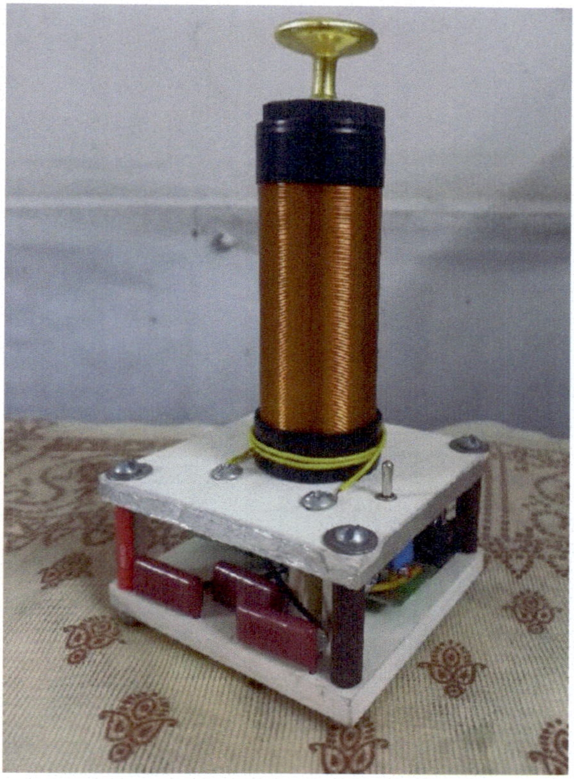

Fig. 18 The completed Tesla coil setup

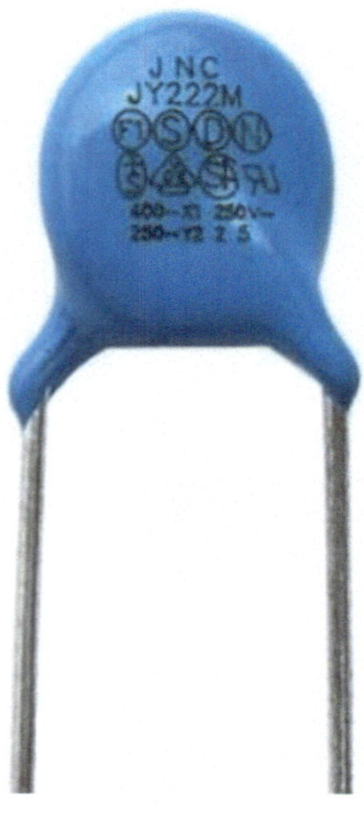

Fig. 19 High voltage capacitor for AC–DC conversion

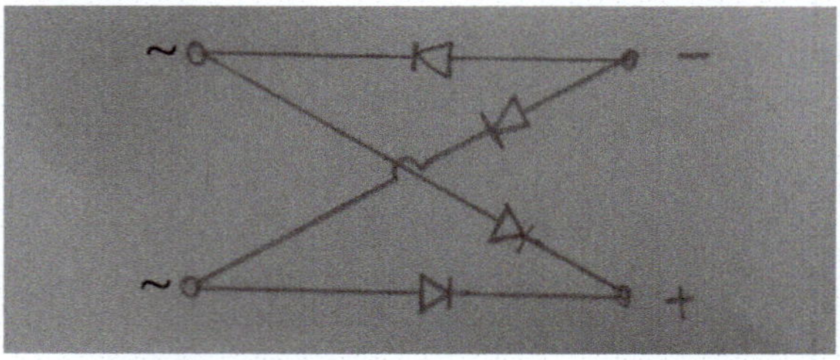

Fig. 20 MB10M circuit diagram

The MB10M circuit diagram is shown in Fig. 20.

The circuit first completes a cycle through the polypropylene capacitors, as shown in Fig. 21.

The other cycle is through spark gap (Fig. 22).

The size of the spark gap is adjusted to control the frequency and output.

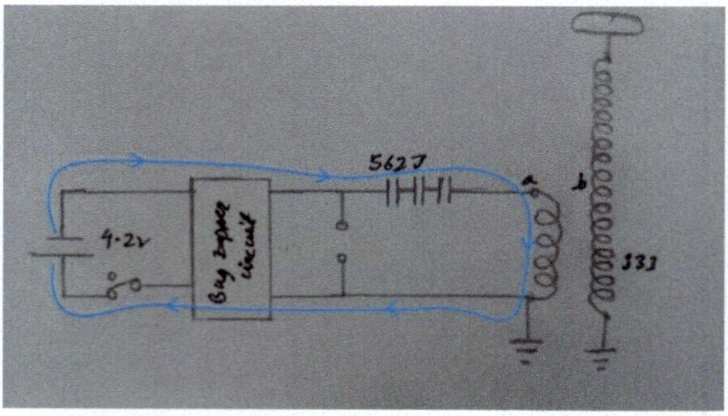

Fig. 21 Direction of current flow (before spark)

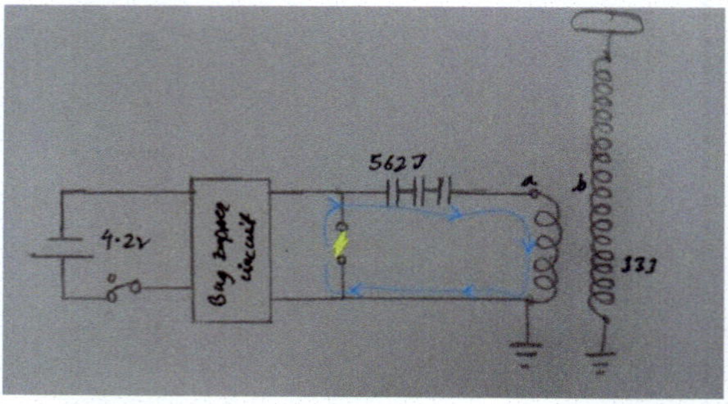

Fig. 22 Direction of current flow (during spark)

3 Results

The pictures are taken first in a dark room since the sparks are not much visible in the light.

The discharge at 3 mm spark gap (Fig. 23).

Discharge at 1.5 mm spark gap (more current density) is shown in Fig. 24.

Discharge to hand showing with a small Vernier caliper for distance measurement. The one cm spark is at the spark gap of 3 mm thus reducing the frequency. The number of sparks almost becomes countable. A ticklish feel can be sensed on the fingers.

With the spark gap at 1.5 mm, nothing was felt on my finger because the frequency is higher than previous (3 mm) gap.

In the dark, a fluorescent bulb lit up at a distance of 20 cm or less (Fig. 25).

More spectacular is the Filament bulb discharge, with a 200 W filament bulb held on top (Figs. 26, 27 and 28).

Fig. 23 Image in dark at gap of 3 mm

Fig. 24 Image in dark at spark gap of 1.5 mm

Fig. 25 The fluorescent lamp lights up not from the RF from the coil, but from the induced currents in the mercury vapor

Tiny neon bulbs (not LED) were found to glow (wirelessly) while kept on paper held above the head of Tesla Coil (Fig. 29).

The battery charge at the start was 3.77 V which dropped to 3.56 V after around 4 min use.

3.1 Change of Capacitors

Many trials were done by changing the capacitor with different capacitance and voltage capacity, as seen in Table 1.

When three 33 nF capacitors were used (connected in series), the spark length was seen to be longer than any of the sparks with three 5.6 nF capacitors (in series).

It was difficult to take pictures because it will require a camera with higher frame rates.

Some images are taken from a video recording (Figs. 30 and 31).

Fig. 26 Formation of plasma in a 200 W size filament bulb

Fig. 27 Very strong discharge through the thin filaments

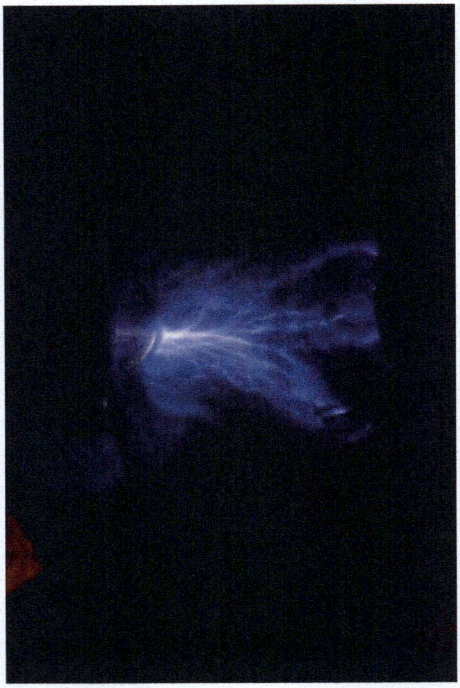

Fig. 28 Massive concentrated discharge through the bottom of the bulb to our hand

Thus, changing the capacitor turned out to be a promising upgrade, as the arc lengths doubled (Figs. 32 and 33).

The primary coil is seen in this image which has 2.5 turns. The number of turns was reduced to 1.5 turns (Fig. 34).

28 mm was the longest possible discharge we could record. The primary coil is visible with 1.5 turns. Three capacitors of 333 J were in series.

4 Comparison of Three Constructions

Three different Tesla coils were constructed, leading to progressively increasing discharge lengths (Figs. 35, 36; Table 2).

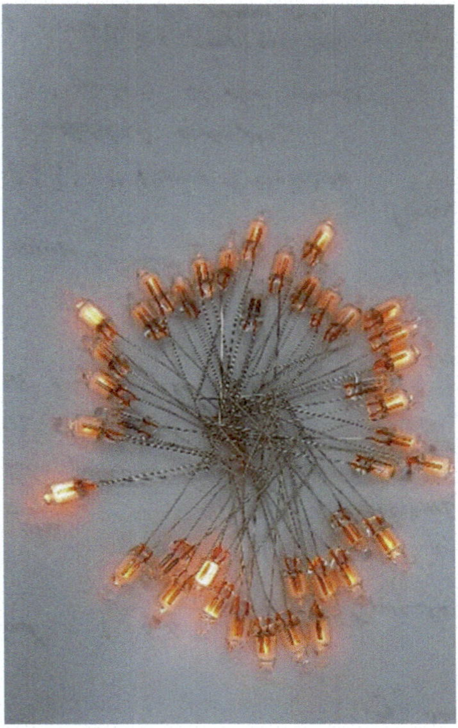

Fig. 29 Glows the neon bulbs (Tesla coil is under the paper)

Table 1 Different capacitors, primary turns, and arc discharge lengths of the Tesla coil, showing progressively increasing discharge length

Capacitor code	Capacitance (nF)	Primary turns	Spark gap size (mm)	Arc discharge length (mm)
562 J 3 kV	1.9	2.5	1.5	7
			2	3
333 J 2 kV	11	2.5	1.5	20
			2	16
		1.5	1.5	28

Fig. 30 Arc length 7 mm (with spark gap adjusted to 1.5 mm)

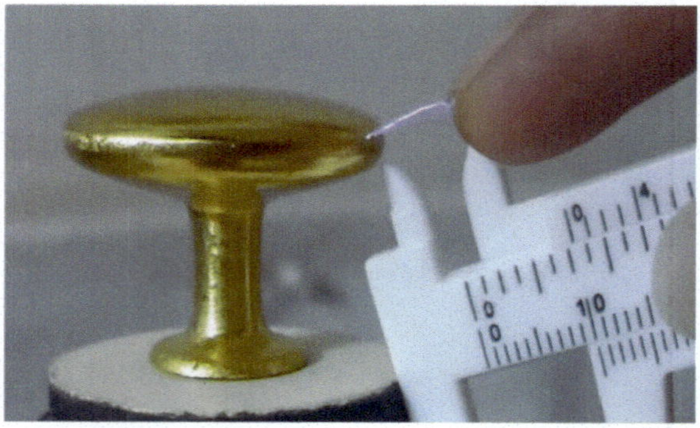

Fig. 31 Discharge of 3 mm (with spark gap adjusted to 2 mm)

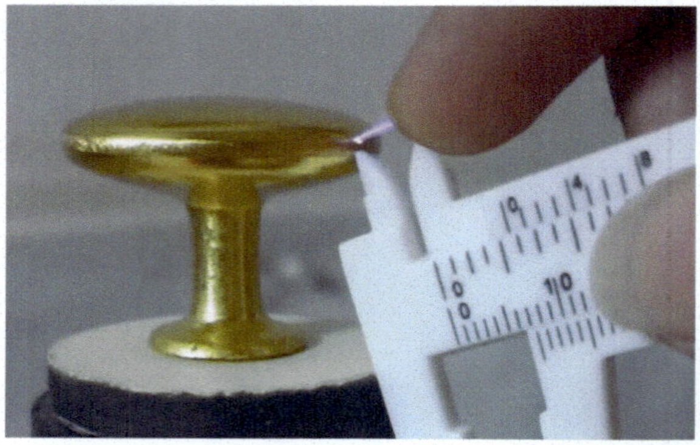

Fig. 32 Discharge of 20 mm (with spark gap adjusted to 1.5 mm)

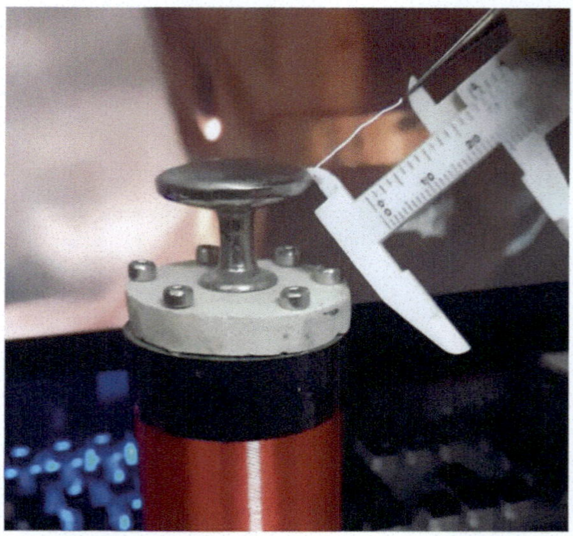

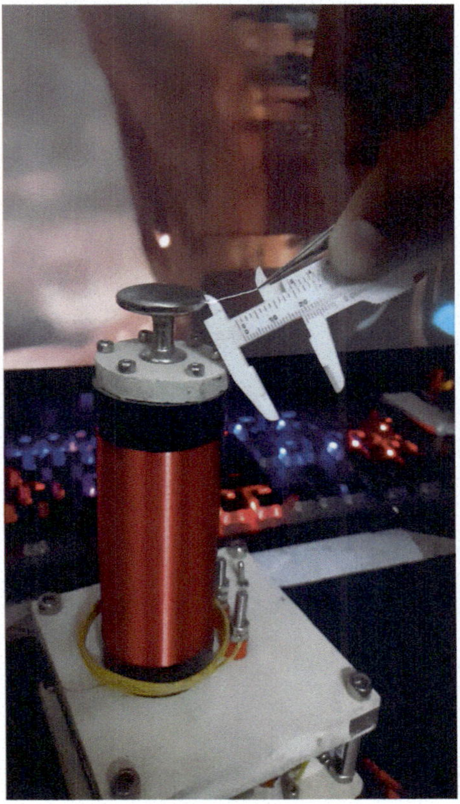

Fig. 33 Spark of 16 mm (spark gap 2 mm)

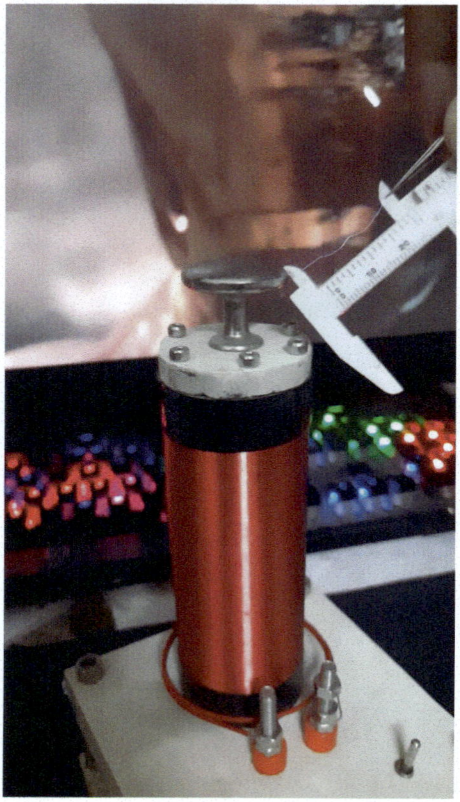

Fig. 34 Spark of 28 mm (spark gap 1.5 mm)

Fig. 35 The first miniature Tesla coil constructed by the authors around 2021, powered by a computer USB cable

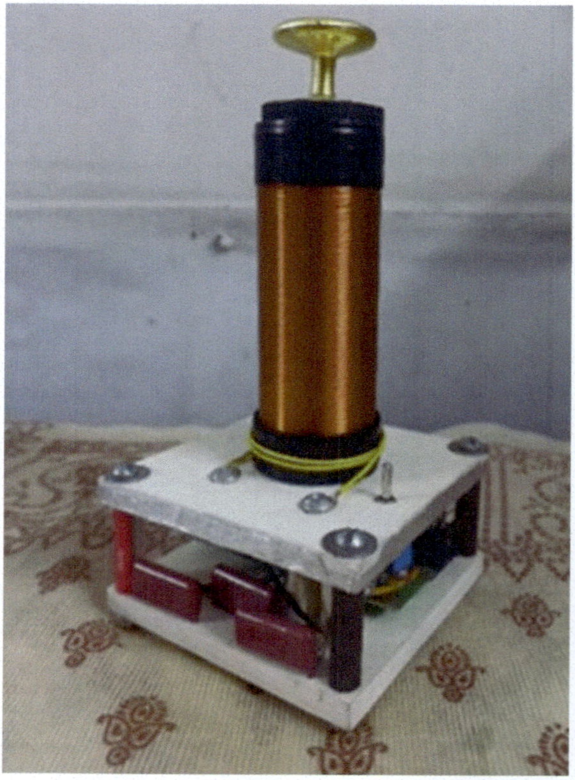

Fig. 36 Spark of 28 mm (spark gap 1.5 mm)

Table 2 Comparison of the three Tesla coils constructed by the authors from 2021 to 2023

Parameters	Construction 1	Construction 2	Construction 3
Secondary turns	300	300	400
Capacitors (× 3)	562 J 3 kV	562 J 3 kV	333 J 2 kV
Series capacitance (nF)	1.9	1.9	11
Secondary diameter (in.)	1.5	1.5	1.7
Secondary length (in.)	4	4	4.5
Base $L \times B$	9.5 cm × 9.5 cm	9.5 cm × 9.5 cm	9.5 cm × 9.5 cm
Base height (in.)	1.5	1.5	1.2
Primary turns	1.5	2.5	1.5
Spark gap (mm)	2	2	1.5
Arc length (mm)	5	7	28

5 Conclusion

The detailed adjustments described above have contributed to the effectiveness and increased power of a Tesla coil running on a 4.2 V battery. Progressively improved versions were constructed, demonstrating longer arcs to the finger, the greater distance at which a CFL glowed, and the greater lighting for neon bulbs. Many adjustments proved to be learning experiences during the constructions. Many of the distances had to be just right and the spark gap had to be closely adjusted.

Intriguing is the imperceptible 1 cm discharge through the finger, showing how centimeters-long discharge is hardly perceptible to the finger.

The device consists of high voltage operation and yet doesn't heat up much.

With some simple precautions, this easily constructed low-voltage, Tesla coil with $20 of components, shows much promise for promoting teaching and research at schools and universities.

References

1. Rahman S, Khan S (2022) The USB powered miniature Tesla coil, with filament bulb, fluorescent lamp and discharge to body. In: 2022 IEEE international IoT, electronics and mechatronics conference (IEMTRONICS), Toronto, ON, Canada, pp 1–7
2. Krbal M, Siuda P (2015) Design and construction solution of laboratory Tesla coil. In: 2015 16th international scientific conference on electric power engineering (EPE), pp 311–314
3. Kolchanova VA (2002) Computational modeling of the Tesla coil parameters. In: Proceedings of the 8th international scientific and practical conference of students, post-graduates and young scientists modern technique and technologies, 2002. MTT 2002, pp 32–33
4. Khan S (2019) AC circuits, 3rd edn. Dhaka, Bangladesh. ISBN: 978-984-33-5146-6
5. Bruns DG (1992) A solid-state low-voltage Tesla coil demonstrator. Am J Phys 60:797
6. Farriz MB, Din A, Rahman AA, Yahaya MS, Herman JM (2010) A simple design of a mini Tesla coil with DC voltage input. In: 2010 international conference on electrical and control engineering, pp 4556–4559
7. Khan S (2018) Semiconductor devices and technology, 3rd edn. Dhaka, Bangladesh. ISBN: 978-094-33-5983-4
8. Ghilinţă C, Stegaru SC, Popeea T, Ţăpuş N (2015) Portable audio-modulated Tesla coil for demonstrative actions. In: 2015 14th RoEduNet international conference—networking in education and research (RoEduNet NER), pp 238–241
9. Skeldon KD, Grant AI, MacLellan G, McArthur C (2000) Development of a portable Tesla coil apparatus. Eur J Phys 21(2):125
10. Tompkins N (2019) Miniature Tesla coil teaching lab. Phys Teach 57(6):390–392

Analysis of an Inexpensive Robotic Arm Implemented with Arduino

Sajid Ahmed Chowdhury, A. F. M. Afnan Uzzaman Sheikh, and Shahriar Khan

Abstract Robotic arms are popular worldwide, but are often prohibitively expensive, especially for developing countries. This paper implemented an inexpensive robotic arm with five degrees of freedom, controlled with variable resistances and an ATmega328P on an Arduino UNO board. Methodology, challenges, and solutions of such a robotic arm have been explored. Oscillations and instability have been brought under control with added resistors. The model can be an educational tool and a basis for other specialized robotic arms . Mechanisms to control the servomotors and writing code for the microcontroller are discussed. The final product is inexpensive at USD 43, with supply of 5 V and 2 A . Lastly, the robotic arm is compared with robotic arms used in surgery and in harsh environments.

Keywords Robotic arm · Arduino · Variable resistance · Servomotors · Education · ATmega328P

1 Introduction

Robotic arms have been widely implemented, such as for welding car bodies, performing surgery, and making prosthetic arms for the handicapped. Other applications have been implemented in factories, and on rovers sent into harsh environments. However, these arms are often prohibitively expensive, especially for developing countries, where their use remains low.

Present Address:
S. A. Chowdhury · A. F. M. A. U. Sheikh · S. Khan (✉)
Department of EEE, Independent University, Bangladesh, Dhaka, Bangladesh
e-mail: skhan@iub.edu.bd

S. A. Chowdhury
e-mail: 2010385@iub.edu.bd

A. F. M. A. U. Sheikh
e-mail: 2130635@iub.edu.bd

P. G. Bradford et al. (eds.), *Proceedings of IEMTRONICS 2024*, Lecture Notes in Electrical Engineering 1229, https://doi.org/10.1007/978-981-97-4780-1_29

Methods used to implement robotic arms normally have a frame, servomotors, and a microcontroller, all of which may be expensive [1, 2]. Coding for the microcontroller requires testing and additional labor. Several publications address these issues, but cost reduction and simplicity of the design has rarely been a major objective [3].

The ATmega328P (as in this paper) is a common choice for the robotic arm [4, 5]. Of the various options for control, low-cost variable resistances for a potentiometer (as in this paper) have been used [6]. The servomotor presents a problem of electrical control of a mechanical system [7].

The purpose for such a robotic arm and their control mechanism may be educational [8], ease of design and production, and low-cost implementation.

The constructed robotic arm illustrates the functionality of potentiometers and servomotors. It is seen how to write a simple program for the controller. The signals sent through the potentiometers can be replaced with signals from sensors or data from communication devices depending on the specific purpose of the robotic arm. The program can be used to learn how the servomotor angles are controlled using signals sent to the microcontroller. Simple additions can be made to the program to suit the hardware's purpose. If a different microcontroller is used other than an ATmega328P, the code in this paper can be used as a foundation for writing code for other microcontrollers.

2 Methodology

The communication between the potentiometers and the motor in the robotic arm is done using the microcontroller ATmega328P on an Arduino Uno board. This control mechanism allows the robotic arm to quickly respond to changes made in the potentiometers.

The constructed arm chassis had five different joints; therefore, its mobility or degrees of freedom (DoF) was five. Arduino programming was used to interface the potentiometers with the actuators used to move our arm.

The five degrees of freedom were controlled using five separate servomotors at the rotation of five joints: shoulder, elbow, wrist, wrist's axial rotation, and the gripper (Fig. 1).

2.1 Servomotors

In accordance with its datasheet [9], servomotors MG90S are placed in each joint of the robotic arm chassis. The servomotor supply voltage was 5 V (Fig. 2).

The servomotors have a restricted movement that allows them to rotate 180° (Fig. 3).

The V_{CC} and ground connections are in parallel with the Arduino UNO. The PWM (Pulse Width Modulation) orange pin is connected to one of the digital pins of

Fig. 1 The frame, servomotors, and five degrees of freedom of the robotic arm

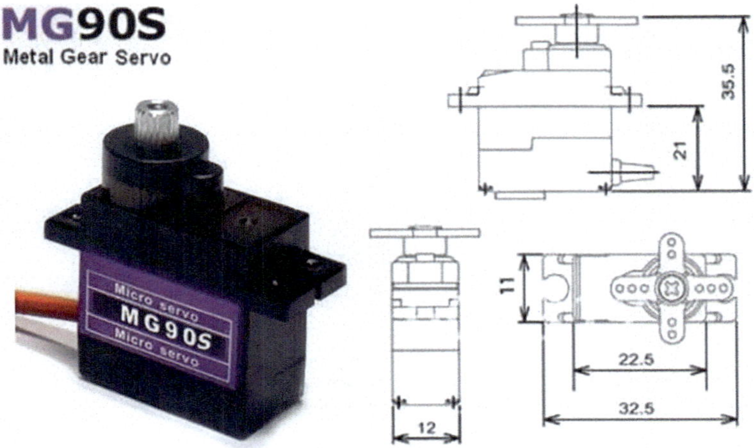

Fig. 2 1 MG90S servomotor [7]

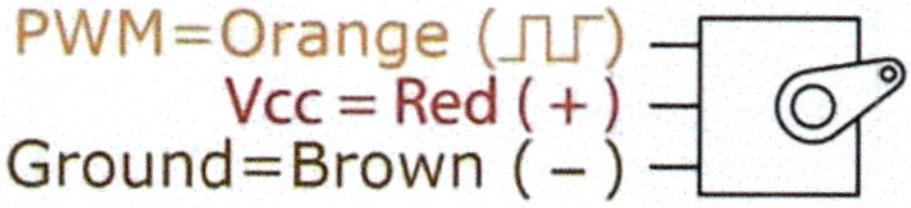

Fig. 3 The pin diagram of the servomotor [7]

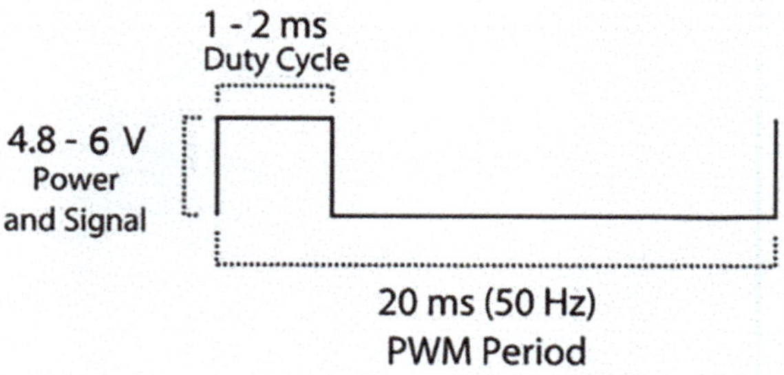

Fig. 4 PWM signal to control the direction of rotation [2]

the Arduino UNO board. Changes in the pulse width of a 50 Hz astable oscillating signal (0–5 V) can control the rotation of the servos. When the pulse width, or the time when the signal is high (logical high, or 5 V), is 1 ms then the servo rotates all the way to the left. When the pulse width is 1.5 ms, the servo is in the middle of its entire rotational path, and when the pulse width is 2 ms, the servo rotates all the way to the right. The overall duty cycle is kept short so most of the time the signal is mostly logic low or 0 V, for greater energy efficiency. The data transfer is of greater concern, rather than power via the orange PWM wire (Fig. 4).

2.2 *Potentiometers*

The servomotors had potentiometers with linear variable resistance. When a potentiometer's knob is rotated, the resistance changes linearly, and in turn, the voltage across the potentiometer changes from 0 V to its maximum value linearly (Fig. 5).

This linear change in voltage can be mapped into a rotational motion via the microcontroller, ATmega328P on the Arduino UNO board.

2.3 *The Arduino UNO*

The ATmega328P microcontroller has 14 digital input/output pins (of which 6 support PWM outputs), 6 analog inputs, a 16 MHz quartz crystal, a USB connection, a power jack, an ICSP header, and a reset button [10, 11]. The operating voltage for the Arduino is 5 V (Fig. 6).

The analog input reads the voltage value and converts it to a digital 10-bit representation with the help of its built-in analog to digital converter (ADC)). The voltage

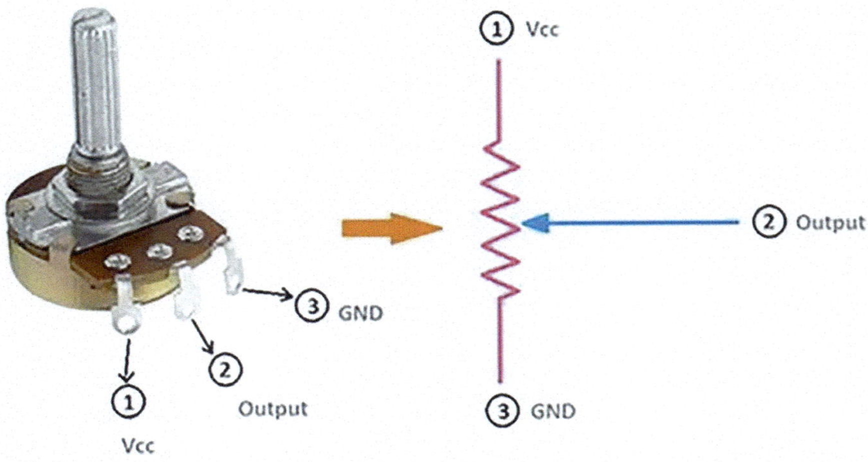

Fig. 5 The standard potentiometer pinout

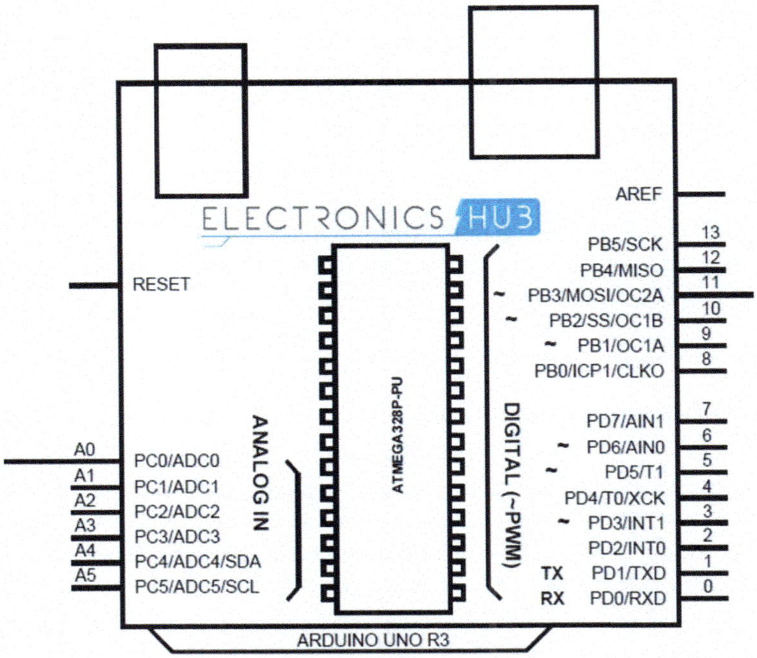

Fig. 6 Arduino UNO R3 pinout [10]

range is converted to integer levels ranging from 0 to 1023. The code is written such that the angle of a servomotor changes from 0 to 180° according to the value at the analog input.

The servomotors are run using a DC power supply providing a voltage of 5 V and approx. 2 A current. When the potentiometer's knob is rotated the analog input signal value changes and the corresponding voltage is used to change the angle of the servomotor. The 0–1023 analog to digital converted values are mapped from 0 to 180° angle of rotation of the servos. This mechanism is used by 5 servomotors that control each of the 5 degrees of freedom.

2.4 The Chassis

The robotic arm consists of a base motor that controls the rotation of the "shoulder" joint. Two motors are attached to the gripper: One controlling the angle at which the gripper holds objects and the other controlling the grip itself. Additionally, two other motors are controlling the "elbow" and "wrist" joints.

3 System Design, and Arduino Code

The potentiometers are connected to the Arduino UNO from pins A1 to A5 (Fig. 7). The digital I/O pins that relate to the signal pin of the servos are pins 10, 9, 6, 5, and 3. The reader may use any pins to their liking though the Arduino IDE sketch (program file) needs to be appropriately updated.

3.1 Hardware Implementation

After getting satisfactory simulation results, a prototype of the robotic arm was implemented with the Arduino UNO. The servomotors are attached to the digital pins of the Arduino board. They can achieve high-precision angular positions based on the input provided by the Arduino board. The potentiometers are attached to the analog pins. Each of the analog pins connected to the potentiometers was grounded through a 10 kΩ resistor (Figs. 8, 9, 10 and 11).

Since high-rated power supplies are hard to find, a standard lab bench power supply was used, as the servos need adequate current to run smoothly. This was done to allow the robotic arm to be run using cheap equipment. Since the servos are powered in parallel to each other and the power supply, the current is divided equally among each other. From experimentation, it was found that a 5 V and approximately 2 A power supply gives the best performance. The inability to provide a proper

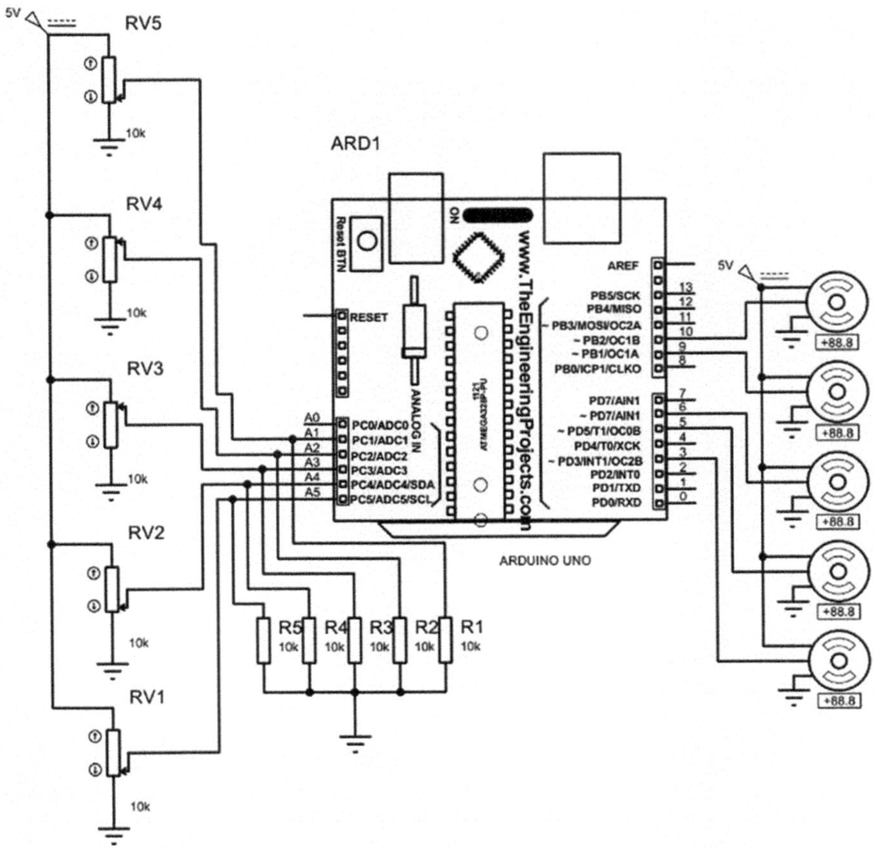

Fig. 7 Circuit diagram of robotic arm

current flow results in servo malfunction, i.e., fidgeting of the servo rotor. However, the servo will still work afterward.

An overflow of the potential supplied to the servomotors is very dangerous for the motors. Hence a great solution would be to use voltage regulator ICs, such as the LM7805 which has a max rated current of 1 A and outputs 5 V no matter what. Due to the parallel connections of the servomotors, the current delivered to the motor is about 0.5 A so the voltage regulator ICs connected with each of the motors will work without issues. A heatsink should also be added as these ICs tend to heat up.

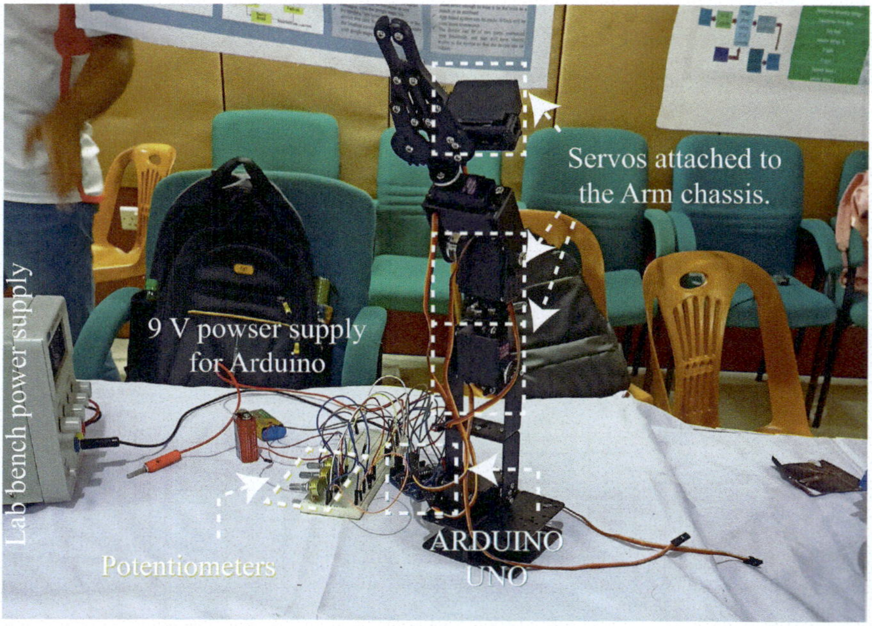

Fig. 8 Basic hardware prototype of the robotic arm

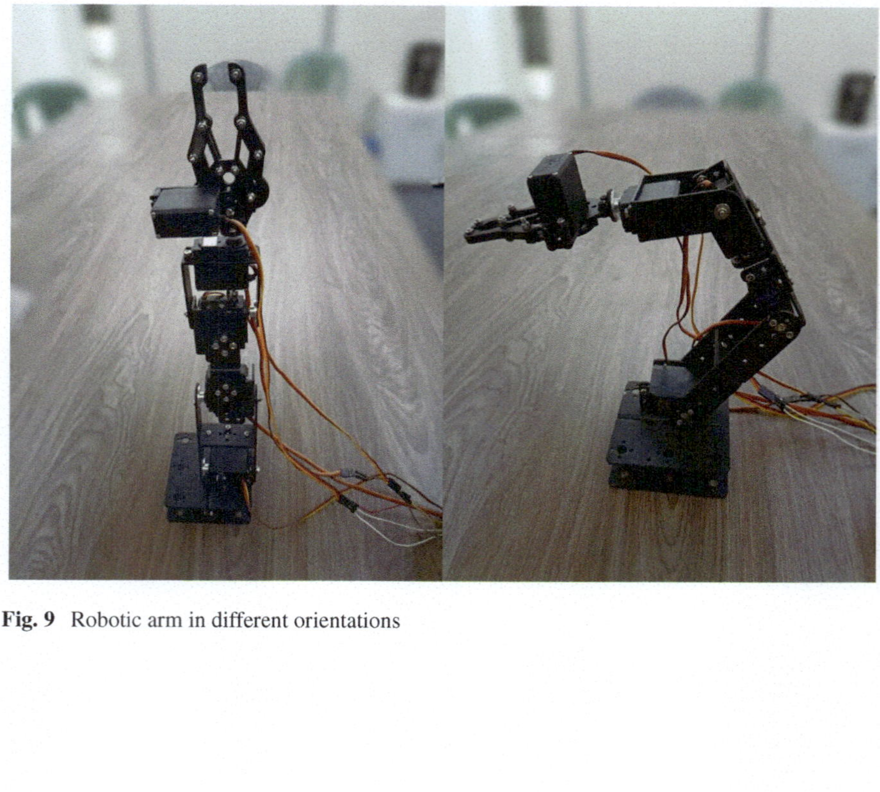

Fig. 9 Robotic arm in different orientations

Fig. 10 Simple repeating unit of the potentiometer control system (1. Potentiometer pin 1 to + V_{CC}, 2. 10 kΩ resistor from Arduino analog pin to ground, 3. Arduino analog pin to potentiometer pin 2, 4. Potentiometer pin 3 to ground)

Fig. 11 Robotic arm holding a marker pen

3.2 Arduino Code

Arduino code is used to send the data taken from the potentiometers to drive the servo.

```
  #include<Servo.h>
Servo  servo1;
Servo  servo2;
Servo  servo3;
Servo  servo4;
Servo servo5;
int servo1Pin = 10, pot1 = A1;
int servo2Pin = 9, pot2 = A2;
int servo3Pin = 6, pot3 = A3;
int servo4Pin = 5, pot4 = A4;
int servo5Pin = 3, pot5 = A5;
void setup() {
  servo1.attach(servo1Pin);
  servo2.attach(servo2Pin);
  servo3.attach(servo3Pin);
  servo4.attach(servo4Pin);
  servo5.attach(servo5Pin);
}
void loop() {
  int reading1 = analogRead(pot1);
  int angle1 = map(reading1, 0, 1023, 0, 180);
  servo1.write(angle1);
  delay(15);
  int reading2 = analogRead(pot2);
  int angle2 = map(reading2, 0, 1023, 0, 180);
  servo2.write(angle2);
  delay(15);
  int reading3 = analogRead(pot3);
  int angle3 = map(reading3, 0, 1023, 0, 180);
  servo3.write(angle3);
  delay(15);
  int reading4 = analogRead(pot4);
  int angle4 = map(reading4, 0, 1023, 0, 180);
  servo4.write(angle4);
  delay(15);
  int reading5 = analogRead(pot5);
  int angle5 = map(reading5, 0, 1023, 0, 180);
  servo5.write(angle5);
   delay(15);
}
```

In this Arduino code, the Servo library is first included. The part of the code declares 5 servomotors with the names servo1–servo5. Another set of variables is declared which correspond to the pins where the servos and potentiometers are attached.

Inside the void setup (), the servo.attach () method is used to specify the pin where the servos are attached.

The void loop portion is the most important part of the code, and this is the part that always repeats. The analog values from the potentiometers are read using analogRead(), and the integer value is stored inside the values reading1, reading2, and so on. The angle the servo moves is specified using the variables angle1, angle2, etc. and this angle is determined using the mapping function that converts the integer values from the analog to digital converter to a value of angle in degrees ranging from 0 to 180°. The servo is then rotated by this angle. The delay of 15 ms that comes afterward is for keeping the servo movement smooth.

4 Results

The robotic arm developed through the methods mentioned above demonstrated a smooth response to changes with the variation in the input signal provided by the potentiometers. It can lift lightweight objects such as pens and markers or objects of similar shape and mass, although this can be greatly improved.

A problem faced was instability in movement and lack of response to changes in the input signal. This problem is caused by floating voltage acting as a noise that distorts the input signal due to lack of current.

The 10 kΩ resistors were connected from the Arduino analog pins to the ground to discharge the floating voltage and overcome the problem of instability. If the servomotors of the arm fail to move completely, then increasing the current may solve the problem.

5 Cost Breakdown

The cost breakdown is given below, adding up to only about US $45. The main costs were the chassis and the servomotors (Table 1).

The cost of the components can be further reduced, such as by using an ATmega328P microcontroller directly, without an Arduino board. Using a 3D-printed plastic chassis will be cheaper than a ready-made metal chassis. Instead of using a

Table 1 Cost breakdown for the robotic arm

Component	Qty	Price/USD (Sept 2023)
5 DOF robotic arm chassis with servomotors	1	34.49
Arduino Uno	1	5.87
Breadboard	1	1.45
Potentiometers	5	0.77
Jumper wires (male to male)	40	0.91
Resistors	10	0.09
Total		43.58

breadboard to connect the controller to the Arduino board and the robotic arm, soldering the components on a Veroboard will reduce the cost even further.

6 Comparative Study

To understand the quality of the robotic arm in this paper, a comparative study was conducted through literature review. The comparison was made between experimental robotic arms made to conduct surgeries and ones that are tailored to function in harsh environments.

Robotic arms that are used for surgery need to have extremely precise control and stability. Moreover, there has to be an effective method for the doctor performing the surgery to communicate with the robotic arm. To make it safe for surgery, it also requires sensors that can warn the doctor and stop the robotic arm's movement whenever a dangerous movement is detected. The sensors, microprocessors, and other components required to achieve this precision and efficient communication can prove to be expensive. A simple experimental robotic arm made for surgery would require a voltage of approximately 12 V and current of 1.2 A for every motor [12].

Robots sent to harsh environments need to be prepared for situations that can be very unpredictable. A harsh environment can be a location that is remote, unknown, unstructured, and dynamic [13]. A robot needs to have the physical robustness to tackle these situations. Their frames can be heavy requiring high power to move. Generally, they require about 4–6 degrees of freedom. The components used for such robotic arms are also manufactured to withstand harsh environments. The above reasons make robotic arms for harsh environments very expensive.

Both comparisons show that in terms of responsiveness and stability, robotic arms made for surgery are better. In terms of robustness, robots made for harsh environments are better. However, achieving the responsiveness, stability and robustness makes them extremely expensive, making them unsuitable as an educational tool. An educational tool needs to be easy to understand, show the foundation of the system, have the robustness to withstand being handled by inexperienced users and has to be sufficiently cheap. The robotic arm mentioned in this paper contains the qualities necessary to be an educational tool. It can easily be made within $50 and can be powered with a regular lab bench power supply. Overall, this robotic arm can be used as a reference to other robotic arms that have more specific purposes.

7 Discussion and Conclusion

This paper has implemented an inexpensive robotic arm with five degrees of freedom, controlled simply with a controller made of an Arduino and variable resistances. With components costing only US $45, this constructed arm can be used for demonstration and educational purposes.

Challenges and solutions of such a robotic arm have been explored. Experimentation showed that a 5 V, 2 A power supply was adequate for driving the arm. Problems such as unstable movement were minimized by experimentation and added resistors.

Limitations of the robotic arm include the shape and low weight of the objects it can hold and lift.

Instead of signals from potentiometers, the movement of the arm can be automated using input automated signals from sensors.

If human control is required, a gesture-controlled system can be implemented where the robotic arm moves according to the movement of a person's hand.

7.1 Future Scope

The robotic arm in this paper can be manually controlled through potentiometers. For future works, two different control systems can be kept, one that uses potentiometers and another that uses switches, making this more suitable for education. For further improvements, sensors can be used and algorithms that automate the system can be implemented. Although it makes the robotic arm more expensive it can still prove to be a valuable tool for practical education. Lastly a neural network system that takes voice as an input and gives motor movements will have voice recognition functions. This may be difficult and require high computational resources and time, but it is certainly worth exploring.

Acknowledgements The authors would like to thank Dr. Dewan A F K Choudhury, CEO of Bioforge Health Systems Limited, for funding this project and the Department of Electrical and Electronic Engineering, Independent University, Bangladesh (IUB) for technical support and hardware debugging.

References

1. McKerrow PJ (1991) Introduction to robotics. Addison Wesley. ISBN: 0-201-18240-8
2. Khan S (2018) Semiconductor devices and technology, 3rd edn. Dhaka, Bangladesh. ISBN: 978-094-33-5983-4
3. Moran ME (2007) Evolution of robotic arms. J Robot Surg 1(2):103–111
4. Bhargava A, Kumar A (2017) Arduino controlled robotic arm. In: 2017 international conference of electronics, communication and aerospace technology (ICECA), vol 2. IEEE

5. Mohammed Ali H, Hashim Y, Al-Sakkal GA (2022) Design and implementation of Arduino based robotic arm. Int J Electr Comput Eng 12(2):1411
6. Mathew EB, Khanduja D, Sapra B, Bhushan B (2015) Robotic arm control through human arm movement detection using potentiometers. In: 2015 international conference on recent developments in control, automation and power engineering (RDCAPE), Noida, India, pp 298–303
7. Khan S (2012) Systems and control, 3rd edn. Dhaka, Bangladesh. ISBN: 978-984-33-3561
8. Jahnavi K, Sivraj P (2017) Teaching and learning robotic arm model. In: 2017 international conference on intelligent computing, instrumentation and control technologies (ICICICT), Kerala, India, pp 1570–1575
9. "MG90S servo, metal gear with one bearing" datasheet. Accessed Sept 2023
10. Arduino (2023) Arduino Uno R3, modified Sept 2023. [Online]. Available: UNO R3 | Arduino documentation. Accessed Sept 2023
11. ElectronicsHub, 15 Feb 2016. [Online]. Accessed 14 Sept 2023. Available: https://www.electronicshub.org/arduino-light-sensor/
12. Sahoo AK, Brahma B, Pattanaik A (2019) Design & development of robotic arm for medical surgery. In: 2019 2nd international conference on power and embedded drive control (ICPEDC), Chennai, India, pp 105–110. https://doi.org/10.1109/ICPEDC47771.2019.9036490
13. Wong C, Yang E, Yan X-T, Gu D (2017) An overview of robotics and autonomous systems for harsh environments, pp 1–6. https://doi.org/10.23919/IConAC.2017.8082020

ChatGPT for Writing Literature and Songs: End of the Road for Poets and Songwriters?

Shahriar Khan

Abstract The emergence of ChatGPT is revolutionary, but its operation, depth, and capabilities are largely unknown. There is concern for the potential of ChatGPT and similar programs to displace or atrophy human intelligence. ChatGPT can write poetry, literature, and songs, but how much can readers and listeners identify computer-generated literature? How can ChatGPT change the need and future for poets and songwriters? What is the future of literature in schools and universities? This study created poetry, literature, and songs from ChatGPT, and surveys whether readers can recognize them as computer-generated. People were found to have some ability to identify computer-generated writings, implying that ChatGPT for now, is not the end of the road for poets and songwriters. Rather ChatGPT is another technological advancement that must be understood and overcome by poets, so that they can produce literature and songs that appeal more to people than computer-generated literature. But with the ongoing rapid improvement of platforms like ChatGPT, we are not far away from when computer-generated literature can no longer be identified by readers. In some years, software will be able to give tunes, music, and voice to their self-generated lyrics. There can be computer-generated video of attractive humans lip synching and dancing to the songs. Will future rock concerts have all computer-generated lyrics, voice, music, and video?

Keywords ChatGPT · AI · Computer generated · Literature · Poetry · Poets · Students · Songs · Lyrics · Audio

---

S. Khan (✉)
Department of EEE, Independent University, Bangladesh, Dhaka, Bangladesh
e-mail: skhan@iub.edu.bd

P. G. Bradford et al. (eds.), *Proceedings of IEMTRONICS 2024*, Lecture Notes in Electrical Engineering 1229, https://doi.org/10.1007/978-981-97-4780-1_30

1 Introduction

ChatGPT and similar software like Gemini, Ernie Bot, LLaMA, Claude, Grok, Copilot, Bard, etc., have changed our concept of creativity and originality. ChatGPT is capable of writing literature and songs, although it is unclear how it does so [1]. Quite unexpectedly, ChatGPT had learned languages like Bangla, and even creates literature in the language like great poets. ChatGPT is changing from day-to-day as its creators modify existing features and add new features.

Considering the improving creativity of such software, there is now a very real question of the future of literature and poetry [2]. Are poets and songwriters nearing the end of the road? How much can students use ChatGPT for writing literature and doing homework [3, 4]? Can teachers distinguish between human-written literature by students and computer-generated literature [5]? What is the future of literature departments in universities across the world [6, 7]?

The key issue worthy of investigation is whether readers and listeners can detect computer-generated literature and song lyrics, as surveyed in this paper.

Anything that listeners perceive as computer-generated will have little appeal to them. What listeners perceive as written by a human, will help in forming a human connection, inspiring the listeners.

Teachers can become better aware of how problems for students can be solved with ChatGPT. Users can better create literature from ChatGPT. The makers of ChatGPT can use the findings of this paper to create better literature and song lyrics.

2 Procedure and Survey

In keeping with the objectives of this paper, a survey was conducted online in forms.google.com, and was advertised non-commercially on Facebook and some WhatsApp groups. 12 persons of various profiles responded.

The author also conducted the surveys in his classes of EEE 233 (Energy Conversion) with 16 student attendees, and in EEE 422 (Economics, Management, and Professional Ethics) with 25 student attendees, and his graduate class of 561 with 7 attendees (as below).

	Age group	Course	Date	No. of responses
General	All age groups	Connected in Facebook and WhatsApp	End of Sept., 2023	12
EEE 233	Undergraduate students	Energy conversion	Oct. 1, 2023	16 students
EEE 422	Undergraduate students	Economics, management	Oct. 2, 2023	25 students

(continued)

(continued)

	Age group	Course	Date	No. of responses
EEE 561	Graduate students	Energy conversion	Oct. 3, 2023	7 students

ChatGPT-generated poetry in English and Bangla were shown to respondents, who were asked whether they were written by a poet or computer-generated.

The author chose the first computer-generated lines from ChatGPT, rather than choosing from a wide variety of outputs by ChatGPT.

3 Results of Survey

In general, respondents showed some ability to distinguish ChatGPT-generated quotes from those actually written by people.

3.1 Age and Education of Respondents

The online respondents were from various age groups. The students of EEE 233 and EEE 422 were all in their 20s.

	General (%)	EEE 233 (%)	EEE 422 (%)	EEE 561 (%)
19 and below	0	0	0	0
20–29	33	100	100	86
30–39	17	0	0	14
40–49	17	0	0	0
50 and above	33	0	0	0

All respondents had passed high school (SSC/HSC), meaning they had some familiarity with the poets and songs from their literature classes.

	General (%)	EEE 233 (%)	EEE 422 (%)	EEE 561 (%)
High school (SSC/HSC)	17	100	100	0
Bachelor's degree	58	0	0	100
Master's and higher	25	0	0	0

3.2 Interest in Literature and Music

Most respondents appeared less interested in reading poetry and literature and more interested in listening to music and songs.

Question: Are you interested in poetry and literature?

	General (%)	EEE 233 (%)	EEE 422 (%)	EEE 561 (%)
Very interested	33	0	22	14
Interested	67	50	39	72
Not interested	0	50	39	14

Most respondents were found to be quite interested in poetry and literature.

Question: Do you listen to music and songs?

	General (%)	EEE 233 (%)	EEE 422 (%)	EEE 561 (%)
Very much	50	75	61	86
Often	17	19	35	0
Sometimes	25	6	0	14
Rarely	8	0	0	0

Most respondents frequently listened to music and songs.

3.3 British Poets of the Nineteenth and Twentieth Centuries

The respondents were mostly Bangladeshi, but had some familiarity with British literature from high school (SSC/HSC). Undergraduate students could better recognize the writings below as computer generated than the graduate students. The reasoning for this may be that younger students were more familiar with the capabilities of ChatGPT than graduate students.

Question: Who wrote this?

"Under the azure sky, dreams take flight,
In whispered winds, secrets of the night."

	General (%)	EEE 233 (%)	EEE 422 (%)	EEE 561 (%)
A British poet from nineteenth century	55	25	0	100
Computer-generated	45	75	100	0

This was a good response, considering that all respondents were not native speakers of English.

Question: Who wrote this?

"Amidst the shadows, whispers of time's embrace,
In the dance of stars, our fleeting lives find grace."

	General (%)	EEE 233 (%)	EEE 422 (%)	EEE 561 (%)
A British poet from twentieth century	45	87	61	86
Computer-generated	55	13	39	14

3.4 English Rock Music

Although the respondents are native speakers of Bangla, most are familiar with English songs. The undergraduates were better able to recognize the lines below as computer generated than the graduate students.

Question: Where are these lines from?

"Moonlit grooves set the stage alight,
Thrilling beats bring the world to the night."

	General (%)	EEE 233 (%)	EEE 422 (%)	EEE 561 (%)
Song by Michael Jackson	50	69	0	71
Computer-generated	50	31	100	29

Question: Who wrote this?

"Thunderous riffs, a metallic crusade,
Shattered silence, in the maelstrom we wade."

	General (%)	EEE 233 (%)	EEE 422 (%)	EEE 561 (%)
Rock song by band Metallica	55	31	25	86
Computer-generated	45	69	75	14

3.5 Bangladeshi Poets Tagore and Nazrul

Most respondents were native Bangla speakers familiar with the works of famous poets Tagore and Nazrul. A respondent in her 80s, familiar with Nazrul, said she did not find any semblance below with Tagore.

Question: Who wrote this?

"মন আমার ভাােলাােবেস চাােঁ দর আেলায়,
িবম এক হেব ভাষা ও ভাােলাবাসার সো যুদ্ধ বািজ Fছড়ায়।"

	General (%)	EEE 233 (%)	EEE 422 (%)	EEE 561 (%)
Rabindranath Tagore	18	0	0	43
Computer generated	82	100	100	57

Question: Who wrote this?

"এক্ষু িতর সো িমেল েস্রাােবর গান, িআম পাই সান্না,
FযMিল সংকট দর করেব □াণ।"

	General (%)	EEE 233 (%)	EEE 422 (%)	EEE 561 (%)
Rabindranath Tagore	45	75	12	86
Computer generated	55	25	88	14

The respondent in her 80s, familiar with poet Nazrul, said she did find some semblance with poet Nazrul.

Question: Who wrote this?

"জীবেনর আMােন েউQজনা
েধর,নেয়ন িফের যায় অধীর
সেময়।"

	General (%)	EEE 233 (%)	EEE 422 (%)	EEE 561 (%)
Kazi Nazrul Islam	27	60	9	71
Computer generated	73	40	91	29

Question: Who wrote this?

"ধরা যাক বায়র তেরার েঝড়,
মানষ আসেব জীবেনর সো িমলােন"

	General (%)	EEE 233 (%)	EEE 422 (%)	EEE 561 (%)
Kazi Nazrul Islam	55	40	19	0
Computer-generated	45	60	81	100

3.6 *Bangla Modern (Adhunik) Music (From Last 40 Years or So)*

Another important genre in Bangla music is "Adhunik" or modern, meaning written in the last 40 years or so.

Most respondents, all native Bangla speakers, were unable to recognize the lines as computer-generated (ChatGPT).

Question: Who wrote this?

"জ◌ীবেনর র◌েঙ লকা◌েন◌া, ম◌ু খ একট◌া হ◌া◌িস,
◌ে◌র বাত◌া◌েস ওড়◌া, প◌া◌েব◌া নতন দশা।"

	All (%)	EEE 233 (%)	EEE 422 (%)	EEE 561 (%)
Adhunik (modern) Bangla song	50	25	10	14
Computer-generated	50	75	90	86

3.7 *Bangla Rock Music*

Bangla rock music usually has drums and guitarists and keyboardists doing their own thing. Warfaze and James are two of the best known among rock bands. Most respondents were of the younger generation, likely to listen to Bangla rock songs, but were unable to identify the lines as computer-generated.

Question: Who wrote this?

"ব◌ু র F◌ে◌ম আধঁ ◌ার
◌িফ◌ের, বাধা Fেছড় মন
উড়ায় ◌িফের।"

	General (%)	EEE 233 (%)	EEE 422 (%)	EEE 561 (%)
James, Bangla rock singer	67	20	0	86
Computer-generated	33	80	100	14

Question: Who wrote this?

"আIM◌েনর ধারায় □◌েল মন, Fজা◌ের
হা◌িরেয়,মহাকা◌েলর নকশা রেচ রেে়র
নীলা□◌েন।"

	General (%)	EEE 233 (%)	EEE 422 (%)	EEE 561 (%)
Warfaze, Bangla rock band	60	20	0	100
Computer-generated	40	80	100	0

4 ChatGPT in the History of Literature

Spoken language has been in existence for primitive Man and predecessors for perhaps millions of years. Songs and music must have been in existence for hundreds or tens of thousands of years, when songs were passed on through the generations by word-of-mouth. To this day, Indian classical music is passed on without being written. The absence of written literature and songs must have allowed room and space for many poets.

Then came written language, when it became possible to write down literature for future generations. We find today that most Western classical music (Beethoven, Bach, Mozart, etc.) is written down. In the presence of a wide variety of written poetry and literature, newcomer poets must have had a harder time to compete.

Sometime around 1920, recordings of sounds and voice became possible, meaning that well-recorded songs by the best singers became available to all. This made it harder for amateurs and others to enter the market for music.

4.1 The New Challenge to Poets and Songwriters

Today's ChatGPT will be a similar challenge for poets and songwriters, as they have to compete with ChatGPT. The challenge to the Poets will be that ChatGPT may be able to write better than them. The challenge to poets would be to prove the human nature of their own writings.

The advantage to the writers may be that their human voice for their writings and songs will be far more powerful than the computer-generated voice available today.

ChatGPT appears to not be the end of the road for poets and songwriters. Rather ChatGPT must be understood and overcome by them, so that they can continue to produce literature and songs that appeal more to people than ChatGPT-generated literature.

4.2 Tunes, Voice, and Music to Accompany Computer Lyrics

Technology will continue with the trends started by ChatGPT. Already, text-to-voice converters are hard to distinguish from humans. These voices can be made to speak in tune, or sing in tune with a beat, which is the building block for a song. Guitars, drums, base, and keyboards can be added, meaning the songs will be much like the rock songs of today. This raises the question of whether rock concerts of tomorrow could have all computer-generated lyrics, voice, and music. Would computer-generated graphics show attractive men and women lip-synching computer-generated songs? The answer would be that an audience always looks for the human element. The more human elements are shown and heard, the more the audience may be attracted to the songs.

4.3 Paintings and Visual Art

We notice some parallels with visual art or paintings. Around the 1990s, computer art became popular with MS PowerPoint and Paint. But painters are continuing their profession of painting, and computer art has gained only limited popularity.

5 Conclusion

Literature by ChatGPT has proven worthy of investigation, as shown in this paper. The online survey and the surveys in the author's class were instructive, and permitted some far-reaching conclusions about the future of literature and ChatGPT.

All respondents had some familiarity with English and Bangla literature from their high-school literature classes.

Undergraduate students could better recognize computer-generated writings than graduate students. This could be because younger students were more familiar with the capabilities of ChatGPT than graduate students.

Regardless of how good ChatGPT can mimic poets, people still had some ability to distinguish the writings of ChatGPT.

ChatGPT and similar softwares will evolve so that humans can less recognize their writings as computer generated. But readers will continue their search for what is written by a human.

But how much longer will it be until humans can no longer distinguish between computer-generated work and that written by a poet?

In future continuation of this work, we can interview larger numbers of those who appreciate art and music, both English and Bangla. The makers of ChatGPT and similar AI software can use the findings of this paper to improve their writings.

In the coming years, software will be able to give tunes, voices, and music to their self-generated lyrics. This raises the question whether rock concerts of the future could have an audience listening to all computer-generated lyrics, voice, and music. Would computer-generated graphics show attractive dancing men and women lip-synching the computer generated songs? The answer would be that an audience will be seeking and be attracted to the realistic human element in computer generated audio and video.

Acknowledgements Gratitude is expressed to the respondents including students and professionals who filled out the survey.

References

1. Saini N (2023) What Google CEO Sundar Pichai said about software engineers losing jobs due to ChatGPT-like tools. Livemint.com, Apr 2023
2. Kirmani AR (2022) Artificial intelligence-enabled science poetry. ACS Energy Lett 8(1):574–576
3. Khalil M, Er E (2023) Will ChatGPT get you caught? Rethinking of plagiarism detection. arXiv preprint arXiv:2302, p 04335
4. Kangasharju A, Ilomäki L, Lakkala M, Toom A (2022) Lower secondary students' poetry writing with the AI-based poetry machine. Comput Educ Artif Intell 3:100048
5. Köbis N, Mossink LD (2021) Artificial intelligence versus Maya Angelou: experimental evidence that people cannot differentiate AI-generated from human-written poetry. Comput Hum Behav 114:106553
6. Lo CK (2023) What is the impact of ChatGPT on education? A rapid review of the literature. Educ Sci 13(4):410
7. Baidoo-Anu D, Ansah LO (2023) Education in the era of generative artificial intelligence (AI): understanding the potential benefits of ChatGPT in promoting teaching and learning. J AI 7(1):52–62

ChatGPT for Solving Engineering and Math: Challenges for Education

Shahriar Khan

Abstract The emergence of ChatGPT and competing software like Bard and Copilot is revolutionizing education today, but their functioning and capabilities are largely unknown. University level problems in physics and engineering can be solved, but the nature and limitations of this capability of ChatGPT are worthy of investigation, as attempted in this study. The author used his three-decades long experience in universities to analyze implications of ChatGPT in education. Differentiation, integration, matrices, Fourier and Laplace transforms could be easily solved, but questions are still misinterpreted, giving wrong answers, as seen in this study. The challenge to students was to better formulate the query as text, understandable to ChatGPT. Entering phasor angles, such as for AC circuits, was problematic. Figures are not recognized, meaning it is difficult to input a series parallel or multi-loop circuit. The freely used version of ChatGPT was investigated, and undergraduate engineering students were surveyed. The findings of this study can be used by developers to improve ChatGPT. Teachers can adapt to the software, and homework may now be assigned in forms more difficult to enter into ChatGPT. ChatGPT should not be interpreted as a problem for teachers assigning homework, nor should we fear the atrophy of the intellect of students. Rather, just like the calculator, simulations and Google Search changed assignments over the last few decades, ChatGPT should be taken as a tool for making education and homework more sophisticated than ever. Education can move towards assigning open-ended and complex-engineering problems, which form the basis of today's Outcome-Based education. A new generation of students will approach problems far more effectively than their predecessors. This study provided some insight and appreciation for what lies in the next few years for education in the light of such AI problem-solving softwares.

Keywords ChatGPT · Bard · Copilot · Electrical · Problems · Solving · Education · Tool · OBE

S. Khan (✉)
Department of EEE, Independent University, Bangladesh, Dhaka, Bangladesh
e-mail: skhan@iub.edu.bd

P. G. Bradford et al. (eds.), *Proceedings of IEMTRONICS 2024*, Lecture Notes in Electrical Engineering 1229, https://doi.org/10.1007/978-981-97-4780-1_31

1 Introduction

Introduced in November, 2022, ChatGPT and competing softwares Copilot, Bard, etc. have shown great capabilities in creating literature [1, 2] and solving mathematical and scientific problems [3]. However, much of how ChatGPT works is unknown, and merits further investigation. Quite unexpectedly, ChatGPT has learned Bangla language, and even creates excellent literature in Bangla. The software and its capabilities are undergoing constant improvements, leading to even greater capabilities [4].

1.1 Implications for Homework in Education

Assignments constitute an important part of education, and ChatGPT can easily solve assignments on differentiation, integration, Fourier and Laplace transforms, matrices, etc. Teachers are now challenged with forming homework hard to enter into ChatGPT. The nature of homework and education in universities has changed [5–7], and the question is how this trend will continue.

1.2 Objectives of this Study

This study investigated challenges and limitations of ChatGPT, especially in undergraduate engineering education. The findings can be used to better enter text into the software. Teachers can become better aware of how to formulate problems and assign homework. The developers of ChatGPT and similar AI software, can use the findings to improve their software.

Finally, there can be some appreciation of what lies in the next few years for such problem-solving AI software.

1.3 Procedure

The author uses his three-decades long experience in universities to analyze some implications of ChatGPT in education. We entered various problems of increasing complexity into ChatGPT, and examined the answers. We submitted problems where the text was hard to enter, and tried to identify situations and problems, where ChatGPT was wrong or of borderline effectiveness and accuracy.

Undergraduate engineering students were surveyed, regarding the new homework assigned to them, and how they used ChatGPT when possible.

Based on the findings, we reached some conclusions about the challenges and implications of ChatGPT on the future of education.

2 Results

The simplest problems, such as with Ohm's law were easily entered and solved by ChatGPT, which even explained the formula and deliberations.

2.1 Recognized Formulas Entered as Text

ChatGPT was able to recognize equations and mathematical expressions entered as text. For example, entering

$$(40^2 + 20^2 - 30^2)/(2 * 40 * 20)$$

was correctly recognized as

$$\frac{40^2 + 20^2 - 30^2}{2 \times 40 \times 20}$$

Conversely, the formula entered from MS Word is correctly converted to the appropriate formula as plain text. For example, entering the following expression as cut-and-paste

$$\frac{40^2 + 20^2 - 30^2}{2 \times 40 \times 20}$$

is correctly converted by ChatGPT to

(〖40〗 ^2+ 〖20〗 ^2- 〖30〗 ^2)/(2*40*20) = 0.6875

In the above case where it was part of problem on phasors, ChatGPT continues, attempting to find arc cos (0.6875) = 44.75 degrees.

The new bracket " 〖40〗 " was introduced by ChatGPT.

2.2 Recognizing Matrices

An input such as [1, 2, 3; 4, 5, 6; 5, 3, 4] was correctly identified as a matrix, and ChatGPT could correctly find the transpose, determinant, and inverse.

2.3 Solving Differentiation, Integration, Laplace and Fourier Transforms

When given simple problems on Differentiation, Integration, Fourier and Laplace transforms, ChatGPT easily answered correctly.

A question like "What is the Laplace transform of 1?" gives a correct answer (=1/*s*) complete with derivation.

"What is the inverse Laplace of 1/*s* ?" also gets the correct answer of the step function *u*(t).

"What is the inverse Laplace of $1/(1 + s)$?" is also correctly recognized and solved as e^t.

"What is the inverse of $1/(1+s^2)$?" is correctly recognized and solved as $\sin(t)$. "What is the Fourier transform of 1?" gets the correct and full derivation and answer of $2\pi\delta(t)$.

"What is the inverse Fourier transform of $2\pi\delta(t)$?" gets the full derivation and answer of 1.

3 AC Circuits

Ohm's law with DC current were easily solved easily.

AC circuit expressions were usually recognized and answered, except that phasor angles were not well-understood.

Question: "Impedances of **j**6 and −**j**3 are placed in parallel with a current source of 12 A. What is the voltage?".

We got the correct answer of −**j**72 (Fig. 1).

3.1 Angles not well Understood

An attempt to input angle of a phasor was not well-recognized. In many cases we found that ChatGPT gives us a complex formula, but asks us to enter the formulas on a calculator.

"When voltage is 200/50, and current is 40/30, what is the impedance?" This gave the wrong answer of 3, as the angles were interpreted as fractions.

Question: An inductor of impedance 4 has an AC current of 5 amps flowing through it. What is the voltage across the inductor?

The answer was

$$\mathbf{V} = \mathbf{I} \times \mathbf{Z} = 20\text{V}$$

where there is no mention of phasor angle.

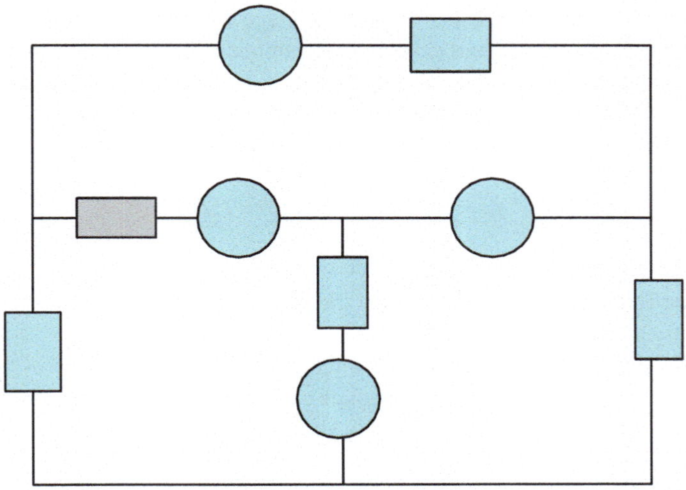

Fig. 1 A figure and circuit which cannot be easily entered into ChatGPT. Rather a system of equations can be written and then entered

It was not easy to enter the angles as formulas, so we enter angles as text.

When entering "50 at 30 degrees × 20 at 40 degrees", the answer is wrongly interpreted as

$$50 \sin(30) \times 20 \sin(40) = 321.395$$

Question: "20 at an angle of 30 × 40 at an angle of 30"

Answer: ChatGPT fails to give us the formula and answer for the magnitude (=1200). Also it wrongly gives us two (meaningless) answers for the angle

$$\theta_1 = \cos^{-1}\frac{40^2 + 20^2 - 30^2}{2 \times 40 \times 20} \qquad \theta_2 = \cos^{-1}\frac{30^2 + 20^2 - 40^2}{2 \times 30 \times 20}$$

It was not easy to enter phase angle understandable by ChatGPT, but power factor was better recognized instead.

Question: An impedance has a voltage of 200 V, and a current of 10 A at pf 0.5 lagging. What is the value of the impedance?

Answer: $\mathbf{Z} = \mathbf{V}/(\mathbf{I}.\text{pf}) = 40$ ohms

3.2 *Incorrect Answers*

In the following problem, the problem is stated correctly, but ChatGPT misunderstands and gives the wrong formula and answer.

Question: A series circuit has a source voltage of 100 V, across a resistor = 4 ohms, inductor of 6 ohms, and capacitor of 3 ohms. What is the current?

Answer (wrong): To calculate the current in a series circuit, we'll use the total impedance (**Z**) of the circuit, which is the sum of the resistance (R), inductive reactance (XL), and capacitive reactance (XC):

$$\mathbf{Z} = R + X\mathrm{L} + XC\,\Omega = 7\Omega$$

So current = 100 V/7 = 14.29 A

In the above problem, when we added the word "AC series circuit," the problem was correctly solved.

Question: An AC series circuit has a source voltage of 100 V rms, across a resistor = 4 ohms, inductor of 6 ohms, and capacitor of 3 ohms. What is the current?

The formulas were used correctly this time to give the right answer of 20 A. However, no angle was given.

4 Transcendental Equations

Transcendental equations have no closed form of solution, and ChatGPT sometimes did not give numerical answers in spite of repeated requests.

$$x \sin(x) = 0.5$$

For example, when the above was entered, the response was lengthy procedures and computer code, including in Python, for solving equations. In spite of repeated requests, no numerical solution was given.

However, entering the following transcendental equation gave a solution as $x =$ 0.584.

$$\log(x) + \sin(x) = 0.5$$

5 Open Ended Problems

Open-ended problems don't have a single correct answer but allow for multiple solutions or interpretations. These problems encourage creativity, critical thinking, and exploration. Solutions may vary based on individual, experiences, or interpretations and often relate to real-world scenarios.

Examples of open-ended problems include,

Design a new product that solves a common electrical engineering problem

How can we improve access to clean water in a slum?

Design a new mousetrap.

Clearly, these types of assignments which cannot be solved well by ChatGPT.

As teachers move more towards open-ended problems and assignments, they will promote higher-order thinking skills, and creativity. Students will think critically, analyze information, and generate innovative solutions.

5.1 Complex Engineering Problems

Complex Engineering Problems (C.E.P.), are a part of Outcome Based Education and Accreditation, the new buzzwords of today. C.E.P. has more features than Open-ended problems, such as having conflicting requirements, requiring depth of thought, etc. ChatGPT will find it even more difficult to address CEP than Open Ended Problems.

The trend of the future may be to assign CEP, and promote creative and critical thinking in students.

6 From Calculator to Google Search to Simulation, then ChatGPT

The emergence of ChatGPT for solving problems does not mean the end of the road for assignments and education. Rather, ChatGPT may even be viewed as beneficial, making education more sophisticated. Just like the calculator and Google Search changed assignments over the last few decades, assignments requiring ChatGPT will be far more advanced than the assignments of today.

Before the popularization of the slide rule in 1900s, students and practitioners were spending much time getting their multiplication, division, and square roots right. With their slide rules and log tables, they could focus on the problem itself rather than the arithmetic.

During the popularization of the calculator in the 1970s and 1980s, it was feared the ease of calculations would make lazy and atrophy the minds of students. But the bypassing of menial and tedious arithmetic proved to be beneficial. Finding sine, cosine and log became easy, and students could focus more on the solutions to problems.

After the invention of computers and simulation softwares, homework and assignments have adapted to the change. The emergence of Google search meant yet another tool for students and professionals, letting them focus on difficult scientific, and open-ended problems.

With the emergence of ChatGPT, homework and curriculums must adapt to adjust to its availability. So teachers can move towards giving problems which ChatGPT cannot solve, such as complex-engineering problems.

Teachers and instructors must continue adapting to the technology available to students. Online exams and take-home exams may lose their popularity, as an examinee can easily access ChatGPT at home.

ChatGPT should be taken as much more than a shortcut for solving problems. Rather ChatGPT should be taken as a tool for learning, meaning that new generations of students may have capabilities far beyond the imagination of today's teachers and instructors.

There may be a trend towards assigning complex-engineering problems, which ChatGPT mostly fails to answer.

7 Conclusion

This study has explored the capabilities and limitations of ChatGPT in solving scientific problems, and discussed the implications on education. Close-ended problems are solved well, at least for the simpler problems where references are found on the internet. Problems like differentiation, integration, Fourier and Laplace transforms are well-solved, at least for simpler problems available online. Angles entered for AC circuits were not well-understood. What is not available online, cannot be solved.

In many cases we found that ChatGPT gave a complex formula, but asks us to enter the values ourselves. Answers may be wrong, even to correctly input questions, as ChatGPT may misunderstand the question. The challenge to students is to input the question in a more precise manner avoiding ambiguity. The challenge to the creators of ChatGPT is to better understand text input by the user.

Softwares like Matlab and PSim can solve AC circuits and matrices, Fourier and Laplace transforms, etc. But Matlab requires entry in very specific form, whereas ChatGPT can understand natural-style language by humans. Plus ChatGPT considers follow-up questions, which are not entertained by Matlab.

The makers of ChatGPT and similar software can use the findings of this paper to improve their software. Entry of data and problems into ChatGPT can be made more user-friendly.

7.1 Impact on Education

Teachers are now modifying how they assign homework. Assignments are increasingly involving figures, such as circuit networks, which ChatGPT cannot recognize. A multi-loop circuit would require the student to convert it to a system of equations to be input into ChatGPT or MATLAB. Errors in arithmetic can no longer be an excuse for the student getting wrong answers.

Assignments and education should now move towards open-ended problems and complex-engineering problems, which will greatly stimulate the intellect of students.

A new generation of students will see approach problems far more effectively than their predecessors.

Past practices of take-home exams and online exams have become problematic, as the examinee can easily take the help of ChatGPT. So the future trend may be to shift from take-home tests/exams and assignments.

References

1. Baidoo-Anu D, Ansah LO (2023) Education in the era of generative artificial intelligence (AI): understanding the potential benefits of ChatGPT in promoting teaching and learning. J AI 7(1):52–62
2. Khan S (2024) ChatGPT for writing literature and songs: end of the road for poets and songwriters?. IEMTRONICS 2024, London
3. Wardat Y, Tashtoush MA, AlAli R, Jarrah AM (2023) ChatGPT: a revolutionary tool for teaching and learning mathematics. Eurasia J Math, Sci Technol Educ 19(7)
4. Khan S (2024) ChatGPT for design of transformers and machines: implications for open-ended problems. IEMTRONICS 2024, London
5. King MR (2023) ChatGPT.: a conversation on artificial intelligence, chatbots, and plagiarism in higher education. Cell Mol Bioeng 16(1):1–2
6. Oguz FE, Ekersular MN, Sunnetci KM, Alkan A (2023) Can chat GPT be utilized in scientific and undergraduate studies?. Ann Biomed Eng 1–3
7. Lo C (2023) K: what is the impact of ChatGPT on education? A rapid review of the literature. Educ Sci 13(4):410

ChatGPT for Design of Transformers and Machines: Implications for Open-Ended Problems

Shahriar Khan

Abstract ChatGPT is revolutionizing technology today, but its depth, capabilities, and mode of operation are not well-understood. It is good at solving problems in physics and engineering as encountered in education. But the question arises how it would approach open-ended problems with no single solution, as encountered in manufacture and Outcome-based education. This study explored transformer and machine design, as examples of open- ended problems. When asked to design transformers and machines of progressively increasing complexity, ChatGPT responded initially with words of advice and guidance. ChatGPT did not attempt to perform the iterative steps and optimization required for solving open-ended problems. Often it gave a range of solutions, from data on the web. Once it asked us to choose from two very different ranges of answers. Or it made assumptions and assumed unrealistic values, and used correct formula to calculate often incorrect solutions. When these errors were identified in follow-up questions, ChatGPT drastically changed the answers. Only free software was used, meaning that paid versions are likely to perform better. The findings can be used for better solving open-ended problems such as power transformer and machine designs. Users can take the advice and formulas from ChatGPT, but they have to do the remaining iterative calculations by themselves. ChatGPT was found to tune itself from our questions and follow-ups, and it may be difficult for other users to replicate the same responses. This study will help their developers to further improve GhatGPT. As it is changing fast, this paper will provide a record of ChatGPT's performance during date of submission of this paper.

Keywords ChatGPT · Bard · Artificial intelligence · Inductor · Transformer · Machine · Single phase · Three phase · Generator · Induction motor · Synchronous motor · Open-ended problem · Complex engineering problem

---

S. Khan (✉)
Department of EEE, Independent University, Bangladesh, Dhaka, Bangladesh
e-mail: skhan@iub.edu.bd

P. G. Bradford et al. (eds.), *Proceedings of IEMTRONICS 2024*, Lecture Notes in Electrical Engineering 1229, https://doi.org/10.1007/978-981-97-4780-1_32

1 Introduction

ChatGPT and competing software such as Gemini, Ernie Bot, LLaMA, Claude, Grok, Copilot, Bard have raised concern about their potential to displace or atrophy human intelligence, and enable plagiarism. But just like the invention of the calculator revolutionized technology in the 1970s, ChatGPT will allow promote technology in unexpected ways, as has been investigated in this paper.

In spite of their great ongoing impact, its depth, capabilities, and mode of operation of these softwares are not well-understood. ChatGPT unexpectedly learned Bangla [1], and even created Bangla literature hard to distinguish from the works of Nobel laureate poet Tagore [2].

ChatGPT is changing fast, responding to the needs of consumers. It can only learn from what exists on the web, and cannot answer what it cannot learn from the web. It has been sued for using content from Game of Thrones, meaning it has to become less dependent on outside content. In many cases, such as found in this paper, it gives inaccurate or outright wrong answers [3].

1.1 Objectives of this Study

It is worth investigating ChatGPT (as in this paper) for the added reasons below.

(a) for users to better utilize and ask questions of ChatGPT (and similar AI programs).
(b) to help provide a tool for manufacturing, education, Complex Engineering Problems, and Outcome Based Education.
(c) for developers of ChatGPT to improve their software
(d) to help converge towards a manual for ChatGPT, which would be easily—understood, and yet continuously changing.
(e) to provide a record of the performance of ChatGPT in the date of submission of this paper.
(f) to have some prediction of what lies in the future for ChatGPT and similar AI programs.

1.2 Homework and Assignments for Education

Mathematical problems, typical of homework and assignments in educational institutions, can be done easily by ChatGPT [4]. The homework of the future should be in a form which cannot be easily solved by ChatGPT [5–9]. Students can easily take the help of ChatGPT during take-home exams and online exams, meaning we must shift towards supervised exams in class where ChatGPT is not accessible.

1.3 Transcendental Equations

ChatGPT correctly approached the solution of transcendental equations like $x \sin x = 0.5$.

It wrote Python code to find the answers, but stopped short of giving a value for the answer.

1.4 Open-ended Problems

Open-ended Problems (OEP) are characterized by conflicting requirements and no single solution. A special case of OEP are Complex engineering problems, used for manufacturing [10] and which is an important component of Outcome-Based Education in universities [11, 12].

1.5 Design of Transformers and Machines

The design of transformers falls well within an open-ended problem or complex engineering problem, requiring compromise in conflicting requirements. Designing a transformer requires minimizing core size, minimizing copper and iron losses, minimizing weight, cost, etc. The objective is to have minimum iron and copper, and yet have minimum loss, maximum linearity, and efficiency (Fig. 1).

Some universities even assign their final year project to design and build a transformer for welding purposes.

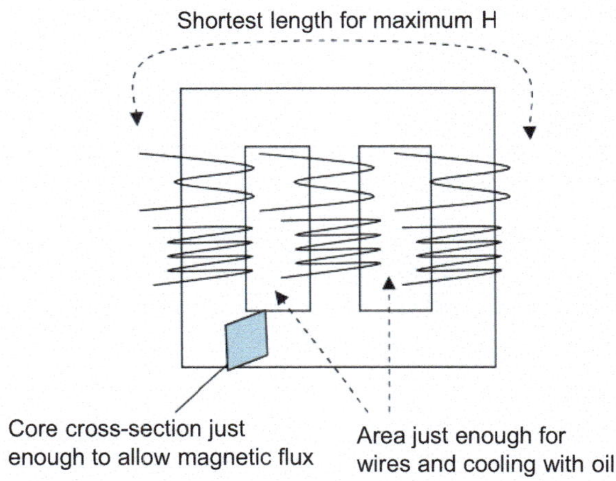

Fig. 1 The conflicting considerations for designing a transformer

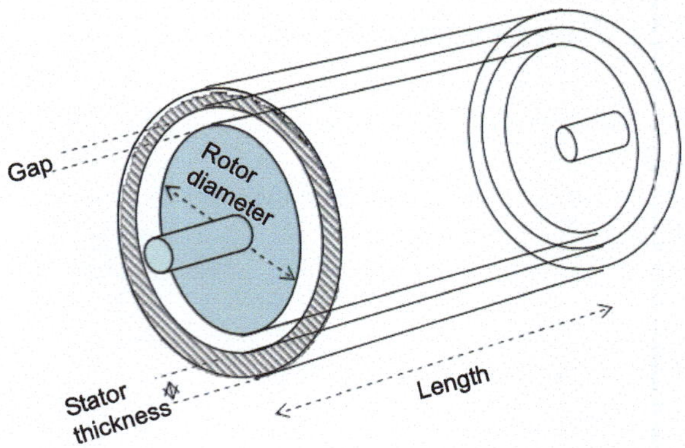

Fig. 2 Conflicting requirements of minimizing iron, rotor diameter, and air gap in design of an electric machine

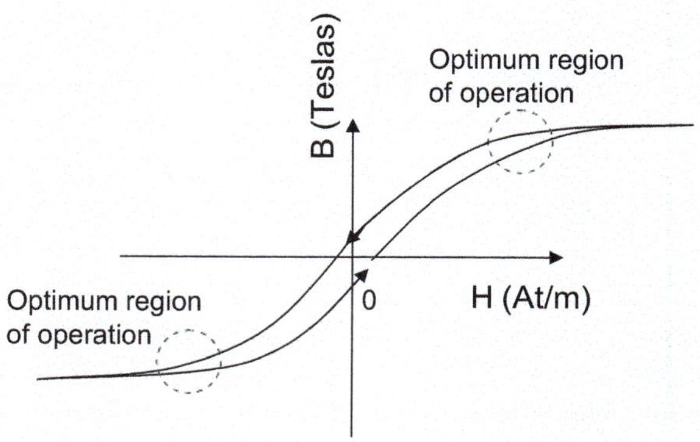

Fig. 3 Operating region in the borderline of linearity optimizes usage of the iron

The design of an electric machine is also a compromise in conflicting requirements, with no single solution. A minimum amount of iron is required in the rotor and stator for supporting the magnetic field. Too much iron means too much weight, volume, and cost.

Too great a diameter means too much centrifugal forces, which can damage generators running at the practical 50 or 60 rotations/second (Fig. 2).

The compromise for both transformers and machines is to operate in the borderline linear section of the B-H curve (Fig. 3).

1.6 Calculations, Simulations, Prototype, Testing, Commercialization

Faced with these conflicting requirements, the steps followed for manufacturing (and Outcome Based Education) are basically (a) calculating on paper, (b) computer simulation, (c) building and testing prototype/s (Fig. 4). Feedback and adjustments in the design may be made during each step. Once a satisfactory prototype has been

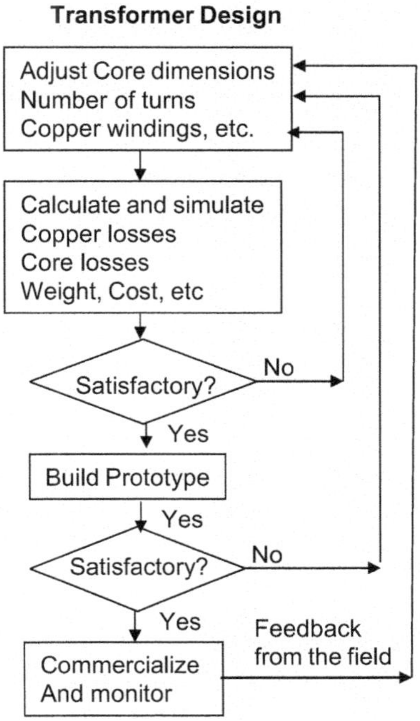

Fig. 4 Iterative and feedback processes for commercial design of transformers and generators

built, it can be commercialized and sent to the market, from where feedback on its performance can be used to make additional changes to the transformer or machine.

2 Procedure

We started with the inductor and single-phase transformer, as they are the simplest, then progressed to increasing complexity of three-phase transformers. We compared our answers to models on the web, and with our proprietary designs of transformers.

For motors, we started with the common three-phase induction motor, moving to the three-phase synchronous motor. Then we attempted Single-phase induction motors, which are of smaller size, but have an added capacitor increasing their complexity.

We only used the free version of ChatGPT, meaning that paid versions may function a little better.

3 Results for Transformer Design

When asked about transformer and machine design, we initially received words of advice and guidance. We got numerical answers, when we persisted in our follow-up questions. The designs by ChatGPT seemed very impractical and did not come close to the proprietary designs available to the author.

3.1 Inductor (single phase)

Single-phase inductors provided a good starting point for our investigation of transformers.

When asked to design an inductor of 1 mH, 1 A, it assumed the core to be ferrite, not the iron used in power transformers. The answers were mostly unrealistic:

$$\text{Ferrite core, No of turns} = 178.51$$
$$\text{Cross - sectional area} = 1.571 \times 10^{-6}\text{m}^2$$
$$\text{Diameter of the coil}(d) := 0.00141 \text{ m}$$

Next, we specified iron-core in our question.

Question: "Design an iron-core inductor of 1 mH, 1 amp."

In response, formulas were scrolled for many minutes, but we could not get numerical formulas. This was a bug that we had encountered and identified.

3.2 Single Phase Transformers

When asked for size of a single-phase transformer, the initial reaction was only words of advice and suggestions, with no numerical answer.

Question: What is the core size of a 1 KVA single phase 11 KV to 415 step-down transformer?

Even when we replaced "core size" with "dimensions" and reduced the KVA rating to 20 VA, and step-down from 220 V to 5 V, the first response was words of advice, without any numerical answers.

We followed up and received the following numerical answer for 1 KVA, single phase 11 KV to 415 V transformer:

Answer:

$$\text{Core} : 100 \text{ mm} \times 100 \text{ mm} \times 50 \text{ mm}$$
$$\text{Primary winding diameter} : 30 \text{ mm}$$

$$\text{Secondary winding diameter} : 90 \text{ mm}$$
$$\text{Insulation thickness} : 5 - 10 \text{ mm}$$

We received the following specifications for a 20 VA single-phase step-down transformer from 220 V to 5 V

$$\text{Core dimensions} : 50 \text{ mm} \times 50 \text{ mm} \times 25 \text{ mm}$$
$$\text{Primary winding diameter} : 15 \text{ mm}$$
$$\text{Secondary winding diameter} : 50 \text{ mm}$$
$$\text{Insulation thickness} : 2 - 5 \text{ mm}$$
$$\text{Number of turns given as} : 44,000 : 2252$$

When we entered that these are far too many turns for the small transformer specified, ChatGPT apologized profusely (!) and recalculated with a 1 cm^2 core area, to a more realistic turns ratio.

$$1000 \text{ turns} : 225 \text{ turns}$$

So ChatGPT was making wrong assumptions on core area = 1 cm^2, instead of finding a value from the web, or calculating an optimum value.

When asked a new question about an unusually rated 0.73 KVA single-phase step-down 11 kV to 415 V transformer, the assumption again was an unrealistic 1 cm^2 core area, and we repeatedly got the wrong turns ratio of 100: 2656 for a step-up transformer.

3.3 *Three-phase Transformers*

Having investigated single-phase inductors and transformers, we moved to three-phase distribution transformers, which are very common on the web.

Question: Find the dimensions of a three-phase 1 KVA distribution transformer, with output of 415 line to line.

Initial answers were words of advice and numerical answers appeared during follow-up questions. In this case, we were asked to choose from two choices (Fig. 5).

Answer:

Choice 1: The Core Size may range from 6 in × 6 in to 8 in by 8 in
Another answer was
10 in to 12 in in height,
8 in to 10 in in width, and 6 in to 8 in in depth.

$$N_{\text{primary}} = 1019, \text{ and } N_{\text{secondary}} = 26,506$$

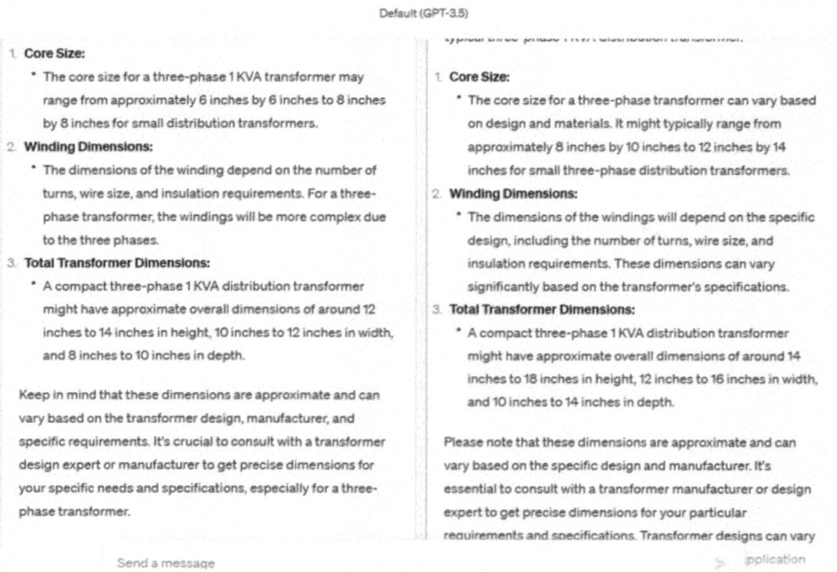

Fig. 5 User is asked to choose between two widely varying ranges of transformer design

The turns ratio was a complete reversal, meaning it was applicable for a step-up transformer and not the given step-down transformer.

4 Results for Machine Design

Questions on machine design resulted initially in words of advice. Follow-up requests were answered with formulas and numbers.

4.1 Three-Phase Generators

We first examine three-phase generators, as they are frequently encountered in the web, and there are a large number of advertised generators.

Question: Find the dimensions of a 1 KVA, 0.8 pf, 415 V generator

ChatGPT made an error at first on recalculating apparent power as 1.25 KVA, which contributed to follow-up errors in the answer. Also, the sizes were too large.

$$\text{Width} = 1.25 \text{ m}$$
$$\text{Height} = 1.25 \text{ m}$$

$$\text{Depth} = 1.25\text{ m}$$
$$\text{Length} = 2.5\text{ m}$$

With a follow-up question on the difference between depth D and length, ChatGPT revised its answers to the following still unrealistic values:

$$\text{Width} = 1.25\text{ m}$$
$$\text{Height} = 1.25\text{ m}$$
$$\text{Depth} = 1.25\text{ m.}$$

When the KVA was increased from 1 KVA to 100 KVA, all the dimensions were increased by a factor of 100 also, which was very unrealistic.

$$\text{Width} : 125\text{ m}$$
$$\text{Height} : 125\text{ m}$$
$$\text{Depth} : 125\text{ m}$$
$$\text{Length} : 25\text{ m}$$

With a follow-up statement that the above were too large, ChatGPT apologized profusely (!) and reduced the answer drastically:

$$\text{Width} : 1\text{ m}$$
$$\text{Height} : 0.5\text{ m}$$
$$\text{Length} : 0.5\text{ m}$$

Clearly we users cannot rely on such hugely varying answers.

4.2 Three Phase Induction motors

When asked to design a three-phase induction motor, initially we received a realistic-looking range of dimensions for length, depth, and height.

Question: Find the dimensions of a 1 KW, 415 V three-phase induction motor

Answer:

$$\text{Length} : 200 - 300\text{ mm}$$
$$\text{Width} : 100 - 200\text{ mm.}$$
$$\text{Height} : 150\text{ to }250\text{ mm.}$$
$$\text{Diameter of stator} = 200\text{ mm}$$
$$\text{Diameter of rotor} = 160\text{ mm}$$

This was somewhat within the commercially manufactured ABB IE3 model 1.1 KW, which had published dimensions of

$$(L \times W \times H)L = 386 \text{ mm} \times W = 195 \text{ mm} \times H = 217 \text{ mm}$$

4.3 Three-Phase Synchronous Motors

When asked for the dimensions of a synchronous motor of the above power and voltage, ChatGPT gave the same dimensions as the induction motor above.

Question: Find the dimensions of a 1 KW three-phase 415 V synchronous motor:

Answer:

$$L = 200 \text{ to } 300 \text{ mm} \times W = 100 \text{ to } 200 \text{ mm} \times H = 150 \text{ to } 250 \text{ mm}.$$

This was well outside the published dimensions of the Terco MV1008 synchronous machine of 465 × 300 × 310 mm

4.4 Single-phase Induction Motors

When asked for dimensions of a Find the dimensions of a 1 KW single phase 220 V induction motor

Answer:

$$\text{Length} = 180 \text{ to } 250 \text{ mm}.$$
$$\text{Width} = 100 \text{ to } 150 \text{ mm}.$$
$$\text{Height} = 120 \text{ to } 200 \text{ mm}.$$

These did not compare well with the dimensions of the commercially available single-phase induction motors.

5 Conclusion

This study investigated transformer and machine design as examples of open-ended problems, characterized by conflicting requirements and no single best solution.

When asked about specifications for transformer and machine design, the outputs were words of advice, formula, and numerical answers, which were usually incomplete, inaccurate, or outright wrong.

ChatGPT did not attempt to perform the iterative steps required for solving open-ended problems. Once it gave a range of solutions, asking us to choose from two sets of data. Or it made usually unrealistic assumptions and values, and used formula to calculate a usually incorrect solution. For example, power transformers were designed with impractical iron cross-section of 1 cm^2. When these errors were identified in follow-up questions, ChatGPT apologized profusely (!), and drastically changed the answers.

We also identified a bug during our design of an inductor, where the response was the same formulas being scrolled repeatedly for some minutes, before it stopped (aborted) without showing any values.

While it is clear that ChatGPT is very good at solving problems which fall into a clear formula, it is not able to balance conflicts, and solve open-ended problems. So users can try to start with the formula and guidance given by ChatGPT, and do the rest of the iterative calculations by themselves.

These indicate assignments and homework of the future in schools and universities will show a shift towards open-ended problems and Complex Engineering Problems.

5.1 Improving ChatGPT for Open-ended Problems

For the software developers of ChatGPT, this paper has identified ways of better improving its capabilities for open-ended problems. ChatGPT may someday be taught how to perform the iterative steps for open-ended problems.

ChatGPT can learn more deeply from the available designs of transformers and machines on the web. A requested design of a transformer of new specifications can be interpolated on the basis of existing designs.

In general, ChatGPT can be taught to learn better from the web, for the purpose of Open-ended problems.

Only the free version of ChatGPT was used in this study. The paid version can be the subject of a future study. As ChatGPT is changing fast, this study will provide a record of its performance at the date of submission of this paper.

References

1. Google AI teaches itself Bangla, The business standard, April 19, 2023
2. ChatGPT for writing literature and songs: end of the road for poets and songwriters? Submitted for publication
3. Deng J, Lin Y (2022) The benefits and challenges of ChatGPT: an overview. Front Comput Intell Syst 2(2):81–83

4. ChatGPT for solving mathematical problems: challenges and some implications for education, submitted for publication
5. Baidoo-Anu D, Ansah LO (2023) Education in the era of generative artificial intelligence (AI): understanding the potential benefits of ChatGPT in promoting teaching and learning. J AI 7(1):52–62
6. King MR (2023) ChatGPT, a conversation on artificial intelligence, chatbots, and plagiarism in higher education. Cell Mol Bioeng 16(1):1–2
7. Oguz FE, Ekersular MN, Sunnetci KM, Alkan A (2023) Can chat GPT be utilized in scientific and undergraduate studies? Ann Biomed Eng 1–3
8. Lo CK (2023) What is the impact of ChatGPT on education? A rapid review of the literature. Educ Sci 13(4):410
9. Fergus S, Botha M, Ostovar M (2023) Evaluating academic answers generated using ChatGPT. J Chem Educ 100(4):1672–1675
10. Wang X, Anwer N, Dai Y, Liu A (2023) ChatGPT for design, manufacturing, and education. Procedia CIRP 119:7–14
11. Akir O, Eng TH, Malie S (2012) Teaching and learning enhancement through outcome-based education structure and technology e-learning support. Procedia Soc Behav Sci 62:87–92
12. Zhu L, Mou W, Yang T, Chen R (2023) ChatGPT can pass the AHA exams: open-ended questions outperform multiple-choice format. Resuscitation 188:109783

Discharging Inductors and Capacitors with Superconductivity, Analyzed with Duality of Charge

Shahriar Khan

Abstract Circuit theory is well established for inductors and capacitors, but becomes problematic in extreme cases of zero resistance, such as in superconductivity. The concept of duality has been extended to power circuits, switching, semiconductors, superconductivity, and other varied fields. However, the concept of duality of charge has rarely been explored and applied. This paper develops the concept of the dual of charge, and uses it in multiple inductances discharging through a resistor. Attempts to calculate individual inductor currents are either tedious, or impossible in the limiting case of zero resistance. Duality concepts are used for discharging inductors, where other methods are either tedious or impossible. Discharging through zero resistance, as in superconductivity, violates the law of conservation of energy, as there is a loss of energy without a resistor to absorb the loss. The law of conservation of momentum during mechanical collisions implies an inexplicable loss of kinetic energy, and is seen to be analogous to the laws of conservation of charge and conservation of flux linkage. The analysis also helps to handle situations of zero resistance, arising from superconductivity.

Keywords Duality · Charge · Inductors · Capacitors · Zero · Infinite · Superconductivity · Momentum

1 Introduction

Duality has been a fundamental tool for analyzing electrical circuits [1, 2]. The dual of voltage is current, the dual of inductance is capacitance, and vice versa. In recent years, the concept of duality has been extended to areas such as semiconductors and superconductivity [3], switching circuits, and in power electronics [4–6].

However, the duals of parameters like charge are less often encountered and explored in the literature. One reason is that the charge ($V \times C$) of a capacitor remains mostly unchanged, whereas the dual flux linkage ($I \times L$) of an inductor

S. Khan (✉)
Independent University, Bangladesh, Dhaka, Bangladesh
e-mail: skhan@iub.edu.bd

P. G. Bradford et al. (eds.), *Proceedings of IEMTRONICS 2024*, Lecture Notes in Electrical Engineering 1229, https://doi.org/10.1007/978-981-97-4780-1_33

usually decreases owing to decreasing current. The flux linkage of a superconducting inductor is a good tool for analysis, as the current continues indefinitely without decreasing.

Circuit theory is well established for inductors and capacitors, but becomes problematic in extreme cases of zero resistance, such as in superconductivity. In these cases, the laws of physics break down, as there is a loss of energy, without any resistance to absorb the loss. The methods in this paper have been extended for limiting cases of infinite and zero resistance, as for superconductivity.

This paper develops the concept of duality for discharging inductors and capacitors, using methods which have seldom been used before. Cases of infinite and zero resistance, such as for superconductivity, have been included.

2 Two Discharging Inductors

We first consider two or more inductances discharging with a resistance (Fig. 1).

Initially, the inductor currents are bypassed through the switch. When the switch is opened, the inductor currents must pass through the resistor. The resistor currents and voltages are easily calculated as

$$i_R(t) = (I_1 - I_2)e^{-t/\tau}$$

$$v_R(t) = R(I_1 - I_2)e^{-t/\tau}$$

Here τ is calculated from the Thevenin's equivalent inductance across the resistor terminals.

$$L_{\text{eq}} = \frac{L_1 L_2}{L_1 + L_2}$$

From elementary theory, $\tau = LAB/R$. Hence

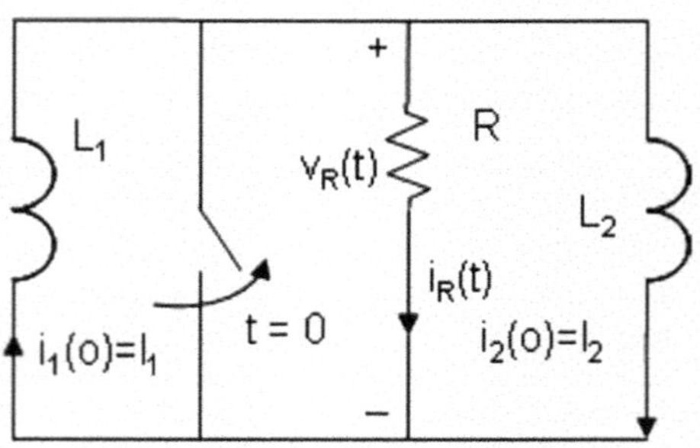

Fig. 1 Inductor currents decaying through resistor upon opening of switch

$$\tau = \frac{L_1 L_2}{(L_1 + L_2)R}$$

2.1 Final Circulating Current in Inductor

The above are relatively straightforward, but the current in the individual inductors, and final current *If* are somewhat more tedious to calculate. Without using duality concepts, these currents have been calculated by equating the voltage across the inductors and the resistor.

$$L_1 \frac{di_1}{dt} = L_2 \frac{di_2}{dt} = R(i_1 + i_2) = R(I_1 - I_2)e^{-t/\tau}$$

Hence

$$\frac{di_1}{dt} = \frac{1}{L_1}(I_1 - I_2)e^{-t/\tau}$$

Integrating, we get

$$i_1(t) = -\frac{L_2}{(L_1 + L_2)R}(I_1 - I_2)e^{-t/\tau} + K$$

Using the boundary condition, $i_1(0) = I_1$, the constant of integration K, is found to be

$$K = \frac{I_1 L_1 + I_2 L_2}{L_1 + L_2}$$

Hence, current $i_1(t)$ is found to be

$$i_1(t) = -\frac{L_2(I_1 - I_2)}{L_1 + L_2} e^{-\tau/\tau} + \frac{I_1 L_1 + I_2 L_2}{L_1 + L_2}$$

The final circulating current *If*, can be found by putting $t = \infty$

$$I_f = \frac{I_1 L_1 + I_2 L_2}{L_1 + L_2}$$

The above process for finding final circulating current is understandably quite tedious. Duality concepts have not been used so far (Fig. 2).

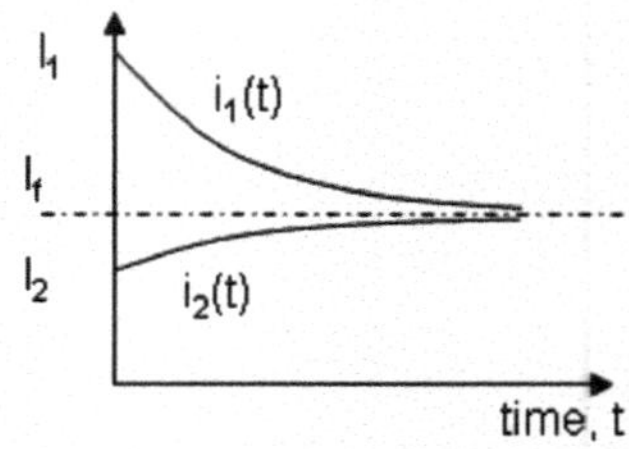

Fig. 2 Inductor currents $i_1(t)$ and $i_2(t)$ discharging to final current I_f

2.2 *Case When* R = ∞

The above equations break down, when $R = \infty$, as there is a sudden loss of energy, with no energy-absorbing resistor (Fig. 3).

Putting $R = \infty$, results in a $0 \times \infty$ situation, with no easy resolution of this indefinite value. Differential equations are no longer applicable because of the unresolved discontinuity in the stored energy.

We illustrate by putting $L1 = 1, L2 = 3, I1 = 2, I2 = 4$, we find

$$\text{Initial energy} = W_{\text{initial}} = 0.5 \times 1 \times 2^2 + 0.5 \times 3 \times 4^2 = 26\ \text{J}$$

$$I_f = 13/4 = 3.25\ \text{amps}$$

$$W_{\text{final}} = 0.5L \times I_f^2 = 0.5 \times 4 \times 3.25^2 = 21.125\ \text{J}$$

The law of conservation of energy is violated, as there is no resistance to absorb the change in energy from 26 to 21.125 J.

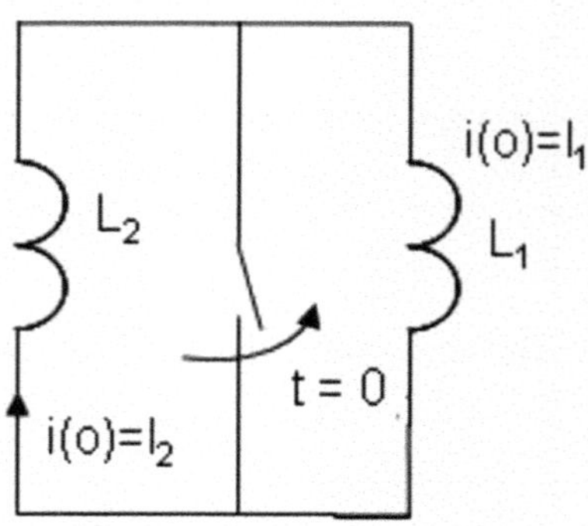

Fig. 3 Limiting case with no resistance, such as in superconductivity

3 Two Discharging Capacitors

The final circulating current can be calculated more simply, if we use the concept of dual of charge, such as from the well-known case of two capacitors, discharging through a resistor (Fig. 4).

This circuit is within the scope of an undergraduate course on electrical engineering. It is simpler to analyze than the discharging inductors, because here we can use the well-known law of conservation of charge Q (Fig. 5).

As in the case of discharging inductors,

$$v_R(t) = (V_1 - V_2)e^{-t/\tau}$$
$$i_R(t) = \frac{V_1 - V_2}{R}e^{-t/\tau}$$

Here τ is calculated from Thevenin's equivalent inductance across the resistor terminals.

$$C_{\text{eq}} = \frac{C_1 C_2}{C_1 + C_2}$$

From elementary theory, $\tau = R\, C_{\text{AB}}$, giving:

$$\tau = \frac{R C_1 C_2}{C_1 + C_2}$$

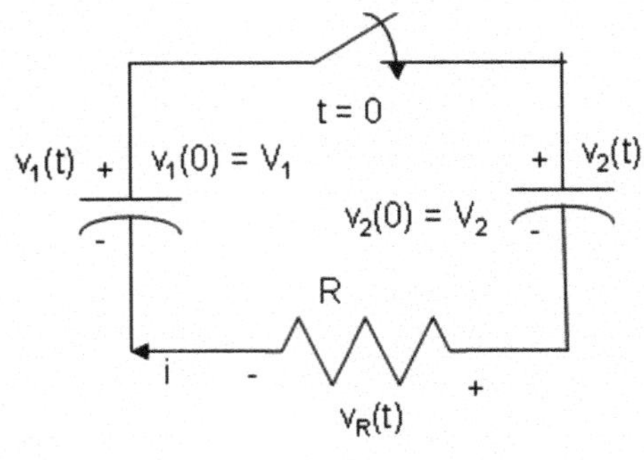

Fig. 4 Two capacitors discharging through resistor R

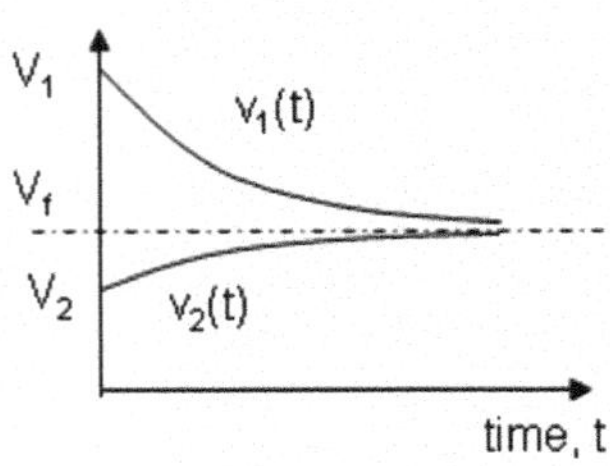

Fig. 5 Capacitor voltages $v_1(t)$ and $v_2(t)$ discharging to final current V_f

3.1 Using Law of Conservation of Charge

The Law of conservation of charge tells us that the charge will be redistributed so that both capacitors come to the same final potential Vf. Using the law of conservation of charge, at the final point, the combined capacitance can be taken as $C_1 + C_2$ holding charge below

$$(C_1 + C_2)V_f = V_1C_1 + V_2C_2.$$

This gives

$$V_f = \frac{V_1C_1 + V_2C_2}{C_1 + C_2}$$

We notice that the final voltage is independent of the value of R. The energy dissipated in the resistor is independent of the value of R. The above equation is applicable even when $R = 0$, or the law of conservation of energy is not applicable.

4 Applying Dual of Charge to Inductors

This paper develops the concept that the dual of charge $Q = V.C$ should be $I.L$, also known as flux linkage. This paper further postulates that the combined $I.L$ of two inductors should be conserved after the switch is opened. Hence:

$$(L_1 + L_2)I_f = I_1L_1 + I_2L_2.$$

$$I_f = \frac{I_1L_1 + I_2L_2}{L_1 + L_2}$$

The above agrees with the more tedious derivation shown earlier. This also shows us that the final current is independent of the value of R, and that the energy dissipated in the resistor is independent of the value of R.

5 Mechanical Analogy: Conservation of Momentum

The discharging inductors and capacitors discussed earlier can be compared to colliding mechanical bodies (Fig. 6). The law of conservation of charge can be compared to the law of conservation of momentum in the mechanical domain. Both of these can be compared to the law of conservation of flux linkage.

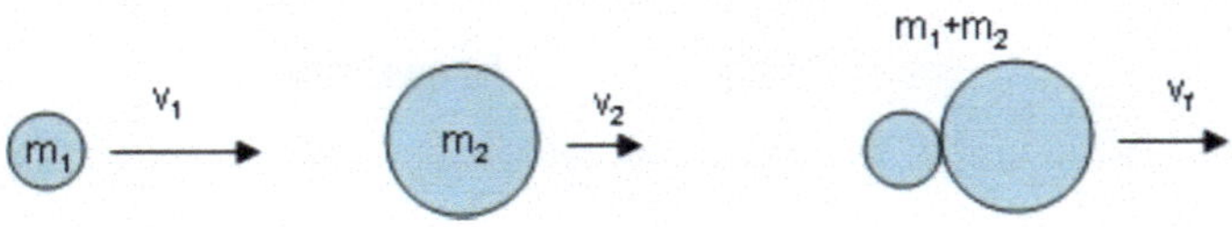

Fig. 6 Mechanical analogy of two colliding bodies moving away with common velocity

In figure above, two bodies of masses M_1 and M_2, moving with initial velocities $V1$ and $V2$ collide and stick and move away with a common velocity of Vf. Although kinetic energy is lost in the process, momentum is conserved.

$$M_1 V_1 + M_2 V_2 = V_f (M_1 + M_2)$$

$$V_f = \frac{M_1 V_1 + M_2 V_2}{M_1 + M_2}$$

The similarity of the above with inductors and capacitors is readily visible. Although momentum is conserved, the law of conservation of energy is violated.

6 Generalized Case of Multiple Capacitors

The above reasoning is now continued with multiple capacitors, $C1$, $C2$, … CN discharging through a resistor R (Figs. 5 and 7).

Here Thevenin's equivalent capacitance across the resistor is

$$C_{\text{eq}} = \left[\frac{1}{C_1} + \frac{1}{C_2} \cdots \frac{1}{C_N}\right]^{-1}$$

Hence time constant $\tau = RC_{\text{eq}}$. The current through the resistor equals

$$i(t) = \frac{V_1 + V_2 + \ldots V_N}{R} e^{-t/\tau}$$

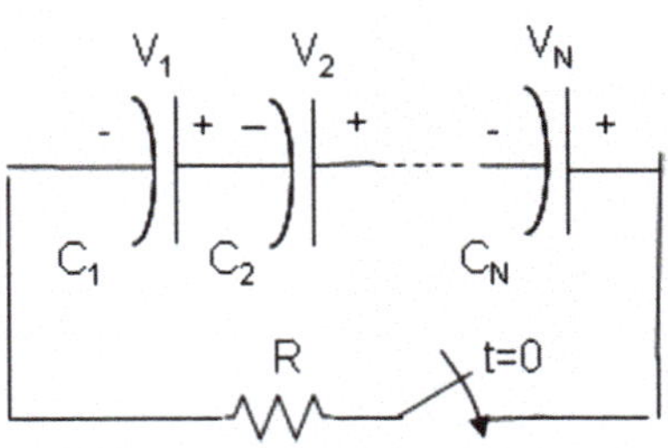

Fig. 7 Generalized case of N capacitors discharging through a resistor R

Integrating $i(t)$ from 0 to ∞ gives the total charge ΔQ that flows through the resistor after the switch is closed.

$$\Delta Q = (V_1 + V_2 + \ldots V_N)\left[\frac{1}{C_1} + \frac{1}{C_2} \cdots \frac{1}{C_N}\right]^{-1}$$

Hence the final voltage of the nth capacitor Vnf is

$$V_{nf} = \frac{C_n V_n - \Delta Q}{C_n}$$

It is the concept of conservation of charge, which allows the simple calculations above.

7 Generalized Case of Multiple Inductors

The generalized case of multiple capacitors can be now extended to the case of multiple inductors (Fig. 8)

It is readily apparent that the voltage across the resistor is

$$v(t) = (I_1 + I_2 + \ldots I_N)Re^{-t/\tau}$$

Here,

$$\tau = \frac{1}{R}\left[\frac{1}{L_1} + \frac{1}{L_2} \cdots \frac{1}{L_N}\right]^{-1}$$

From the generalized case of multiple capacitors (Sect. 5), we can infer that the change in the total $L_{eq}I_{total}$, or $\Delta\lambda$ is

$$\Delta\lambda = (I_1 + I_2 + \ldots I_N)\left[\frac{1}{L_1} + \frac{1}{L_2} \cdots \frac{1}{L_N}\right]^{-1}$$

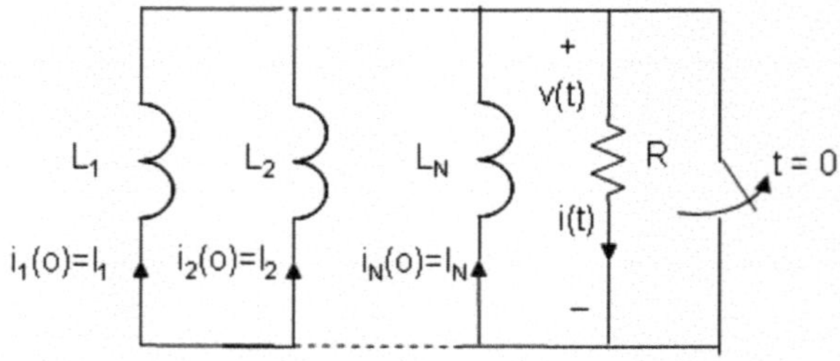

Fig. 8 Generalized case of multiple inductors discharging through resistor *R*

From the capacitance analogy shown above, the final current *Inf* in the nth inductor can be found as

$$I_{nf} = \frac{L_n I_n - \Delta\lambda}{L_n}$$

The calculation of final circulating current in each inductor is hence greatly simplified by analyzing it as the dual of multiple capacitors. We notice that the circulating currents are independent of the value of the resistor, as predicted by the case of capacitors. The energy lost in the resistor is independent of the value of the resistor. Unlike other methods, this analysis is applicable in the limiting case of $R = \infty$.

8 Law of Conservation of Inductance-Current Product (Flux Linkage)

The Law of Conservation of Charge is well-recognized, and it also forms the basis for Kirchoff's current law. According to the duality concepts developed in this paper, current-inductance product (flux linkage) should arguably be conserved. Unlike the conservation of charge, current-inductance product is normally not conserved. The simple case of current decaying in an *R–L* circuit has flux linkage being lost, or being converted to heat energy. Also, conservation of flux linkage may be more applicable in cases of linearity, such as for air-core inductors, rather than for iron-core inductors.

Another analogy for the laws of conservation of charge and conservation of flux linkage may be the Law of conservation of momentum, applied to the collision of two bodies. During the mechanical collisions, both the conservation of momentum and conservation of energy are followed. Similarly, the law of conservation of energy must be followed in addition to the laws of conservation of charge and conservation of flux linkage, for cases shown in this paper.

9 Conclusion

Although the duality of current, voltage, inductance, and capacitance are well understood, the duality of charge ($C.V$) has seldom been explored and developed. In this paper the dual of charge, which is $L.I$, or flux linkage, is shown to be useful in giving insight into multiple inductance behavior. A clear application of the duality of charge is the case of discharging inductances, which is analyzed in this paper as the dual of discharging capacitors. From the duality concept, it is apparent that the final circulating currents are independent of the value of the resistor R. The law of conservation of inductance-current product still holds, and can be used to find the circulating currents in each inductor.

Duality concepts are applicable in the limiting case of $R \rightarrow \infty$ or 0, where energy is lost without a resistance, as is the case for superconductivity. In the case of mechanical collision of two bodies, the law of conservation of momentum may be seen as analogous to the laws of conservation of charge, and conservation of flux linkage.

References

1. Freeland SD (1992) Techniques for the practical application of duality to power circuits. IEEE Trans Power Electron 7:374–384
2. Khan S (2022) Electrical circuits In: Khan S. Dhaka, Bangladesh
3. Davidson A, Beasley MR (1979) Duality between superconducting and semiconducting electronics. IEEE J Solid-State Circuits 14(4):758–762
4. Wolfs PJ (1993) A current-sourced DC-DC converter derived via the duality principle from the half-bridge converter. IEEE Trans Ind Electron 40(1):139–144
5. Khan S (2018) Semiconductor devices and technology. In: Khan S. Dhaka, Bangladesh
6. Cheng KWE, Ye YM (2015) Duality approach to the study of switched-inductor power converters and its higher-order variations. IET Power Electron 8(4):489–496

Infinite and Zero May Have Similar Applicability for Engineering Analysis

Shahriar Khan

Abstract Zero "0" is widely used in calculations and has great acceptability, whereas infinite "∞" remains vague and is seldom used. But zero is questionable too, because an absolute zero cannot be attained for temperature, vacuum, and even resistance. The solution of general relativity for a black hole gives absolute zero of volume and infinite density. The number zero taught in primary school, implies an absolute zero, which is not encountered in physics. So an alternative system of numbers is proposed here, where something infinitely large, such as the universe is represented by alternative-number zero, and an infinitesimally small quantity (temperature) is represented by infinite. The coldest temperature attained till today would be 2.06 $\times$ 10^{10} units (1/T) in this alternate infinite-based numbering system. Human civilization has a "bias" towards assigning increasing numbers to increasing quantities. The alternative to the bias of human civilization is assigning increasing numbers to decreasing quantities. It is proposed here how infinite and zero may be two sides of the same coin, and may have the same usefulness in engineering calculations. The simple cases of discharging inductors and capacitors without resistances produce inexplicable changes in energy, and zeroes and infinite, which must be solved by incorporating resistances. Colliding bodies which attach and conserve momentum produce similar inexplicable changes in energy, involving zeroes and infinite.

Keywords Infinite · Absolute · Zero · Engineering · Physics · Temperature · Superconductivity · Infinitesimal · Resistance

1 Introduction

The analysis of zero and infinite should have some significance in physics and/or engineering. It is not sufficient to just have numbers which have no meaning in engineering or physics. Zero "0" is widely used in calculations and has great acceptability, whereas infinite "∞" is almost never used for calculations.

S. Khan (✉)
Department of EEE, Independent University, Bangladesh, Dhaka, Bangladesh
e-mail: skhan@iub.edu.bd

P. G. Bradford et al. (eds.), *Proceedings of IEMTRONICS 2024*, Lecture Notes in Electrical Engineering 1229, https://doi.org/10.1007/978-981-97-4780-1_34

The limitless nature of infinite is widely understood. However, the uncertainty and limitless nature of zero is seldom thought of, and as a result, we do not associate zero with uncertainty and impracticality. Zero is usually taken to be exactly zero, and seldom associated with infinitesimally small.

1.1 Infinite Quantities

The first application of infinite is the size of the universe. For example, we do not know whether the universe is infinite or not [1].

Another infinite density of a black hole (with zero volume), as derived from General relativity.

Other examples include infinite voltage across an inductor when trying to change current in zero time. Similarly, the current through capacitors become zero when discharging in zero time. Another is the infinite force on bodies colliding and attaching while saving momentum. All these problems have been approached in later sections of this paper.

1.2 Procedure

We investigate the role of infinite and zero in various engineering problems. From there, we try to come to conclusions about the similar nature of zero and infinite.

1.3 Literature Review

The nature of infinite has been investigated for long. Well-known is the investigation of numbers by the famous Russell [2] and others [3]. The author has investigated infinite situations arising from discharging capacitors and inductors in zero time [4].

2 Absolute Zero of Temperature, Pressure, and Resistance

We are familiar with the unattainability of the absolute zero of temperature, as stated in the third law of thermodynamics. The absolute zero of temperature can only be approached, but never attained. This raises questions about the zero we state for similar physical quantities.

We often speak of a high-vacuum, but it is intuitive and self-evident that an absolute vacuum cannot be attained. We can visualize the few hydrogen atoms in 1 m^3 of outer space, but could there be subatomic particles remaining? Clearly, we

Table 1 Zero and infinite in physics and engineering

	Zero	Infinite
Applicability and acceptability	Zero enjoys many applications and acceptability in analysis from the simplest school calculations to advanced physics and engineering calculations	Infinite is controversial and seldom used even in the simplest engineering analysis. Requires modification of the theory to replace the infinite
Temperature	Absolute zero of temperature cannot be attained	
Black hole	Relativity says the volume of a black hole is zero, which is thought to be a problem	Relativity says the density of a black hole is infinite, which must be overcome using the quantum theory
Vacuum	It is intuitive that a zero vacuum cannot be attained	
Universe		It is not known whether the universe is finite or infinite. This would be zero in the alternative numbering system
Quantum theory	Quantum theory looks deep into matter, but we cannot look into infinitesimally tiny space and matter	

cannot remove all infinitesimally small yet—undiscovered particles in the space, supporting the unattainability of an absolute vacuum.

The quantum theory investigates matter and energy at the nanoscale [5], but what if we continue to investigate say at the pico scale, or much deeper? Clearly, quantum mechanics cannot investigate and explain at an infinitesimally small scale.

We summarize our applications of zero and infinite below (Table 1).

2.1 Absolute Zero of Resistance

If absolute zero of temperature and vacuum cannot be attained, how would it be possible to have an absolute zero of resistance, as implied by superconductivity?

Zero resistance for superconductivity is well-known, which raises the question of whether the resistance is actually zero, or extremely small. According to the literature, once a current was started in a superconducting coil, there was no measurable change in resistance after about 3 years [6]. A current may be started in a superconducting coil by moving a pre-existing magnet after the superconductivity has been initiated (Fig. 1).

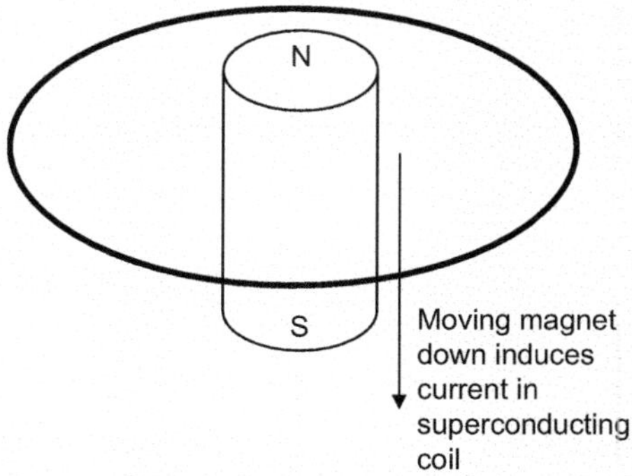

Fig. 1 A current may be started in a superconducting coil by moving a pre-existing magnet in the coil

Absolute zero of resistance is counter-intuitive. Would there be no measurable loss in resistance in the above loop after hundreds of years? Millions of years? Billions of years? Clearly, there will be a point in time, when the current will start decaying measurably. Values for resistance in superconductive coils have been proposed [7].

We demonstrate how to calculate the resistance in the superconducting coil, with a 1% measurable decay of current after three years.

$$3 \text{ years} = 3 \times 60 \text{ sec} \times 60 \text{ min} \times 24 \text{ hours} \times 365 \text{ days} = 10^8 \text{sec}$$

The decay of current in an inductive coil with resistance is given by the following formula:

$$i(t) = I_0 \exp(-L/R)t$$

In case of $L = 1$ H, with 1% perceptible drop in the current, the resistance can be calculated:

$$\begin{aligned}
0.99\, I_0 &= I_0 \exp(-R)t \\
0.99 &= \exp\,(-R)10^8 \\
\text{In } 0.99 &= -10^8 \times R \\
0.01 &= 10^8 \times R \\
R &= 10^{-10}\text{ohm} = 100 \text{ pico - ohm} \\
1/R &= 10^{10}\text{alternative units}
\end{aligned}$$

3 Alternative Infinite-Based Numbering System

In primary school, we are taught about numbers (integers), such as

$$0, 1, 2, 3, \ldots$$

This has promoted from an early age that the zero is an absolute zero, which is about impossible in physics. If we show an apple, and then remove it (integer 1 to 0), does it mean that there is no remaining atomic or subatomic particle of an apple left?

If we show a person and then remove him/her (integer 1 to 0), does that mean that there is no atomic or subatomic particle of the person left behind?

Clearly, the concept of integers detracts from the reality that a physical quantity can be extremely small, but not absolutely zero.

3.1 *Alternative Infinite-Based Numbering System*

If our conventional numbers cannot adequately deal with the impossibility of an absolute zero, how can we have an alternative system of numbers based on infinite?

Since the number zero implies infinitesimally small, we can attempt a new system of numbers where zero would imply infinitely large, and infinite would imply zero, or infinitesimally small. This numbering system can better deal with the quantities which are infinitely large (Table 2).

Table 2 Comparison of conventional and Infinite-based numbering systems

Numbers	Conventional meaning	Meaning according to our alternative infinite-based integers
0	A quantity infinitesimally small, which is an impossibility for temperature, pressure, electrical resistance, etc.	Infinitely large, such as what the universe may be
1	A finite quantity, such as resistance, temperature, and pressure	A finite quantity, such as resistance, temperature, and pressure
2	A finite quantity, such as resistance, temperature or pressure	A finite quantity, such as resistance, temperature, or pressure
10^{10}		Possible resistance of a superconducting coil in $1/R$, as calculated earlier in this paper
2.06×10^{10}		The lowest temperature obtained in 2018, with temperature unit expressed as $1/T$ (K)
1.5×10^{14}		The highest vacuum ever created in terms of $1/P$ (in pascal)
∞	A quantity infinitely large, such as the infinitely large space, infinite density of a black hole	A quantity infinitely small, such as infinitesimally small temperature, pressure, or electrical resistance

The lowest temperature that matter has been cooled to is 38 pico Kelvin (−273.149999999962 C), just 38 trillionths of a degree above absolute zero. (by QUANTUS Team in Germany, in August 2018). In the alternative infinite-based numbering system, this would be the

$$1/T = 1/\left(38 \times 10^{-12}\right) = 2.0636 \times 10^{10}\text{units}$$

in our alternative infinite-based numbering system.

4 Discharging Inductors with Zero Resistance

The simple case of interaction of two or more inductors without any resistance leads to violation of the law of conservation of energy, and a zero and infinite situation (Fig. 2).

We can try to resolve the conflict in physics by adding a resistance in parallel (Fig. 3).

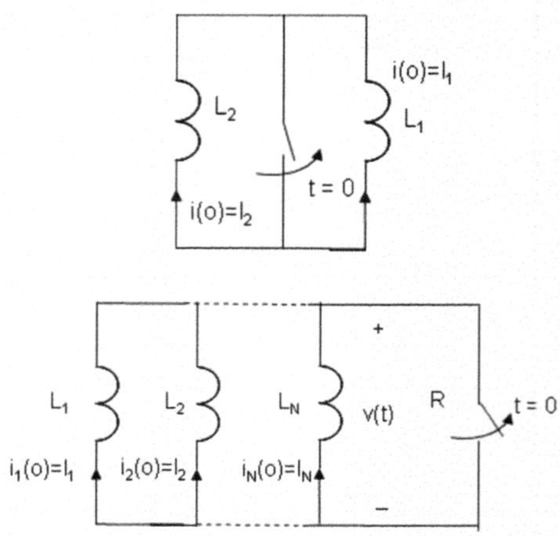

Fig. 2 Inductors discharging without a resistor, when most equations break down because of the discontinuity of energy

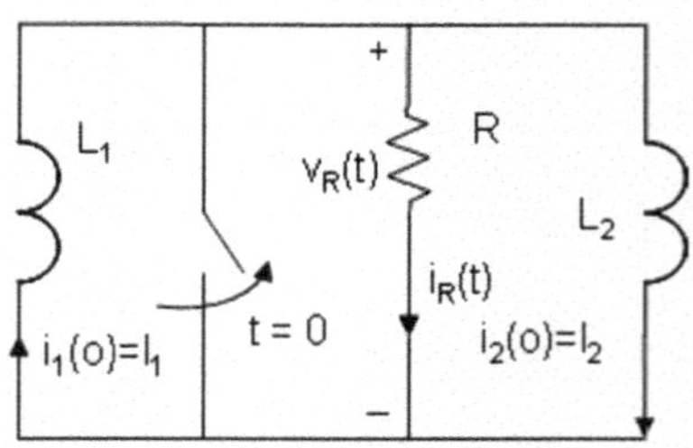

Fig. 3 We try to overcome the zero-infinite conflict by incorporating a resistance

The adding of a resistance is distantly similar to overcoming the zero volume, infinite density singularity of a black hole by incorporating the quantum theory for a black hole.

The resistor currents and voltages are easily calculated as

$$i_R(t) = (I_1 - I_2)e^{-t/\tau}$$

$$v_R(t) = R(I_1 - I_2)e^{-t/\tau}$$

Here τ is calculated from the Thevenin's equivalent inductance across the resistor terminals.

$$L_{eq} = \frac{L_1 + L_2}{L_1 + L_2}$$

From elementary theory, $\tau = L_{eq}/R$. Hence

$$\tau = \frac{L_1 L_2}{(L_1 + L_2)R}$$

The currents have been calculated by equating the voltage across the inductors and the resistor.

$$L_1 \frac{di_1}{dt} = L_2 \frac{di_2}{dt} = R(i_1 + i_2) = R(I_1 - I_2)e^{-t/\tau}$$

Hence,

$$\frac{di_1}{dt}\frac{1}{L_1}(I_1 - I_2)e^{-t/\tau}$$

Integrating, we get

$$i_1(t) = -\frac{L_2}{(L_1 + L_2)R}(I_1 - I_2)e^{-t/\tau} + K$$

Using the boundary condition, $i_1(0) = I_1$, the constant of integration K, is found to be

$$K\frac{I_1 L_1 + I_2 L_2}{L_1 + L_2}$$

$i_2(t)$ can be found just as above, and has been shown with $i_1(t)$ in figure below. The final circulating current If, can be found by putting $t = \infty$ (Fig. 4)

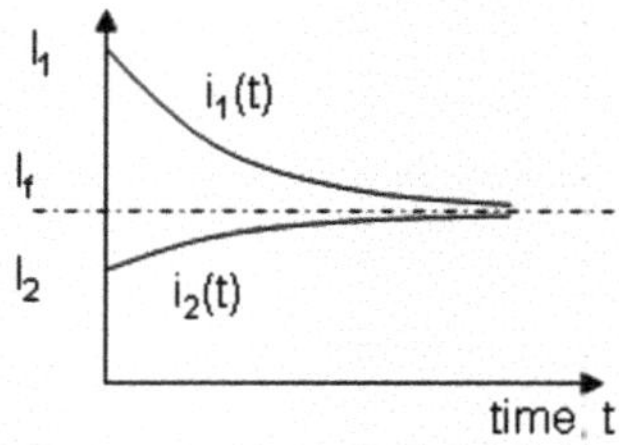

Fig. 4 Inductor currents $i_1(t)$ and $i_2(t)$ discharging to final current I_f

$$I_f = \frac{I_1 L_1 + I_2 L_2}{L_1 + L_2}$$

4.1 *Special case with no resistor, or when R = ∞*

The above equations break down in the absence of a resistor, or when $R = \infty$ (Fig. 2). This is because of the sudden loss in stored energy before and after the opening of the switch, with no resistor to absorb the energy. Putting $R = \infty$, results in a $0 \times \infty$ situation, with no easy resolution of this undefined value. Differential equations are no longer applicable because of the unresolved discontinuity in the stored energy.

We now illustrate the above with a numerical example.

Although charge is conserved, the law of conservation of energy is violated. To illustrate inexplicable loss of energy, we consider

$$L_1 = 3, L_2 = 2, I_1 = 4, I_2 = 1$$

Before the closing of the switch, total energy, considering for inductor $W = 0.5 \times L \times I^2$

$$W_{\text{initial}} = 0.5 \times 3 \times 4^2 + 0.5 \times 2 \times 1^2 = 25 \text{ J}$$

This gives

$$I_{\text{final}} = 14/5 = 2.8, \text{ with energy } W_{\text{final}} = 0.5 \times 5 \times 2.8^2 = 19.6 \text{ J}$$

There is no explanation for the 25–19.6 = 5.4 J loss in energy.

We can only explain the loss by considering a finite resistance which absorbs the energy.

4.2 *Generalized Case of Multiple Inductors*

The generalized case of multiple capacitors can be now extended to the case of multiple inductors (Fig. 5).

It is readily apparent that the voltage across the resistor is

$$v(t) = (I_1 + I_2 + \ldots I_N)R\,e^{-t/\tau}$$

Here,

$$\tau = \frac{1}{R}\left[\frac{1}{L_1} + \frac{1}{L_2} \cdots \frac{1}{L_N}\right]^{-1}$$

From the generalized case of multiple capacitors, we can infer that the change in the total $L_{\text{eq}}I_{\text{total}}$, or $\Delta\lambda$ is

$$\Delta\lambda = (I_1 + I_2 + \ldots I_N)\left[\frac{1}{L_1} + \frac{1}{L_1} \cdots \frac{1}{L_N}\right]^{-1}$$

From the capacitance analogy shown above, the final current I_{nf} in the nth inductor can be found as

$$I_{nf} = \frac{L_n I_n - \Delta\lambda}{L_n}$$

We notice that the circulating currents are independent of the value of the resistor, as predicted by the case of capacitors. The energy lost in the resistor is independent of the value of the resistor. Unlike other methods, this analysis is applicable in the limiting case of $R = \infty$.

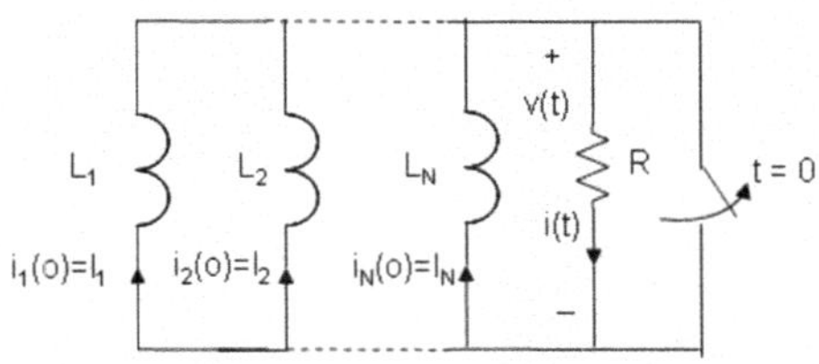

Fig. 5 Generalized case of multiple inductors discharging through resistor R

5 Capacitors Discharging Through Resistance

We now look at discharging capacitors, which is the dual of discharging inductors. In the absence of a resistor, there is no place for loss of energy, and the laws of physics break down. This conflict in physics can be overcome by the addition of resistors as shown below.

As in the case of discharging inductors,

$$v_R(t) = (V_1 - V_2)e^{-t/\tau}$$

$$i_R(t) = \frac{V_1 - V_2}{R} e^{-t/\tau}$$

Here τ is calculated from Thevenin's equivalent inductance across the resistor terminals.

$$C_{\text{eq}} = \frac{C_1 C_2}{C_1 + C_2}$$

From elementary theory, $\tau = RC_{\text{AB}}$, giving:

$$\tau = \frac{RC_1 C_2}{C_1 + C_2}$$

The equations above are equivalent to the case for discharging inductors.

From the above final voltage, Vf, the decay of the individual capacitor voltages $v_1(t)$ and $v_2(t)$ can be calculated (Fig. 6).

$$v_1(t) = \left(V_1 - V_f\right)e^{-t/\tau} + V_f$$
$$v_2(t) = \left(V_2 - V_f\right)e^{-t/\tau} + V_f$$

As for the discharging inductors, the final voltage is (Fig. 7)

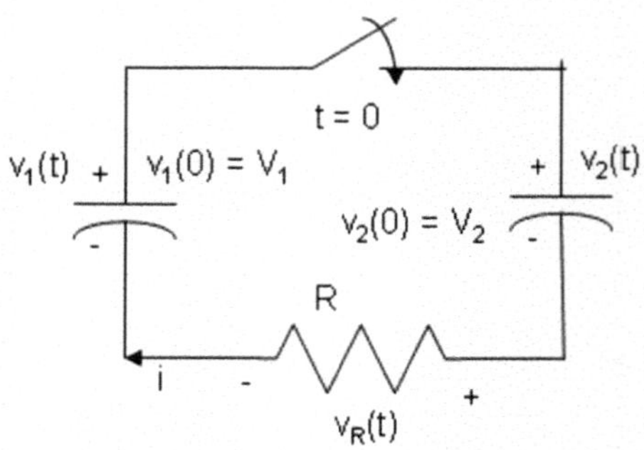

Fig. 6 Discharging resistors must have a resistor, so that there is a place for the absorption of energy

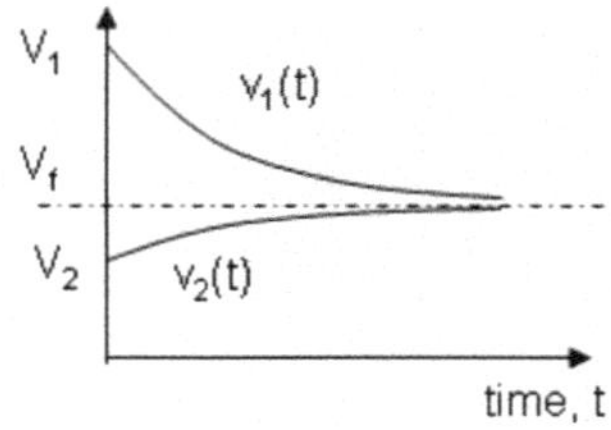

Fig. 7 Capacitor voltages $v_1(t)$ and $v_2(t)$ discharging to final value V_f

$$V_{\text{final}} = \frac{V_1 C_1 + V_2 C_2}{C_1 + C_2}$$

5.1 *Zero Resistance, as in Superconductivity*

We now consider the case of zero resistance, as in superconductivity.

Although charge is conserved, the law of conservation of energy is violated. To illustrate inexplicable loss of energy, we consider

$$C_1 = 3, C_2 = 2, V_1 = 4, V_2 = 1$$

Before the closing of the switch, total energy

$$W_{\text{initial}} = 0.5 \times 3 \times 4^2 + 0.5 \times 2 \times 1^2 = 25 \text{ J}$$

This gives

$$V_{\text{final}} = 14/5 = 2.8, \text{ with energy}$$
$$W_{\text{final}} = 0.5 \times 5 \times 2.8^2 = 19.6 \text{ J}$$

There is no explanation for the 25–19.6 = 5.4 J loss in energy.

We can only explain the loss by considering non-zero resistance which absorbs the energy.

5.2 *Multiple Discharging Capacitors*

The above reasoning is now continued with multiple capacitors, $C1$, $C2$, …CN discharging through a resistor R (Fig. 8).

Here Thevenin's equivalent capacitance across the resistor is

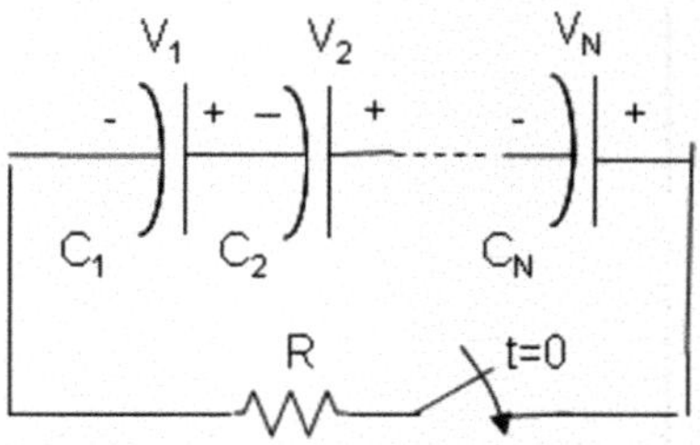

Fig. 8 Generalized case of N capacitors discharging through a resistor R

$$C_{\mathrm{eq}} = \left[\frac{1}{C_1} + \frac{1}{C_2} \cdots \frac{1}{C_N}\right]^{-1}$$

The time constant $\tau = RC_{\mathrm{eq}}$. The current through the resistor equals

$$i(t)\frac{V_1 + V_2 + \ldots V_N}{R}e^{-t/\tau}$$

Integrating $i(t)$ from 0 to ∞ gives the total charge ΔQ that flows through the resistor after the switch is closed.

$$\Delta Q = (V_1 + V_2 + \ldots V_N)\left[\frac{1}{C_1} + \frac{1}{C_2} \cdots \frac{1}{C_N}\right]^{-1}$$

Hence the final voltage of the nth capacitor V_{nf} is

$$V_{nf} = \frac{C_n V_n - \Delta Q}{C_n}$$

6 Mechanical Analogy: Conservation of Momentum

We now consider the school-level problem of two bodies of masses m_1 and m_2, moving with initial velocities v_1 and v_2 collide and stick and move away with a common velocity of v_f (Fig. 9).

$$M_1 V_1 + M_2 V_2 = V_f (M_1 + M_2)$$

$$V_{\mathrm{final}} = \frac{M_1 V_1 + M_2 V_2}{M_1 + M_2}$$

The similarity of the above with inductors and capacitors is readily visible.

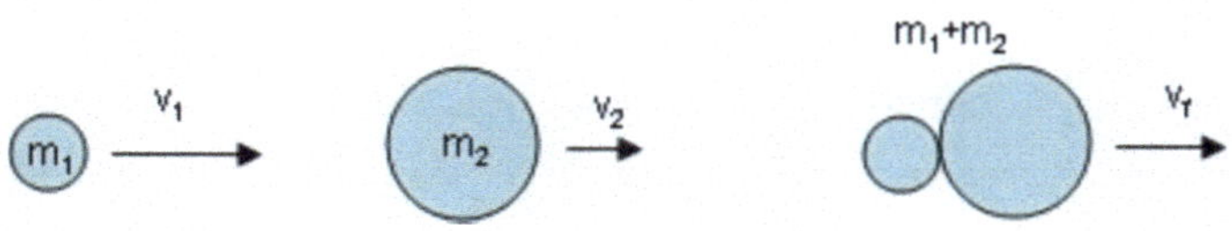

Fig. 9 Two colliding bodies moving away together with common velocity

6.1 *Violation of Conservation of Energy*

Although momentum is conserved, the law of conservation of energy is violated. To illustrate inexplicable loss of energy, we consider

$$M_1 = 3, M_2 = 2, V_1 = 4, V_2 = 1$$

Before the collision, total energy

$$W_{\text{initial}} = 0.5 \times 3 \times 4^2 + 0.5 \times 2 \times 1^2 = 25 \text{ J}$$

This gives

$$V_{\text{final}} = 14/5 = 2.8, \text{ with energy}$$
$$W_{\text{final}} = 0.5 \times 5 \times 2.8^2 = 19.6 \text{ J}$$

There is no explanation for the 25 – 19.6 = 5.4 J loss in energy.

We can only explain the loss by considering balls which absorb energy by deformation, which is analogous to electrical resistance.

7 Conclusion

Human civilization is seen to have a “bias” towards assigning increasing numbers to increasing quantities. But this does not address very small quantities like low temperatures, pressures, and resistances, and very large quantities like the vastness of space and the density of a black hole. The solution proposed here is to have an alternative decreasing numbering system where the infinite of space is represented by zero, and an infinitesimally low temperature, pressure, or resistance is represented by infinite.

In the case of inductors and capacitors discharging through zero resistance, the laws of physics break down because of the sudden loss of energy. This is not very far from the zero-volume, infinite-density case for a black hole. We try to solve this problem by including a resistor (like incorporation of quantum theory with relativity for black hole). We then try to find some solutions in the limiting case of $R \to \infty$ or zero, where differential equations are no longer applicable.

It is discussed here how Infinite and Zero may be two sides of the same coin, and may have the same usefulness in engineering calculations.

References

1. Bruno G (2011) On the infinite universe and worlds
2. Russell B, Potter M (2022) Introduction to mathematical philosophy. Routledge
3. Lazerowitz M (1979) The infinite. London and New York, George Allen and Unwin, Ltd
4. Khan S (2024) Discharging inductors analyzed using concept of duality of charge, accepted for IEMTRONICS 2024, London, UK
5. Aguirre A, Tegmark M (2011) Born in an infinite universe: a cosmological interpretation of quantum mechanics. Phys Rev D 84(10):105002
6. Hassenzahl WV, Hazelton DW, Johnson BK, Komarek P, Noe M, Reis CT (2004) Electric power applications of superconductivity. Proc IEEE 92(10):1655–1674
7. Halperin BI, Refael G, Demler E (2010) Resistance in superconductors. Int J Modern Phys B 24(20–21):4039–4080

Mobile Application Security: Malware Threat and Defenses

Fahad Mira

Abstract Regular mobile phones to smart phones are phenomenal because massive leap in capabilities. The abundance of applications available for download and installation is one of the main selling points of smart phones. But it also means that malicious software may be readily distributed to smart phones, allowing hackers to launch a variety of attacks. Effective detection procedures and preventative measures should be taken to deal with this problem. The reasons why cell phones are susceptible to security breaches are initially covered in this article. After then, it shows signs of malicious behavior and virus threats. After that, it takes a look back at the methods currently used to identify and stop malware. It highlights the need for additional research in these areas, as well as the efforts of app developers, app store authorities, and users to protect themselves from this type of malware.

1 Introduction

Smart phones are multi-purpose portable computers and, communications devices may handle communications through multimedia and apps for play and work, in contrast conventional mobile phones that primarily offer mobile telephone capabilities. The exponential growth in smart phone adoption can be attributed to the dramatic improvement in functionality compared to older mobile phones. A record-breaking one billion smart phones were shipped globally in 2023 [1].

Business use of smart phones is expected to accelerate the global smart phone market's rapid rise in the coming years. The new BYOD [2] paradigm, which holds employees accountable rather than the corporation, is rapidly replacing the old corporate-liable model in businesses around the globe. According to IDC's research, there were 132.3 million employee-liable devices and 61.4 million corporate-liable smart phones in use in 2013. When compared to shipments in 2012, this represents a growth rate of 50.3% for the first model and 18.5% for the second. Transportation

F. Mira (✉)
Department of Computer Science and Technology University of Bedfordshire, Bedfordshire, UK
e-mail: Fahad.Mira@beds.ac.uk

P. G. Bradford et al. (eds.), *Proceedings of IEMTRONICS 2024*, Lecture Notes in Electrical Engineering 1229, https://doi.org/10.1007/978-981-97-4780-1_35

of smart-forecasted to reach 88 million units in 2017 and 328.4 million units in 2016 [2].

Installing and running third-party application programmes, also known as apps, is one of the distinguishing aspects of smart phones. They greatly expand the capabilities of smart phones, making the user experience better. These applications are formally disseminated through online marketplaces called app stores. The iOS platform's equivalent is the Apple App Store, and the Android platform's equivalent is the Google Play Store. App creators can easily distribute their programmes to these marketplaces, and app users can easily discover and download new apps. Because of this, the pace of app creation has been absolutely skyrocketing in the past several years. As an example, there were over 700,000 apps available in the Apple App Store by September 2012 and over 650,000 in the Google Play Store.

Malicious software, or malware, can penetrate any cyber system and execute on infected devices secretly, even cell phones. Hackers have found an easy way to infect cell phones with malware through app markets, even if people love installing apps from these platforms. They might, for instance, repackage well-known games with malware and then distribute them through app stores. The infected programmes are commonly downloaded by unsuspecting users. With 254,158 apps detected on Android alone, the number of malware-infected apps has reached 267,259 [3]. A 614% growth in malware in apps since 2023 was also hinted at. Numerous additional entry points exist for malicious software to infiltrate targets [4]. Malicious software sometimes masquerades as file macros. Some install themselves by taking advantage of security holes in computers, smart phones, and other networked devices. Some infiltrate victims' cell phones through email attachments or multimedia messaging service (MMS) messages. Whatever the situation may be, malware has the potential to significantly impact users' and businesses' data privacy and information security.

Malicious behavior and threats posed by malware are presented in the next sections of this article, following a discussion of the reasons why cell phones are susceptible to security attacks. The current methods for detecting and preventing malware are subsequently reviewed. We contend that everyone from app store managers to app developers to researchers to end users must work together to combat this kind of virus. As a last point, we bring out a number of unresolved security concerns that necessitate additional study.

2 Why Smartphones Are Vulnerable?

Below, we'll go over some of the things that make smart phones susceptible to security breaches. To begin, many people's private information is kept on their smart phones. In instance, some data can be extremely sensitive since an increasing number of users conduct financial transactions, such online banking and buying, from their cell phones. Because of the large sums of money that hackers may make off of such sensitive data, cell phones are attractive targets for them.

Additionally, Android version is being used by new handsets. Malware authors can learn more about Android thanks to the company's open-source kernel policy. In order to increase their market share, Google promotes the creation of third-party apps and makes it easy to publish them. Therefore, hackers have a lot of room to manoeuvre when it comes to building and sharing malware. Malware installation also seems to rise in tandem with the frequency with which people install apps on their cell phones.

Third, the general public tends to think of smart phones as nothing more than regular mobile phones with a bunch of apps on them for all their communication and entertainment needs. They don't care that their cell phones are effectively portable computers that can be breached by cybercriminals. The outcome is that safety precautions are disregarded. Furthermore, malware developers are less limited in their ability to execute their harmful operations with the introduction of Smart phone hardware and operating systems. Additionally, there are instances when it is practical for attackers to create mobile malware; they may easily transfer virus from PCs to Smart phone.

3 Threats of Malware and Malicious Behavior

Malicious attack behavior, remote control behavior, and propagation behavior are the defining characteristics of mobile malwares [5]. The propagation behavior describes the potential ways in which malware can infect victims. The mobile malware uses a remote server to further abuse the infected device, as seen by the remote control behavior. After infecting a victim's devices, the malware will target those devices through various communication channels, such as Bluetooth. This is known as the attack behavior. What follows is a more in-depth analysis of the dangers presented by malware.

After infecting a smart phone, malware will attempt to access the data stored within, disrupt the device's usual operations, or introduce new security holes, including allowing unauthorized remote access. In general, malware can conduct a variety of attacks. As seen in Table 1, common forms of malware include worms, spyware, phishing, surveillance, dialer-ware, financial malware.

Attacks from Phishing Websites—PC users are often targets of phishing attacks. This kind of assault is appropriate on smart phones and other platforms since it does not require any kind of attack on the users' systems. Malware can collect sensitive information, including credit card numbers, by simply posing as legitimate-looking websites. About a quarter of malicious software includes questionable URLs.

Smart phones are attractive targets for phishing attacks for a number of reasons. The first problem is that malicious programmes can be easily distributed in app stores by masquerading as legal ones. To add insult to injury, the small screen size of most smart phones makes it easy to hide trust cues that users use to determine the safety of submitting sensitive information, such as the presence or absence of Secure

Table 1 Typical attacks launched by malware [6]

Phishing	Credentials including account details and credit card numbers are gathered through credible apps, emails, or SMS
Spyware	Monitoring users' smartphone activity may result in the extraction or inference of personal information. Unlike surveillance assaults, spyware does not target specific individuals
Surveillance attacks	The built-in sensors of an infected smartphone are used to monitor a specific user
Diallerware attacks	Malware steals users' money by making disguised calls to premium numbers or SMS services
Financial malware attacks	Attacks on cellphones or financial apps try to steal user credentials ordo man-in-the-middle attacks
Worm-based attacks	A worm is a malware that replicates and spreads from one device to another via existing networks
Botnets	Virus-infected zombie machines, known as botnets, allow hackers to remotely control them

Sockets Layer. Thirdly, hackers can utilize numerous channels in cell phones, such as instant messaging, short of such permission, even when an app appears to have a valid requirement to transmit data to the outside world. As an example, a weather app with spyware could misuse its authorization to send location data to ad servers for spamming marketing information, even though it has the necessary authorization to send location data to weather information servers [7], message service (SMS), and others, to conduct phishing attacks. Fourth, phishing is a real threat on cell phones, but most users don't realize it. More people put more faith in their cell phones than in their computers.

The term "spyware" describes a type of malicious software that secretly gathers data from infected devices. Smart phones are prime targets for spyware due to the large amounts of personal data and sensitive information they store and process. Also, smart phones include hidden routes that hackers might use to get their hands on the data they've collected. Permission settings on smart phones aren't always fine-grained enough to prevent misuse.

The severity of the harm can vary according to the nature of the data being gathered. In the above scenario, users are only irritated since spam messages are triggered by their location data. But worse things can happen if more sensitive data is gathered. For instance, compared to other prevalent forms of spyware, the new Android spyware known as Zitmo poses a greater threat. It receives confirmation texts issued by financial institutions. This kind of text message can include the monitored phone's login information for online banking. The hacker can then use this data to commit fraud [8].

Threats to Privacy—Sensors like the GPS, accelerometer, microphone, and camera found on most smart phones make them vulnerable to surveillance attacks. Smart phones infected with appropriate spyware can be used to keep targeted users

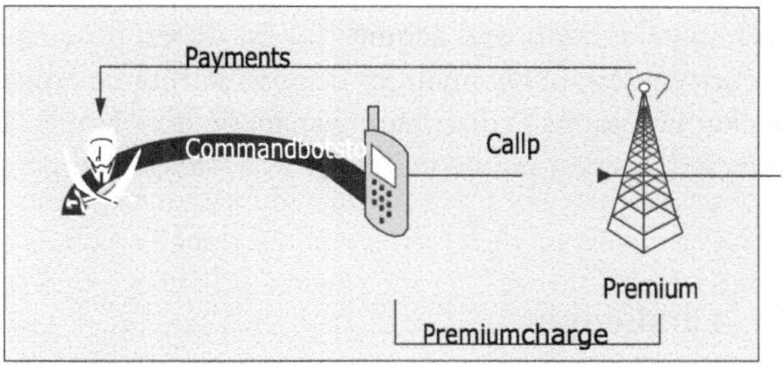

Fig. 1 Diallerware attacks

under surveillance, especially when they are strongly associated with their owners [9]. Because it may reveal extremely private information, the GPS sensor is especially helpful. Hackers are already taking advantage of real apps in order to keep their intended victims under constant surveillance. Also, it is possible to secretly modify even non-spyware apps to enable tracking.

Attacks using dialerware—As illustrated in Fig. 1, hackers can secretly send premium- rate SMS messages to smart phone users, causing them to suffer financial expenses. The initial intent of premium-rate text messages and phone calls was to offer supplementary services, including stock quotes and news, to customers for an additional fee that would be charged to their phone bills. A hacker might make money by abusing premium-rate calls under this assault. Hackers trick people into subscribing to premium services that they control after infecting their handsets. One example of Android malware is HippoSMS, which uses a premium-rated number to send SMS messages. It prevents customers from receiving SMS notifications from service providers, allowing them to remain unaware of any unwelcome additional payments [9].

Attacks by Financial Malware—The goal of financial malware is to execute man-in- the-middle attacks on financial applications or steal passwords from cell phones.

Smart phones are just as susceptible to financial malware as PCs. A key logger that steals credit card info is one example of financial malware. Worst case scenario: it's an app pretending to be a legitimate banking app. After users install and use the programme, the hacker gains the ability to conduct man-in-the-middle attacks on financial transactions.

Cyber attacks Involving Worms—Smart phones are vulnerable to attacks that utilize worms. In addition, it can self-propagate, meaning it can move from one device to another across an existing network even when users aren't actively involved. Worms, in reality, can infect smart phones worldwide with a high probability of success with only one click. Additionally, worm-based assaults on the virtualization environment, and by extension, smart phones, are anticipated to rise when network function virtualization is integrated into next-generation mobile networks in an effort to decrease operating and capital costs [10].

"Botnets" refer to networks of infected "zombies" that hackers may command from afar. An example of a mobile botnet would be a group of infected and remotely controlled cell phones. The majority of botnets are utilized by organized crime to initiate assaults and acquire funds, which poses a significant danger to Internet security. It could be anything from gathering information that could be utilized for unlawful reasons to distributing spam or launching a Denial-of-Service attack. Once infected, a smart phone might act as a walking zombie in the face of cyber assaults.

4 Challenges

Smartphone security is much different from PC security. Security issues with smart phones, in particular, stem from the fact that they use so many different technologies to connect to the web. What sets mobile security apart from desktop security are these three aspects:

Portability: users may take their devices with them anywhere, giving them great mobility. So, they could be physically or intellectually tampered with or stolen. Typically, the individual who owns a device is also its only user, which allows for strong customization. Reliable connection: people can use their cell phones to access a wide range of online services. Therefore, malware can infiltrate devices through several ways. Furthermore, the constrained capacities of mobile devices stand out from a personal computer in the clearest way. Possible security solutions are limited in intricacy because to limited CPU power, memory, and battery life. Complex intrusion detection methods, for instance, demand too much processing power from cell phones to be practical.

5 Defense Methodology

A two-level approach can be used to protect against malware. Stopping malware from infecting cell phones is the first level's goal. The second tier depends on instruments that proactively identify the presence of malicious software. As soon as it is found, the smart phone systems are cleared of it and cleaned up. First, we go over this tactic for generic situations below. Then we will concentrate on a few particular kinds of attacks.

6 General Scenarios

Preventive Measures: As Fig. 2 illustrates, cooperation amongst the stakeholders is crucial to preventing assaults from malware that resides in cell phones [11].

Application Developers: It is the responsibility of application developers to make sure that their apps respect privacy and secures coding principles, and that they don't

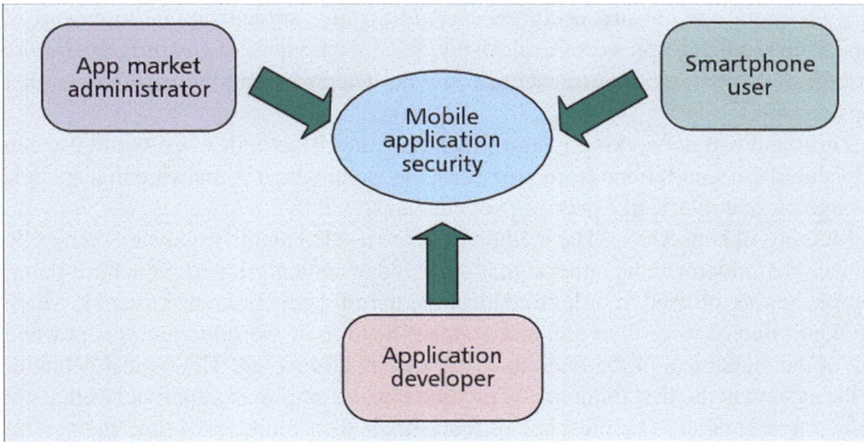

Fig. 2 Co-operation among the stake holders

access superfluous data. Malware would thus have a more difficult time launching assaults by exploiting the vulnerabilities in another application, which would make it more effective. Alternately, developers have the option of substituting a one-of-a-kind identity for the IMEI number. Not only should sensitive data be encrypted before being transported to remote systems, but it should also be held locally. Third-party libraries should go through the proper channels for approval before being used in app development.

Furthermore, although while Android apps come with roughly 100 built-in permissions that regulate functions like making phone calls and sending quick messages, you should utilize these rights as little as possible. Application developers should take extra care when utilizing built-in permissions because most smart phone users just use the default settings. Moreover, application developers might offer supplemental security services to counteract device flaws or fend off malware attacks.

App Market Administrators: It is imperative for app market administrators to thoroughly review and eliminate any dubious applications that are posted. With differing degrees of effectiveness, server-side screening procedures have recently been created to identify and eliminate fraudulent programs from app marketplaces [12]. Additionally, having a clearly established security policy by administrators is beneficial to developers. For instance, before an app can be made available through the App Store, it must comply with Apple's security guidelines. Apple uses encryption keys to sign code before approving apps. Installing apps on an iPhone can only be done through the App Store. This guarantees that iPhone distribution can only occur for apps that comply with Apple's security policies. Google has released Bouncer, a new app scanner for mobile malware.

Users of Smartphones: It is recommended that users establish a robust anti-malware system (personal firewalls, for example) that can guard against malicious activity and notify users of any unusual occurrences. Additionally, students ought to download apps from reputable app stores exclusively. Read the reviews of any program before installing it. It is important to exercise caution when granting the permissions that apps require.

Turning down networking features like WiFi and Bluetooth when not in use can help shield the smartphone from proximity malware, which is malware that spreads through close contact, like peer-to-peer chat apps.

Methods of Detection—The methods that are used to identify malware on mobile devices are fundamentally either signature-based or anomaly-based. Signature-based approaches are utilised in order to collect the harmful acts of known viruses, which are then referred to as their signatures. Identification of the infection occurs when one of the signatures of the malicious software is discovered. The typical behavior of the system is the first thing that is predicted when employing approaches that are based on anomalies. The presence of the virus is then found each time the system exhibits behavior that deviates from the typical behavior that is anticipated.

As illustrated in Fig. 3, malware detection techniques can be classified into two domains: client detection and network detection, depending on where the malware detection is being performed. In general, host-based or cloud-based methods can be used for client detection. We refer to the methods that operate locally on cell phones as host-based methods. This type of detecting technology is referred to as cloud-based, and it is possible that the heavy computation that is required could be transferred to a server located in a remote location in order to improve efficiency.

There is a significant majority of mobile antivirus programmes that are currently being sold by security companies. These programmes generally perform the same functions as their desktop counterparts. As a result, a restricted capacity for detection is provided, but at the expense of a significant demand on resources. It's possible that such an approach won't be successful. In the realm of mobile malware detection technology, signature-based detection methods constitute the majority of approaches. The availability of a recently updated signature database is a necessary condition for the success of these methods. It is frequently necessary for the device to maintain a substantial signature database in order to perform static scanning.

Utilising might be one method that can be utilised to reduce the size of the database. A fundamental drawback of signature-based detection systems is that they are susceptible to obfuscation. This is true regardless of the circumstances. Metamorphosis and polymorphism are two ways that may be utilised by programmers who create malicious software in order to escape being identified by static detection methods.

The cloud-based solution should be applied to get around this. Efficient detection for heterogeneous devices is possible by offloading intensive computations to the cloud [12].

A cloud-based malware detection server has numerous scan engines for anomaly or signature detection, as illustrated in Fig. 4. For instance, second, a cloud-based security solution for cell phones was created by [13]. By continuously sending the device's inputs and network connections to the cloud, it simulates a registered smart phone within the system. Resource-intensive security studies on the simulated device become possible in this way.

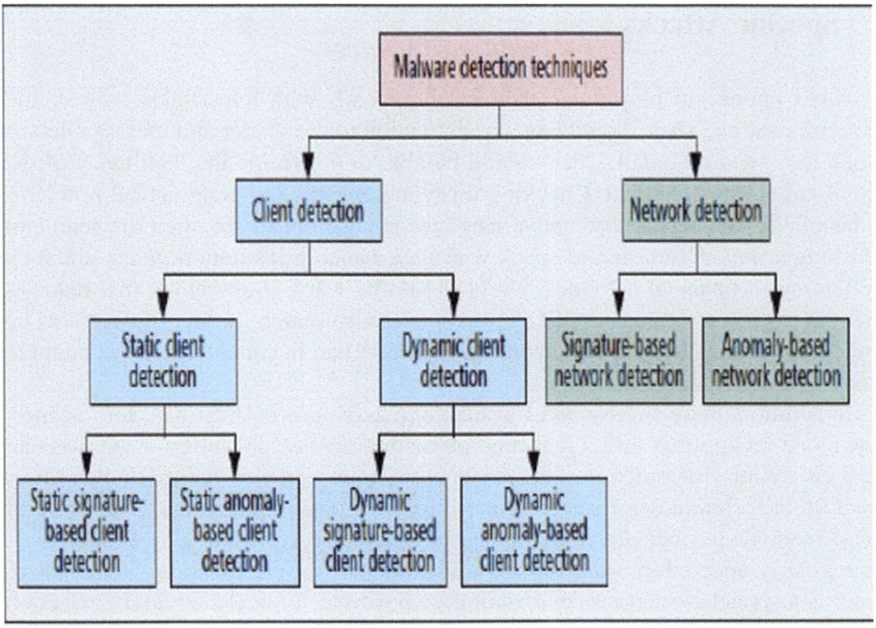

Fig. 3 Malware detection techniques

Alternatively, network detection strategies can be used to provide security if the user is unwilling or unable to install mobile malware detection software. Through the capture of network traffic and the monitoring of malicious events originating from smart phones, these techniques seek to detect malware in the mobile Internet. For instance, a prototype named Airmid that can automatically identify and react to malware infection in cell phones was created and implemented by [14].

Either dynamic analysis or static analysis can be used to accomplish detection. Static analysis analyzes programs or codes without running them. Three steps make up this process: unpacking, disassembly, and analysis. In general, it is quick and easy. With dynamic analysis, an isolated environment is used to continuously observe the activity of programs. This method gathers and examines an application's runtime data, such as system calls and events. While dynamic analysis concentrates on the reasons behind and frequency of specific suspicious events, static analysis techniques concentrate on what is being accessed [15].

7 Specific Attacks

In smart phone phishing when they receive emails with hyperlinks, they should exercise caution. They should pay close attention to the sender and message details when they read an email. They should be able to determine the legitimacy of the email based on its content. Checking for typographical and grammatical problems is beneficial. Senders whose native language is not English are often the source of phishing scams. Users should speak with their banking institution in any situation pertaining to financial activities. Users should be wary of any email that requests them to log in via the linked URL. Users should also search for security symbols on their browsers, such as the lock icon in the address bar. In conclusion, users ought to exercise extreme caution.

In addition, there are resources available to assist users. Users can, for instance, download an app that offers real-time protection against phishing for any website they visit. Anti-virus software, anti-phishing software, and browser and toolbar filters are further techniques for reducing the impact of phishing assaults. Every one of these methods has benefits and drawbacks. For instance, phishing signatures must be regularly updated in order for anti- phishing software to function. Nonetheless, their widespread use and ease of updating are positives. Typical methods of detection include:

- Content-based screening, which has the ability to identify phishing attempts in emails with success.
- The blacklist approach, which necessitates human verification. This technique is extensively used as anti-phishing in the toolbar because of its extremely low false positive rate.
- Whitelist: this approach differs from blacklist in that it requires record-keeping of all websites available online.

Mobile network providers can also take part in the fight against phishing scams. Using anti-malware software, they may filter SMS and MMS messages that contain phishing URLs by scanning both incoming and outgoing messages in the mobile network. By sharing database information, they can also form international alliances with other operators to stop the spread of mobile viruses. Like phishing, consumers can actively guard against spyware by being vigilant. Modify the security settings of your browser: Most browsers let you change the level of security, typically on a range from "high" to "low." Users should make use of these options so the browsers can stop unwanted operations.

Watch out for pop-ups: Pop-up windows frequently include advertisements that may serve misleading objectives. Some can pose as offering a warning about a viral infection. Users must never select "yes." Rather, users ought to adopt a skeptical computing approach and presume that any new software may be hazardous unless demonstrated differently. Users should be aware that clicking "yes" on a popup that they do not understand can lead to the loading of spyware. Peruse the terms and conditions at all times: In their terms and conditions, reputable software providers would provide information on how they gather and handle user data. Unfortunately, very few users really take the time to read them. Users should be aware of what they are signing up for if they are especially concerned about safeguarding their online privacy.

Advertising in pop-up windows is typically deceiving. Some resemble viral illness alerts. User must never choose "yes." Instead, customers should suspect each new application is hazardous unless proved otherwise. To prevent infection, avoid clicking "yes" on unexpected popups. Always read the T&Cs: Reputable software vendors declare data collection and usage in their terms and conditions. Unfortunately, few read them. Internet consumers should know what they're getting if privacy is vital.

Anti-spyware programs may identify phone monitoring malware. While quarantineable, smart phone malware is difficult to eradicate. Most anti-spyware programs monitor incoming messages and prevent threats. Anti-spyware software needs periodic updates [16].

Use anti-spyware scanners: Users can use anti- spyware apps to find dangerous tracking software on their cell phones. Spyware on smart phones can be difficult to remove, but it can always be placed in quarantine to stop being utilized. Through the examination of incoming traffic and the blocking of any potential dangers, the majority of anti-spyware solutions offer ongoing protection. Anti-spyware programs undoubtedly require regular updates in order to function properly.

Botnets: There are indicators that suggest a smart phone has been turned into a zombie. It's possible that the smart phone seems a little slower than normal.

Sometimes it might freeze or restart by itself. The users should be aware of these indicators. Examining if the device is actively sending or receiving data even when no relevant apps are open is another method to identify botnet activity. Unwanted behaviors and a decline in system performance are other indicators. Zombies can use up storage space, CPU power, and network traffic. Consumers must be aware of any strange behavior their cell phones may display.

In recent years, there has been a lot of study on botnets, with a primary focus on botnet detection, measurement, tracking, mitigation, and future prediction. Effective methods should be available soon to assist smart phone owners in preventing their devices from becoming zombie-compromised, according to this research. Yazdinejad et al. [17] for instance, suggested a simple method for identifying malware. To intercept communications delivered by malicious software, it installs phantom contacts on the device. Additional analysis is done based on the messages that were captured in order to determine the message's signature or even the malware's signature.

The idea of a "zombie smartphone." in cellphone-dependent world is alarming. Once vibrant and responsive, these devices reveal minor but alarming signs of degradation. Zombie-like slowness, freezes, and restarts may occur. The observant eye must delve deeper to recognize these irregularities symptoms of more nefarious network penetration. Despite smartphone botnets, botnet research remains promising. Researchers are developing zombie-prevention diagnostics and predictions. New knowledge may help smartphone users battle this silent invasion and defend their digital environments from the undead [18].

8 Results and Next Work

In conclusion, smart phones will become more widely used in the market. Malware distribution into cell phones is also anticipated to increase. To protect against malware, consumers, researchers, app developers, and app store authorities must all put out effort. The main attacks against smart phones brought on by malware have been compiled in this article, along with potential defenses and detection techniques.

Malware behavior is constantly evolving, and hackers are constantly coming up with new ways to compromise cell phones, even though certain attacks may still be avoided with the aid of detection tools and preventive measures already in place.

Therefore, it is necessary to develop more advanced tools. Moreover, when designing such solutions, the constrained resources of mobile devices should be considered. Future malware attack mitigation on smart phones is expected to be a shared responsibility between the cloud and the device. While the device may do local detection using simplified classification algorithms, computationally demanding operations should be executed in the cloud. We offer the following recommendations for future research directions as we wrap up this essay.

First, greater focus needs to be placed on developing a methodical approach to collecting data on new malware, as it is usually hidden in apps that are distributed through different third-party markets. Because of the competition, security companies are reluctant to release their malware database to the public. As a result, researchers have access to relatively few malware samples. For this reason, it's critical to find solutions to automate a systematic process for compiling these dangerous programmes.

Second, as mentioned, secret of signature- based detection is knowing which malware to link to which in order to perform security analysis. These days, package names and cryptographic hashes are frequently employed as identifiers.

Unfortunately, because hackers may readily alter them, these are ineffective methods. Security analysts often need to perform time-consuming reverse engineering procedures to identify malicious functions and structures. Therefore, future research should focus on finding efficient ways to link newly discovered malware with existing malware in the database.

Third, mobile malware is changing quickly to evade the detection methods now in use. Novel techniques are therefore required for the prompt detection of emerging infections. Recently, some researchers proposed employing machine learning techniques for Android malware detection that are based on Bayesian classification [11].

Fourth, a paradigm change is required. To achieve flawless security, a lot of work has been put into finding and fixing every potential vulnerability when designing security systems up to this point. Otherwise, enemies can take down the entire system with just one vulnerability. This puts a lot of challenges on the design process. The enemy must combat any obstacle preventing it from accomplishing its attack objectives, yet, if the blame for this single point of failure may be transferred from defenders to advertisers. Therefore, next research should examine enemies' single point of failure.

References

1. Aldauiji F, Batarfi O, Bayousef M (2022) Utilizing cyber threat hunting techniques to find ransomware attacks: a survey of the state of the art. IEEE Access 10:61695–61706
2. Aldhaban F, Daim T, Harmon R, Basoglu N (2020) Technology adoption in emerging regions: case of the smartphone in Saudi Arabia. Int J Innov Technol Manag 17:2050003
3. Ashawa M, Morris S (2021) Analysis of mobile malware: a systematic review of evolution and infection strategies
4. Aslan ÖA, Samet R (2020) A comprehensive review on malware detection approaches. IEEE Access 8:6249–6271
5. Bank D (2023) Problematic monetization in mobile games in the context of the human right to economic self-determination. Comput Hum Behav 149:107958
6. He D, Chan S, Guizani M (2015) Mobile application security: malware threats and defenses. IEEE Wirel Commun 22(1):138–144
7. Buffa S, Fouladfar MH, Franchini G, Lozano Gabarre I, Andrés Chicote M (2021) Advanced control and fault detection strategies for district heating and cooling systems—a review. Appl Sci 11:455
8. Chen L, Xia C, Lei S, Wang T (2021) Detection, traceability, and propagation of mobile malware threats. IEEE Access 9:14576–14598
9. Chowdhury MM, Del Toro JM, Kambhampaty K (2022) Active cyber defense by benevolent worms. In: 2022 IEEE international conference on electro information technology (eIT), pp 580–585
10. Gamba J, Rashed M, Razaghpanah A, Tapiador J, Vallina-Rodriguez N (2020) An analysis of pre-installed android software. In: 2020 IEEE symposium on security and privacy (SP) pp 1039–1055
11. Mat SRT, Ab Razak MF, Kahar MNM, Arif JM, Firdaus A (2022) A Bayesian probability model for android malware detection. ICT Express 8:424–431
12. Muhammad Z, Anwar Z, Javed AR, Saleem B, Abbas S, Gadekallu TR (2023) Smartphone security and privacy: a survey on APTs, sensor-based attacks, side-channel attacks, google play attacks, and defenses. Technologies 11:76
13. Pan Y, Ge X, Fang C, Fan Y (2020) A systematic literature review of android malware detection using static analysis. IEEE Access 8:116363–116379
14. Yazdinejad A, HaddadPajouh H, Dehghantanha A, Parizi RM, Srivastava G, Chen M-Y (2020) Cryptocurrency malware hunting: a deep recurrent neural network approach. Appl Soft Comput 96:106630
15. Ruggia A, Possemato A, Dambra S, Merlo A, Aonzo S, Balzarotti D (2023) The dark side of native code on android. Authorea Preprints
16. Sharma T, Rattan D (2023) Android Malwares with their characteristics and threats. In: Mobile radio communications and 5g networks: proceedings of third MRCN 2022, ed. Springer, pp 1–12
17. Zhang Z, Wang N, Wu H, Tang C, Li R (2021) MR-DRO: a fast and efficient task offloading algorithm in heterogeneous edge/cloud computing environments. IEEE Internet Things J
18. Adomako S, Abdelgawad SG, Ahsan M, Amankwah-Amoah J, Liedong TA (2023) Nonmarket strategy in emerging markets: the link between SMEs corporate political activity, corporate social responsibility, and firm competitiveness. J Bus Res 160:113767

AquaLynk: Connected Pisciculture Using Internet of Things (IoT)

Arkaprava Paul and Dibyadarshi Das

Abstract The Internet of Things (IoT) is a ground breaking technological advancement that allows objects to digitally represent themselves. It has brought a new level of automation to various fields, modernizing everything. In addition, IoT is playing a significant role in pisciculture and agriculture including crop monitoring and soil management. In this context, the authors have designed a smart system for monitoring and predicting fish farming conditions based on IoT technology. The system focuses specifically on water quality, using three sensors pH and temperature sensors to determine the pH and temperature values of the water, and a water level sensor to detect the water level. The proposed system displays the relevant information regarding quality of water through a mobile application and predicts the best fish to be farmed based on the water quality.

Keywords Pisciculture · Internet of things (IoT) · Water quality monitoring

1 Introduction

Fish are important natural resources that contribute to increased national income, improved nutrition, reduced unemployment, and the generation of foreign currency. In comparison to red meat, it is a source of low-cost high-protein dietary element having other health-beneficial effects [1]. Nonetheless, the existing infrastructure and conventional fish farming have failed to meet the predicted demand for fish in order to meet the expanding population's needs. Traditional techniques for monitoring water quality included taking water samples and sending them to a chemical laboratory for analysis of harmful materials or using simple use-and-throw testing kits. The system's disadvantage was all of the manual processing, such as measurements, system maintenance and control, and so on.

A. Paul · D. Das (✉)
Institute of Engineering Management Y-12, Salt Lake, Sector 5, Kolkata, West Bengal 700091, India
e-mail: dasdibyadarshi2021@gmail.com

P. G. Bradford et al. (eds.), *Proceedings of IEMTRONICS 2024*, Lecture Notes in Electrical Engineering 1229, https://doi.org/10.1007/978-981-97-4780-1_36

Internet of Things (IoT) can enable people to control objects remotely by connecting through internet. Data from the surroundings can be collected through sensors on the objects and sent to the cloud database. Use of IoT in fish farming can be done to detect dissolved oxygen, water temperature, alkalinity, acidity and level of water from floating point each moment. This will help fishermen to get real-time information about the water condition and take necessary steps to produce more fish.

The present project aims to develop an IoT based system which can monitor the condition of the water of pond so that preventive measures to maintain the water quality can be taken and also help in determining the type of fish suitable for farming.

2 Objectives

2.1 Development of an IoT-Based Device and Web-Application

The primary aim is to design and implement an Internet of Things (IoT) based system comprising both hardware and software components. This system will facilitate remote measurement of key parameters such as pH, temperature, and water level in a fish pond. Additionally, a user-friendly web application will be developed to visualize and interpret the data collected from the IoT device.

2.2 Real-Time Monitoring of Water Parameters

The system aims to provide real-time monitoring of essential water quality parameters to ensure optimal conditions for fish farming. By deploying sensors capable of measuring pH, temperature, and water level, the system will continuously collect data from the fish pond and transmit it to the cloud-based database.

2.3 Generation of Alerts for Abnormal Water Conditions

To enhance proactive management of fish pond environments, the system will incorporate alert mechanisms to notify users when water parameters deviate from predefined healthy ranges. Alerts will be generated based on thresholds set for each parameter, allowing farmers to take timely corrective actions to maintain water quality.

2.4 *Support for Preventive Measures and Fish Species Selection*

The IoT-based system will not only monitor water quality but also provide insights to support preventive measures and fish species selection. By analyzing the collected data, the system will offer recommendations for maintaining optimal water conditions and suggest suitable fish species based on the observed water quality parameters.

2.5 *Ease of Use and Accessibility*

The development process will prioritize the creation of a user-friendly interface for both the IoT device and the web application. The system should be intuitive to operate, allowing fish farmers to easily access real-time data and receive alerts on their preferred devices, such as smartphones or computers. Accessibility considerations will ensure that the system can be utilized effectively by fish farmers of varying technical backgrounds.

3 Literature Review

The intersection of population growth, food security, and environmental sustainability has placed a significant emphasis on the role of aquaculture in meeting global dietary needs. As elucidated by the Food and Agriculture Organization of the United Nations (FAO) in their report" The State of World Fisheries and Aquaculture 2016," aquaculture contributes substantially to global fish production, providing essential protein sources for human consumption [1]. However, as demand for fish continues to rise, there is a pressing need for sustainable aquaculture practices and effective management strategies to ensure the long-term viability of fish farming operations.

Studies such as those conducted by Jumatli and Ismail underscore the importance of promoting sustainable aquaculture practices to meet the increasing demand for fish products [2]. Sustainable practices encompass not only the responsible stewardship of aquatic resources but also the enhancement of aquatic animal health and the conservation of natural habitats. Despite the potential of aquaculture to address food security challenges, fishermen often face obstacles in increasing fish production due to a lack of knowledge and inadequate management of water quality parameters [3].

Water quality is a critical determinant of fish health and productivity in aquaculture systems. Parameters such as dissolved oxygen, temperature, pH, and ammonia levels play pivotal roles in maintaining optimal aquatic environments for fish growth and survival. Deviations from ideal water quality conditions can lead to stress, disease outbreaks, and mortality among fish populations [4]. To address these challenges, researchers and practitioners have increasingly turned to technological innovations, particularly IoT-based solutions, to enhance monitoring and management of aquaculture systems.

The advent of IoT technology has revolutionized the way fish farming operations are monitored, controlled, and managed. IoT-based systems offer real-time monitoring capabilities, enabling continuous data collection on water quality parameters such as dissolved oxygen, temperature, and pH [5]. These systems utilize sensors and connectivity to transmit data to centralized databases, where it can be analyzed and interpreted in real-time. By leveraging IoT technology, fish farmers can proactively identify and address issues related to water quality, thereby optimizing fish health and productivity.

Numerous studies have demonstrated the efficacy of IoT-based systems in improving aquaculture management practices. For instance, IoT solutions enable remote monitoring and control of fish tank parameters, allowing farmers to adjust environmental conditions in real-time to meet the specific needs of their fish stocks [6]. Additionally, IoT-based approaches utilizing wireless sensor networks enable continuous monitoring of fish cultivation, ensuring optimal conditions for fish health and productivity [7].

Empirical analyses have further highlighted the growing adoption of IoT technology in the fisheries and aquaculture sectors. Studies such as those conducted by Kiranmayi and Sharma provide insights into the development and deployment of IoT-based solutions for aquaculture management [8]. These solutions leverage advanced technologies such as artificial intelligence (AI) to enhance the efficiency and effectiveness of fish farming operations. By integrating IoT and AI technologies, researchers have developed real-time monitoring systems capable of analyzing complex datasets to optimize fish farming practices [9].

In addition to real-time monitoring capabilities, IoT-based systems offer predictive analytics and remote access functionalities, empowering fish farmers with actionable insights for sustainable fish farming practices [10]. These systems enable farmers to anticipate and mitigate potential risks associated with water quality fluctuations, thereby promoting the long-term sustainability of aquaculture operations [11].

4 Proposed Solution

The proposed solution offers a comprehensive approach to revolutionize fish farming practices through the integration of IoT technology. Central to this solution is the AquaLynk device, which employs an Arduino-based platform and integrates sensors to monitor crucial water quality parameters in real-time. These sensors include pH sensors for acidity/alkalinity measurement, temperature sensors for thermal monitoring, and water level sensors for assessing pond water levels. Through advanced data processing algorithms and cloud connectivity, the AquaLynk device provides fish farmers with remote access to real-time monitoring data and alerts, enabling timely intervention to maintain optimal water conditions. Predictive analytics further empower farmers to anticipate potential issues and optimize farm management practices. Rigorous validation procedures ensure the accuracy and reliability of sensor data, while key performance metrics evaluation underscores the system's robust functionality and suitability for real-world applications. Overall, this solution aims to enhance productivity, cost-efficiency, and sustainability in the aquaculture industry by leveraging IoT technology for proactive water quality management and predictive analytics.

5 Methodology

The AquaLynk device operates through a series of integrated components and sophisticated algorithms, ensuring seamless data acquisition, processing, and transmission. The following elucidates the detailed working mechanism of each key component.

5.1 Sensor Integration and Data Acquisition

pH Sensor: The pH sensor consists of an electrode immersed in the water, which undergoes chemical reactions in response to changes in hydrogen ion concentration. These reactions generate electrical signals proportional to the pH level, which are then converted into digital data by the device's microcontroller.

Temperature Sensor: The temperature sensor employs a thermistor to detect variations in water temperature. As temperature changes, the resistance of the thermistor alters leading to voltage fluctuations that are interpreted by the microcontroller to derive precise temperature readings.

Data Transmission: Data collected by the sensors are transmitted to the cloud database via internet connectivity. This seamless data transmission enables remote access to real-time water quality metrics, empowering fish farmers with actionable insights and decision-making capabilities.

Mobile Application Interface: The AquaLynk system features a user friendly mobile application interface, allowing fish farmers to remotely monitor water quality parameters and receive instant alerts on their smartphones or tablets. The application provides visual representations of data trends, facilitating intuitive interpretation and timely intervention.

5.2 Data Processing and Analysis

Upon acquiring data from the sensors, the AquaLynk device initiates data processing algorithms to analyze the incoming information. Advanced algorithms are employed to interpret pH and temperature readings, identifying deviations from optimal ranges indicative of potential water quality issues.

Real-time data analysis enables the device to generate actionable insights and alerts, facilitating timely intervention by fish farmers to maintain optimal water conditions and mitigate risks to fish health.

5.3 Cloud Database and Remote Access

Processed data are transmitted to a cloud-based database through internet connectivity, ensuring remote access and monitoring capabilities. Cloud storage facilitates seamless data management, enabling historical data analysis, trend identification, and predictive analytics. Fish farmers can remotely access the cloud database via the AquaLynk mobile application or web interface, empowering them with real-time insights into water quality parameters and actionable recommendations for optimal fish farm management.

5.4 Alert Generation and Notification

The AquaLynk device is equipped with an alert mechanism that triggers notifications in response to deviations from predefined water quality thresholds. Alerts are generated instantaneously upon detecting anomalies in pH or temperature levels, ensuring prompt attention and intervention.

Notifications are delivered to fish farmers via the mobile application, enabling timely response to critical water quality fluctuations and mitigating potential risks to fish health and farm productivity.

5.5 Validation and Performance Evaluation

Comparative Analysis: To validate the accuracy and reliability of the AquaLynk device, data generated by the device were compared with simultaneous manual measurements using established water quality testing equipment. Discrepancies between device-generated data and manual measurements were analyzed to assess the device's performance and calibration accuracy.

Performance Metrics: The AquaLynk device underwent rigorous performance evaluation to assess key metrics such as response time, data accuracy, system stability, and connectivity reliability. Performance tests were conducted under various environmental conditions and operational scenarios to ensure robust functionality and suitability for real-world fish farming applications.

Through the intricate synergy of sensor integration, data processing algorithms, cloud-based infrastructure, and alert mechanisms, the AquaLynk device emerges as a transformative solution in modernizing fish farming practices. Its ability to provide real-time insights, predictive analytics, and remote monitoring capabilities revolutionizes fish farm management, enhancing productivity, sustainability, and profitability in the aquaculture industry.

6 Feasibility Analysis

6.1 Technical Feasibility

Project AquaLynk is an IoT-based web application that leverages various technologies and tools to monitor and manage fish farming environments effectively. The key technologies and tools involved in AquaLynk include microcontroller boards, physical sensors, HTML, CSS, JSP, MySQL, JavaScript, Figma, and diagram drawing tools. The technical skills required for utilizing these technologies are readily available and manageable. Additionally, these technologies are freely accessible, facilitating ease of implementation. The development timeline is well-coordinated, and the application's initial hosting on a free web hosting service ensures minimal hosting costs. As AquaLynk does not entail multimedia aspects, the required bandwidth is modest, further reducing operational expenses.

6.2 Financial Feasibility

AquaLynk entails hosting costs for web application deployment, which are mitigated by its minimal bandwidth requirements. The hardware components necessary for the application are affordable, contributing to its financial feasibility. Adopting freeware software standards eliminates licensing expenses for potential customers. While there are associated costs for bug fixes and maintenance tasks, AquaLynk's target market of local fisheries stands to benefit significantly from its functionalities. Therefore, despite these costs, the project remains financially feasible.

6.3 Operational Feasibility

The operational feasibility of AquaLynk primarily hinges on its deployment in fish farming environments, particularly in India. Critical operational considerations include:

(i) Electrical Power Supply: The system's functionality relies on the availability of nearby electrical power sources to operate sensor units within fish farms
(ii) Internet Connectivity: A consistent albeit low-bandwidth internet connection is essential for transmitting data from microcontrollers to the server.
(iii) Environmental Factors: Given the deployment of AquaLynk in outdoor water bodies housing fish, hardware components must be weatherproof and capable of withstanding environmental conditions. Testing to validate resilience to natural forces is imperative.

Addressing these operational challenges ensures the effective implementation of AquaLynk in fish farming contexts, enhancing its operational feasibility and utility.

7 Results

The AquaLynk device was experimented in a pond of a fishery in Midnapore district of West Bengal. Analysis of transmitted data was done for a period of one month. Table 1 shows the summary statistics of measurement of pH and temperature of water.

The mean pH of water measured by AquaLynk and manual pH meter were 6.63 and 32.62 respectively, but the difference was not statistically significant (p¡0.05). There was also no significant difference in the mean temperature displayed by the device and water thermometer (p¡0.05). Hence, the results provided by the device corroborated with physical measurement done by standard instruments (Figs. 1 and 2).

Table 1 Measure of pH and water temperature for 30 days (N = 30)

	pH		Temperature (°C)	
	AquaLynk	Manual measurement	AquaLynk	Manual measurement
Mean	6.63	6.61	32.62	32.65
Standard deviation (SD)	0.73	0.12	0.51	0.56
Standard error of Mean (SE)	0.13	0.20	0.94	0.10
Independent t-test	$t = 0.683$; $df = 58$; $p = 0.498$		$t = 2.34$; $df = 58$; $p = 0.816$	

Fig. 1 Device prototype

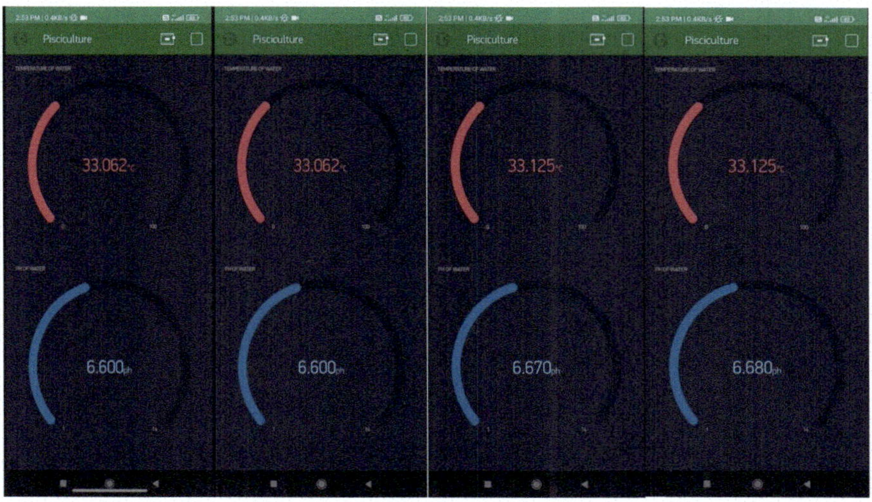

Fig. 2 Real time test data

8 Conclusion

The suggested Pond Monitoring System Device only relies on the Internet of Things. This real-time monitoring tool will aid fish farmers in increasing fish production. Through the Android application, the owner can gain knowledge about the surroundings of his or her developed pond and the different elements in the water and take the appropriate action. Farmers will also be aware of the best fish for their pond. Lastly, the initiative will result in advancement in fish farming.

9 Future Scope

Project AquaLynk exhibits promising potential for future enhancements and expansions to further augment its functionality and effectiveness in fish farming management. The following avenues represent key areas for future development:

1. Expansion of Sensor Capabilities: Integrating additional sensors for measuring a broader range of water quality parameters beyond pH, temperature, and water level will enrich AquaLynk's monitoring capabilities. Parameters such as dissolved oxygen, turbidity, and ammonia levels are critical indicators of water quality that can significantly impact fish health and productivity.
2. Integration of Machine Learning Models: Implementation of machine learning algorithms presents an exciting opportunity to enhance AquaLynk's predictive capabilities. By analyzing historical data collected from sensors, machine learning models can predict variations in water quality, facilitating proactive management strategies and optimizing fish farming practices.
3. Anomaly Detection and Root Cause Analysis: Incorporating anomaly detection algorithms enables AquaLynk to identify deviations in water quality parameters from expected norms. Coupled with root cause analysis functionalities, the system can provide insights into the underlying factors driving these anomalies, empowering fish farmers to address potential issues promptly and effectively.
4. Automated Water Health Correction Setups: Developing automated systems for water health correction represents a significant advancement in aquaculture management. By integrating actuators and control mechanisms, AquaLynk can dynamically adjust water parameters to maintain optimal conditions for fish growth and survival. These setups can include mechanisms for pH adjustment, oxygenation, and nutrient supplementation, ensuring continuous optimization of water quality.

By embracing these future directions, AquaLynk can evolve into a comprehensive and intelligent solution for fish farming management, offering enhanced predictive capabilities, proactive anomaly detection, and automated corrective actions. These advancements not only streamline operations for fish farmers but also contribute to the sustainable growth and development of the aquaculture industry.

10 Discussion

The IoT based smart system has been developed to measure temperature and pH levels of the water. The hardware system has been integrated with the mobile application. The developed system has been tested placing the sensors inside the water and the whole system is working fine. By using the device users will get real time information about the water quality factors on their mobile phone and can take proper steps to make the water quality better so that fish can live there without facing any problems.

The application of this proposed system is not only limited to fish farming. Our future plan is to scale up this system and apply to other areas as well to monitor water pollution.

Bibliography

1. FAO (2016) The state of world fisheries and aquaculture 2016. Contributing to food security and nutrition for all. Rome, 200 pp
2. Jumatli A, Ismail MS (2021) Promotion of sustainable aquaculture, aquatic animal health and resource enhancement in Southeast Asia. Proceedings of the international workshop on the promotion of sustainable aquaculture, aquatic animal health, and resource enhancement in Southeast Asia
3. Ullah A (2020) IoT based smart pisciculture monitoring and prediction system in Bangladesh. J Adv Res Mob Comput 2(1):1–19
4. Khan DMM (2021) An IoT based smart water monitoring system for fish farming in Bangladesh. In: 5th international electronic conference on water sciences
5. Harun Z, Reda E, Hashim H (2018) Real time fish pond monitoring and automation using Arduino. IOP Conf Ser Mater Sci Eng 340(1)
6. Paulin NS, Anupriya N, Prasanthi S (2017) Pisciculture environment control using automated monitoring system. Asian J Appl Sci Technol 1(2):60–65
7. Agossou BE, Toshiro T (2021) IoT & AI based system for fish farming: case study of Benin. In: GoodIT '21: conference on information technology for social good, pp 259–264
8. Tavares M, Mendes R (2019) IoT solutions for fish farming: a review. In: 2019 IEEE international conference on industrial technology (ICIT)
9. Chakraborty D, Das D, Saha D, Dey N (2020) IoT-based real-time water quality monitoring system for aquaculture. In: Intelligent computing and IoT: proceedings of ICICI 2019, vol 2
10. Kundu S, Roy S, Dey N, Shankar B (2021) Development of IoT based real-time aquaculture monitoring system for sustainable fish farming. Sustain Prod Consumption 27:532–542
11. Singh P, Chaudhary A, Singh S (2019) Internet of things (IoT) based smart aquaculture monitoring system. In: 2019 5th international conference on advanced computing and communication systems (ICACCS)
12. Garg R, Sharma S, Agarwal M (2020) IoT-based intelligent system for water quality monitoring and control in aquaculture. SN Comput Sci 1(1):1–8
13. Biswas P, Choudhury A (2017) Fish farming using IoT. Int J Comput Appl 164(10):33–36
14. Subramanian K, Sivanandan R (2016) IoT based real time monitoring and controlling of fish tank parameters. Int J Adv Res Comput Sci 7(5):83–87
15. Gupta RP, Sood SK, Sharma S, Kumar A (2017) An IOT based approach for monitoring fish cultivation using wireless sensor networks. Int J Sci Eng Res 8(8):866–871
16. Kiranmayi B, Sharma R (2019) Empirical analysis of IoT in fisheries and aquaculture. In: Proceedings of the 10th international conference on computing, communication and networking technologies (ICCCNT)
17. Agossou BE, Toshiro T (2018) Development of a real-time monitoring system for fish farming using IoT and AI technologies. Aquacult Int 26(4):1149–1162
18. Ghosh S, Sengupta A (2018) An IoT based intelligent aquaculture management system. In: 2018 international conference on communication and signal processing (ICCSP)

Opinion Mining on Offshore Wind Energy for Environmental Engineering

Isabele Bittencourt, Aparna S. Varde, and Pankaj Lal

Abstract Renewable energy sources are vital to help mitigate the effects of climate change, and reducing the carbon dioxide emissions of fossil fuels, e.g. the state of New Jersey has a goal of producing 100% clean energy by 2050. However, the plans for offshore wind energy by the shore of the state still brings much controversy between residents due to the wind farms' impact on wildlife, coastline, and the people's view from the beaches. In this context, we perform sentiment analysis on social media data to investigate people's opinions and concerns regarding offshore wind energy. We adapt 3 machine learning models, i.e. TextBlob, VADER and SentiWordNet for sentiment analysis because different functions are provided by each model, all of which are useful in our work. Techniques in NLP (natural language processing) are harnessed to gather meaning from the textual data in social media. Data visualization tools are suitably deployed to display the overall results. Despite the controversy surrounding this topic, our findings indicate some positive reception, suggesting potential support for modern-day renewable energy goals. However, there are neutral and negative comments as well, thus potentially helping to find areas for further improvement. The results of this work can be thus useful in a variety of decision-making contexts by governmental organizations and companies, hence aiding and enhancing offshore wind energy policy development. Hence, this work is much in line with citizen science and smart governance via involvement of mass opinion in decision support. In our paper, we highlight the role of sentiment analysis from social media in this aspect.

I. Bittencourt
Computational Linguistics Program, Montclair State University (MSU), Montclair, USA
e-mail: bittencourti1@montclair.edu

A. S. Varde (✉)
School of Computing, Montclair State University, Montclair, USA
e-mail: vardea@montclair.edu

P. Lal
Department of Earth and Environmental Studies, Montclair State University, Montclair, USA
e-mail: lalp@montclair.edu

I. Bittencourt · A. S. Varde · P. Lal
Clean Energy and Sustainability Analytics Center, Montclair State University, Montclair, NJ 07043, USA

P. G. Bradford et al. (eds.), *Proceedings of IEMTRONICS 2024*, Lecture Notes in Electrical Engineering 1229, https://doi.org/10.1007/978-981-97-4780-1_37

Keywords Environmental management · Clean energy · Offshore wind · Machine Learning · Natural language processing · Sentiment analysis · Smart governance

1 Introduction

Renewable energy creates shared value, circular economy approaches and commitment to the United Nations Sustainable Development Goals (UN SDG). Producing more renewable energy and abandoning conventional sources, such as coal, natural gas and petroleum, is a need shared by all countries in the world, especially as global warming grows more intense. Renewable energy has become more popular over the years; and hence many countries, e.g. the United Kingdom, Germany and China have adopted it as a source of clean energy.

Offshore wind energy is a type of renewable energy; it is obtained from the use of the force of the wind on the high seas, where it reaches much higher and more constant speed compared to onshore wind, due to the lack of barriers. According to Environmental America [1], the United States has the potential to achieve the production of more than 7200 terawatt-hours (TWh) of electricity from offshore wind, which is almost twice the amount of electricity the nation consumed in 2019, and almost 90% of the amount of electricity it would consume in 2050.

Many regions in the United States are coastal states, such as New Jersey, which makes them an ideal location for offshore wind farms. Figure 1 is a group of pictures taken at ACUA, a wastewater treatment facility in Atlantic City, New Jersey.

During our own site-visit to the facility, we have learned that the plant requires 2.5 megawatts of power each day, and when the speed of the wind is above 12 miles per hour, the electricity they need for a day can be generated by 2 of the 5 turbines they have on site [2]. Considering the potential of these turbines in producing high amounts of clean energy, the state of New Jersey has established a goal of having 100% clean energy by 2050, and 7500 megawatts coming from offshore wind by 2035 [3].

Although this type of energy is clean and renewable, it raises a lot of concerns regarding wildlife, tourism, fisheries, and coastal communities. For instance, in the state of New Jersey, the Protect Our Coast NJ website is an initiative from the community to stop the building of wind farms, as can be seen in Fig. 2. Accordingly, Fig. 3 is a sample of two posts from a New Jersey Facebook group, where people can post and debate on the topic of offshore wind energy. The wind farm process plan includes "create participation opportunities and resources that address resident concerns in relation to livelihood, landscape, and property/ownership types" [4].

Hence, given this background and motivation, our study in this paper investigates people's sentiment towards offshore wind energy through sentiment analysis, offering insights on media opinion to help in fostering the effectiveness of various communication strategies, and also to identify controversies and assess them, thereby supporting the growth of the offshore wind energy initiatives. The main focus of our study in this paper is the state of NJ where we have made site visits, e.g. as seen in

Fig. 1 Pictures taken during the offshore wind energy fellowship meeting at ACUA, wastewater treatment facility in Atlantic City. On the top, working wind turbines. On the bottom left, solar panels. On the bottom right, clarifier tank used for sedimentation

Fig. 2 Poster taken from a New Jersey community website w.r.t. renewable energy

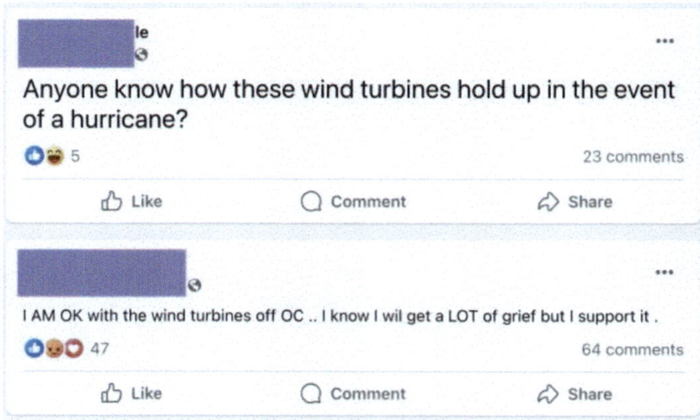

Fig. 3 Examples of social media posts from a community group on Facebook

Fig. 1 here. Note that the opinion mining conducted here can have broader implications because similar views can be expressed by masses elsewhere in response to renewable energy sources.

2 Related Work

Sentiment analysis has surfaced as a very popular method of analyzing mass opinion; and consequently, a significant amount of work has been done on urgent and emerging subjects using machine learning models.

In [5, 6] sentiment analysis is used to analyze sentiment towards climate change on Twitter. Sentiment analysis is also performed in [7], where the authors analyze the legitimacy of wind power in Germany using newspaper articles from 2009 to 2018. In [8], sentiment analysis is used for analyzing a large dataset of geo-tagged tweets to investigate the main topics of discussion and sentiment of the population regarding climate change in different countries over time. In [9], the authors perform sentiment analysis and topic modelling on online reviews for policy amendments, showing how air traffic perceptions were impacted by COVID. In [10], opinion mining is used to investigate public reactions towards urban ordinances for smart governance.

Work has also been conducted towards the policies regarding offshore wind energy to investigate major concerns that were already previously known. In [11], the authors provide an overview of the current development of offshore wind power in different countries, and explore issues around its development. The authors in [12] review three areas that can help with reducing energy consumption, IoT, cloud computing and opinion mining, the latter being significant for its contribution when the aim is to understand feelings and demands from energy consumers and stakeholders to help in the creation of better policies. Additionally, in [13], the authors present an overview of the main issues associated with the economics of offshore wind. In [14], the costs and benefits of offshore wind are discussed relative to onshore wind power and conventional electricity production, with a review of cost estimates.

3 Data Description

The data used in this project is gathered mainly from Facebook groups. It has become very common for towns to create groups for discussing topics about their community. We aptly target communities from the NJ coast, e.g. groups from Atlantic City, Ocean City, The Wildwoods and Cape May. We only collect data from publicly accessible groups, so that there are no privacy concerns. We use the Apify web scraper platform [15] to collect and create the dataset. Comment samples can be seen in Figs. 4 and 5.

Fig. 4 Sample 1 of comments from a Facebook NJ community group

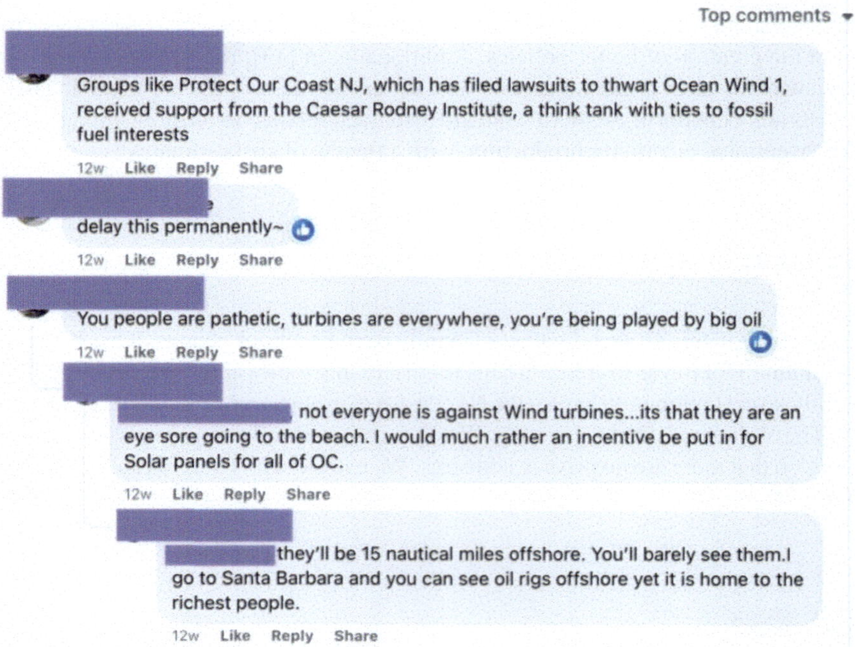

Fig. 5 Sample 2 of comments from a Facebook NJ community group

We use standard search engines to find pertinent comments related to the keywords "offshore wind energy", "wind farms" and "wind turbines". A total of 6569 comments are collected, constituting the main source of data in our analysis. Note that we focus our analysis on Facebook posts, however our models and algorithms can be used for other social media sources for opinion mining in related contexts.

4 Details of Methods

Our overall approach for sentiment analysis is synopsized with an illustration in Fig. 6. After collecting the data, our first step is pre-processing. We use the Python NLTK (natural language toolkit) library [16] to remove punctuation, html links and stop words, and to perform lemmatization and stemming. It is important to use these pre-processing techniques before analyzing text with sentiment analysis models, in order to discard irrelevant text, e.g. pronouns, articles, that do not add useful meaning to the analysis. The pseudocode for pre-processing can be seen in Algorithm 1.

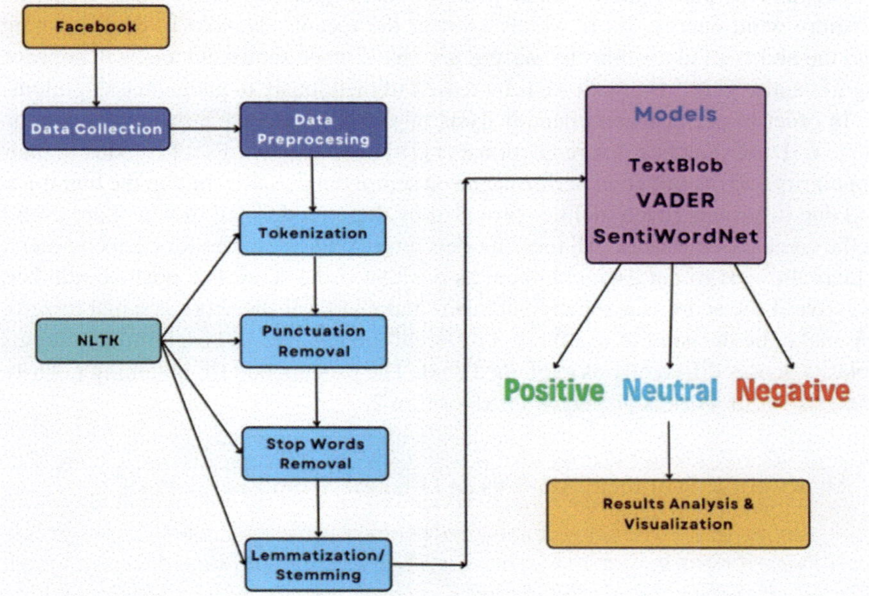

Fig. 6 Approach for sentiment analysis

Algorithm 1: Pre-processing Social Media Data on Offshore Wind

```
INPUT: Facebook comments data-frame δ
FOR comment α in δ DO:
    IF α is null:
        Remove α from δ
    END IF
    Make α lowercase
    Delete punctuation in α
    Delete URLs from α
    Delete # from α
    DO stop-word elimination on α
    DO Lemmatization on α
    IF α is < 3:
        Remove α from δ
    END IF
END FOR
OUTPUT: pre-processed text data as cleaned text β
```

Once the data is pre-processed, we commence the core of the analysis. We deploy the paradigm of sentiment analysis, i.e. a series of methods, techniques and tools used to detect and extract subjective information, such as opinion and attitudes, from language. This paradigm suits our goals of identifying subjective information on offshore wind energy. As is widely known, the recent advances in deep learning and the ability of algorithms to analyze text, have made sentiment analysis improve significantly. Thus, its practice has increased tremendously to gauge mass opinions.

In order to perform sentiment analysis on our data, we use 3 models: TextBlob [17], VADER (Valence Aware Dictionary) [18] and SentiWordNet [19], due to their popularity and overall good performance on sentiment classification in the literature, and due to various functionalities provided by these models, all of which are useful in the context of our work. All these models return polarity scores for each comment, which allow us to tag each comment as positive if the score is a positive number, negative if the score is a negative number, and neutral if the score is equal to zero. We make the decision to use these 3 different models because each model outputs polarity scores differently as explained next. The pseudocode for obtaining polarity analysis in our work is displayed in Algorithm 2.

Algorithm 2: Sentiment Analysis on Offshore Wind Energy Posts

```
INPUT: Pre-processed Facebook comments data-frame δ
FOR comment α in δ DO:
   IF polarity score φ > 0 DO:
      Add positive label
   IF φ = 0 DO:
      Add neutral label
   ELSE DO:
      Add negative label
   END IF
END FOR
OUTPUT: polarity scores φ for each comment α
```

TextBlob gives polarity scores by aggregating individual words' sentiment scores. Besides returning polarity scores, TextBlob also returns subjectivity scores. The subjectivity scores lie between 0 and 1 and show if the text expresses subjective or objective content. The higher the subjectivity means that the text has more opinions, emotions or personal judgements (rather than factual or more neutral content which yields a lower subjectivity score). For example, if a post contains the text, "I do not like offshore wind energy, it's boring!", it is really subjective and opinionated, and would thus get a very high subjectivity score. On the other hand, if it has the text, "Offshore wind energy costs us 10% more than our current usage which we cannot afford due to our profits being 20% lower this year", the information is more objective and factual, hence its subjectivity score would be lower. Although we are looking into people's opinions towards offshore wind energy, TextBlob's subjectivity feature helps us understand the type of information that is being disseminated in the comments, thereby providing a clearer insight into the authenticity of the comments.

Besides opinionated comments, factual comments can show us that people are actually discussing facts about the topic. This is our main reason for deploying TextBlob to get a good idea of the subjective versus objective content in the posts.

VADER (Valence Aware Dictionary and Sentiment Reasoner) returns normalized polarity scores. VADER's compound scores are calculated using the sum of positive, negative, and neutral scores, which are normalized between -1 (most negative) and 1 (most positive). For instance, if a user comments that "Renewable energy sources maybe a bit expensive but are much healthier", it can get a highly positive score for "much healthier" but a slightly negative score for "bit expensive" yielding a compound score that tilts more towards the positive side. Hence, we harness VADER for compound scores to enable a more cumulative analysis of the posts.

SentiWordNet thrives on the well-known lexical source WordNet and considers the given context using "synsets". These synsets are sets of synonyms that are grouped together by their semantic equivalence, which are useful for analyzing the semantic context of the textual data. For example, the word "estimable" would get a neutral score when referring to an item because it implies an item that can be estimated (e.g. its cost); while the same word "estimable" would receive a positive score when referring to a person because it typically refers to a person worthy of esteem or respect. This justifies our choice of SentiWordNet to facilitate adequate analysis with reference to context in the posts considered.

5 Experimental Results

The sentiment analysis performed in this work shows that the overall sentiment on offshore wind energy seems fairly positive on the whole across all the 3 models, TextBlob, VADER, and SentiWordNet; however, there are many users posting negative and neutral comments as well. Figures 7, 8 and 9 present the sentiment distribution as obtained in TextBlob, VADER and SentiWordNet respectively. All these figures include bar plots and pie plots for at-a-glance visualization of the results derived from each model. Delving deeper into the posts pertaining to the negative and neutral comments can pave the way to identify loopholes in existing policies based on their reception by the masses, thus presenting potential areas for improvement. Likewise, the positive posts can offer the scope for continuation of the respective policies and further encouragement along similar lines.

As discussed earlier, we are able to get subjectivity scores from TextBlob. In Fig. 10, we show the subjectivity scores within the dataset in a bar plot. It is possible to see that the content of the dataset is mostly factual, which indicates that besides expressing their opinion, people are discussing real facts within their comments.

Moreover, it is possible to visualize the most frequent negative and positive words present in the comments. We extract the most frequent negative words in the negative comments and the most frequent positive words in the positive comments from each model. These are plotted in Figs. 11 and 12 for TextBlob, Figs. 13 and 14 for VADER,

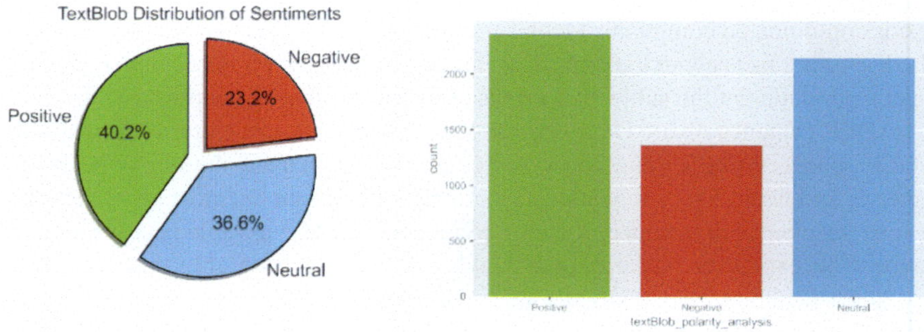

Fig. 7 TextBlob plots for distribution of sentiments visualization

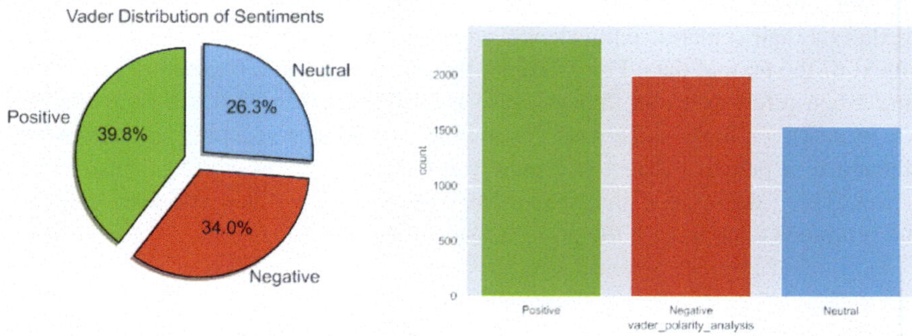

Fig. 8 VADER plots for distribution of sentiments visualization

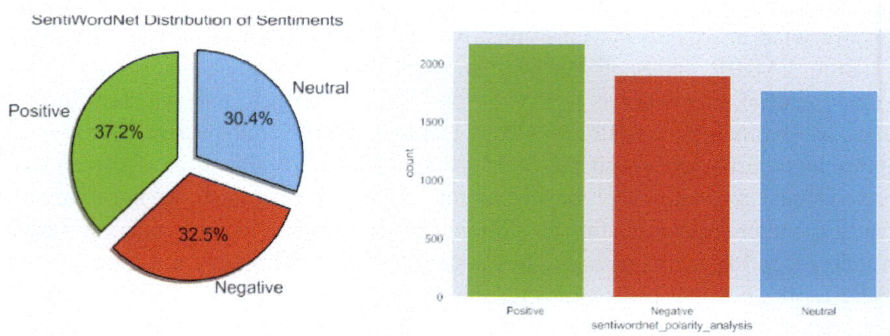

Fig. 9 SentiWordNet plots for distribution of sentiments visualization

and Figs. 15 and 16 for SentiWordNet, respectively. These serve to provide good visual depictions of the main terms of interest in the social media posts.

Frequent words reveal major topics of interest in the data. Some highly significant topics of discussion in this analysis can be seen in the positive and negative WordClouds in Figs. 17, 18, 19, 20, 21 and 22, created using the most frequent positive

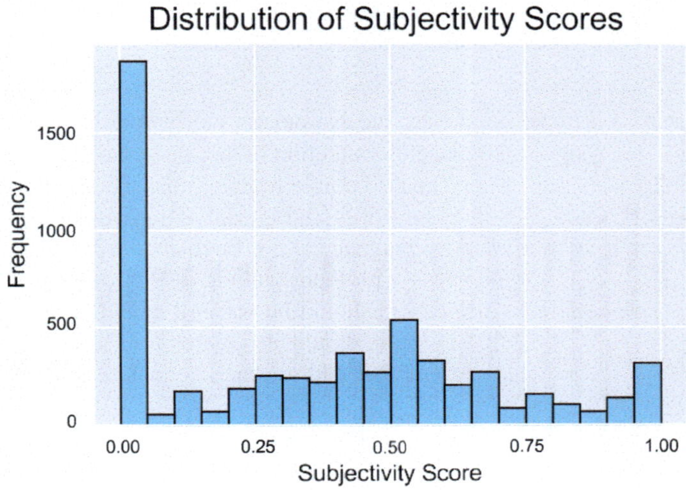

Fig. 10 TextBlob plots for distribution of subjectivity visualization

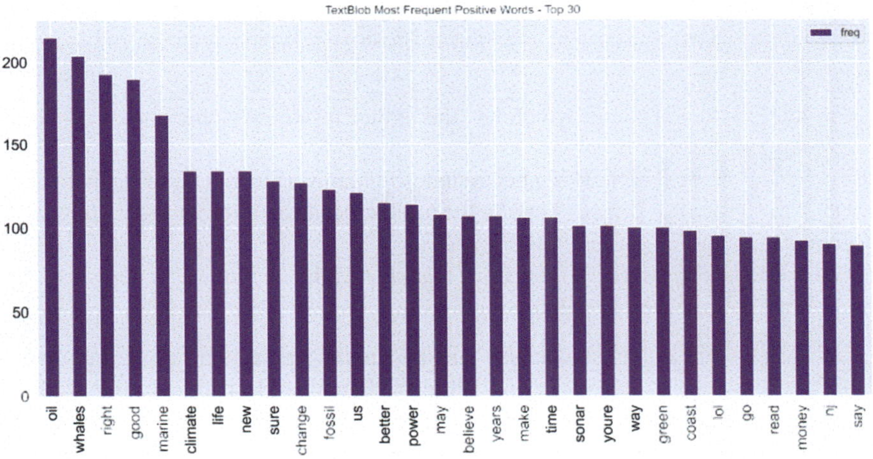

Fig. 11 TextBlob's top 30 positive words

and negative words from each model. The bigger the word appears in the image, the more frequent it is. A few significant words are marine mammal, Osterd, whale, fossil fuel, propaganda, climate change, wildlife, money, democrats, sonar mapping, death etc.

Fig. 12 TextBlob's top 30 negative words

Fig. 13 VADER's top 30 positive words

6 Conclusions and Future Work

This work harnesses methods in machine learning and natural language processing to address issues in environmental management, more specifically, clean energy. In particular, offshore wind energy is targeted here, in order to gauge its reception by the masses. Sentiment analysis is performed on social media data, more specifically Facebook posts geo-located in NJ, to investigate people's opinions. The need for renewable sources of energy is high due to the severe effects of climate change on earth. This study provides support to the commitment of producing 100% clean

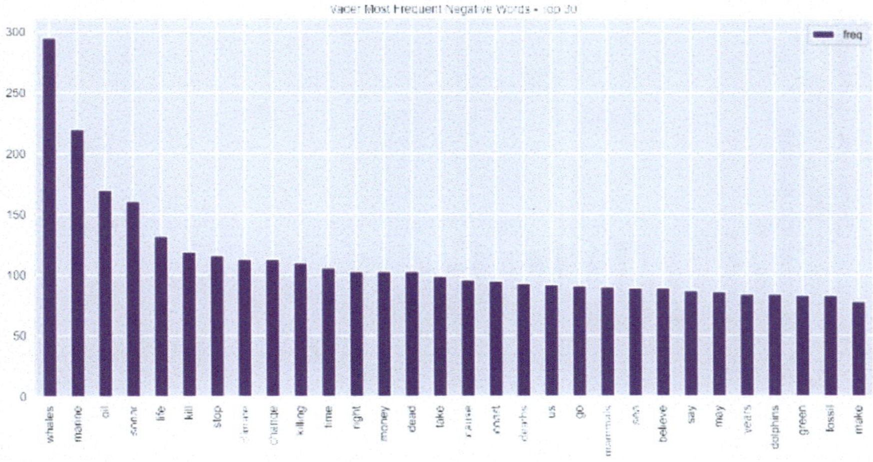

Fig. 14 VADER's top 30 negative words

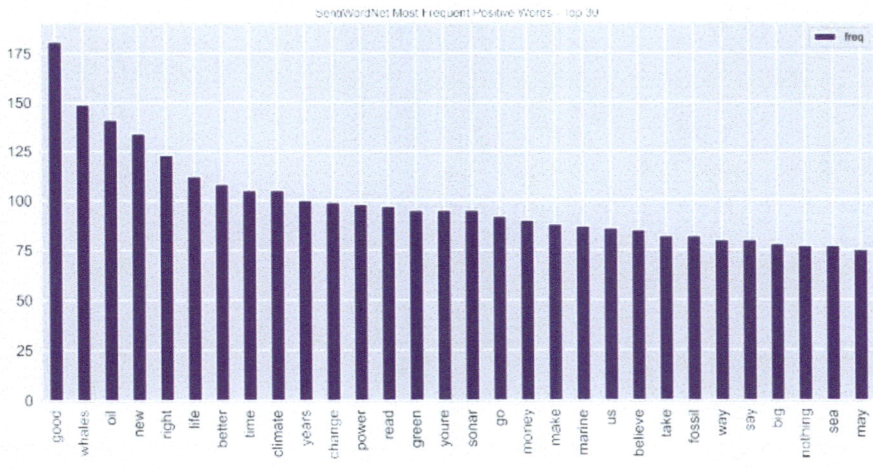

Fig. 15 SentiWordNet's top 30 positive words

energy by 2050 that is being targeted across many regions in the world today, e.g. in the state of New Jersey here. Although the overall opinion of the population in this dataset is somewhat positive, the neutral and negative counts are not far behind due to concerns over the impacts of offshore wind energy on wildlife and beaches, the state's budget money, etc. Subjectivity scores show that people have a considerable extent of concrete and factual discussions in their social media posts, which is an interesting finding.

The results presented here can be used for future decisions by government leaders and interested companies on the topic, i.e. wind farms' location, projects budget,

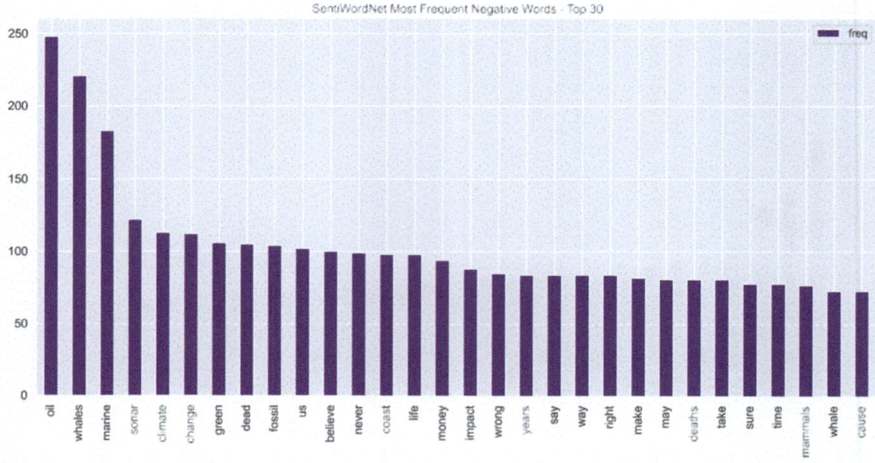

Fig. 16 SentiWordNet's top 30 negative words

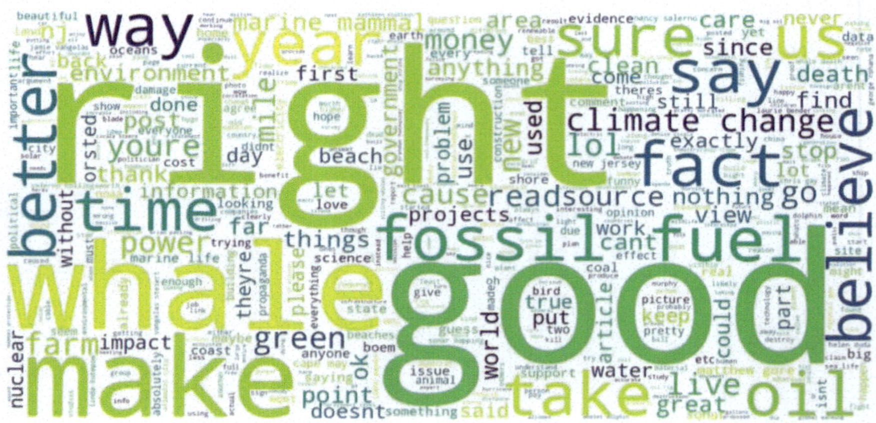

Fig. 17 Positive WordCloud from TextBlob

environmental impacts etc. Tailoring policies to public opinion and identifying key concerns are an important part of the process of investing in the build of wind farms by the coast of New Jersey.

Likewise, similar analyses can be performed on social media posts in other regions to gain deeper insights on the reception of the policies by the common masses. This is in line with paradigms such as citizen science and smart governance that entail active involvement of the masses with more transparency and openness, providing the foundation for more adequate decision-making. Our work in this paper thus makes a modest contribution to these paradigms.

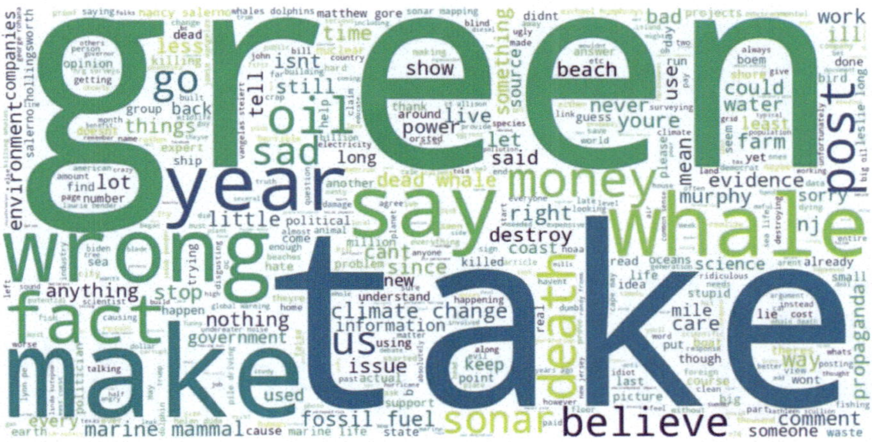

Fig. 18 Negative WordCloud from TextBlob

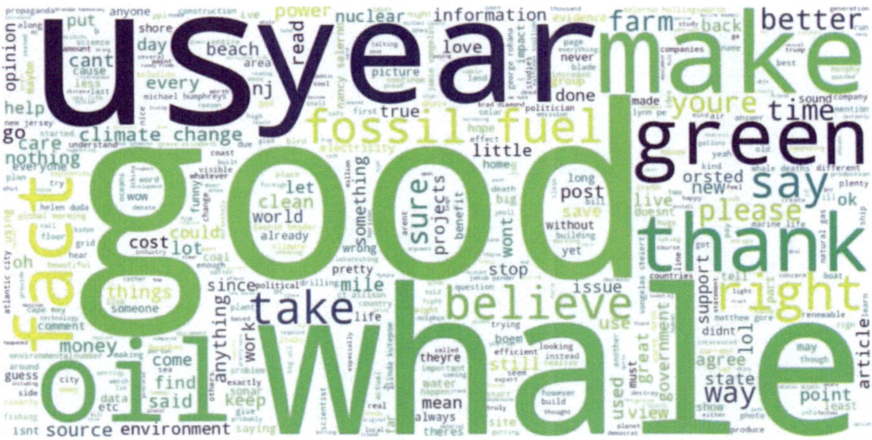

Fig. 19 Positive WordCloud from VADER

Future work emerging from our research includes investigating other sources of text such as online news, corporate and governmental websites, user blogs, and research articles to discover knowledge from text. This can be achieved via machine learning techniques in conjunction with natural language processing. It can include topic modelling using LDA (Latent Dirichlet Allocation), for instance as in [20] where LDA is used to explore climate change, energy and food security trends in newspapers and public documents. It can also entail analyzing more information with Large Language Models (LLMs), such as in [21], where the authors use LLMs for topic modelling and perform quite well. Furthermore, it is possible to map text to structured data to enable the usage of data mining techniques such as association rules for enhanced knowledge discovery. More research on these lines can be conducted

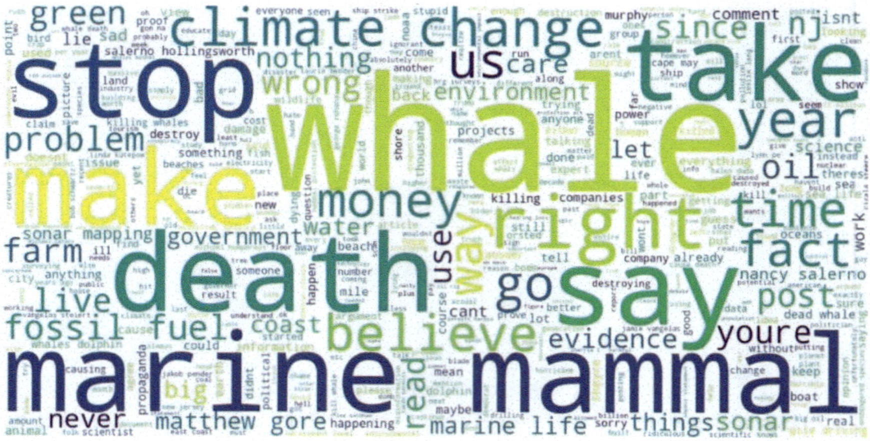

Fig. 20 Negative WordCloud from VADER

Fig. 21 Positive WordCloud from SentiWordNet

in the broad realm of computational linguistics such that it can be useful in studies on the environment.

In sum, our paper provides the ground for more work on machine learning and NLP techniques to tackle environmental issues, e.g. in clean energy and related areas. As we move forward towards a greener future, there has been an ample amount of research done towards analysis of environmental topics, deployment of environment-friendly applications, scientific data mining in general, and machine intelligence on the whole [22–31] much of which focuses on work done in our research group. Hence, our work in this paper is orthogonal to such research. Overall, our study in this paper stands on a notable bridge between computational linguistics and environmental

Fig. 22 Negative WordCloud from SentiWordNet

management, helping in the achievement of more sustainable practices for a better and greener world.

Acknowledgements All the authors acknowledge the New Jersey Wind Institute Fellowship Program by NJEDA, a state wide initiative in NJ. We also acknowledge the Clean Energy and Sustainability Analytics Center (CESAC) at MSU, NJ. In addition. Dr. Aparna Varde acknowledges NSF MRI grants 2018575 and 2117308 as well.

References

1. ENVIRONMENT AMERICA, offshore wind for America. https://environmentamerica.org/center/resources/offshore-wind-for-america-3/#:~:text=The%20United%20States%20has%20the,2050%20if%20we%20electrified%20our. Accessed 1 Dec 2023
2. ACUA, Jersey-Atlantic wind farm. https://www.acua.com/Projects/Jersey-Atlantic-Wind-Farm.aspx. Accessed 1 Feb 2024
3. Department of Environmental Protection, Offshore wind. https://dep.nj.gov/offshorewind/. Accessed 1 Dec 2023
4. Elmallah S, Rand J (2022) After the leases are signed, it's a done deal: exploring procedural injustices for utility-scale wind energy planning in the United States. Energy Res Soc Sci 89:102549
5. Rosenberg E, Tarazona C, Mallor F, Eivazi H, Pastor-Escuredo D, Fuso-Nerini F, Vinuesa R (2023) Sentiment analysis on Twitter data towards climate action. Results Eng 19:101–287
6. McNamee B, Varde A, Razniewski S (2022) Correlating facts and social media trends on Environmental quantities leveraging common sense reasoning and human sentiments. In: Proceedings of the 2nd workshop on sentiment analysis and linguistic linked data, pp 25–30
7. Dehler-Holland J, Okoh M, Keles D (2022) Assessing technology legitimacy with topic models and sentiment analysis—the case of wind power in Germany. Technol Forecast Soc Chang 175:121–354

8. Dahal B, Kumar SAP, Li Z (2019) Topic modeling and sentiment analysis of global climate change tweets. Soc Netw Anal Min 9:24
9. Field A, Varde A, Lal P (2022) Sentiment analysis and topic modeling for public perceptions of air travel: COVID issues and policy amendments. Language resources and evaluation conference, LREC, pp 2–8
10. Puri M, Varde A, De Melo G (2018) Smart governance through opinion mining of public reactions on ordinances. IEEE 30th international conference on tools with artificial intelligence. ICTAI, pp 838–845
11. Sun X, Huang D, Wu G (2012) The current state of offshore wind energy technology development. Energy 41:298–312
12. Shrestha S, Varde A (2023) Roles of the web in commercial energy efficiency: IoT, cloud computing, and opinion mining. Association for computing machinery special interest group on hypertext, hypermedia and web, Article 5. ACM SIGWEB Autumn, pp 1–16
13. Green R, Vasilakos N (2011) The economics of offshore wind. Energy Policy 39:496–502
14. Snyder B, Kaiser M (2009) Ecological and economic cost-benefit analysis of offshore wind energy. Renew Energy 34:1567–1578
15. Web Scraper, Apify, apify.com. Accessed 15 Dec 2023
16. Bird S, Klein E, Loper E (2009) Natural language processing with Python: analyzing text with the natural language toolkit. O'Reilly Media, Inc
17. Textblob, textblob.readthedocs.io/en/dev/index.html. Accessed 1 Dec 2023
18. Hutto CJ, Gilbert EE (2014) VADER: a parsimonious rule-based model for sentiment analysis of social media text. Eighth international conference on weblogs and social media
19. Baccianella S, Esuli A, Sebastiani F (2010) SentiWordNet 3.0: an enhanced lexical resource for sentiment analysis and opinion mining. In: Calzolari N, Choukri K, Maegaard B, Mariani J, Odijk J, Piperidis S, Rosner M, Tapias D (eds) Language resources and evaluation conference. LREC
20. Benites-Lazaro LL, Giatti A, Giarolla A (2018) Topic modeling method for analyzing actor discourses on climate change, energy and food security. Energy Res Soc Sci 45:318–330
21. Stammbach D, Zouhar V, Hoyle A, Sachan M, Ash E (2023) Revisiting automated topic model evaluation with large language models. Conference on empirical methods in natural language processing. Association for Computational Linguistics, ACL, pp 9348–9357
22. Pawlish M, Varde A, Robila S (2012) Cloud computing for environment-friendly data centers. In: Proceedings of the 4th international workshop on cloud data management, pp 43–48
23. Pawlish M, Varde A (2010) A decision support system for green data centers. In: Proceedings of the 3rd workshop on Ph.D. students in information and knowledge management, pp 47–56
24. Shrestha S, Buckley B, Varde A, Cwynar D (2023) Hybrid CNN-LSTM and domain modeling in climate-energy analysis for a smart environment. IEEE 35th international conference on tools with artificial intelligence, pp 229–233
25. Singh A, Yadav J, Shrestha S, Varde A (2023) Linking alternative fuel vehicles adoption with socioeconomic status and air quality index. The 37th AAAI conference on artificial intelligence. https://doi.org/10.48550/arXiv.2303.08286
26. Varde A, Pandey A, Du X (2022) Prediction tool on fine particles pollutants and air quality for environmental engineering. SN Comput Sci 3:article number 184
27. Gonzalez-Moodie B, Daiek S, Lorenzo-Trueba J, Varde A (2021) Multispectral drone data analysis on coastal dunes. IEEE International conference on big data, pp 5903–5905
28. Prasad A, Varde A, Gottimukkala R, Alo C, Lal P (2021) Analyzing land use change and climate data to forecast energy demand for a smart environment. 9th international renewable and sustainable energy conference, pp 1–6
29. Suchanek FM, Varde AS, Nayak R, Senellart P (2011) The hidden Web, XML and the semantic web: scientific data management perspectives. In: Proceedings of the 14th international conference on extending database technology, pp 534–537

30. Varde AS, Takahashi M, Rundensteiner EA, Ward MO, Maniruzzaman M, Sisson RD Jr (2004) Apriori algorithm and game-of-life for predictive analysis in materials science. Int J Knowl-Based Intell Eng Syst 8(4):213–228
31. Tandon N, Varde AS, de Melo G (2018) Common sense knowledge in machine intelligence. ACM SIGMOD Rec 46(4):49–52

Author Index

P. G. Bradford et al. (eds.), *Proceedings of IEMTRONICS 2024*, Lecture Notes in Electrical Engineering 1229, https://doi.org/10.1007/978-981-97-4780-1

The manufacturer's authorised representative in the EU is Springer Nature Customer Service Centre GmbH, Europaplatz 3, 69115 Heidelberg, Germany. If you have any concerns regarding our products, please contact ProductSafety@springernature.com

Printed and bound by CPI Group (UK) Ltd, Croydon, CR0 4YY
30/09/2026
02225298-0017